Geodynamics Series

Geodymanics Series

Ice Sheets, Sea Level and the Dynamic Earth

Jerry X. Mitrovica
Bert L. A. Vermeersen
Editors

Geodynamics Series   Volume 29

American Geophysical Union
Washington, D.C.

Library of Congress Cataloging-in-Publication Data
Ice sheets, sea level, and the dynamic earth / Jerry X. Mitrovica, Bert L. A. Vermeersen, editors.
      p.cm.-- (Geodynamics series ; v. 29)
   Includes bibliographical references.
   ISBN 0-87590-531-5
   1. Glacial isostasy. 2. Sea level. 3. Earth--Crust. 4. Geodynamics. I. Mitrovica, Jerry X.,
1960-. II. Vermeersen, Bert, 1961-. III. Series.
QE511.I34 2002
551.1'3--dc21   2002025425

ISBN 0-87590-531-5
ISSN 0277-6669

CONTENTS

PREFACE

In this monograph, we present recent progress in geophysical modeling and observational tools related to the process of glacial isostatic adjustment (GIA). Rather than a retrospective view, however, we have been led by one overarching mission: to gather significant contributions that present the state-of-the-art in the field and beyond, just as it is being reshaped by new space-geodetic technologies. In this light, the monograph includes discussion on new progress in a number of long-standing problems: the modeling of the Earth's viscoelastic response; the prediction and analysis of sea-level changes and anomalies in the Earth's rotation and gravity field; and the inference of mantle viscosity. Such contributions are complemented by papers that focus on results obtained by GPS and constraints expected from impending satellite missions, as well as predictions of geophysical observables (e.g., present-day 3-D deformations, gravity signals and fault instability) related to these efforts. In these many applications it is important to understand recent progress in GIA research and the limitations that currently impact that research.

Although the GIA community is not large, relatively speaking, the continuing influence of the field reflects the remarkably broad set of research disciplines that must contend with a GIA signal. Accordingly, the audience for this volume extends beyond the immediate community of scientists active in GIA studies to include students and researchers within a variety of cognate disciplines. Geologists and oceanographers, for example, will learn how markers of Holocene sea-level change are being adopted in modern GIA analyses and will also be guided through the often subtle connection between ice mass variations and site-specific trends in sea level. Solid-earth geophysicists will be interested in updated constraints on the radial profile of mantle viscosity, which exerts a fundamental control on the dynamics of the Earth's interior. Furthermore, geodesists will be reminded of the influence of GIA on the long-term rotational state of the planet. They, together with glaciologists, will also find a detailed account of the role modern satellite missions play both in the evolution of GIA studies and in ongoing efforts to constrain the mass of present-day ice reservoirs. All of these disciplines will be impressed by the astonishing accuracy of a GPS survey of Fennoscandian GIA, and be motivated by questions raised by the published maps of 3-D crustal deformation.

The review process significantly improved the collection, and we thank all of those involved, reviewers and authors especially. Finally, we dedicate this volume to William M. Kaula, who served as the original oversight editor for the volume. He enthusiastically endorsed our proposal and argued that it had the potential for becoming, in his words, an excellent book. We hope that we have realized the potential for excellence Dr. Kaula initially pointed out to us.

Jerry X. Mitrovica
Bert L. A. Vermeersen

Glacial Isostatic Adjustment and the Earth System

Jerry X. Mitrovica

Department of Physics, University of Toronto, Toronto, Canada

Bert L. A. Vermeersen

DEOS, Faculty of Aerospace Engineering, Delft University of Technology, The Netherlands

Historical perspective makes it relatively easy to identify transformative contributions to the study of glacial isostatic adjustment (GIA). However, even in the absence of such perspective, it is clear that GIA research has entered, to adopt the term of *Smith and Turcotte* [1993; p. xi], the "next great step." This revolution is being accompanied by unprecedented and widespread interest in the GIA problem, and it is being fueled by the application of a suite of remarkably accurate space-geodetic techniques (e.g., surveying using the global positioning system, or GPS, and satellite-based gravity field mapping). The present volume of papers is a product of this renewed interest.

The analysis of data related to the deformation of the Earth in consequence of the waxing and waning of Late Pleistocene ice sheets is a long-standing field of geophysical study. Traditionally, GIA studies have focussed on two primary applications. The first, which dates back to *Haskell* [1935], *Vening-Meinesz* [1937] and their contemporaries, involves the inference of mantle viscosity from GIA observables related to Holocene sea-level change and anomalies in both the Earth's rotational state and gravitational field. The second encompasses efforts to constrain the space-time history of Late Pleistocene ice cover.

Over the last decade, these "traditional" GIA applications have broadened dramatically. Indeed, GIA studies now routinely include applications as diverse as: (1) the correction of tide gauge data for ongoing GIA in order to estimate global sea-level rise over the last century; (2) the analysis of post-glacial stress regimes and their connection to fault instabili-

ty and seismicity; (3) the study of GIA-induced perturbation to the Earth's orbital elements (e.g., precession and obliquity) and the impact of this signal on paleoclimate proxies; (4) constraining excess ice volumes during the last glacial maximum; and (5) the prediction of present-day three-dimensional crustal deformations. Modern GIA research has clearly evolved into a multi-disciplinary Earth system science.

The focus on 3-D crustal deformations first emerged in the GIA literature in the early 1990's [e.g., *James and Morgan*, 1990; *James and Lambert*, 1993; *Mitrovica et al.*, 1993, 1994a,b], and it was motivated by the promise of a new generation of space-geodetic measurements based on GPS and very-long-baseline-interferometry technology. This promise was soon realized. The first-ever detection of post-glacial deformation using GPS was reported in the mid 1990's [*BIFROST*, 1996]. The BIFROST GPS network, which is composed of a dense array of continuously operating and permanent GPS receivers within Fennoscandia, has more recently yielded a map of GIA-induced 3-D deformation with sub-mm/yr accuracy [*Milne et al.*, 2001; *Johansson et al.*, 2002; *Scherneck et al.*, this volume].

In addition to maps of crustal deformation, satellite missions such as GRACE, CHAMP and GOCE are expected to deliver spatial and temporal variations of the Earth's global gravity field that are orders of magnitude (on a global scale) more precise than existing constraints. For example, the GRACE mission will detect secular geoid variations up to harmonic degree and order ~40 [see *Wahr and Davis*, this volume], and may be able to separate contributions of Greenland and Antarctic melting. Furthermore, the GOCE mission is expected to yield a global (minus two "polar gaps") gravity field with an accuracy of ~1 mgal for gravity anomalies and 1 cm for the geoid at a resolution better than 100 km [see *Visser et al.*, this volume].

Ice Sheets, Sea Level and the Dynamic Earth
Geodynamics 29
Copyright 2002 by the American Geophysical Union
10.1029/029GD01

Space-geodetic measurements of 3-D crustal deformations and global gravity variations will also be applied to constrain ongoing mass flux from present-day ice reservoirs [e.g., *Hager*, 1991; *Wahr et al.*, 1995]. In some regions, for example the Antarctic, modeling of the deformation associated with both GIA and any ongoing mass flux will be important [see *Ivins et al.*, this volume]. In this regard, it has been shown that some separation of the two processes is possible by invoking gravity measurements [*Wahr et al.*, 1995].

As the accuracy of observational constraints on the GIA process improves, the theory underlying numerical predictions of these observables must keep pace. This volume includes a number of articles outlining improvements in the prediction and analysis of post-glacial sea-level change, as well as the underlying theory governing the response of viscoelastic Earth models. In general, GIA predictions have been based on spherically symmetric (that is, radially stratified "onion skin") models, and the extension of these models to the case of laterally varying structure represents a major current goal of a number of research groups [e.g., *Tromp and Mitrovica*, 1999; *Kaufmann and Wu*, this volume].

The first part of this volume comprises three overview articles that cover recent and impending geodetic constraints on GIA, the modern prediction and analysis of GIA-induced sea-level variations using Australia as a case study, and a reanalysis of anomalies in Earth rotation associated with GIA. This is followed by a series of articles sampling a wide range of the applications described above.

In some areas of geophysics it is possible to supply "standard" numbers for use by scientists whose expertise lies elsewhere. A suite of uncertainties related to ice history, Earth structure and even the underlying mathematical and physical theory makes it impossible to provide robust (i.e., unique) numbers for the GIA process. This uncertainty poses a significant challenge for scientists outside the GIA mainstream and an opportunity for those within the field. Our primary motivation for compiling this volume has been to establish a bridge between these two sets of scientists. Specifically, the articles are written in a voice suitable for the general reader, and they highlight the advances as well as the limitations of present research in the field. Understanding the dynamic range of the process will help those at the periphery to make informed judgements required for their research, and it will guide "experts" through the evolving and always lively world of GIA.

REFERENCES

BIFROST Project Members, GPS measurements to constrain geodynamic processes in Fennoscandia, *Eos Trans. AGU*, 77, 337, 341, 1996.

Hager, B. H., Weighing the ice sheets using space geodesy: A way to measure changes in ice sheet mass, *Eos Trans. AGU*, 72, 91, 1991.

Haskell, N. A., The motion of a fluid under a surface load, 1, *Physics*, 6, 265-269, 1935.

James, T. S., and W. J. Morgan, Horizontal motions due to post-glacial rebound, *Geophys. Res. Lett.*, 17, 957-960, 1990.

James, T. S., and A. Lambert, A comparison of VLBI data with the ICE-3G glacial rebound model, *Geophys. Res. Lett.*, 20, 871-874, 1993.

Johansson, J. M., J. L. Davis, H.-G. Scherneck, G. A. Milne, M. Vermeer, J. X. Mitrovica, R. A. Bennett, B. Jonsson, G. Elgered, P. Elosegui, H. Koivula, M. Poutanen, B. O. Ronnang, and I. I. Shapiro, Continuous GPS measurements of post-glacial adjustment in Fennoscandia, 1. Geodetic results, *J. Geophys. Res.*, in press, 2002.

Milne, G. A., J. L. Davis, J. X. Mitrovica, H.-G. Scherneck, J. M. Johansson, M. Vermeer, and H. Koivula, Space-geodetic constraints on glacial isostatic adjustment in Fennoscandia, *Science*, 291, 2381-2385, 2001.

Mitrovica, J. X., J. L. Davis, and I. I. Shapiro, Constraining proposed combinations of ice history and Earth rheology using VLBI determined baseline length rates in North America, *Geophys. Res. Lett.*, 20, 2387-2390, 1993.

Mitrovica, J. X., J. L. Davis, and I. I. Shapiro, A spectral formalism for computing three-dimensional deformations due to surface loads, 1. Theory, *J. Geophys. Res.*, 99, 7057-7073, 1994a.

Mitrovica, J. X., J. L. Davis, and I. I. Shapiro, A spectral formalism for computing three-dimensional deformations due to surface loads, 2. Present-day glacial isostatic adjustment, *J. Geophys. Res.*, 99, 7075-7101, 1994b.

Smith, D. E., and D. L. Turcotte, Preface, in *Contributions of Space Geodesy to Geodynamics: Crustal Dynamics*, AGU Geodyn. Ser., Vol. 23-25, ed. by Smith, D. E., and D. L. Turcotte, AGU (Washington, DC), p. xi, 1993.

Tromp, J., and J. X. Mitrovica, Surface loading of a viscoelastic Earth - I. General theory, Geophys. *J. Int.*, 137, 847-855, 1999.

Vening-Meinesz, F.-A., The determination of the Earth's plasticity from the post-glacial uplift of Scandinavia, Proc. Kon. Ned. Acad. Wetensch., 40, 654-662, 1937.

Wahr, J., A. Trupin and D. Han, Predictions of vertical uplift caused by changing polar ice volumes on a viscoelastic Earth, *Geophys. Res. Lett.*, 22, 977-980, 1995.

Jerry X. Mitrovica, Department of Physics, University of Toronto, 60 St. George Street, Toronto, Canada M5S 1A7 (e-mail: jxm@physics.utoronto.ca)

Bert L. A. Vermeersen, DEOS, Faculty of Aerospace Engineering, Kluyverweg 1, 2629 HS, Delft, The Netherlands (e-mail: b.vermeersen@lr.tudelft.nl)

Geodetic Constraints on Glacial Isostatic Adjustment

John M. Wahr

Department of Physics, University of Colorado, Boulder, Colorado

James L. Davis

Harvard-Smithsonian Center for Astrophysics, Cambridge, Massachusetts

Geodetic observations are playing an increasingly important role in the study of glacial isostatic adjustment (GIA). Observations of the Earth's gravity field, fluctuations in the Earth's rotation, and motion of the Earth's surface yield information on ongoing GIA. Satellite altimetric observations of ice sheet elevations are on the verge of revolutionizing our knowledge of mass imbalance of the polar ice sheets. In this paper we discuss some of the ways in which these measurement techniques have contributed to our knowledge of GIA and the present-day mass imbalance of the ice sheets. We discuss improvements in measurement techniques that are expected to occur in the near future, and that should lead to notable improvements in our understanding of both GIA and polar ice mass imbalance.

1. INTRODUCTION

The enormous ice sheets of the last glacial period left their imprint throughout northern North America and Scandinavia. Their effects can be seen in numerous geological features, which hold clues not only to the ice sheet extent and eventual retreat, but also to how the Earth has responded to these varying loads. This response depends critically on the Earth's viscosity profile. In fact, observations of this glacial isostatic adjustment (GIA) provide what are arguably the best existing constraints on the global-scale viscosity of the Earth's mantle. These viscosity values have important implications for mantle convection, at least to the extent that the mantle can be represented as a Newtonian fluid during the convective process.

Geodetic observations are playing an increasingly important role in constraining GIA. The geological evidence, especially observations of relative sea level (RSL) variations during the Late Pleistocene and Holocene (i.e., over the last 20,000 years, or so), has proven exceptionally useful for learning about upper mantle viscosity and lithospheric thickness. But RSL is relatively insensitive to deep-mantle viscosity. Furthermore, RSL variations are strongly dependent on details of the ice sheet retreat which are poorly known. There are certain types of geodetic observations, on the other hand, that are primarily sensitive to lower mantle viscosity. Furthermore, although all geodetic observables are also sensitive to the history and spatial distribution of the ice sheets, they are sensitive in different ways than are the RSL data. Thus, combining geodetic data with RSL observations can help resolve the trade-off between viscosity and ice sheet models.

Ice Sheets, Sea Level and the Dynamic Earth
Geodynamics Series 29

Various types of geodetic measurements are sensitive to the effects of GIA. The first to be used to constrain GIA models were observations of the large negative free-air gravity and geoid anomalies centered over the original ice sheet locations [e.g., *O'Connell*, 1971], although there is now evidence that sources other than GIA may be playing a large part in generating these anomalies. More recently, with the development of remarkably precise geodetic measurement techniques, have come measurements of time-varying quantities: of the Earth's gravity field, of positions of points on the surface, and of changes in the Earth's rotation. These time-varying quantities reflect the ongoing shift of mass within the Earth as the Earth continues to adjust to the removal of the ice load during the Late Pleistocene. These geodetic constraints are becoming increasingly important as more data are being accumulated, and are likely to become even more useful once there are a few years of data from GRACE: a dedicated-gravity satellite mission scheduled for launch in early 2002.

There is also likely to have been GIA caused by changes in the polar ice sheets (i.e. Antarctica and Greenland) over the last several thousand years. Detecting the effects of GIA in those regions is especially difficult. Local geological evidence is scarce, because the land is still mostly ice-covered. Geodetic measurements of time-variable quantities in these areas is complicated by the possible contributions from present-day changes in ice thickness. For example, any ongoing variations in ice mass could contribute to changes in the gravitational field, and could induce elastic motion of the Earth's crust. These signals cannot easily be separated from the effects of GIA.

Airborne- and, especially, satellite-altimetry provide a means of constraining present-day changes in ice, so that its effects can be removed from other geodetic observations to extract the GIA signal. Conversely, the GIA signal can contaminate the altimetric measurements of present-day mass imbalance of the polar ice sheets. By combining the altimetry with other types of geodetic observations, it is conceivable that this contamination can be reduced, resulting in improved altimetric mass-balance estimates.

In this paper we describe various geodetic measurements (i.e. space- and surface-based gravity observations, crustal motion measurements, and radar and laser altimetry) that are sensitive to GIA and/or polar mass balance. We discuss these measurements both in terms of what they have provided so far and what the future might hold. We discuss the impact these measurements might have on studies of both GIA and polar mass imbalance.

2. THE STATIC GRAVITY FIELD

The first geodetic results recognized as being relevant to GIA were free air gravity anomalies inferred from surface gravimeter measurements [see, e.g., *Walcott*, 1973]. Those measurements showed that there are anomalous gravity lows over the central Canadian shield (with an amplitude of about -50 mgals) and, to a lesser extent, over Scandinavia (with an amplitude of about -10 mgals). These features also show up prominently, of course, in today's global gravity solutions that are obtained largely from satellite laser ranging observations. Global gravity solutions are often displayed in the form of a map of the geoid, where the gravity anomaly over Canada shows up as a geoid low of about -40 m.

Maps of geoid anomalies and of gravity anomalies are different expressions of the same information. For example, it is usual to expand the geoid height, N, obtained using satellite laser ranging measurements as a sum of associated normalized Legendre functions, $\tilde{P}_{lm}$, in the form [see, e.g., *Chao and Gross*, 1987].

$$N(\theta,\phi) = a\sum_{l=1}^{\infty}\sum_{m=0}^{l}\tilde{P}_{lm}(\cos\theta)\big[C_{lm}\cos m\phi \\ + S_{lm}\sin m\phi\big], \qquad (1)$$

where θ and ϕ are co-latitude and eastward longitude, a is the Earth's radius, and the C_{lm} and S_{lm} are dimensionless Stokes' coefficients. These same C_{lm} and S_{lm} appear in the expansion of the free air gravity anomaly on the geoid:

$$g_{FA}(\theta,\phi) = g\sum_{l=1}^{\infty}\sum_{m=0}^{l}(l-1)\tilde{P}_{lm}(\cos\theta)\big[C_{lm}\cos m\phi \\ + S_{lm}\sin m\phi\big], \qquad (2)$$

where g is the mean value of gravity on the Earth's surface. For each $\tilde{P}_{lm}$ term in these expansions, the index l is inversely proportional to the horizontal scale: with the scale (half-wavelength) $\approx 20,000/l$ km. The index m determines, in an approximate sense, the spatial orientation of this horizontal variability: i.e., how much of the variability is east-west as opposed to north-south.

Plate 1 shows both the geoid height (top panel) and the free air gravity anomalies (middle panel) over North America inferred from the gravity field model EGM96, which is based on a combination of satellite laser ranging, satellite radar altimetry over the oceans, and data from gravimeters taken on land, on ships, and from airplanes [*Lemoine et al.*, 1998]. Here, only C_{lm} and S_{lm} values for l and m up to 30 have been included in the

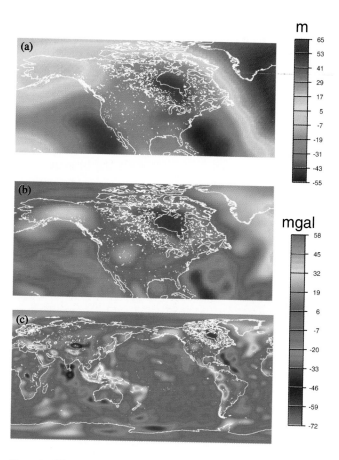

m

65
53
41
29
17
5
-7
-19
-31
-43
-55

mgal

58
45
32
19
6
-7
-20
-33
-46
-59
-72

Plate 1. The geoid height over North America (a); and the medium- to large-scale free air gravity anomalies over North America (b), and over the entire globe (c). Results inferred from gravity field model EGM96 [*Lemoine et al.*, 1998], which is based on a combination of satellite- and ground-based observations. All Stokes' coefficients up to degree and order 30, corresponding to spatial scales of about 670 km and larger, are included when constructing these maps.

sums (1) and (2). Note that the negative anomaly centered over Hudson Bay is evident in both the geoid and the gravity maps. The fact that this anomaly is located about where it would be expected if it were the result of GIA, and that it has an amplitude and a spatial extent that are consistent with plausible assumptions about the Earth's viscosity and ice sheet history and extent, suggests that it is caused by GIA. The amplitude of the anomaly has been frequently used as a constraint on GIA models, where it has proven particularly useful for constraining lower mantle viscosity.

But there has long been uneasiness about whether this anomaly is caused primarily by GIA, or whether it partially reflects some deep-seated density anomaly of either thermal or compositional origin [see, e.g., *Cathles*, 1975]. Note from the bottom panel of Plate 1 that the Canadian anomaly is still quite prominent on a global free air gravity map (again, shown truncated to $l \leq 30$), but that there are several other negative anomalies elsewhere on the globe that are just as large and that are certainly not related to GIA (note that the gravity low over Scandinavia is not particularly prominent in the bottom panel of Plate 1 at this resolution.)

This global gravity map can be compared to Plate 2, which shows global seismic tomographic results for the shear wave velocity, v_s, at a depth of 200 km, computed using spherical harmonic model MK12WM13 of *Su and Dziewonski* [1997]. This model includes spherical harmonics only for degrees $l \leq 12$, corresponding to spatial scales ≥ 1700 km. Shorter-scale features cannot be resolved. Note the large $\approx 4\%$ positive anomaly under the central Canadian Shield. The amplitude of this anomaly is far too large to be related to GIA. For example, GIA models suggest that constant density surfaces at this depth in the upper mantle should presently be uplifted relative to their isostatic position by on the order of 100 m or so [*Tamisiea and Wahr*, 2001]. (The reason the constant density surfaces beneath the lithosphere are uplifted at present, rather than depressed, is that they were reasonably smooth before the deglaciation began; though the lithosphere was depressed beneath the ice load. The entire column beneath the ice then uplifted when the ice began to melt. The lithosphere is still uplifting to fill in the initial depression. But the material below the lithosphere is now subsiding to flatten out the constant density surfaces again.) The lateral v_s anomalies due to this GIA uplift would then be on the order of 100 m $\times dv_s/dr$, which (taking dv_s/dr from the spherical seismic model PREM of *Dziewonski and Anderson* [1981]) is only about 0.01% of v_s: far smaller than the anomalies shown in Plate 2.

Presumably, then, this Canadian v_s anomaly is due to either thermal or compositional anomalies. It is far from clear, but at depths this large or larger, thermal anomalies could well dominate. In that case the positive v_s anomaly would imply a decreased temperature in the region, which would be associated with an increased density. The direct effect of this increased density would be to cause an overlying gravity high and a positive geoid anomaly. But the increased mass would also cause a depression of the overlying outer surface, as well as of all internal boundaries in the surrounding region, and these depressions would cause a gravity low and a negative geoid anomaly. Whether this latter, indirect effect is larger or smaller than the direct effect depends on the Earth's viscosity profile. *Peltier et al.* [1992] found that by using a then-current seismic model for v_s and adjusting the viscosity profile until it gave the best fit to the global geoid up to degree and order 8, they could obtain a gravity low over Northern Canada that agreed reasonably well with that shown in Plate 1. Their results do not prove conclusively that the Canadian gravity-geoid anomaly is solely the consequence of a deep-seated thermal density anomaly. In fact, *Simons and Hager* [1997] concluded that probably about half of this gravity anomaly is, indeed, caused by GIA. But all these results do suggest that caution should be exercised when using this anomaly to constrain GIA.

3. ONGOING CRUSTAL MOTION

If the gravity and geoid anomalies described above were representative of GIA only, they would be a measure of the total amount of compensation that must still take place in order for the Earth to be fully compensated. In contrast, the RSL data are observations of surface displacements as functions of time. The motion in that case occurred long ago, during or soon after the deglaciation phase. The displacements were large, in some places averaging several hundred meters over several thousand years, so that the geological measurements, while somewhat imprecise, are relatively unambiguous. Motion that large in those regions could only have been due to GIA.

Modern geodetic techniques permit the identification of ongoing crustal deformation at the mm or sub-mm level. Moreover, space geodetic techniques can be used to measure three-dimensional site motions, giving the prospect of using determinations of horizontal as well as vertical GIA deformations [*James and Morgan*, 1990]. Probably the most accurate space geodetic technique for this purpose is dual-frequency very-long-baseline interferometry (VLBI), which has been in use since the late 1970's [e.g., *Clark et al.*, 1985]. VLBI obtains its accuracy in part from the use of very large, and very expensive, radio telescopes. Because of the high cost involved in building and maintaining such facilities, the number of VLBI stations is small and their geographic density is not great. Early studies that used VLBI data to constrain GIA models [*James and Lambert*, 1993; *Mitrovica et al.*, 1993] found that significant sensitivity of VLBI-determined site velocities to mantle viscosity was limited to a small number of VLBI sites that happened to be located in or near areas undergoing GIA, such as sites Westford (Massachusetts) or Onsala (Sweden). This unfortunate situation renders VLBI data unsuitable for GIA studies when used alone. Inclusion of satellite-laser ranging sites does not alleviate this situation [*Argus et al.*, 1999].

A more recent applicable technique is geodesy with the Global Positioning System (GPS). The equipment required of the user is several orders of magnitude less expensive than for a VLBI site, and the cost continues to decrease. Used in the "continuous GPS" mode, the equipment is emplaced at a single site for a long period. Data from a continuous or "permanent" GPS site are typically downloaded and analyzed once per day, so a large number of position estimates are used to determine site velocities. The resulting velocities can be highly accurate, and the low equipment cost enables high site density to be obtained.

3.1. Sensitivity of Three-Dimensional Velocity Data

Crustal deformation due to GIA is a consequence of the specific time-dependent ice geometry ("ice model") as well as the viscoelastic response of the Earth. Uncertainties in either or both of these components will contribute to our uncertainty of ongoing crustal deformation. The goal of measuring ongoing crustal deformation due to GIA is therefore to reduce these uncertainties, and a geodetic network must be designed with these goals in mind. In addition to these geophysical goals, one must attempt to obtain as much redundancy (in the statistical sense) as the budget allows.

The first observational geodetic experiment having the explicit goal of measuring GIA deformations was the BIFROST Project [*BIFROST Project*, 1996; *Johansson et al.*, 2002]. With initial observations in 1993, BIFROST installed 31 permanent GPS receivers throughout Finland and Sweden, augmenting the three permanent GPS receivers already in Fennoscandia. (An additional GPS

site was later installed at Riga by another group.) The location of these sites, along with a numerical prediction of the radial motion due to GIA, is shown in Plate 3.

These predictions are based on a model for the Earth that is spherically symmetric, compressible, and with Maxwell viscoelasticity. The model has three viscosity layers including an elastic lithosphere with thickness 70 km, an upper mantle viscosity of 4×10^{20} Pa s and a lower mantle viscosity of 5×10^{21} Pa s. The elastic structure of our Earth model is taken from the seismically constrained PREM [*Dziewonski and Anderson*, 1981]. The ocean component of the surface load is computed via a revised sea-level algorithm that solves the sea-level equation [*Farrell and Clark*, 1976] in a gravitationally self-consistent manner while incorporating the effects of GIA-induced perturbations to the Earth's rotation vector [e.g., *Milne*, 1998] and the postglacial influx of ocean water/meltwater to once-ice-covered regions [*Milne*, 1998]. Predictions of the load-induced three-dimensional deformation rates are calculated via the theory of *Mitrovica et al.* [1994a].

The ice model is a hybrid of the high-resolution Fennoscandian ice model of *Lambeck et al.* [1998], which provides a good fit to local sea-level observations for our choice of Earth model, and the lower resolution global ice model Ice-3G [*Tushingham and Peltier*, 1991]. The hybrid was created by replacing the Fennoscandian and Barents Sea components of Ice-3G by the *Lambeck et al.* model. (The ice outside of Fennoscandia, especially within Laurentia, contributes significantly to GIA deformations within Fennoscandia. See, for example, *Mitrovica et al.* [1994b].)

In this section, we use an inverse approach [e.g., *Peltier*, 1976] to conduct a sensitivity analysis for the inference of the viscosity profile from three-dimensional site velocity information. For these calculations, we assume that we are solving for mantle viscosity only. (We adjust for the log of the viscosity.) We divide the mantle into 23 shell-like layers, with thinner layers in the upper mantle where greater sensitivity may be obtained. Elements of the design matrix for the inversion are obtained from Frechet kernels. An unconstrained inversion yields a covariance matrix with high correlations and large uncertainties, so constraints are required to stabilize the solution. In our case, we use very light constraints in the form of smoothing constraints. We apply these constraints in the form of observations that require equality between the log of viscosity for adjacent levels, with a standard deviation of 0.4. The uncertainties are assumed to be 1 mm/yr for the radial rates and 0.3 mm/yr for the horizontal. No constraint is im-

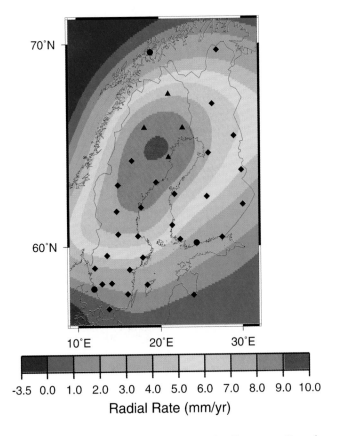

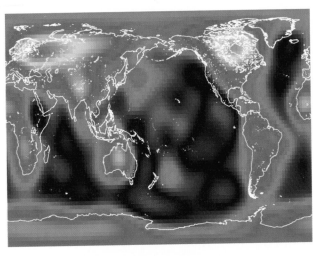

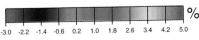

Plate 2. Global seismic tomographic results for the laterally varying component of the shear wave velocity at a depth of 200 km, plotted as a percentage of the spherically symmetric component at that depth. Results are from *Su and Dziewonski* [1997], and include all spherical harmonic coefficients up to degree and order 12, corresponding to spatial scales of about 1700 km and larger.

Plate 3. Predicted radial crustal rates for Fennoscandia calculated as described in the text. The locations of permanent GPS stations in this area that are used for scientific geodesy are shown. (The different symbols are used in the covariance analysis.) Most of these sites are part of the BIFROST network [*Johansson et al.*, 2002], the specific goal of which is to study GIA.

posed across the 670 km upper/lower mantle boundary. (A discussion of the implications of this boundary for modeling is given by *Peltier* [1986].)

We present the results of the sensitivity analyses for three different networks in Figure 1. In Figure 1a, we perform the analysis assuming we have data from only a small network of sites near the center of uplift, indicated by triangles in Plate 3. Our analysis indicates that the viscosity profile determined using radial rates only is relatively uncertain, especially for the lower mantle. The viscosity is much better determined from the horizontal components of velocity, although there is still a loss of sensitivity across the 670 km boundary. The greater sensitivity associated with the horizontal components of velocity arises from the direction information. A change in the viscosity that changes the radial rates by 10% may move the location of the center of uplift. Since the horizontal components of velocity radiate away from this center, such a viscosity change may result in a ~100% change in the horizontal components near the center of uplift. (This argument is similar to that made below regarding the sensitivity to ice model errors.) The "U" shape to the sensitivities results from the form of the constraints.

Figure 1b shows the analysis for three sites near the periphery of uplift, indicated by circles in Plate 3. The sensitivity of the radial rates for the lower mantle is somewhat better than that of Figure 1a, because the location of the peripheral bulge is very sensitive to the lower mantle viscosity [e.g., *Davis and Mitrovica*, 1996]. However, as with Figure 1a, the sensitivity associated with the radial rates is not commensurate with that of the horizontal rates. From an experimental viewpoint, in which we must consider that the errors will be to some extent unknown and must be assessed, it is not advantageous to have one subset of data dominate a solution.

The viscosity sensitivity from the full GPS network of Plate 3 is shown in Figure 1c. In this solution, the combined sensitivity is nearly constant except for the uppermost and lowermost mantle regions. The "U" shape is nearly gone, indicating that the constraint equations play a minimal role (although not negligible), and the sensitivities for the radial and horizontal components contribute more nearly equally, except perhaps in the deep lower mantle. These results suggest that a relatively dense geodetic network (the intersite spacing of the BIFROST network is ~100 km) that measures horizontal as well as radial motions could contribute significantly to the determination of mantle viscosity at depth.

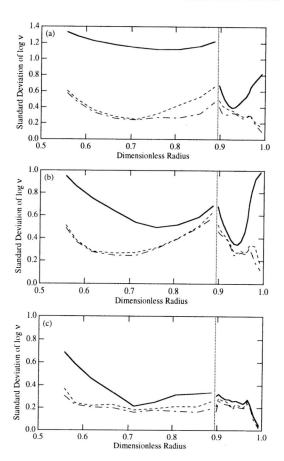

Figure 1. Results of sensitivity analyses for estimation of viscosity profile from Fennoscandian GPS data. Shown are the calculated standard deviation of $\log \nu$, where ν is the viscosity in Pa s. The sensitivity analyses are described in the text. Data consist of three-dimensional site velocities. Shown in each plot are the results for radial rates only (solid line), horizontal rates only (dashed line), and combined (long-short dashes). The vertical dotted lines indicate the 670 km upper/lower mantle boundary. (a) Network consists of four GPS sites located near the center of uplift (triangles in Figure 3). (b) Network consists of three GPS sites located near the periphery of uplift (circles in Figure 3). These GPS sites existed prior to the BIFROST project. (c) Network consists of all GPS sites shown in Figure 3. Note that the vertical scale is different for (a).

3.2. Ice Model Errors

In the previous section and throughout this article, we generally assume that we have perfect knowledge of the time-dependent geometry of the ice load (the "ice model"). In fact, the ice model is difficult to determine. Some measure of the global ice volume can be obtained from the $\delta^{18}O$ record [e.g., *Crowley and North*, 1991], and there is even fragmentary geologic evidence for pre-Quaternary glaciation (summarized, for example by *Benn and Evans* [1998]). A more detailed ice model is obtained from geologic evidence, and is therefore limited to the last glacial period, in the Late Pleistocene. Geomorphological data yield information regarding the time-dependent geometry of the ice margins and flow directions during the last retreat of the Late Pleistocene ice sheets. Such data may also be used to infer the location of the margins during stadials. Although these data constrain (within their limits of accuracy) the broader glacial morphology, the details of the timing and the ice thickness cannot be determined. Reconstruction of ice sheets therefore depends on a model for the interpretation of the geological data [e.g., *Kleman*, 1990].

Ice sheet reconstructions for the Late Pleistocene can also incorporate the sea level record. The Ice-3G [*Tushingham and Peltier*, 1991] model and its descendants, for example, use a global database of sea level records from the Late Pleistocene to infer the deglaciation history. The global ice model at any epoch is represented by several hundred discrete elements of ice, each of which can be varied until an adequate match is obtained, in a statistical sense, to the global sea level record. The geologic record of glaciation is used to guide the placement of ice. A consistent solution is found for the sea level and the vertical deformation due to the viscoelastic response of the Earth to the changing ice load [e.g., *Peltier and Andrews*, 1976]. The criticism of such global models is that they are non-unique, matched to a particular global mantle viscosity profile, and that there is no physics built into the ice sheets. In addition, since the goal is to match the sea level record globally, the ice history in any local area may be seriously deficient.

Lambeck et al. [1998] used a similar approach for northern Europe only. The geology and glacier physics were roughly incorporated by using the ice sheet history of *Denton and Hughes* [1981], who used a parabolic profile for ice sheet thickness. (This profile represents the solution for a steady-state perfectly plastic ice sheet on a horizontal bed [*Paterson*, 1998].) *Lambeck et al.* [1998] found, however, that this assumption produces ice sheets that are too thick to match the sea level record, and so they reduced the maximum thickness from ~3.4 km to ~1.5 km, with no attempt to impose a parabolic profile. *Milne et al.* [2001] demonstrated that the ice model of *Lambeck et al.* [1998] provides a better fit to the Fennoscandian GPS data [*Johansson et al.*, 2002] compared to that provided by the *Tushingham and Peltier* [1991] model.

A similar result for Fennoscandian ice sheet thickness was found by *Davis et al.* [1999], who introduced an ice model parameter that could be directly estimated from the data set. *Davis et al.* [1999] used modern tide-gauge records to estimate a parameter that scaled the thickness of the Fennoscandian ice sheet, based on the Ice-3G model [*Tushingham and Peltier*, 1991]. Their estimates for the ice-thickness scaling parameter ranged from 0.58 to 0.88, with the lower values being used with a mantle viscosity profile consistent with that of *Davis and Mitrovica* [1996].

The results of *Davis et al.* [1999] indicate that it is possible, at least under some conditions, to simultaneously estimate viscosity and ice model parameters. The data type used, sea-level rates from tide gauges, corresponds roughly in informational content to the radial velocity from a geodetic site. The horizontal motions may therefore be important in further separating viscosity and ice-model effects. To understand why this is so, consider an ice model error that can be described as a horizontal shift in the position of the ice sheet. We will make use of a simplified load based on a circular disk, after *Mitrovica et al.* [1994b]. At glacial maximum the disk has a basal radius of 8° and provides a rough approximation to the maximum geographic extent of the Fennoscandian ice sheet. The time dependence of the disk mass is a simple sawtooth function with a 90-kyr growth phase followed immediately by a 10-kyr deglaciation phase. Within each cycle the radius and height of the ice disk are varied in order to maintain plastic equilibrium [*Paterson*, 1998]. The final deglaciation event for the ice disk models is assumed to have ended 6 kyr B.P. The Earth model is chosen to yield a peak present-day uplift rate of 10 mm/yr.

The resulting radial and horizontal deformations are shown in Figure 2, along with the error caused by a mislocation of the center of the ice disk by 2°. (The mislocation is along the direction for which the horizontal rates are plotted.) The maximum error for the radial rates is ~35% of the peak radial motion, whereas the maximum error for the horizontal rates is ~82% of the peak horizontal motion. This peak horizontal error occurs near the center of uplift, where the expected horizontal velocities should be close to zero. Thus, the horizontal velocities should be highly sensitive to positions errors of the ice model.

Ice model errors may yield information on the Earth's climate, for the relationship between the climate and its great ice sheets is an intimate one. The primary cyclical climate variations during the late Quaternary are associated with the glaciation cycles of 100,000 year period (i.e., the Milankovitch cycles), during which glaciers of maximum thickness 3–4 km covered large parts of both hemispheres. These cycles are driven by the periodic changes associated with orbital precession and eccentricity, and in the Earth's obliquity and axial precession, all of which influence solar insolation [*Imbrie and Imbrie*, 1979; *Imbrie et al.*, 1992].

It would be a mistake, however, to view the ice-age glaciers as simply a response to climate changes. The response of the climate and the growth of glaciers further modifies the climate [*Imbrie et al.*, 1993]. General Circulation Models (GCMs) must take into account the response of and the interaction between the atmosphere, the oceans, the cryosphere, and the biosphere. One example of this interaction is the increase in the Earth's albedo for glaciated areas. Fresh snow has an albedo that exceeds 0.8 [e.g., *Paterson*, 1998]. This increased albedo causes a reduction in the absorbed solar radiation for glaciated areas causing, in turn, an extreme lowering of temperature, which *Williams* [1979] demonstrated is more critical to glacier growth than precipitation.

Glaciers are a huge sink of mass and energy. Precipitation that results in the formation of glaciers is not immediately returned to the climate system, but is stored within a glacier. Sea water can also contribute to the growth of glaciers. At the time of the last glacial maximum, 22–14 kyr B.P. [*Crowley and North*, 1991], the global sea level was reduced by ~120 m relative to that of today [*Fairbanks*, 1989]. Especially in the far northern latitudes, the lowered sea levels caused the glaciers to dam in large amounts of sea and fresh water, further accelerating glacier growth [*Hughes*, 1998].

The extent to which information on the ice model and the climate that drives the glaciers can be extracted from geodetic data remains to be determined. We will comment on some possible techniques for this in Section 6.

3.3. Sea-Level Observations

Strictly speaking, tide gauges are not geodetic instruments. Their intended purpose, as their name suggests, is to measure the change in sea-surface height. (In bodies of water without tides, these instruments are often referred to as "mareographs.") In fact, they measure changes in the sea-surface height relative to changes in the land height:

$$\Delta T(\theta, \phi) = \Delta S(\theta, \phi) - \Delta R(\theta, \phi) \qquad (3)$$

where ΔT is the change in the tide-gauge reading (the "apparent" sea-level change) at co-latitude θ and east longitude ϕ, ΔS is the "absolute" change in the sea-

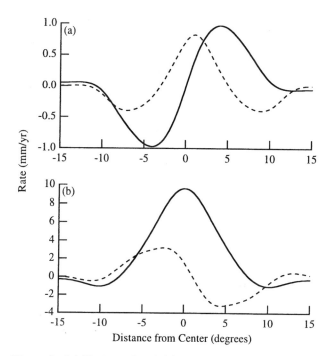

Figure 2. (a) Horizontal and (b) radial deformation rates associated with an idealized "ice disk" load (solid lines), and the error associated with a 2° offset in the position of the disk (dashed lines).

surface height, and ΔR is a change in the radial position of the crust to which the tide gauge is attached. The expression (3) may of course be differentiated with respect to time to yield a similar relationship between rates:

$$\dot{T}(\theta, \phi) = \dot{S}(\theta, \phi) - \dot{R}(\theta, \phi) \qquad (4)$$

where $\dot{T}$ is the rate of apparent sea-level registered by the tide gauge, and so on. $\dot{S}$ (or ΔS) may have contributions from a number of processes, including tides, pressure loading (inverted barometer), surface wind stresses, and thermal expansion. These changes represent a rearrangement of the existing volume of water. Similarly, $\dot{S}$ may have a component due to the response to the changing geoid caused by GIA. On the other hand, $\dot{S}$ may have a component due to changes in the volume of water caused, for example, by ongoing melting of glaciers. This last contribution will not itself produce a spatially uniform signal in $\dot{S}$, since it will be accompanied by spatially variable changes in the geoid as well as a spatially variable contribution to $\dot{R}$ [*Mitrovica et al.*, 2001]. Globally averaged, the signature of ongoing melting is believed to be between 1 and 3 mm/yr.

$\dot{R}$ (or ΔR) may have contributions from local processes like subsidence due to groundwater extraction,

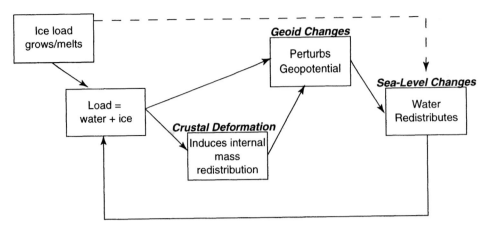

Figure 3. Relationship among changes in sea level, solid surface, and the geoid for GIA.

or from global or regional effects such as present-day changes in glacier loading and tectonic processes. GIA will also induce a spatially coherent global and regional signal. The GIA signal will contribute not only to $\dot{R}$ (with a maximum of 10–12 mm/yr in the northern hemisphere), but also contribute to $\dot{S}$ through changes in the geoid as the Earth adjusts.

Thus generally speaking, the forms of (3) and (4) notwithstanding, the contributions to apparent sea-level change (as observed by tide gauges) cannot be neatly separated into phenomena that affect the sea surface only and phenomena that affect the solid surface only. Because the ocean represents a surface load to which the solid Earth responds viscoelastically, and because redistributions of both the solid Earth and the oceans perturb the geoid to which the ocean surface conforms, the problem must be treated in a gravitationally self-consistent manner. This approach leads to the "sea-level equation" introduced by *Farrell and Clark* [1976]. The interrelationship between changes in sea level, solid surface, and the geoid, illustrated in Figure 3, is discussed in detail by *Milne* [1998].

The recent prospect of high-accuracy space-geodetic velocities has generated a re-examination of the sea-level theory of *Farrell and Clark* [1976]. As a result several significant improvements to the theory have come about. A time-dependent continent margin (TDCM) allows for the varying shape of the ocean load as the sea level changes [*Johnston*, 1993; *Peltier*, 1994; *Milne and Mitrovica*, 1998]. The sea-level equation is solved iteratively, and it has been assumed that the change in the sea level at any point is incremental and reflects the increments in water volume, solid-surface motion, and gravitational attraction. However, a glacial retreat in a region of the ocean previously covered by glacier will be accompanied by a relatively large inrush of water

that could not properly be described in terms of such increments. Therefore, the traditional sea-level theory must be modified for this "water dumping" effect [*Milne*, 1998; *Peltier*, 1998; *Milne et al.*, 1999]. Finally, the redistribution of the ocean load and solid Earth will induce true polar wander, to which the sea level will respond [*Han and Wahr*, 1989; *Milne and Mitrovica*, 1996; *Bills and James*, 1996].

GIA produces a global sea-level signature. In areas that were previously glacier covered, such as Fennoscandia and Laurentia, near the regions of maximum uplift $|\dot{R}| \gg |\dot{S}|$ so that $\dot{T}(\theta, \phi) \simeq -\dot{R}(\theta, \phi)$. Near the edge of uplifting areas, however, this approximation cannot be assumed, since $\dot{R}$ and $\dot{S}$ can be of the same order of magnitude. In the "far field" (areas far from previously glaciated areas), $\dot{R}$ can be near zero and $\dot{S}$ may actually dominate [e.g., *Mitrovica and Davis*, 1995].

The main aspect of sea-level observations that limits their usefulness is their accuracy. When straight lines are fit to annual means of tide-gauge observations, for example, the RMS residual is ~60 mm for Fennoscandia and ~20 mm for the east coast of North America. In some cases, regional common-mode errors may be filtered out to increase accuracy, but assumptions must be made which may introduce systematic errors [*Davis et al.*, 1999]. Nevertheless, the spatial pattern of the GIA signal may be distinct enough to obtain useful information from these data. *Davis and Mitrovica* [1996] used tide gauge data from the east coast of North America to estimate lower mantle viscosity. This was possible since the peripheral bulge (the area of subsidence surrounding the main uplift peak) makes a distinct pattern along the line of the North American east coast and the location of the peripheral bulge is controlled by the lower mantle viscosity.

A number of studies [e.g., *Peltier and Tushingham*,

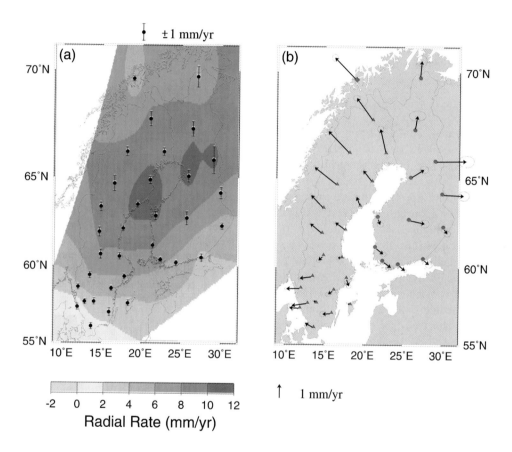

Plate 4. (a) Radial and (b) horizontal rates from the *Johansson et al.* [2002] study (after *Milne et al.* [2001]). The error bars in (a) show the 1-σ uncertainties. The error ellipses (b) are 1-σ. Reprinted with permission from *Milne et al.* [2001], Copyright 2001, American Association for the Advancement of Science.

1989; *Trupin and Wahr*, 1990; *Douglas*, 1997] have used numerical GIA predictions to remove the GIA signal from tide-gauge rates with the goal of determining the component of present-day glacier melting. A global increase in sea level is widely considered to be a sensitive indicator of climate change [*Houghton et al.*, 1990; *Woodworth et al.*, 1992; *Warrick et al.*, 1993]. The explicit assumption in all such analyses has been that present-day variations in the volume of polar ice complexes, which are thought to dominate the residual tide gauge trends [*Peltier and Tushingham*, 1989], are accompanied by a uniform, "eustatic," sea-level change [*Peltier and Tushingham*, 1989], and that any scatter in the GIA-corrected tide gauge trends reflects unmodeled effects or noise. In such an analysis, we rewrite (4) as

$$\dot{T}(\theta, \phi) = \dot{S}_{\text{GIA}}(\theta, \phi) - \dot{R}_{\text{GIA}}(\theta, \phi) + \dot{\mu} \qquad (5)$$

where $\dot{\mu}$ is the (assumed) eustatic sea-level rate, $\dot{S}_{\text{GIA}}$ is the component of sea-level rate due to the adjusting geoid, and $\dot{R}_{\text{GIA}}$ is the rate of adjustment of the solid surface.

In fact, this procedure does significantly reduce the geographic variation of apparent sea-level rates, resulting in much greater confidence in the estimate of present-day sea level rise [*Douglas*, 1997]. However, the assumption that the glacial-melt signature is geographically uniform is grossly invalid [*Mitrovica et al.*, 2001; *Plag and Jüttner*, 2001], since it ignores the readjustment of the solid surface and the geoid upon glacier melting. Moreover, the estimate of the sea-level rise is highly sensitive to the Earth model used to correct for GIA [*Mitrovica and Davis*, 1995]. Because of this problem, until the BIFROST GPS studies [*Johansson et al.*, 2002], the important Fennoscandian tide gauges have never been used in a study of global sea level rise. *Milne et al.* [2001] combined the radial rates from the BIFROST network with these tide-gauge data to obtain a regional Fennoscandian sea-level rate of $\dot{\mu}_{\text{Fenno}} = 2.1 \pm 0.3$ mm/yr (a positive value indicating a sea-level rise). These results are shown in Figure 4.

3.4. Results from Crustal Deformation Studies

As reviewed in the beginning of this section, the first studies to use space geodetic crustal deformation measurements to infer GIA signals [*James and Lambert*, 1993; *Mitrovica et al.*, 1993] depended on the rather sparse VLBI networks, and so the results were rather limited. *James and Lambert* [1993], for example, found that of 18 baselines examined only three exhibited significant motions, and that of these three only two seemed to be correlated with predicted GIA motions. *Mitrovica et*

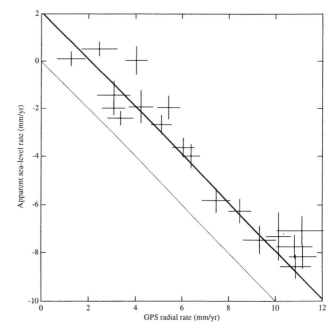

Figure 4. Rates of sea-level change determined from tide-gauge records at 20 sites in Fennoscandia corrected for regional geoid variations due to GIA versus GPS-determined radial velocities at the same (or nearby) sites. The dotted line is, following (5), the result for $\dot{\mu} = 0$. The solid line is the best-estimate through the data and it yields 2.1 ± 0.3 mm/yr of regionally coherent sea-surface rise. Reprinted with permission from *Milne et al.* [2001], Copyright 2001, American Association for the Advancement of Science.

al. [1993] were the first to attempt to constrain Earth models using these data. However, they also concluded that the sensitivity was limited to a few baselines, and that, in fact, a no-GIA "null model" could not be ruled out with high statistical confidence.

BIFROST Project [1996] used radial rates from 19 of the then-longest operating BIFROST sites to rule out a no-GIA "null model" for Fennoscandia with 99% confidence. A more detailed study from the full BIFROST network appeared in *Johansson et al.* [2002], and a discussion of this network appears elsewhere in this book [*Scherneck et al.*, 2001]. *Johansson et al.* [2002] examined the radial as well as the two components of horizontal velocity of 33 BIFROST sites (Plate 4) and demonstrated conclusively the correlation between geodetically measured velocity and predicted GIA velocities (calculated using the same model as in Section 3.1). Figure 5 (from *Johansson et al.* [2002]) shows these correlations. Excellent correlations for each velocity component are clearly evident, but systematic differences can be observed. Assuming a simple linear model for

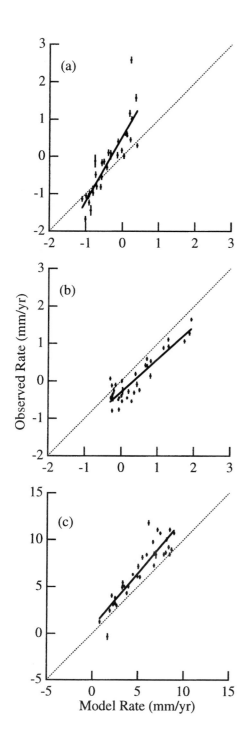

Figure 5. Comparison between the rates observed from the BIFROST GPS data and those predicted from GIA calculations for the (a) east, (b) north, and (c) radial components of velocity. From *Johansson et al.* [2002]. The dotted lines show the "model equals observed" relationship, whereas the solid lines are the best fit lines.

the error in the predicted rates we can determine the best-fit lines shown in Figure 5. The root-mean-square (RMS) residuals to this line are then a measure of the accuracy of the BIFROST rate determinations. *Johansson et al.* [2002] found a value of 0.4 mm/yr for east, 0.3 mm/yr for north, and 1.3 mm/yr for up. (These RMS residuals provided the basis for the uncertainties used in Section 3.1.)

The good correlation that was demonstrated by the *Johansson et al.* [2002] study indicated that the data would be of high enough quality to yield useful constraints on the viscosity profile. The resulting study [*Milne et al.*, 2001] was based on forward calculations only. Plate 5 shows contours of the χ^2 residual between the observations of *Johansson et al.* [2002] and GIA predictions based on a range of upper and lower mantle viscosities. The combined results (Plate 5c) indicate a χ^2 minimum at $\nu_{UM} = 8 \times 10^{20}$ Pa s and $\nu_{LM} = 10^{22}$ Pa s. Similar viscosity profiles have been reported using a variety of data [*Hager and Clayton*, 1989; *Davis and Mitrovica*, 1996; *Mitrovica and Forte*, 1997; *Simons and Hager*, 1997; *Lambeck et al.*, 1998].

The differing sensitivities between horizontal and radial observables discussed in Section 3.1 is clearly seen in the difference in the χ^2 contours of Plates 5a and b. However, there is also a difference between the minima for these data types. This perhaps is indicative of the inappropriateness of a radially symmetric three-layer (lower mantle, upper mantle, and elastic lithosphere) viscosity assumption, of an error in the effective thickness of the lithosphere used for these calculations (see Section 3.1), or perhaps of ice-model errors.

More complicated parametrizations of mantle viscosity can be obtained using inverse techniques. Generally, as in Section 3.1, the mantle viscosity is overparametrized, so that some type of constraints are required. Resolution studies are then required to determine the ranges in the mantle that can be independently determined (or nearly so) [*Mitrovica and Peltier*, 1991].

In the near future, data from other geodetic networks will be available to study this problem. In North America, for example, the PANGA network [*Miller et al.*, 1998] and Western Canadian Deformation Array [*Chen et al.*, 1996], both in the Pacific Northwest, will be sensitive to GIA deformation, although other sources of deformation are present in this area as well. A geodetic network on the east coast of the U.S. [*Nerem et al.*, 1997] has grown out of a focus on the Chesapeake Bay region. Some forward-analysis studies have been performed on these data, but it is too early in this project to yield high-accuracy results [*Park*, 2000]. A project

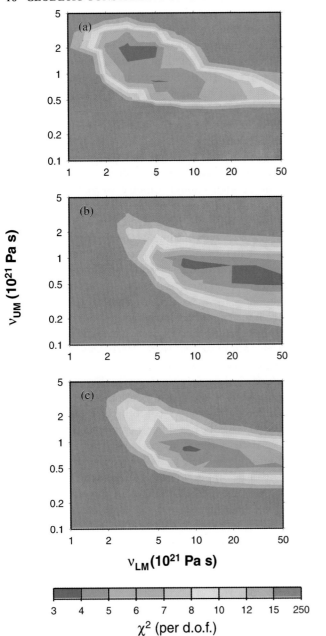

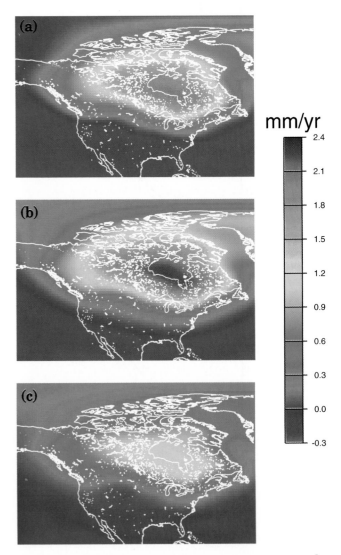

Plate 5. Contours for the χ^2 residual per degree-of-freedom (d.o.f.) between the observed Fennoscandian three-dimensional GPS site velocities [*Johansson et al.*, 2002] and GIA predictions based on various values for upper and lower mantle viscosities (ν_{UM} and ν_{LM}, respectively). The χ^2 was calculated using three different subsets of the observed velocities: (a) radial components; (b) horizontal components only; and (c) radial and horizontal components combined. Reprinted with permission from *Milne et al.* [2001], Copyright 2001, American Association for the Advancement of Science.

Plate 6. Predicted rate of change of the geoid over North America for three plausible GIA models, differing only in their assumed values for the lower mantle viscosity: (a) 4.5×10^{21} Pa sec, (b) 10×10^{21} Pa sec, and (c) 50×10^{21} Pa sec.

known as SEAL led by the Hermann von Helmholtz Association of German Research Centers will include a densified continuous GPS network in Canada (see URL http://op.gfz-potsdam.de/seal/.) Finally, the proposed Plate Boundary Observatory (PBO), if funded, will result in a dense network of GPS sites throughout western North America [*Silver et al.*, 1999]. Although the main goals of the PBO are related to tectonic deformation, the unique pattern of three-dimensional GIA deformation, the wide aperture, and the density of sites may make the PBO an extremely useful tool for studying GIA.

Geodetic measurements of crustal motion are also being used to study present-day motion, growth, and ablation of ice sheets. These are discussed in Section 6.

3.5. Gravimetry

Locally acquired gravity data have been an important source of information regarding GIA. The change in gravity at a site in an unglaciated area is due to the radial motion of the site as well as to the redistribution of mass in the Earth beneath the site. The former contribution (akin to the free-air correction) is theoretically about a factor of three larger than the latter contribution (akin to the Bouger correction). *Ekman and Mäkinen* [1996] used gravity data in part to investigate the role that a mantle phase change may have on GIA. They estimated a mantle flow parameter c whose value would be $c = 1$ if the gravity changes were due completely to viscous inflow associated with GIA and $c = 0$ for no inflow. Their estimated value of $c = 0.67 \pm 0.41$ (where the uncertainty represents a 95% confidence interval) seems to prefer the pure inflow model but, given the uncertainty, seems able to admit some intermediate result.

Although local gravity measurements are still being acquired, these have mainly given way to GPS observations, which are easier and less expensive to perform, especially over a dense network [*Larson and vanDam*, 2000]. However, one area where gravity measurements are quite important is in currently glaciated areas such as Antarctica and Greenland. Here, gravity measurements have been shown to be useful in discriminating between the elastic response of the Earth to present-day glacier load variations and GIA. This is discussed further in Section 6.2.

4. SATELLITE MEASUREMENTS OF TIME-VARIABLE GRAVITY

The ongoing redistribution of mass that is still occurring deep within the Earth as part of the GIA process,

causes temporal variations in the gravity field. These variations are part of the signal measured by surface gravimeters. But, as mentioned in Section 3.5, the measured signal depends more on the vertical displacement of the crustal surface beneath the gravimeter, than on the redistribution of mass within the Earth. The effects of mass redistribution are not only smaller than the effects of vertical motion of the instrument, but they can also be obscured by local gravitational effects that are unrelated to GIA, such as those caused by local changes in soil moisture and ground water beneath or near to the meter.

Satellite gravity measurements offer a method of determining changes in the gravity field that are insensitive to crustal motion (other than the extent to which that motion causes an actual change in gravity) and to very local non-GIA gravity effects. Furthermore, a satellite, in principle, is far better at mapping gravity fluctuations over large (hundreds to thousands of kms) spatial scales than are surface gravimeters. In effect, a gravity satellite can be envisioned as a large gravimeter network spread along the satellite orbit; though with the weakness that these "gravimeters" only record data at the instant the satellite passes through that location. This latter issue means that short-period gravity signals can conceivably alias into long-period satellite averages of the gravity field. Though with enough passes over or near the same locations, those aliased effects will tend to cancel.

The value of long wavelength, GIA-induced gravity fluctuations is that they are sensitive to viscosity deep inside the Earth; more deeply, for example, than are crustal motion observations. As a general rule of thumb, the depth sensitivity to viscosity is roughly the same order as the horizontal scale of the gravity fluctuation.

Plate 6 shows the predicted rate of change in the geoid over North America, for three plausible GIA models that differ only in their lower mantle viscosity (i.e., the viscosity between the core/mantle boundary and the seismic discontinuity at 670 km depth). Each model assumes that the viscosity of the upper mantle (i.e. the region between 670 km depth and the bottom of the lithosphere) is 10^{21} Pa sec, that the thickness of the elastic lithosphere is 120 km, that the Earth's elastic parameters (i.e., the density, and the elastic v_p and v_s velocities) are consistent with the seismic model PREM [*Dziewonski and Anderson*, 1981], and that the ice deglaciation history is adequately described by the *Tushingham and Peltier* [1991] Ice-3G model. Note from Plate 6 that the pattern of the signal is roughly the same for each of these three lower mantle viscosity values, a consequence largely of the fact that each model is driven with the same ice model; but that the amplitude of the signal depends quite strongly on that viscosity. At issue is whether it is possible to distinguish between these three GIA predictions, and others like them, using satellite measurements of time variable gravity.

One of the most serious obstacles is that satellite observations of GIA gravity fluctuations are not entirely free of contamination from other secularly-varying mass anomalies. Particularly troublesome for existing measurements have been the possible effects of on-going changes in the mass of the Antarctic and Greenland ice sheets. An unambiguous determination of the GIA signal requires that the satellite results have a short enough spatial resolution to be able to separate the secular gravity signals over northern Canada and Scandinavia from those over Antarctica and Greenland. Although this has proven difficult for existing satellites, the situation should soon improve dramatically.

4.1. Results from Satellite Laser Ranging

Results from satellite laser ranging (SLR), and especially from LAGEOS, have been used to determine secular changes in the longest-wavelength zonal gravity coefficients: the C_{l0} in (1) and (2). LAGEOS was launched in 1976, and so there are almost 25 years of available data that can be used to separate the secular variability from signals that are long-period but not truly secular. Measurement accuracy has steadily improved over this time, but even the early data are sufficiently accurate to permit the identification of secular variability. After just 5 years of LAGEOS measurements, for example, *Yoder et al.* [1983] and *Rubincam* [1984] were able to infer rates of secular change in C_{20} (1.3×10^{-11} yr^{-1} for *Yoder et al.* and 1.2×10^{-11} yr^{-1} for *Rubincam*) that are about equal to recent estimates that use the full time period of data [see, for example, *Eanes*, 1995; *Cheng et al.*, 1997]. What the more recent SLR data have permitted, however, are estimates of secular changes in C_{l0} for l larger than 2: up to $l = 6$ in some cases [e.g. *Cheng et al.*, 1997]. So far, the SLR results have not provided secular coefficients for any non-zonal (i.e. $m \neq 0$) term. It is more difficult to obtain those terms, because they represent gravity signals that change sign as the satellite orbits from west to east. Thus, unlike for the zonal terms which have the same sign at every longitude, the effects of a non-zonal term on a satellite orbit tend to cancel during any single revolution, and so do not build up as quickly.

The early secular C_{20} results were interpreted as being caused solely by GIA. Because C_{20} describes a global-scale variation (half-wavelength of 10,000 km), it is

particularly sensitive to deep-mantle viscosity. For example, Figure 6 shows the predicted values of $\dot{C}_{20}$ (the secular change in C_{20}) as a function of lower mantle viscosity. To generate this prediction, we used visco-elastic Green's functions computed as described by *Han and Wahr* [1995], for an Earth model with elastic constants taken from PREM, and assuming an upper mantle viscosity of 10^{21} Pa s, an elastic lithospheric thickness of 120 km, and a uniform lower mantle viscosity that is variable but that can take on any of the range of values shown on the x-axis in Figure 6. These Green's functions were convolved with the Ice-3G [*Tushingham and Peltier*, 1991] ice deglaciation model. Also shown in Figure 6 is a range of observed SLR values for $\dot{C}_{20}$ taken from the studies referenced above.

Note from Figure 6 that over this range of lower mantle viscosities, the predicted values pass through the observed values either for small ($\approx 2 \times 10^{21}$ Pa sec) or for large ($\approx 95 \times 10^{21}$ Pa sec) viscosity values. The explanation is that for small viscosity values the rebound is rapid, but since it has been rapid during the entire deglaciation history, there is not much uncompensated mass remaining. Whereas for large viscosity values the flow is slower, but there is a considerably larger mass of material that is still relaxing. Thus, either limit results in the observed $\dot{C}_{20}$ signal.

The difficulty with using the $\dot{C}_{20}$ observations to constrain GIA, is that there are other physical processes, most notably present-day changes in polar ice sheet mass, that could produce competing secular C_{20} signals. The present-day mass imbalance of the polar ice sheets is not well known, but recent estimates [*Warrick et al.*, 1996] suggest that Antarctica could have been thickening over the last century at an average rate of anywhere between -40 to $+40$ mm/yr (corresponding to a rate of global sea level rise of between 1.4 and -1.4 mm/yr), and Greenland from between -85 and 85 mm/yr (corresponding to a rate of sea level rise of between 0.4 and -0.4 mm/yr). If the upper bounds on the thickening rates are used to estimate the Antarctic and Greenland contributions to $\dot{C}_{20}$, the results (see Figure 6) are on the same order as or even larger than the effects of GIA.

Attempts can be made to separate the GIA and polar ice contributions to the secular gravity field observed with SLR. For example, it is possible, in principle, to separate northern hemisphere signals (Canadian, Scandinavian, and Greenland GIA + present-day Greenland mass imbalance) from southern hemisphere signals (Antarctic GIA + present-day Antarctic mass imbalance) by using all existing SLR secular zonal coefficients: up to $l = 6$, for example. Though this is complicated, in practice, by apparent systematic effects in the LAGEOS orbit that affect the odd degrees. Even if this separation were successful, it is not possible to use only the zonal terms (the C_{l0}) to meaningfully separate the Canadian or Scandinavian GIA signals from each other, or from either the GIA or present-day signals over Greenland. For example, the difference in longitude between the center of Greenland and the center of Hudson Bay is about 45°. This implies that the separation of the Canadian GIA signal from the Greenland GIA + present-day signals would require solutions for the nonzonal coefficients up to at least $m = 360/45 = 8$. The implication of all this is that it is dangerous to use the existing secular C_{20} measurements to constrain lower mantle viscosity.

4.2. Expected Contributions from GRACE

Satellite measurements of time-variable gravity will become much less ambiguous after the launch of the Gravity Recovery and Climate Experiment (GRACE) in early 2002, under the joint sponsorship of NASA and the Deutsches Zentrum für Luft- und Raumfahrt. Scheduled for a nominal 5-year lifetime, GRACE will consist of two satellites in identical low-Earth orbits (an initial altitude in the range of 450–500 km) and a few hundred kilometers apart, that range to each other using microwave phase measurements. Onboard GPS receivers will determine the position of each spacecraft in a geocentric reference frame. The low altitude compared with SLR satellites (LAGEOS, for example, has an altitude of about 6000 km) plays a large part in the ability of GRACE to deliver the gravity field at high spatial resolution (see below). But the low altitude also means greatly increased problems from atmospheric drag. To deal with these problems, onboard accelerometers will be used to detect the nongravitational acceleration so that its effects can be removed from the satellite-to-satellite distance measurements. The residuals will be used to map the gravity field.

The gravity field will be determined by GRACE orders of magnitude more accurately and to considerably higher spatial resolution than by any prior satellite mission. It will permit temporal variations in gravity to be determined down to scales of a few hundred kilometers and larger, every 30 days. These temporal variations can be used to study a large number of problems in a number of disciplines, from monitoring changes in water, snow, and ice on land, to determining changes in seafloor pressure. Comprehensive descriptions of these and other applications are given by *Dickey et al.* [1997] and *Wahr et al.* [1998].

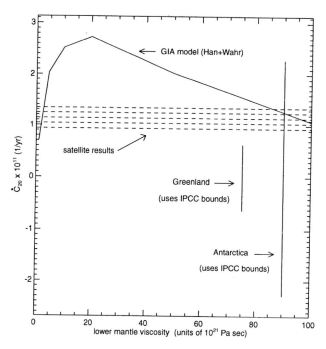

Figure 6. Predicted and observed values of $\dot{C}_{20}$. Shown are predicted results from the GIA models of *Han and Wahr* [1995] as a function of lower mantle viscosity, as well as the possible range of contributions from present-day changes in Antarctic and Greenland ice. The range of observational results obtained using satellite laser ranging is shown with the horizontal lines. Note that the contributions from Greenland and, especially, Antarctica could be as large as the GIA contributions.

The improved gravity field available from GRACE will extend the role time-variable gravity plays in GIA studies for two reasons. One is that the GRACE measurements will be accurate enough to be sensitive to the GIA signal to much higher degrees and orders than is now available with SLR.

For example, Figure 7a compares the expected errors for GRACE (solid line) with the size of the GIA signal estimated for a specific viscosity profile (the "default model", represented by the "+" symbols). The secular degree amplitudes plotted in this figure are defined as

$$N_l^{sec} = a\sqrt{\sum_{m=0}^{l}\left(\dot{C}_{lm}^2 + \dot{S}_{lm}^2\right)} \qquad (6)$$

where $\dot{C}_{lm}$ and $\dot{S}_{lm}$ are either the secular rates of change of C_{lm} and S_{lm} caused by the predicted GIA signal (the "+" symbols in Figure 7a), or are the estimated uncertainties in those quantities as delivered by the GRACE measurements (the solid line in Figure 7a). For the

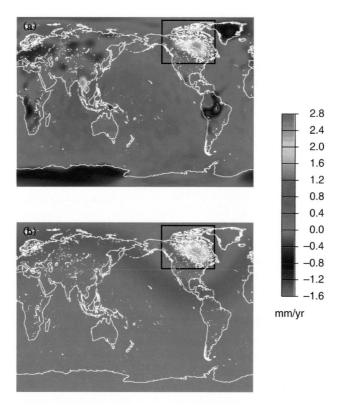

Plate 7. (a) The simulated secular change in the geoid over five years. Obtained by adding the predicted GIA signal for the default model, to predictions of the secular geoid contributions, over a five year period, from continental water storage, from changes in Greenland and Antarctic ice, from the distribution of water in the oceans (including an assumed global sea level rise plus an additional assumed change in Hudson Bay), from errors in the atmospheric mass corrections to the GRACE data, and from a realization of the GRACE observational errors. (b) The predicted contribution to the secular geoid change from the default GIA model over Canada. At issue is how well it will be possible to extract the GIA signal shown in (b) from the total GRACE estimate shown in (a).

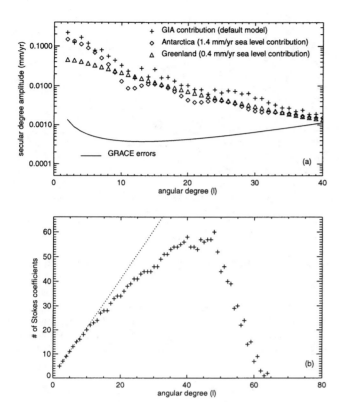

Figure 7. (a) Predictions of degree amplitudes for the secular change in the geoid, as a function of angular degree l. Shown are predictions from our default GIA model (see text), as well as upper bound estimates from present-day changes in Greenland and Antarctic ice. All results are larger than the degree amplitudes of the expected secular GRACE measurement errors (Thomas and Watkins, personal communication; solid line) for degrees up to $l = 40$. Note that the contributions from present-day changes in polar ice can be nearly as large as the GIA contributions. (b) The number of Stokes' coefficients at a given value of l, where the amplitude of the predicted GIA signal is larger than the expected secular GRACE measurement error at that value of l. The GIA signal is predicted using our default model. The dotted line shows the total number of Stoke's coefficients as a function of l. Note that for each l less than about 40, the predicted GIA contributions for over half of the available Stokes' coefficients are larger than the expected GRACE errors. All together, there are a total of 2166 Stokes' coefficients where the predicted GIA signal rises above the expected GRACE errors.

GRACE errors, we use preliminary 30-day GRACE uncertainty estimates provided by Brooks Thomas and Mike Watkins at the Jet Propulsion Laboratory, and assume a 5-year mission length. For the "default" GIA model, we assume upper and lower mantle viscosities of 10^{21} and 10^{22} Pa s, respectively, a lithospheric thickness of 120 km, elastic parameters from PREM, and

the Ice-3G ice model. Note from Figure 7a that the expected GRACE errors are smaller than the predicted GIA signal for all degrees smaller than at least $l = 40$. Since there are $2l + 1$ values of m for every l, and since the GRACE errors are expected to be roughly independent of m, there are likely to be on the order of 40^2 secular geoid coefficients where the GIA signal is larger than the GRACE measurement accuracy. In fact, Figure 7b shows the number of Stokes' coefficients, for each l, where the expected secular errors in the GRACE measurements are smaller than the GIA contributions predicted using the default model. The dotted line in this figure shows the total number of Stokes' coefficients (i.e. $2l + 1$) for each l. In all, there are between 2100 and 2200 Stokes' coefficients where the predicted GIA signal is larger than the expected GRACE measurement errors, as opposed to maybe 5 coefficients—all zonal—which have been detected using SLR.

Figure 8 shows similar comparisons between the expected GRACE errors and the degree variances of the predicted difference between GIA signals for two plausible lower mantle viscosity values (10×10^{21} and 50×10^{21} Pa sec), and for two plausible upper mantle viscosity values (1×10^{21} and 0.6×10^{21} Pa sec). Note that the GRACE errors are smaller than the differences between the two lower mantle models for all degrees less than $l = 18$, corresponding to spatial scales of greater than about 1100 km. This illustrates the sensitivity of GRACE to lower mantle viscosity, and suggests the utility of GRACE for constraining that quantity. The sensitivity to upper mantle viscosity extends out to higher degrees than does the lower mantle sensitivity, but is substantially smaller at low degrees.

Also shown in Figure 7a are the degree variances that could be caused by the present-day mass imbalances of Greenland and Antarctica. To estimate these degree variances, we have used the upper bounds from *Warrick et al.* [1996] (± 40 mm/yr of average ice thickness change for Antarctica, and ± 85 mm/yr for Greenland), and have assumed these changes in ice are distributed uniformly over each ice sheet. The general conclusion from these figures that the polar ice sheet signal is likely to be of the same order as the GIA signal and much larger than the GRACE measurement errors down to degrees of $l = 40$, would be unaffected if the rates of ice thickness change were assumed to be much smaller.

The results in Figure 7 illustrate that GRACE will be highly sensitive to the mass imbalance of the polar ice sheets. But they also raise the concern that the effects of polar mass imbalance could contaminate the GRACE GIA estimates. To assess the value of GRACE for learning about GIA, it is not enough to just consider the sensitivity of GRACE to the GIA signal. It is also important that GRACE provides enough spatial resolution to be able to separate the GIA signal from all other secular signals, especially those from the polar ice sheets.

This is where GRACE has a second advantage over SLR. Because the GRACE degree amplitude errors are smaller than both the GIA and polar ice signals for all degrees ≤ 40, and since for any given l the GRACE errors will be approximately independent of m, every $\dot{C}_{lm}$ and $\dot{S}_{lm}$ with $l \leq 40$ can be used to separate the different secular signals. Since a degree of $l = 40$ corresponds to a spatial scale of about 500 km, this provides sufficient spatial resolution to separate Canada, Scandinavia, Greenland, and Antarctica.

To assess this conclusion, we generated a suite of simulations of GRACE data to obtain a rough estimate of how well GRACE will be able to deliver the GIA signal. For each simulation we constructed five years of monthly, synthetic, C_{lm} and S_{lm} values that included the effects of GRACE measurement errors (using the estimated errors described above), of a particular GIA model, of secular Antarctic and Greenland mass imbalance signals (assuming various values lying within the bounds from *Warrick et al.* [1996] described above), of the secular gravitational effects of a 1.0 mm/yr global sea level rise that we assumed is caused by an increase in oceanic mass, of month-to-month variations in global water storage (computed using global, gridded water storage fields estimated by C. Milly and K. Dunne, personal communication), and of month-to-month variations in sea floor pressure caused by changes in oceanic circulation (rather than by the addition of water to the oceans) and predicted from an ocean general circulation model. (See *Wahr et al.* [1998] for a more complete description.)

We have also added an entirely ad hoc, additional sea level rise in Hudson Bay of between ±1.0 mm/yr, assumed to be caused by an increase in oceanic mass in the region. Although there is no reason to expect an anomalous sea level rise of that magnitude in Hudson Bay, we include it in the hope of obtaining reasonably conservative bounds on the GIA separation error, noting that an unmodeled secular increase in the amount of water in Hudson Bay will cause a gravity signal with a spatial pattern that is similar to that produced by GIA in the region.

The upper panel in Plate 7 shows the secular change in the geoid for one such simulation, truncated to degrees $l \leq 30$. For this simulation sea level contributions

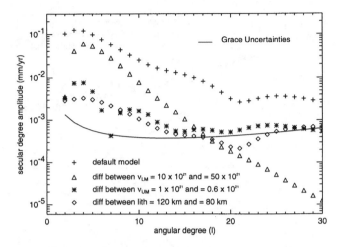

Figure 8. Compares the degree amplitudes of the expected secular Grace measurement errors, to the degree amplitudes of the default GIA model, and to the degree amplitudes of the differences between the default GIA model and three other plausible GIA models, obtained by modifying either the upper or lower mantle viscosities or the lithospheric thickness. The results imply that the GRACE measurements should be sensitive to all three parameters, but particularly to the lower mantle viscosity.

were assumed to be 1.4 mm/yr for the Antarctic and 0.4 mm/yr for Greenland (the upper bounds from Warrick, et al.). The GIA model used in this simulation assumed upper mantle and lower mantle viscosities of 10^{21} and 10^{22} Pa sec respectively, a lithospheric thickness of 120 km, and the Ice-3G deglaciation model, including ice sheets in Canada, Scandinavia, Antarctica, and Greenland. The Antarctic geoid signal is not as pronounced in the simulation shown in Plate 7 as it is for some of the other choices of present-day melting that we consider for this assessment, because in this simulation the effects of present-day Antarctic melting are of the opposite sign of the GIA response to the Holocene Antarctic deglaciation.

The results plotted in the upper panel in Plate 7 thus represent a secular geoid signal that might be extracted from the GRACE data. Note that the Canadian GIA signal (shown alone in the lower panel) is clearly evident in the total signal (upper panel), as are signals over both Greenland and Antarctica, as well as the effects of hydrology, particularly over southeast Asia and the Amazon basin in South America. Also present, though it is not as evident as some of the other signals, are the effects of the assumed 1 mm/yr global sea level rise.

The fact that the Canadian GIA signal can be so clearly identified by simply looking at Plate 7, is an indication that GRACE will be able to deliver this signal reasonably free of contamination from other effects. To obtain a rough estimate of the level of this contamination, we truncate the total GRACE result (top panel of Plate 7) and the GIA signal (bottom panel) to the region inside the box shown in both panels. We remove the spatial mean over this region from both signals, and least-squares fit the residual GIA signal to the total signal over this region. If there were no contamination, we would expect our solution for this fit parameter to be unity.

For the parameters used to construct the simulation shown in Plate 7, we obtain a fit parameter of 0.96. When we repeat this procedure for other simulated data sets for GRACE, obtained by varying the present-day Antarctic and Greenland ice thickening rates and the ad hoc Hudson Bay sea level rise, we obtain fit parameters that vary between about 0.94 and 0.96, which we crudely interpret as implying that GRACE would be able to recover the Canadian GIA signal to an accuracy of about 4–6%. This result is obtained without attempting to simultaneous fit any other parameters (e.g., present-day Antarctic and Greenland mass changes, global sea level rise). For example, a large part of the 4–6% error is due to the effects of the assumed 1 mm/yr global sea level rise. The effects of the sea level rise do not stand out clearly in the upper panel of Plate 7, because of their relatively small amplitude. But their pattern over northern Canada is negatively correlated with the GIA pattern: smallest near the center of the continent because of its distance from the ocean, and larger closer to the coasts. When we simultaneously fit a global sea level rise to the data along with the GIA signal, we obtain GIA fit coefficients that vary between 0.97 and 0.99, suggesting an error in the GIA recovery of 1–3%. We expect that better estimates could be obtained by also including present-day Antarctic and Greenland parameters in the fitting process, though in that case it would probably be desirable to consider other regions in addition to the region inside the Plate 7 box.

This simple simulation does not address the larger issue of estimating the improvement we can expect in our knowledge of the Earth's viscosity profile as a result of GRACE. Resolving that issue would require a simultaneous inversion of simulated GRACE data and other GIA-related observations, and should also take into account the likely uncertainties in the ice deglaciation model. What is clear at this stage is that because of the presence of the various non-GIA signals, the number of independent, useful GIA constraints provided by GRACE will be significantly smaller than the $\approx$2100–

2200 coefficients estimated above. Still, GRACE will provide a large number of new constraints on the GIA signal that have the potential of significantly improving our understanding of the GIA process, the Earth's viscosity profile, and the ice deglaciation history.

5. VARIATIONS IN THE EARTH'S ROTATION

Another type of observation that has occasionally been used to constrain GIA is the measurement of variations in the Earth's rotation. These variations take two forms: changes in the rate of rotation, and variations in the position of the rotation axis.

5.1. Variations in the Rotation Rate

A change in the Earth's rotation rate is often described in terms of the effect it has on the length of a day (LOD). A decrease in rotation rate implies an increase in the LOD, for example.

Ancient eclipse data show that the LOD has been increasing over at least the past 2,500–3,000 years at an average rate of about 1.5–2.0 msec/century. This is illustrated in Figure 9, taken from *Stephenson and Morrison* [1995], which shows ΔT (the negative of the change in Universal Time) between about 700 BC and AD 1990. (The detailed results shown in the inset of Figure 9 from after about AD 1625 represent telescopic observations of lunar occultations.) ΔT is the difference between the time that would have been recorded on an atomic clock, and a measure of time as determined by observing objects in the sky (the Sun and Moon, for example). The time derivative of ΔT is proportional to the change in the LOD.

Stephenson and Morrison [1995] found that the data shown in Figure 9 can be fit reasonably well with a parabolic time dependence of the form

$$\Delta T = A + B \left(t - t_0 \right)^2 \quad , \qquad (7)$$

where t is the time along the x-axis in Figure 9, t_0 is a reference time corresponding to the cusp of the parabola, and A and B are constants, with B found to be positive. Because of the t^2 time dependence in (7), and because B is positive, the LOD is thus inferred to be a linearly increasing function of time. Stephenson and Morrison conclude that the LOD has been increasing over this time period at a rate of 1.7 msec/century.

Most of the secular increase in the LOD is caused by a torque on the Earth from the Moon and Sun, that is indirectly related to tidal friction in the Earth's oceans. This tidal friction effect can be estimated independently of Earth rotation measurements, either by using Lunar

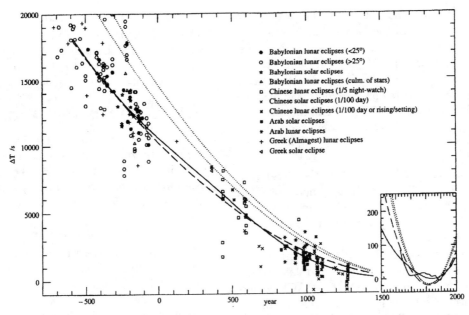

Figure 9. The negative change in Universal Time between about 700 BC and AD 1990, as inferred from ancient eclipse data. The detailed results shown in the inset represent telescopic observations of lunar occultations after about AD 1625. The results imply that the length of a day has been increasing over the past 2,500–3,000 years at an average rate of about 1.5–2.0 msec/century. Reprinted from Figure 5 of *Stephenson and Morrison* [1985] with permission of the authors and the Royal Society.

Laser Ranging (LLR) observations of the evolution of the lunar orbit, or from SLR determinations of the effects of ocean tides on satellite (e.g. LAGEOS) orbits. Those independent estimates suggest that if tidal friction were the only excitation source, the secular increase in LOD rate would be about 2.3 msec/century [*Williams et al.*, 1992]. For example, the top panel of Figure 10 compares the LLR tidal friction estimate from *Williams et al.* [1992] with *Stephenson and Morrison's* [1995] estimates of the LOD (which they inferred by fitting cubic splines to the data shown in Figure 9, and then taking the time derivative of those splines). The results suggest that there is some additional mechanism causing a secular decrease in the LOD of about 0.6 msec/century.

Although the eclipse data show considerable scatter over this 2700 year period, there is indirect, supporting evidence for a non-tidal, secularly decreasing LOD variation. A change in C_{20} can be shown to be directly proportional to a change in the Earth's polar moment of inertia. Thus, the secular change in C_{20} inferred from the ≈ 25 years of SLR observations implies a secular change in the Earth's moment of inertia over that 25-year period. To conserve the Earth's angular momentum, a change in the moment of inertia must be accompanied by a change in rotation rate: a spinning figure skater will spin faster, for example, by raising her hands over her head.

When the $\dot{C}_{20}$ results from SLR are used in this way to estimate the corresponding change in LOD over the last 25 years, the result is found to be in good agreement with the 0.6 msec/century, non-tidal, secular decrease in the LOD (see the bottom panel of Figure 10). The implication is that the non-tidal secular decrease in LOD over the past 2700 years has been caused by a secular change in the Earth's mass distribution, and that this secular change has tended to look about the same over the last 25 years (the time period of LAGEOS) as it has over the last 2700 years. A natural conclusion is that this process is GIA. As a result, this non-tidal, secular LOD decrease has been used to constrain the Earth's viscosity [e.g., *Peltier*, 1983; *Gasperini et al.*, 1986].

In principle, though, this constraint adds no new information to the $\dot{C}_{20}$ constraint already provided by SLR: the non-tidal LOD decrease is directly proportional to $\dot{C}_{20}$. The LOD constraint is subject to the same contamination from the possible effects of changes in polar ice. The fact that the 25-year rate and the 2700-year rate are about the same, does perhaps strengthen the expectation that the effects of GIA may be more important than those of polar ice, given that the GIA rates are probably more likely to have been reasonably constant over a 2700-year time frame than are changes in polar mass. On the other hand, the eclipse data do not really argue for a uniform rate over this time period.

In fact, the LOD results shown in Figure 10 suggest the presence of a large, multi-century-scale, non-secular LOD variation. In short, the issue of polar ice contamination is probably not any better resolved by the addition of the LOD data.

5.2. Secular Motion of the Rotation Axis

In 1891 the American astronomer S.C. Chandler discovered that the Earth's polar axis of rotation moves along an approximately circular path relative to fixed points on the Earth's surface, with about equal power at periods of 12 and 14 months. The 12-month "Annual Wobble" is now known to be a forced motion, driven by seasonal variations in atmospheric pressure and, to a lesser extent, in continental water storage. The 14-month "Chandler Wobble" is a free mode, excited by some quasi-random process either within the Earth or on or above its surface.

In 1899, shortly after Chandler's discovery, the International Latitude Service (ILS) was inaugurated. The ILS was a network of 4–6 optical astronomical observatories distributed along 39° 08' North latitude, that monitored the position of the Earth's rotation axis with the initial goal of improved understanding of the Chandler and Annual Wobbles. It was renamed the International Polar Motion Service in 1962. The optical network effectively ceased operation in 1988, when its role was supplanted by the more accurate techniques of VLBI, SLR, and GPS.

Thus there exists a 100-year record of the motion of the Earth's rotation axis. That record shows that in addition to the Chandler and Annual Wobbles, there is an unexplained decadal wobble (sometimes referred to as the Markowitz wobble) with an irregular period of about 30 years, and a secular drift of the pole of about 3.5 masec/year in a direction lying about midway between Hudson Bay and Greenland [*Dickman*, 1981; *Gross and Vondrak*, 1999].

The secular drift has sometimes been interpreted as the result of GIA [e.g. *Nakiboglu and Lambeck*, 1980; *Sabadini and Peltier*, 1981]. The interpretation is that as mass within the mantle continues to viscously adjust to the deglaciation process, there is a corresponding secular change in the Earth's principal axes of inertia. To conserve angular momentum, the Earth's rotation axis must also move. The observed polar drift has thus sometimes been used as an additional constraint on GIA, and has been found to be particularly sensitive to lower mantle viscosity.

But, like many of the other constraints described above, the interpretation of the secular polar motion

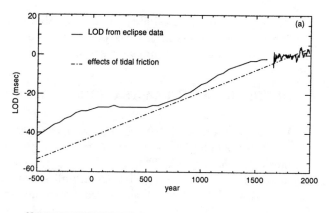

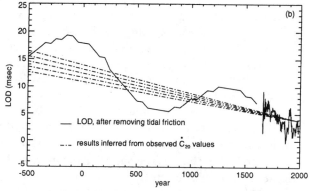

Figure 10. (a) The length-of-day results estimated from *Stephenson and Morrison's* [1995] Universal Time (UT) data (Figure 16), compared with the effects of tidal friction as inferred by [*Williams et al.*, 1992] from Lunar Laser Ranging data. Note that the UT results appear to require some other secular mechanism (besides tidal friction) to decrease the length of day. (b) The length of day results inferred from the UT data, after removing the lunar laser ranging estimate of the effects of tidal friction. These residual results are compared with the rate of change in the length of day implied by satellite laser ranging estimates of $\dot{C}_{20}$. Note that the trend in the eclipse residuals is in good agreement with the $\dot{C}_{20}$ results.

is ambiguous. There could, for example, be substantial contributions from present-day changes in the polar ice sheets. The effects of Antarctica are likely to be relatively small in this case, in contrast to its effects on $\dot{C}_{20}$ and the LOD. The reason is that Antarctica is centered nearly over the Earth's rotation axis, so that a change in ice mass has only a relatively small effect on the Earth's principal axis of inertia. Mass imbalance in Greenland, however, could have a substantial effect. In addition, models have shown that the steady mass redistribution associated with imperfectly compensated tectonic motion could cause secular polar motion that could easily be a sizable fraction of the observed drift [*Vermeersen et al.*, 1994; *Steinberger and O'Connell*, 1997; *Richards et al.*, 1999]. Again, the possibility of these sizable contaminating signals suggests that the observations of the secular polar drift are an unreliable source of information about GIA.

6. GEODETIC CONSTRAINTS ON ICE LOADING VARIATIONS

As discussed above, the effects of ongoing changes in the mass of the polar ice sheets can contaminate geodetic measurements of GIA. The SLR secular geoid coefficients and Earth rotation variations are all sensitive to mass variability on global scales, and so cannot effectively separate the ice sheet and GIA contributions. Geodetic observations of crustal motion in Canada and Fennoscandia (far from existing ice sheets) are not significantly affected by the ice sheets variations, but such variations will have a global sea-level signature. Time variable gravity observations provided by GRACE will have sufficient spatial resolution to allow separation of the effects of the ice sheets from the GIA signals arising from historical glaciation in Canada and Fennoscandia. But it will be difficult, at best, to separate the GIA signal caused by the Holocene deglaciation of Greenland or Antarctica from the effects of present-day ice sheet imbalance. For example, if GRACE data show evidence of secular changes in gravity above Greenland, it will not be clear from the GRACE data alone how much of that signal might be due to viscous flow in the Earth caused by changes in Greenland ice over the last few thousand years, and how much might be due to present-day changes in the direct gravitational attraction of the Greenland ice sheet. The same sort of separability problems occur when interpreting measurements of crustal deformation near the Greenland or Antarctic ice sheets.

In Section 3, we reviewed how errors in the reconstruction of the historical ice load affect the interpretation of geodetic determinations of crustal deformation associated with GIA. Geologic data are useful in performing these reconstructions, although such data cannot yield unique solutions. In previously covered regions, ice model errors are probably the limiting source of error in predicting GIA deformations, especially the horizontal components of deformation.

All of this suggests that improved constraints on present-day ice sheet mass imbalance and historical ice cover could lead to improved geophysical results obtained from GIA data. Mass imbalance has traditionally been estimated by determining the rate of ice flow out of a region by measuring the motion of markers attached to the ice (including, most recently, GPS receivers), and combining those measurements with estimates of the snow accumulation over the region. The recent development of high-precision altimetry, perhaps combined with other types of geodetic observations to separate the GIA effect, is now providing an alternative and improved method of measuring mass imbalance over large regions.

Because crustal deformation measurements are affected by the history of ice loading, there is at least in principle the possibility that these data, also, can be used to improve the ice model. This subject is discussed further below.

6.1. Altimetric Measurements

Radar altimetry from Earth-orbiting satellites has developed into an exceedingly useful oceanographic measurement technique, primarily through its estimates of sea surface height variations. The altimeter generates a radar pulse that travels downward from the satellite, reflects off the ocean surface, and is detected back at the satellite. The round-trip travel time is determined, and transformed into an estimate of the satellite-to-ocean distance, using an estimate of the refractive index of the air the pulse traveled through. The satellite orbit is independently determined from satellite tracking data (e.g. SLR, on-board GPS, DORIS Doppler radio measurements, etc.), so that the satellite's position is known relative to a geocentric reference frame. By combining the orbital results with the altimeter measurements, the sea surface can be mapped in the geocentric reference frame. By mapping the sea surface over successive repeating orbits, time variations in sea surface height can be determined.

It has been found that these same radar altimeters can be used in a similar manner to determine changes in ice sheet elevations [for a review, see *Zwally and Brenner*, 2001]. Altimeter estimates of round-trip travel time are not as precise for the ice sheets as they are for the ocean. Ice sheet surfaces tend to have significant slopes and undulations over the scale of a radar footprint (typically 15 to 40 km in diameter, depending on the altimeter), whereas the ocean is relatively flat (the effects of waves tend to cancel over a radar footprint). This causes degradation of the reflected pulse from an ice sheet surface, which decreases the precision of the travel time. Still, the measurements have proven accurate enough to provide useful mass imbalance estimates over regions with small-to-moderate slopes, which tend to include a substantial portion of the ice sheets.

For example, *Davis et al.* [2000] compared altimeter results from Seasat (data acquired during 1978) and Geosat (data acquired between 1985 and 1988) to estimate the mass imbalance of the southern, interior portion of the Greenland ice sheet between 1978 and 1988. Their estimates only included latitudes below 72° N, since that was the maximum latitude covered by the altimeter ground tracks. And they were able to obtain useful returns only for the ice sheet interior, corresponding approximately to elevations lying above 2000 m, due to degradation of reflected pulses from the rough ice sheet topography outside that region. Their results showed secular changes in ice sheet elevations during this period that varied between about -180 and 240 mm/yr, with a spatial average over this region of only 13 ± 6 mm/yr.

As another example, *Wingham et al.* [1998] used ERS altimeter data from 1992-1996 to estimate changes in Antarctic elevations during this 5-year period. They were restricted to latitudes of less than about 81° S, since that is the maximum ground track latitude, and they could not obtain estimates near the coast due to significant degradation of the reflected pulses from those area, and from a few other regions due to tape-recorder limitations. But they were able to obtain estimates over 63% of the grounded ice sheet. They found that the average rate of increase in ice sheet elevation over this region during 1992-96 was -9 ± 5 mm/yr.

The spatial averages found by *Davis et al.* [2000] and *Wingham et al.* [1998] are far smaller than the upper bounds estimated by *Warrick et al.* [1996] for the century-scale rates of change averaged over the entire ice sheets (± 40 mm/yr for Antarctica, and ± 85 mm/yr for Greenland). But the altimeter results are for only a portion of each ice sheet, and reflect changes over relatively short time periods.

The spatial coverage can be extended using laser altimeters, which have much smaller footprints than radar altimeters and so are less sensitive to surface slope and undulations. *Krabill et al.* [1999] used measurements from an airborne laser altimeter that flew repeat tracks over a portion of the southern Greenland ice sheet during 1993 and 1998, to estimate changes in ice sheet elevations during this time period. Because their laser altimeter footprint was only about 1 m in diameter, their altimeter could obtain useful returns from almost every ground track location. The weakness of the airborne technique is that the measurements are necessarily limited to the relatively sparse set of aircraft flight lines. By interpolating between those flight lines, *Krabill et al.* estimated the rate of elevation change to be 5 ± 7 mm/yr when averaged over about the same region considered by *Davis et al.* [2000]

The advantages of the small footprint available from

lasers and the more complete coverage available from satellites, will be combined in NASA's IceSAT (Ice, Cloud and Land Elevation Satellite), scheduled for an early 2002 launch and a nominal 3–5 year lifetime. The principal instrument aboard IceSAT will be GLAS (Geoscience Laser Altimeter System), a high-precision laser altimeter with a footprint of about 70 m in diameter and designed specifically for ice sheet studies. The satellite will be tracked using on-board GPS receivers to determine its orbit. The IceSAT ground track should pass within 4° of the pole, with off-nadir pointing capable of increasing the coverage by up to one additional degree closer to the pole, which would allow IceSAT to sample all but about 3% of the Antarctic ice sheet and close to 100% of the Greenland ice sheet.

The accuracy with which IceSAT/GLAS can deliver the century-scale trend in ice mass averaged over an entire ice sheet will not be limited by the measurement accuracy of either the altimeter or the satellite tracking. Instead, the dominant errors sources will be the effects of variable snow accumulation and uncertainties in the effects of GIA. The variable snow accumulation errors are of two types: an undersampling error and a compaction error. The undersampling error is simply that the variable accumulation could cause the trend over the 3–5 year IceSAT mission to differ from the century-scale trend. The century-scale trend is of primary interest, since changes in ice flow are thought to be reasonably steady over that sort of time span. On the other hand, mass trends over shorter periods are of interest in their own right, and may even be more appropriate for the limited objective of determining the ice sheet mass balance in order to remove its effects from contemporary geodetic GIA measurements. If the objective is simply to determine the mass trend over the 3–5 year IceSAT mission length, then there is no undersampling error.

The compaction error arises because an altimeter measures ice sheet elevation, not mass. The use of an elevation measurement to deduce the change in mass requires knowledge of the density throughout the column of snow and ice. Newly-fallen snow gradually changes into ice as more snow falls on top of it and compacts the underlying column. Thus the column density depends on the variable accumulation rate that occurs during and prior to the mission time period. It is unlikely that the accumulation will be monitored with sufficient accuracy to allow for an adequate removal of this effect.

A GIA error is present because any vertical movement of the crust, (from GIA, for example) will cause the surface of the ice sheet to also move vertically. The altimeter cannot separate the effects of this elevation change from the effects of a true change in ice sheet thickness.

To estimate the size of these various errors for Antarctica, *Wahr et al.* [2000] used global precipitation fields output from the NCAR Climate System Model CSM-1 [*Boville and Gent*, 1998], along with estimates of the uncertainty in the GIA signal under Antarctica and of the GLAS measurement errors, to construct synthetic GLAS data for Antarctica. They concluded that a 5-year IceSAT/GLAS mission could estimate the century-scale trend in ice thickness, expressed in equivalent water thickness, to an accuracy of about ±9 mm/yr, and could estimate the trend over the 5-year observing period to about ±7 mm/yr. For the 5-year trend, the compaction and GIA errors were estimated to be about equal. These errors are far smaller than the present uncertainty of ±40 mm/yr [*Warrick et al.*, 1996].

By the time IceSAT/GLAS has been in orbit long enough to obtain these high accuracies, GRACE will presumably have provided secular estimates for the C_{lm} and S_{lm} that will be accurate enough to separate the effects of Greenland and Antarctica from the GIA signals over Canada and Scandinavia. But GRACE alone will not be able to separate the gravitational effects of present-day polar mass imbalance from the effects of GIA over the Greenland or Antarctic ice sheets themselves. Data from IceSAT/GLAS could help.

For example, *Wahr et al.* [2000] described a method of combining IceSAT/GLAS and GRACE data to remove the GIA contribution from IceSAT/GLAS mass balance estimates. They found that by using GRACE, the GIA effects on the IceSAT/GLAS estimate of the Antarctic mass trend could probably be determined and removed to an accuracy of about 25%, averaged over the entire ice sheet. *Velicogna and Wahr* [2002] extend this method to look specifically at recovering the GIA crustal uplift rate over Antarctica, and conclude that the spatially-averaged uplift rate can probably be determined to this same 25% accuracy.

6.2. Measurements of Ongoing Crustal Motion

Ice sheet mass imbalance can also be constrained using geodetic measurements of crustal motion. A change in ice sheet mass will load or unload the underlying Earth and cause displacements of the crust in the surrounding region. If those displacements can be measured, they could conceivably serve as constraints on the mass variability.

This method of constraining ice mass has been used, for example, to study the Bagley Ice Field Glacier in southeastern Alaska [*Sauber et al.*, 2000]. Along the

edge of the Greenland or Antarctic ice sheets, however, the crustal displacement would also have contributions from GIA effects caused by any changes in Greenland or Antarctic ice that might have occurred over the last several thousand years. In other words, observations of crustal deformation alone cannot separate the elastic response of the Earth due to ongoing glacier variations from the viscoelastic response associated with GIA. A solution to this problem was proposed by *Wahr et al.* [1995], who described a method of combining measurements of vertical crustal motion with surface gravimeter measurements at the same location, in order to separate the GIA and ice mass imbalance contributions. Briefly, the relative effects of mass imbalance and GIA are different for crustal displacements than they are for surface gravity, and that difference can be exploited to solve for the two effects simultaneously. This method is now being employed at two locations in Greenland [*Wahr et al.*, 2001]. Preliminary results at a location on the western margin of the ice sheet show a substantial rate of GIA subsidence, and support previous suggestions that the nearby ice margin may have actually been advancing over the last 2,000–3,000 years.

6.3. Determination of Ice History-Parameters

As discussed in Section 3.2, reconstruction of the Late Pleistocene and Holocene ice cover depends on a nonunique interpretation of geologic, including RSL, data. Given the sensitivity of space-geodetic determinations of crustal velocity to errors in the ice model, is there a method for using these data to estimate corrections to the ice model?

One of the obstacles in developing such a method is the difficulty in parametrizing the reconstruction. Models like Ice-3G [*Tushingham and Peltier*, 1991] and Ice-4G [*Peltier*, 1994] are characterized by hundreds of independently varied ice elements at each of a number of epochs. Parametrization of such a model would lead to a highly overparametrized problem (which is why these reconstructions are nonunique). Models like that of *Lambeck et al.* [1998], on the other hand, start with an ice sheet that approximates a physically realistic ice sheet that is bounded by the geology, and then modify the thickness in a simple way. Although it has been shown that an overall scaling of the ice thickness is possible and somewhat separable from viscosity parameters [*Davis et al.*, 1999], the resulting "universally scaled" ice sheet may not be reasonable.

A completely different approach to ice sheet reconstruction was developed by *Fastook and Hughes* [1991], who used a finite-element method to simulate the last glaciation cycle on a regional basis. A number of regional climatic and dynamic parameters are required for this approach. For temperature variation the GRIP ice core results [e.g., *Johnson et al.*, 1995] are used as a temperature proxy, but the offset of the mean global temperature from this proxy can be varied. The calculations also require values for parameters for the ice flow and sliding laws, the geothermal heat gradient, amount of calving, and the adiabatic temperature lapse rate. These parameters represent values that, in principle, could be estimated from the geodetic data.

If this approach proves successful, it would not only result in an improved GIA analysis using geodetic data, but also an improved reconstruction of the ice history for the last 20,000 years, and provide constraints on otherwise poorly known climate parameters. Whether these data have the required sensitivity, either alone or in combination with other GIA data types, remains to be seen.

7. SUMMARY

Geological observations of relative sea level (RSL) have long provided useful constraints on GIA. But those constraints alone are not sufficient to fully bound GIA models. They are, for example, relatively insensitive to the viscosity of the deep mantle. Furthermore, like every other GIA observable, they are sensitive both to the Earth's viscosity profile and to the detailed history of the ice load. Reliance solely on RSL data makes it difficult to separate these contributions.

The correlation between the viscosity values and the ice history parameters can be somewhat reduced by dating lateral moraines to bound the ice sheet extent, using far-field sea level rise data to infer the total amount of ice locked up in the ancient ice sheets, and using simplified dynamical models to simulate the shape and history of those ice sheets. Uncertainties exist, however, regarding the detailed structure, strength, and rheology of these ice sheets. To obtain improved information about both the viscosity and the ice sheet evolution, other observations of GIA deformation would be most useful.

Geodetic observations are particularly promising in this regard. Although several apparent geodetic constraints on GIA have proven to be at least partially ambiguous (e.g. the gravity/geoid anomaly over northern Canada; the secular change in the Earth's rotation and in the low-degree zonal geoid coefficients), there are a number of new geodetic measurement techniques that should soon provide enough relatively unambiguous GIA information to lead to substantial improvement

in GIA models. These new techniques include the use of GPS observations to determine the detailed pattern of on-going crustal motion (both vertical and horizontal) in the vicinity of the ancient ice sheets; and the launch, at the end of 2001, of the GRACE dedicated gravity satellite mission, which should determine time-variability in the global geoid down to scales of a few hundred km. For both these types of observations, the accuracy and spatial resolution should be sufficient to remove most of the ambiguities that often arise when interpreting GIA data.

An understanding of GIA is useful for what it might reveal about the Earth's viscosity profile and the history of the ice sheets. But also because good knowledge of the GIA signal makes it possible to remove the GIA effects from various types of observations, in order to obtain better estimates of other quantities. For example, GIA-induced crustal uplift affects tide gauge observations of global sea level rise. By independently modeling and removing the GIA contributions, the sea level rise estimates are improved. This has been well documented in several past studies [e.g., *Peltier and Tushingham*, 1989; *Trupin and Wahr*, 1990; *Douglas*, 1997]. The general problem of GIA contamination will become increasingly important in the future. The GIA signal caused by past changes in Antarctic and Greenland ice can contaminate attempts to use altimeter or time-variable gravity measurements to infer the present-day mass imbalance of the polar ice sheets. Optimal polar ice imbalance estimates will thus require either prior, independent knowledge of the GIA signal; or else will require that both the GIA signal and the ice imbalance signal be solved for simultaneously, perhaps by simultaneously combining various types of data.

Acknowledgments. We thank Dazhong Han for providing viscoelastic Green's functions; Mike Watkins, Brooks Thomas, and Srinivas Bettadpur for their estimates of the GRACE gravity field errors; Mery Molenaar and Frank Bryan for their estimates of the geoid coefficients due to the ocean; Chris Milly and Krista Dunne for their global, gridded water storage data set; Glenn Milne for providing frechet kernels for viscosity; and Bert Vermeersen for helpful comments on the text. This work was partially supported by NASA grant NAG5-7703 to the University of Colorado, and NASA grant NAG5-6068 and NSF grant EAR-9526885 to the Smithsonian Institution.

REFERENCES

Argus, D. F., W. R. Peltier, and M. W. Watkins, Glacial isostatic adjustment observed using very long baseline interferometry and satellite laser ranging geodesy, *J. Geophys. Res.*, *104*, 29,077–29,093, 1999.

Benn, D. I., and D. J. A. Evans, *Glaciers and Glaciation*, Oxford University Press, New York, 1998.

BIFROST Project, GPS measurements to constrain geodynamic processes in Fennoscandia, *Eos Trans. AGU*, *77*, 337–341, 1996.

Bills, B. G., and T. S. James, Late Quaternary variations in relative sea level due to glacial cycle polar wander, *Geophys. Res. Lett.*, *23*, 3023–3026, 1996.

Boville, B. A., and P. Gent, The NCAR climate system model, version one, *J. Clim.*, *11*, 1115–1130, 1998.

Cathles, L. M., *The Viscosity of the Earth's Mantle*, Princeton University Press, New Jersey, 1975.

Chao, B. F., and R. S. Gross, Changes in the Earth's rotation and low-degree gravitational field induced by earthquakes, *Geophys. J. R. Astron. Soc.*, *91*, 569–596, 1987.

Chen, X., R. Langley, and H. Dragert, The Western Canada Deformation Array: An update on GPS solutions and error analysis, *GPS Trends in Precise Terrestrial, Airborne, and Spaceborne Applications*, edited by G. Beutler, G. W. Hein, W. G. Melbourne, and G. Seeber, Springer, Berlin, 70–74, 1996.

Cheng, M. K., C. K. Shum, and B. Tapley, Determination of long-term changes in the Earth's gravity field from satellite laser ranging observations, *J. Geophys. Res.*, *102*, 22,377–22,390, 1997.

Clark, T. A., et al., Precision geodesy using the Mark–III very-long-baseline interferometer system, *IEEE Trans. GARS*, *GE-23*, 438–449, 1985.

Crowley, T. J., and G. R. North, *Paleoclimatology*, Oxford Univ., New York, 1991.

Davis, C. H., C. A. Kluever, B. J. Haines, C. Perez, and Y. Yoon, Improved elevation change measurement of the southern Greenland ice sheet from satellite radar altimetry, *IEEE Trans. GARS*, *38*, 1367–1378, 2000.

Davis, J. L., and J. X. Mitrovica, Glacial isostatic adjustment and the anomalous tide gauge record of eastern North America, *Nature*, *379*, 331–333, 1996.

Davis, J. L., J. X. Mitrovica, H.-G. Scherneck, and H. Fan, Investigations of Fennoscandian glacial isostatic adjustment using modern sea-level records, *J. Geophys. Res.*, *104*, 2733–2747, 1999.

Denton, G. H., and I. J. Hughes (Eds.), *The Last Great Ice Sheets*, Wiley, New York, 1981.

Dickey, J. O., et al., *Satellite Gravity and the Geosphere: Contributions to the Study of the Solid Earth and Its Fluid Envelope*, 112 pp., Natl. Acad. Press, Washington, D.C., 1997.

Dickman, S. R., Investigation of controversial polar motion features using homogeneous International Latitude Service data. *J. Geophys. Res.*, *86*, 4904–4912, 1981.

Douglas, B. C., Global sea level rise: A redetermination, *Surv. Geophys.*, *18*, 279–292, 1997.

Dziewonski, A., and D. L. Anderson, Preliminary reference Earth model, *Phys. Earth Planet. Inter.*, *25*, 297–356, 1981.

Eanes, R. J., A study of temporal variations in Earth's gravitational field using LAGEOS-1 laser ranging observations, Ph.D. thesis, 128 pp., Univ. of Texas, Austin, 1995.

Ekman, M., and J. Mäkinen, Recent postglacial rebound, gravity change, and mantle flow in Fennoscandia, *Geophys. J. Int.*, *126*, 229–234, 1996.

Fairbanks, R. G., A 17,000-year glacio-eustatic sea level record: Influence of glacial melting rates on Younger Dryas event and deep-ocean circulation, *Nature*, *342*, 637–642, 1989.

Farrell, W. E. and J. A. Clark, On postglacial sea level, *Geophys. J. R. Astron. Soc.*, *46*, 647–667, 1976.

Fastook, J. L., and T. J. Hughes, Changing ice loads on the Earth's surface during the last glaciation cycle, in *Glacial Isostasy, Sea-Level, and Mantle Rheology*, edited by R. Sabadini, K. Lambeck, and E. Boschi, pp. 165–201, Kluwer, Norwell, Mass., 1991.

Gasperini, P., Sabadini, R., and Yuen, D.A., Excitation of the earth's rotational axis by recent glacial discharges, *Geophys. Res. Lett.*, *13*, 533-536, 1986.

Gross, R.S., and J. Vondrak, Astrometric and space-geodetic observations of polar wander, *Geophys. Res. Lett.*, *26*, 2085-2088, 1999.

Hager, B. H., and R. W. Clayton, Constraints on the structure of mantle convection using seismic observations, flow models, and the geoid, in *Mantle Convection: Plate Tectonics and Global Dynamics*, edited by W. R. Peltier, pp. 657–763, Gordon and Breach, Newark, N.J., 1989.

Han, D., and J. Wahr, Post-glacial rebound analysis for a rotating Earth, in *Slow Deformations and Transmission of Stress in the Earth*, edited by S. Cohen and P. Vaníček, AGU Mono. Series 49, 1–6, 1989.

Han, D. and J. Wahr, The viscoelastic relaxation of a realistically stratified Earth, and a further analysis of post-glacial rebound, *Geophys. J. Int.*, *120*, 287–311, 1995.

Houghton, J. T., G. J. Jenkins, and J. J. Ephraums (Eds)., *Climatic Change; the IPCC Scientific Assessment*, Cambridge Univ. Press, Cambridge, 1990.

Hughes, T. J., *Ice Sheets*, Oxford University Press, New York, 1998.

Imbrie, J., and K. P. Imbrie, *Ice Ages: Solving the Mystery*, MacMillan, 1979.

Imbrie, J., et al., On the structure and origin of major glaciation cycles, 1. Linear responses to Milankovitch forcing, *Paleoceanography*, *7*, 710–738, 1992.

Imbrie, J., et al., On the structure and origin of major glaciation cycles, 2. The 100,000 year cycle, *Paleoceanography*, *8*, 699–735, 1993.

James, T. S., and A. Lambert, A comparison of VLBI data with the Ice-3G glacial rebound model, *Geophys. Res. Lett.*, *20*, 871–874, 1993.

James, T. S., and W. J. Morgan, Horizontal motions due to post-glacial rebound, *Geophys. Res. Lett.*, *17*, 957–960, 1990.

Johansson, J. M., J. L. Davis, H.-G. Scherneck, G. A. Milne, M. Vermeer, J. X. Mitrovica, R. A. Bennett, B. Jonsson, G. Elgered, P. Elósegui, H. Koivula, M. Poutanen,

B. O. Rönnäng, and I. I. Shapiro, Continuous GPS measurements of postglacial adjustment in Fennoscandia, 1. Geodetic results, *J. Geophys. Res.*, in press, 2002.

Johnson, S. J., D. Dahl-Jensen, W. Dansgaard, and N. Gundestrup, Greenland paleotemperatures derived from GRIP borehole temperature and ice core profiles, *Tellus, 47B*, 624–629, 1995.

Johnston, P., The effect of spatially non-uniform water loads on predictions of sea level change, *Geophys. J. Int., 114*, 615–634, 1993.

Kleman, J., On the use of glacial striae for reconstruction of paleo-ice sheet flow patterns, with application to the Scandinavian ice sheet, *Geograf. Ann., 72*, 217–236, 1990.

Krabill, W., E. Frederick, S. Manizade, C. Martin, J. Sonntag, R. Swift, R. Thomas, W. Wright, and J. Yungel, Rapid thinning of parts of the southern Greenland ice sheet, *Science, 283*, 1522–1524, 1999.

Lambeck K., C. Smither, and P. Johnston, Sea-level change, glacial rebound and mantle viscosity for northern Europe, *Geophys. J. Int., 134*, 102–144, 1998.

Larson, K. M., and T. vanDam, Measuring postglacial rebound with GPS and absolute gravity, *Geophys. Res. Lett., 27*, 3925–3928, 2000.

Lemoine, F. G., et al., The Development of the Joint NASA GSFC and the NIMA Geopotential Model EGM96, *NASA Tech. Paper TP-1998-206861*, NASA Goddard Space Flight Center, 575 pp., July, 1998.

Miller, M. M., D. J. Johnson, C. M. Rubin, E. Endo, J. T. Freymueller, C. Goldfinger, H. M. Kelsey, E. D. Humphreys, J. S. Oldow, and A. Qamar, Preliminary GPS results from the Cascadia margin: The Pacific Northwest Geodetic Array (PANGA), *Eos Trans. AGU, 79*, Fall Meeting Suppl. (abstract), F206, 1998.

Milne, G. A., *Refining Models of the Glacial Isostatic Adjustment Process*, Ph.D. Thesis, Univ. Toronto, 1998.

Milne, G. A., and J. X. Mitrovica, Post-glacial sea-level change on a rotating Earth: First results from a gravitationally self-consistent sea-level equation, *Geophys. J.Int., 126*, F13–F20, 1996.

Milne, G. A., and J. X. Mitrovica, The influence of a time-dependent ocean-continent geometry on predictions of postglacial sea level change in Australia and New Zealand, *Geophys. Res. Lett., 25*, 793–796, 1998.

Milne, G. A., J. X. Mitrovica, and J. L. Davis, Near-field hydro-isostasy: The implementation of a revised sea-level equation, *Geophys. J. Int., 139*, 464–482, 1999.

Milne, G. A., J. L. Davis, J. X. Mitrovica, H.-G. Scherneck, J. Johansson, M. Vermeer, and H. Koivula, Space-geodetic constraints on glacial isostatic adjustment in Fennoscandia, *Science, 291*, 2381–2385, 2001.

Mitrovica, J. X., and J. L. Davis, Sea level change far from the Late Pleistocene ice sheets: Implications for recent analyses of tide gauge records, *Geophys. Res. Lett., 22*, 2529–2532, 1995.

Mitrovica, J. X., and A. M. Forte, Radial profile of mantle viscosity: Results from the joint inversion of convection and postglacial rebound observables, *J. Geophys. Res., 102*, 2751–2769, 1997.

Mitrovica, J. X., and W. R. Peltier, Radial resolution in the inference of mantle viscosity from observations of glacial isostatic adjustment, *Glacial Isostasy, Sea Level, and Mantle Rheology*, edited by R. Sabadini, K. Lambeck, and E. Boschi, Kluwer, Boston, 63–78, 1991.

Mitrovica, J. X., J. L. Davis, and I. I. Shapiro, Constraining proposed combinations of ice history and Earth rheology using VLBI-determined baseline length rates in North America, *Geophys. Res. Lett., 20*, 2387–2390, 1993.

Mitrovica, J. X., J. L. Davis and I. I. Shapiro, A spectral formalism for computing three-dimensional deformations due to surface loads, 1, Theory, *J. Geophys. Res., 99*, 7075–7101, 1994a.

Mitrovica, J. X., J. L. Davis, and I. I. Shapiro, A spectral formalism for computing three-dimensional deformations due to surface loads, 2, Present-day glacial isostatic adjustment, *J. Geophys. Res., 99*, 7075–7102, 1994b.

Mitrovica, J. X., M. Tamisiea, J. L. Davis and G. A. Milne, Recent mass balance of polar ice sheets inferred from patterns of global sea-level change, *Nature, 409*, 1026–1029, 2001.

Nakiboglu, S.M, and K. Lambeck, Deglaciation effects on the rotation of the Earth, *Geophys. J. R. Astr. Soc., 62*, 49–58, 1980.

Nerem, R. S., K. D. Park, T. M. vanDam, M. S. Schenewerk, J. L. Davis, and J. X. Mitrovica, A study of sea level change in the northeastern U.S. using GPS and tide gauge data, *EOS Trans. AGU, 78*, 140–141, 1997.

O'Connell, R. J., Pleistocene glaciation and the viscosity of the lower mantle, *Geophys. J. R. Astron. Soc., 23*, 299–327, 1971.

Park, K.-D., Determination of Glacial Isostatic Adjustment Parameters Based on Precise Point Positioning using GPS, Ph.D. Thesis, 126 pp., Dept. Aerospace Eng. and Eng. Mech., Univ. Texas, Austin, 2000.

Paterson, W. S. B., *The Physics of Glaciers, 3rd Ed.*, Elsevier, Boston, 1998.

Peltier, W. R., Glacial isostatic adjustment, 2, The inverse problem, *Geophys. J. R. Astron. Soc., 46*, 669–706, 1976.

Peltier, W.R., Constraint on deep mantle viscosity from Lageos acceleration data, *Nature, 304*, 434–436, 1983.

Peltier, W. R., Deglaciation-induced vertical motion of the North American continent and transient lower mantle rheology, *J. Geophys. Res., 91*, 9099–9123, 1986.

Peltier, W. R., Ice-age paleotopography, *Science, 265*, 195–201, 1994.

Peltier, W. R., 'Implicit ice' in the global theory of glacial isostatic rebound, *Geophys. Res. lett., 25*, 3955–3958, 1998.

Peltier, W. R., and J. T. Andrews, Glacial isostatic adjustment, 1, The forward problem, *Geophys. J. R. Astron. Soc., 46*, 605–646, 1976.

Peltier, W. R., and A. M. Tushingham, Global sea level rise and the Greenhouse Effect: Might they be connected? *Science, 244*, 806–810, 1989.

Peltier, W. R., A. M. Forte, J. X. Mitrovica, and A. M. Dziewonski, Earth's gravitational field: seismic tomography resolves the enigma of the Laurentian anomaly. *Geophys. Res. Lett., 19*, 1555–1558, 1992.

Plag, H.-P., and H.-U. Jüttner, Inversion of global tide gauge data for present-day ice load changes, *Memoirs of Nat. Inst. of Polar Res., 54*, 301–318, 2001.

Richards, M.A., H.P. Bunge, Y. Ricard, and J.R. Baumgardner, Polar wandering in mantle convection models, *Geophys. Res. Lett., 26*, 1777–1780, 1999.

Rubincam, D. P., Postglacial rebound observed by Lageos and the effective viscosity of the lower mantle, *J. Geophys. Res., 89*, 1077–1088, 1984.

Sabadini, R., and Peltier, W.R., Pleistocene deglaciation and the Earth's rotation: implications for mantle viscosity, *Geophys. J. R. Astr. Soc., 66*, 553–578, 1981.

Sauber, J., G. Plafker, B. F. Molnia, and M. A. Bryant, Crustal deformation associated with glacial fluctuations in the eastern Chugach Mountains, Alaska, *J. Geophys. Res., 105*, 8055–8077, 2000.

Silver, P., et al., The Plate Boundary Observatory: Creating a Four-Dimensional Image of the Deformation of Western North America, white paper based on PBO Workshop, Snowbird, Utah, October 3–5, 1999.

Simons, M., and B. H. Hager, Localization of the gravity field and the signature of glacial rebound, *Nature, 390*, 500–504, 1997.

Steinberger, M. and R.J. O'Connell, Changes of the Earth's rotation axis owing to advection of mantle density heterogeneities, *Nature, 387*, 169–173, 1997.

Stephenson, F. R. and L. V. Morrison, Long-term fluctuations in the Earth's rotation: 700 BC to AD 1990, *Phil. Trans. R. Soc. Lond. A, 351*, 165–202, 1995.

Su, W. and A. M. Dziewonski, Simultaneous inversion for 3-D variations in shear and bulk velocity in the mantle, *Phys. Earth and Planet. Int., 100*, 135–156, 1997.

Tamisiea, M., and J. Wahr. Effects of phase transitions in a conductive Earth on post glacial rebound, submitted to *Geophys. J. Int.*, 2001.

Trupin, A. and J. Wahr, Spectroscopic analysis of global tidal gauge sea level data, *Geophys. J. Int., 100*, 441–453, 1990.

Tushingham, A. M., and W. R. Peltier, Ice-3G: A new global model of late Pleistocene deglaciation based upon geophysical predictions of postglacial relative sea level change *J. Geophys. Res., 96*, 4497–4523, 1991.

Velicogna, I. and J. Wahr, Post-glacial rebound and ice mass balance over Antarctica from ICESat/GLAS, GRACE, and GPS measurements, submitted to *J. Geophys. Res.*, 2002.

Vermeersen, L. L. A., R. Sabadini, G. Spada, and N. J. Vlaar, Mountain building and earth rotation, *Geophys. J. Int., 117*, 610-624, 1994.

Wahr, J. M., D. Han, and A. Trupin, Predictions of vertical uplift caused by changing polar ice volumes on a viscoelastic Earth, *Geophys. Res. Lett., 22*, 977–980, 1995.

Wahr, J., M. Molenaar, and F. Bryan, Time-variability of the Earth's gravity field: hydrological and oceanic effects and their possible detection using GRACE, *J. Geophys. Res., 103*, 30,205–30,230, 1998.

Wahr, J., D. Wingham, C. R. Bentley, A method of combining GLAS and GRACE satellite data to constrain Antarctic mass balance, *J. Geophys. Res., 105*, 16,279–16,294, 2000.

Wahr, J., T. vanDam, K. Larson, and O. Francis, Geodetic measurements in Greenland and their impications, *J. Geophys. Res., 106*, 16,567–16,582, 2001.

Walcott, R.I., Structure of the Earth from glacio-isostatic rebound. *Ann. Rev. Earth Planet. Sci., 1*, 15–37, 1973.

Warrick, R. A., E. M. Barrow, and T. M. L. Wigley (Eds.), *Climate and Sea Level Change: Observations, Projections and Implications*, Cambridge Univ. Press, Cambridge, 1993.

Warrick, R. A., C. Le Provost, M. Meier, J. Oerlemans, and P. L. Woodworth, Changes in sea level, in *Climate Change 1995: The Science of Climate Change: Contribution of Working Group I to the Second Assessment Report of the Intergovernmental Panel on Climate Change*, edited by J. T. Houghton, L. G. Meira Filho, B. A. Callander, N. Harris, A. Kattenberg, and K. Maskell, Cambridge Univ. Press, New York, 359–405, 1996.

Williams, L. D., An energy balance model of potential glacierization of northern Canada, *Arctic and Alpine Res., 11*, 443–456, 1979.

Williams, J. G., X. X. Newhall, and J. O. Dickey, Diurnal and semidiurnal tidal contributions to lunar secular acceleration, *EOS Trans. Amer. Geophys. Union, 73*, 126, 1992.

Wingham, D. J., A. R. Ridout, and C. K. Shum, Antarctic elevation change 1992 to 1996, *Science, 282*, 456–458, 1998.

Woodworth, P. L., D. T. Pugh, J. G. De Ronde, R. G. Warrick, and J. Hannah (Eds.), *Sea Level Changes: Determination and Effects*, American Geophysical Union, Washington, D.C., 1992.

Yoder, C. F., J. G. Williams, J. O. Dickey, B. E. Schutz, R. J. Eanes, and B. D. Tapley, Secular variations of Earth's gravitational harmonic J_2 coefficient from Lageos and the non-tidal acceleration of Earth rotation, *Nature, 303*, 757–762, 1983.

Zwally, H. J. and A. C. Brenner, Ice sheet dynamics and mass balance, in *Satellite Altimetry and Earth Sciences*, edited by L. Fu and A. Cazenave, pp. 351–369, Academic Press, New York, 2001.

J. M. Wahr, Department of Physics, University of Colorado, Boulder, CO. (e-mail: wahr@longo.colorado.edu)

J. L. Davis, Harvard-Smithsonian Center for Astrophysics, 60 Garden Street, MS 42, Cambridge, MA 02138. (e-mail: jdavis@cfa.harvard.edu)

Sea Level Change From Mid Holocene to Recent Time: An Australian Example with Global Implications

Kurt Lambeck

Research School of Earth Sciences, Australian National University, Canberra 0200, Australia

Observed relative sea-level change reflects changes in ocean volume, glacio-hydro-isostasy, vertical tectonics and redistribution of water within ocean basins by climatological and oceanographic factors. Together these factors produce a complex spatial and temporal sea-level signal. For the tectonically stable Australian margin, geological evidence indicates that sea-levels at 7000-6000 years ago were between 0 and 3 m above present level, due primarily to glacio-hydro-isostatic effects of the last deglaciation. The spatial variability of this signal determines the mantle response to the surface loading and leads to an effective lithospheric thickness of 75-90 km and an effective upper mantle viscosity of $(1.5 - 2.5) \times 10^{20}$ Pa s. Compared with results for other regions this is indicative of regional variation in upper-mantle response. Also, ocean volumes continued to increase after 7000 years ago by enough to raise global mean sea level by about 3 m. Much of this increase occurred between 7000 and 3000 years ago. Because of the spatial variability in mantle response, isostatic corrections to tide-gauge records of recent change should be based on regional model-parameters rather than on global parameters. The two longest records from the Australian margin give an isostatically corrected rate of regional sea-level rise of 1.40 ±0.25 mm/year. Comparisons of this rate with rates from other regions indicates that the spatial variability in secular sea-level is likely to be significant, with estimates of regional rates ranging from about 1 mm/year to 2 mm/year. These rates of secular change cannot have persisted further back in time than a few hundred years without becoming detectable in high-resolution geological and archaeological indicators of sea-level change.

1. INTRODUCTION

Changes in sea level are usually expressed as a change in the level of the sea with respect to land. It is therefore a relative measure that is indicative of movement of the land, changes in ocean volume or, in most cases, of both.

As such, sea-level change, both today and in the past, exhibits a complex spatial and temporal pattern that reflects tectonic, isostatic and climate contributions. Future change, likewise, will be geographically variable. Separation of the contributing factors becomes important when examining present change or when predicting future trends from incomplete observations. How much of an observed signal is of local relevance only because of tectonic contamination? How much of the signal is of global relevance, indicative of changes in ocean volume? Under some very special circumstances it becomes pos-

Ice Sheets, Sea Level and the Dynamic Earth
Geodynamics Series 29
10.1029/029GD03

sible to separate these contributions, but the present observational data base is mostly inadequate for this and reliance on numerical model results or on indirect indicators of change is usually required.

This paper examines this question of separability, using geological and tide gauge sea-level evidence from the relatively stable Australian margin. In particular, it examines the separability of glacio-hydro-isostatic contributions to sea-level from changes in ocean volume over the past 7000 years. (Unless otherwise indicated ages are expressed in calendar years. Palaeo sea-level data will be expressed mostly with respect to the radiocarbon ^{14}C time scale. Any conversion to calendar years is based on calibration by *Stuiver et al.*[1998] and *Bard et al.*[1998]. For the interval in question here, and within dating uncertainties, the difference between the two time scales is effectively linear and either is adequate for modelling, provided that all time-dependent functions (sea-level data, ice retreat history and viscosity) are defined in the same system.)

Evidence abounds that sea levels have changed by a few meters over the past few thousand years even though most of the melting of the large ice sheets was complete by about 7000 years ago. This evidence is in the form of raised shorelines, fossil corals above their current growth position, or submerged in-situ tree stumps and other normally sub-aereal vegetation or fauna. Such observations may be indicative of vertical tectonic movement of the land, of on-going changes in ocean volume, or of the on-going isostatic response to the last deglaciation of the high-latitude ice sheets. If the interest in sea-level data is the isostatic rebound or climate signal, then the tectonic component needs to be eliminated. If the interest is in estimating tectonic rates of vertical movement then the rebound and climate contributions must be removed. The importance of tectonic contributions can sometimes be assessed from the geological evidence: is there an independent record of crustal deformation that serves as warning that sea-level data from the locality may be contaminated by a tectonic signal? Can the position of the Last Interglacial (LIg) shorelines be identified and used as an indicator of tectonics? Globally LIg shorelines occur within a few metres of present sea level and if they are found to be well-elevated or submerged they points to tectonic uplift having occurred during the last 125000 or so years.

The margins of the old continents, away from tectonic plate boundaries and away from the former ice sheets, provide some of the best information on recent sea level change largely free of tectonic contributions. Along the Australian margin, for example, there is substantial evidence that sea levels have been up to three meters higher in the past few thousand years and that sea level has been dropping quasi-uniformly since that time (see, for example, papers in *Hopley* [1983]). Two results are illustrated in Figure 1. The first result, based on the past sea level recorded within chenier ridges, is from the Gulf of Carpentaria [*Chappell et al.*, 1982] and indicates that here sea level has fallen by about 2.5 m during the past 5000 (radiocarbon or ^{14}C) years. The second result, from near-shore Orpheus Island in northern Queensland, is based on fossil in-situ corals that have been stranded by a retreating sea and here sea levels have fallen by only about 1 m over a similar period (*Chappell et al.* [1983] and unpublished ANU data). In both instances the evidence points to sea levels having peaked at about 5000 to 6000 ^{14}C years ago and that before this time levels had been rising quite rapidly. This earlier rise can be seen in Figure 2 which summarises age-height data of coral samples from many localities on the Great Barrier Reef and the upper envelope represents a first approximation to the regional sea-level curve for the past 10000 ^{14}C years. Elsewhere along the Australian margin the high-stand occurs up to 3m elevation or higher [*Burne*, 1982]. In Tasmania the mid-Holocene high-stands appears to be absent. The amplitudes also vary across the shelf, reaching their maxima for coastal sites or localities at the head of narrow gulfs, such as Spencer Gulf (see Figure 1 for locations), and decreasing with distance from shore. In northern Queensland, for example, high-stands of about 1.5 m occur at coastal sites such as Yule Point [*Chappell et al.*, 1983], less for the near-shore islands such as Orpheus, and absent from the outer reef about 100 km offshore. At the outer reef corals at the mean-low-water tide level are all much younger than 5000-6000 ^{14}C years [*Hopley*, 1982]. While the precise details of the highstand morphology remains to be established for most locations, a general observation is that the greater its amplitude the more sharply will it be defined and the earlier the occurrence of the maximum level (see *Lambeck and Nakada* [1990] for a discussion of the Australian margin data then available).

Elsewhere, different patterns emerge. For tectonically stable islands in the Pacific the high-stands are usually low, less than about 1 m, and tend to occur later than along the continental margins (e.g. *Pirazzoli and Montaggioni* [1988] and *Moriwaki et al.* [2000]) Figure 3. The interpretation of these results is often complicated by the possibility that some vertical land motion may also occur that is not associated with glacio-hydro isostasy. In particular, Pacific island sites near the tectonic-plate margins should be avoided if the search is for climate or isostatic signals. But even within the

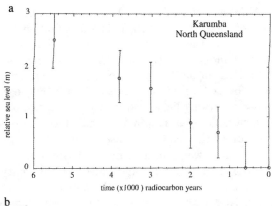

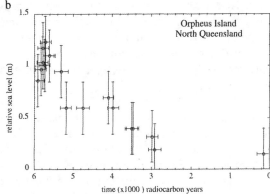

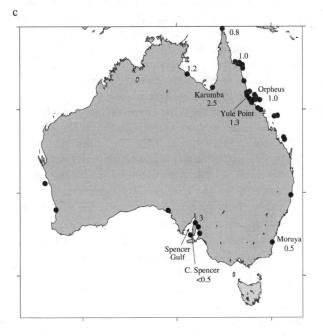

Figure 1. Observed sea level change at two localities in northern Queensland, Australia, for Late Holocene time. The time scale is in ^{14}C years. (a) Near Karumba in the Gulf of Carpentaria [*Chappell et al.*, 1982] and (b) near-shore Orpheus Island (*Chappell et al.*, [1983], and unpublished ANU data). (c) Map showing site locations where late-Holocene highstands have been observed. Numbers indicate the maximum elevations (in metres) observed for some of the sites.

plate interior some vertical movements may result from nearby volcanic loading of the crust, such as some of the islands in the Cook, Society, and Tuamotu groups, or from the motion of the plate over mantle hot spots.

2. THE STRUCTURE OF THE FAR FIELD MID HOLOCENE HIGH STAND.

The small high-stands that develop in mid-Holocene time at locations far from former ice margins - at far-field sites - are largely a result of the planetary response to the last glacial cycle and as such these signals provide significant information on the mantle rheology as well as on any late-Holocene changes in ocean volume. Far from the former ice margins, the first order change in sea level due to the melting of a land-based ice volume V_i is given by the ice-volume-equivalent sea-level function $\Delta\zeta_e(t)$ defined as

$$\Delta\zeta_e(t) = -\frac{\rho_i}{\rho_o}\int_{t_p}^{t}\frac{1}{A_o(t')}\frac{dV_i}{dt'}dt' \qquad (1)$$

where $A_o(t)$ is the ocean surface area and ρ_i, ρ_o are the average densities of ice and ocean respectively. Superimposed on this is the isostatic crustal response to the time-dependent changes in the ice and water loads as mass is exchanged between the ice sheets and the oceans. Thus, in the absence of tectonic contributions,

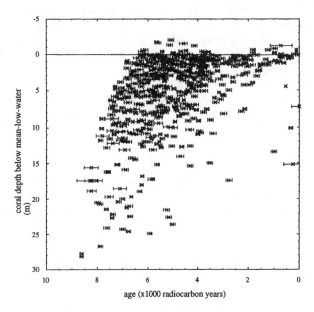

Figure 2. Age-height relationship of corals sampled from reef cores in the Great Barrier Reef region with ages less than 10000 years. The time scale is in ^{14}C years. The height datum is mean-low water. (Data compiled by D. Hopley and D. Zwartz.)

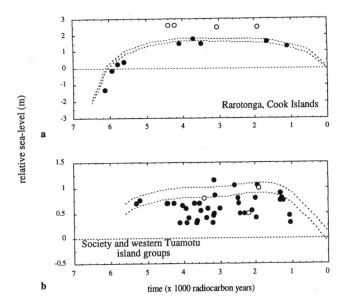

Figure 3. Observed mid- to late-Holocene sea levels in the South pacific. (a) Raratonga, Southern Cook Islands. Solid circles are from microatolls and lagoon and swale deposits and open circles are from sandridge deposits. The estimated mean sea level is shown by the dashed lines (from *Moriwaki et al.*, 2000). (b) Society and western Tuamotu Island groups. Solid circles denote estimates of mean sea level and open circles denote lower limits. The data and the approximate sea-level curve for the region (dashed lines) are from *Pirazzoli and Montaggioni* [1988].

the relative sea-level change can be written schematically as (see *Farrell and Clark* [1976]; *Nakada and Lambeck* [1987]; *Mitrovica and Peltier* [1991]; *Johnston* [1993]; *Milne and Mitrovica* [1998]; *Milne et al.* [1999] for various aspects of this equation)

$$\Delta\zeta_{rsl}(\varphi, t) = \Delta\zeta_e(\varphi, t) + \Delta\zeta_I(\varphi, t)$$
$$= \Delta\zeta_e(t) + \Delta\zeta_i(\varphi, t) + \Delta\zeta_w(\varphi, t) + \Delta\zeta_\Omega(\varphi, t) \quad (2)$$

where $\Delta\zeta_{rsl}(\varphi, t)$ is the height of the palaeo sea surface relative to present sea level and is a function of position φ and time t. The $\Delta\zeta_I = \Delta\zeta_i(\varphi, t) + \Delta\zeta_w(\varphi, t)$ is the total isostatic response contribution to the sea-level change, including contributions from the deformation of the earths surface and from the changes in gravitational potential. The $\Delta\zeta_i$ and $\Delta\zeta_w$ are the components of $\Delta\zeta_I$ representing the glacio- and hydro-isostatic load effects respectively. In formulating these terms ocean-ice mass is conserved and the ocean surface remains a gravitational equipotential surface throughout [*Farrell and Clarke*, 1976; *Nakada and Lambeck*, 1987]. (That is, these terms include the average over the oceans of the response functions to loading and have been included throughout the three generations of models described

by *Nakiboglu et al.* [1983], *Nakada and Lambeck* [1987] and *Johnston* [1993, 1995].) The formulation and programming of the earth-response and sea-level equations has been independently established through these three generations of models and each successive program has been checked against its predecessor. The earth response functions have also been checked against independent code by G. Kaufmann whose formulation of the inversion of the Laplace transformed variables into the time domain was found to be the more robust one on time scales of 10^5 years and longer. Preliminary tests between independent codes developed by M. Nakada and J. X. Mitrovica also indicate agreement of results within the range of expected differences resulting from the use of different ice- and earth-models.

The water depth or terrain elevation at time t and location φ, expressed relative to coeval sea level, is

$$h(\varphi, t) = h(\varphi, t_0) - \Delta\zeta_{rsl}(\varphi, t) \quad (3)$$

where $h(\varphi, t_0)$ is the present-day (t_0) bathymetry or topography at φ. The isostatic terms in (2) are functions of the earth rheology. In addition, the glacio-isostatic contribution is a function of the ice mass through time and the hydro-isostatic contribution is a function of the spatial and temporal distribution of the water load. The ocean surface area A_0 in (1) is a function of time because of (i) the advance or retreat of shorelines as the relative position of land and sea is modified and (ii) the retreat or advance of grounded ice over shallow continental shelves and seas. Thus the relationship between ice volume V_i and equivalent sea level $\Delta\zeta_e$ is weakly model dependent. With these definitions, the ice volume in (1) includes any ice grounded on the shelves that displaces ocean water and the ocean function is defined by the ice grounding line [*Lambeck and Johnston*, 1998; *Milne et al.*, 1999]. Also included in (2) is a term $\Delta\zeta_\Omega$ that allows for the change in the gravitational equipotential surface by deglaciation-induced changes in the Earths rotation [*Milne and Mitrovica*, 1998].

In the calculation of the isostatic term $\Delta\zeta_I(\varphi, t) = \Delta\zeta_i + \Delta\zeta_w$ the two load contributions are not wholly decoupled as implied by the second equality in the schematic representation (2). Near the ice sheet margin, for example, the gravitational attraction of the ice pulls the ocean water up and increases the water load in the neighbourhood of the ice margins. Also, during the ice loading stage, a broad swell forms beyond the ice margins which, when it occurs in an ocean environment, displaces ocean water and will, ignoring the other contributions, cause sea-level to rise and hence result in an increase in the water load elsewhere. When the ice sheet melts the bulge gradually subsides and the sea levels would fall if other contributions are again ignored. These effects are included in the numerical evaluations of the total isostatic term $\Delta\zeta_I$.

Well beyond the former ice margins $\Delta\zeta_i$ is not negligible but varies relatively slowly with position, independently of whether the site lies at or away from a continental margin. This glacio-isostatic term is the sum of the changes in gravitational attraction from the ice and earth deformation as well as the rebound of the crust peripheral to the ice load. The combined effect is a broad zone, out to distances of 3000 km or more from the ice margin, where the Late Holocene glacio-isostatic contribution is negative. Thus, in the absence of the other contributions, sea levels would be rising throughout late- and post-glacial time (see examples in *Lambeck* [1995; 1996]). Beyond this broad zone the glacio-isostatic signal is again positive but of smaller amplitude.

In contrast to the glacio-isostatic term, the water load term $\Delta\zeta_w$ exhibits a much greater variability because of its strong dependence on the spatial distribution of the added melt water. Its geographic form follows the outline of the coastline, appropriately filtered by the elastic or high-viscosity response of the lithosphere. The main reason for this signal is the subsidence of the sea floor under the increased water load which, through the elastic behaviour of the lithosphere, pulls the coastline down partly. This is accentuated by any uplift of the interior of the continent by mantle flow from ocean to continental mantle. A smaller contribution is from the change in gravitational attraction between the water and land as sea level rises.

In the Australian region the glacio-isostatic term is mainly positive in postglacial times and the two contributions combine to produce a fall in relative sea level for the postglacial period as is illustrated in Figure 4 for Queensland and South Australian localities. It is this and the water-load effect that produces the mid-Holocene high-stand at these sites, the peak high-stand indicating when melting ceased. The results in Figure 4 explain qualitatively the spatial variability observed: At Karumba and Orpheus the glacio-isostatic terms are very similar but the hydro-isostatic terms are substantially different because of the different geometries of the nearby water loads: for Karumba on the Gulf of Carpentaria, the Pacific load lies some 500 km to the east and is effectively an inland one such that the hydro-isostatic effect is considerably magnified. The nearer Gulf of Carpentaria water load, distributed through about 90° azimuth only, further magnifies the effect. For the second site, Orpheus Island on the Pacific coast, the high-stand amplitude is smaller because the coast

moves to a greater extent with the ocean floor than it does at Karumba. Likewise, the predicted high-stand amplitude at the head of Spencer Gulf (Port Augusta) exceeds that at the entrance (Cape Spencer) because the former site is effectively 300 km inland.

3. LATE HOLOCENE SEA LEVELS: RHEOLOGY AND OCEAN VOLUME CHANGES.

Because the far-field glacio-isostatic contributions vary only slowly with position, the differential high-stands for near-by sites such as Karumba and Orpheus, or the two ends of Spencer Gulf, are not strongly dependent on details of the ice loads. The differential amplitudes are, however, earth-model dependent as explored by *Lambeck and Nakada* [1990]. Relatively thin lithospheres

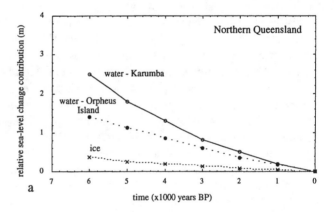

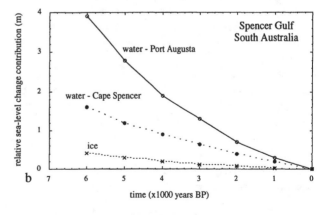

Figure 4. Predicted first-order isostatic contributions to relative sea-level change for the two northern Queensland localities illustrated in Figure 1 and for two localities in South Australia: Port Augusta at the head of the Gulf and Cape Spencer at the entrance to the Gulf. The curves labelled water refer to the hydro-isostatic contribution and the curves labelled ice refers to the glacio-isostatic contribution. This latter contribution is nearly identical for each of the two pairs of sites. (Adapted from *Lambeck and Nakada* [1990].)

and low viscosity upper mantles tend towards enhanced highstand amplitudes and strong gradients of this amplitude across the shelf, whereas thick lithospheres and high viscosity upper mantles reduce both the high-stand amplitude and its spatial gradients. These amplitudes, however, are not a function of rheology alone. They also depend on the melting history of the ice sheets. In the examples illustrated in Figure 4 all melting is assumed to have ceased at 6000 ^{14}C years ago but there is no *a-priori* reason to suppose this. For example, mountain glaciers in the last century have added between 20-30 mm to global sea level rise [*Zuo and Oerlemans*, 1997; *Dyurgerov and Meier*, 1997; *Gregory and Oerlemans*, 1998] and if this had been constant for the past 6000 years the total rise would have been between 1 and 2m. Potentially more important is the ongoing melting of Antarctic ice after cessation of melting of the northern ice sheets at about 6000 ^{14}C years ago. This ice sheet was much larger during the Last Glacial Maximum [*Denton and Hughes*, 1981] and the West Antarctic part, in particular, may have lost up to two thirds of its volume in the subsequent interval [*Bindschadler*, 1998]. In the Antarctic deglaciation model of *Huybrechts* [1994] the melting is largely driven by the rising sea levels that follow from the northern hemisphere deglaciation and much of the Antarctic decay is predicted to have occurred in Holocene time right up to the present. *Conway et al.* [1999] also suggest that the retreat of the West Antarctic ice is still ongoing such that it may contribute up to 0.9 mm/yr to the present sea-level rise [*Bindschadler*, 1998]. Some evidence for recent retreat is seen in new data from Mary Byrd Land where marginal moraines dated at a few thousand years BP occur up to 400 m above the present ice surface (J. O. Stone, private communication).

There is, therefore, no *a-priori* reason for assuming that all melting ceased at 6000 ^{14}C years ago and in interpreting the high-stand amplitudes the possibility of there being also a contribution from a change in ocean volume must be entertained [*Nakada and Lambeck*, 1988]. Thus, in general, the schematic observation equation relating observed sea levels $\Delta\zeta_o$ to the model-dependent predicted values $\Delta\zeta_p$ should be written as

$$\Delta\zeta_o + \epsilon_o = \Delta\zeta_p = \Delta\zeta_e + \delta\zeta_e + \Delta\zeta_i + \Delta\zeta_w \quad (4)$$

where ϵ_o is the observation error and $\Delta\zeta_e$ is a corrective term to the nominal equivalent sea level function $\Delta\zeta_e$ based on the ice sheets used to evaluate the earth-rheology dependent isostatic components $\Delta\zeta_i$ and $\Delta\zeta_\Omega$. (The rotation term $\Delta\zeta_\Omega$ is ignored in this equation but not in the analysis.) (Equation (4) assumes that the

$\delta\zeta_e$ is small such that its neglect in the evaluation of the isostatic terms is negligible. Typically $(|\Delta\zeta_i| + |\Delta\zeta_w|)/|\Delta\zeta_e| \approx (10-15)\%$ so that an error of 10% in $\Delta\zeta_e$ introduces errors of only $1-1.5\%$ in $\Delta\zeta_p$. If $\Delta\zeta_e$ is larger an iterative solution may be required.)

That a solution for both $\Delta\zeta_e$ and the rheological parameters contained in the functions $\Delta\zeta_i$, $\Delta\zeta_w$ is possible can be illustrated as follows. Consider a three-layer mantle model defined by a lithosphere of effective thickness H_l and effective viscosities (η_{um}, η_{lm}) for the lower and upper mantle with the boundary between the two mantle zones at 670 km depth. Consider also observations of the highstand amplitudes $\Delta\zeta_o^{(1)}$ and $\Delta\zeta_o^{(2)}$ at two nearby locations. The difference $\Delta\zeta_o^{(1)} - \Delta\zeta_o^{(2)} \approx \Delta\zeta_w^{(1)} - \Delta\zeta_w^{(2)}$ is independent of the equivalent sea-level function unless the starting ice models from which $\Delta\zeta_w^{(1)}$ and $\Delta\zeta_w^{(2)}$ are evaluated are grossly in error. Thus this differential highstand amplitude is mostly a function of rheology. This dependence is illustrated in Figure 5 for selected pairs of sites as a function of η_{um} and η_{lm} with a lithosphere of H_l= 50 km. The highstands are illustrated here for 6000 years BP (before present) only and superimposed upon the predicted differential values are the observational constraints for the upper and lower limits to the highstands. Shaded areas denote that part of the solution space that is consistent with observational data. For any one pair of sites the response parameters are not well constrained but, because the rheological dependence varies with coastline geometry, the solution based on all the data is well constrained for at least η_{um} (see the last panel of Figure 5). In this case, with an *a-priori* value of $H_l = 50$ km, $\eta_{um} = (2-3) \times 10^{20}$ Pa s and $\eta_{lm} = 10^{22}$ Pa s although the latter is not well constrained. If a lithospheric thickness of 100 km is assumed then the solution space is distinctly different Figure 6 but also less satisfactory: If the Cape Melville data is excluded two solutions for viscosity are possible but neither satisfy some of the other observational evidence that is not included in constructing Figure 6. In particular, the thicker lithospheric value leads to smaller than observed tilt gradients across the Queensland shelf.

Considerable trade-off occurs between the three earth-model parameters and, in particular if an *a-priori* value for lithospheric thickness is assumed, the resulting parameters for the viscosities are only as reliable as the initial choice. There are no relevant *a-priori* estimates for lithospheric thickness appropriate for the loading frequencies and magnitudes of the glacial cycles. Tectonic estimates, corresponding to long load cycles, result in thinner effective lithospheres because the load time-

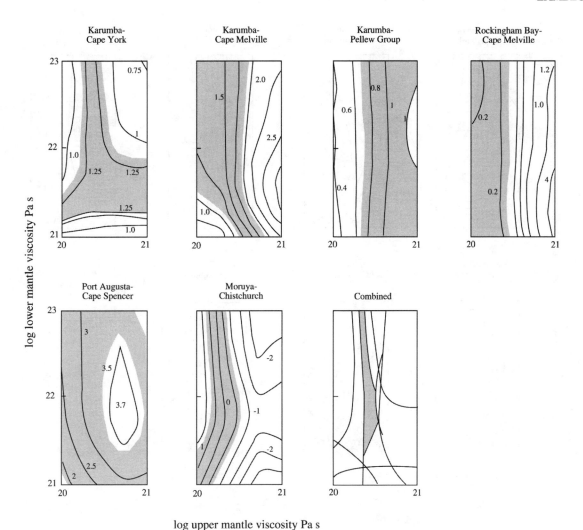

Figure 5. Predicted differential amplitudes for the mid-Holocene highstands at pairs of Australian sites as a function of upper η_{um} and lower η_{lm} mantle viscosities for lithospheric thickness $H_l = 50$ km. The shaded areas correspond to the $\eta_{um} - \eta_{lm}$ solution space in which the predicted values lie within the observed range. (Adapted from *Lambeck and Nakada* [1990]).

constant becomes more comparable to the relaxation times for the lower lithosphere and the load stresses migrate with time into the colder parts of the layer. Seismic estimates, corresponding to very high frequency load cycles, give higher estimates for H_l and one of the reasons for attempting to estimate lithospheric thickness from the sea-level data is to explore this dependence of H_l on the load cycle. Hence, when solving for rheological parameters from rebound data, it is not acceptable to assume an *a-priori* value for H_l and this parameter must be considered as an unknown, along with the mantle viscosity parameters.

Once the rheological parameters are determined from the differential observations, the correction term for the

ice-volume equivalent sea-level follows directly from (4). Some representative results for Karumba and Orpheus Island are shown in Figure 7(a, b) in which the predicted highstands, based on the above mantle solution and zero change in ocean volume after 6000 years ago, are compared with the corresponding observed values. The former are consistently higher than the observed highstands but the difference remains nearly constant from one location to another. It provides, therefore, a measure of the increase in ocean volume ($\delta\zeta_e$) that has occurred in the past 6000 ^{14}C years. The solution for $\delta\zeta_e$ based on all the Late Holocene data is shown in Figure 7(c) and points to the same general conclusion, that ocean volumes have continued to increase be-

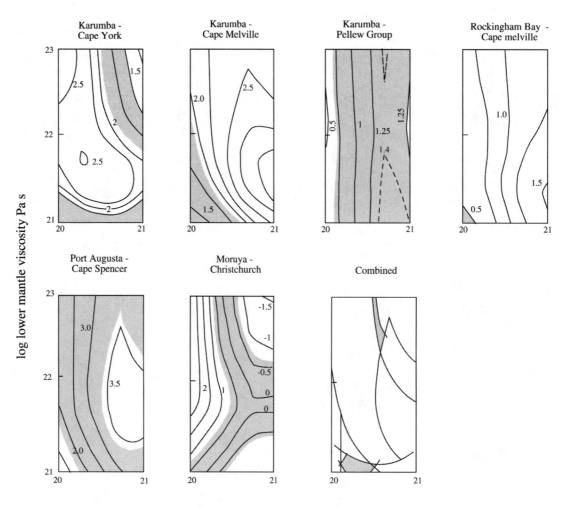

log upper mantle viscosity Pa s

Figure 6. Same as Figure 4 but for $H_l = 100$ km. No single solution is found in this case that satisfies all the observational data used and the combined solution illustrated does not include the Cape Melville data.

tween about 6000 and 3000 ^{14}C years BP. The function $\delta\zeta_e$ will be earth-model dependent but this dependence is small for models within that part of the parameter space in the neighbourhood of the least variance. Also, iterative solutions in which Late Holocene melting is explicitly introduced into the ice models lead to the same results for earth-model parameters and the solutions appear to converge (c.f. examples in *Lambeck et al.* [1996]).

The results illustrated in Figures 5 and 6 are from *Lambeck and Nakada* [1990] and are based on a first-order solution of the sea-level equation in which the time dependence of the shorelines was treated in an approximate manner only. Also, the solution was based on a small subset of the total data available for the last 6000 ^{14}C years. A more comprehensive search through the model space using the complete formulation for the isostatic terms in equation (4) and a larger data base (about 90 sea level estimates with $t \le 6000$ ^{14}C years) has therefore been carried out in which the model-parameter search has been limited to 3 mantle layers of which the lower mantle viscosity has been constrained to 10^{22} Pa s, consistent with the previously found value as well as that found for other localities (see Section 5). For a point in the earth-model parameter space, sea-level values are predicted for the time and location of each observed data point, using the ice models with zero melting after 6000 ^{14}C years BP. The corrective term $\delta\zeta_e$ is estimated along with the corresponding root mean square value of the misfit between

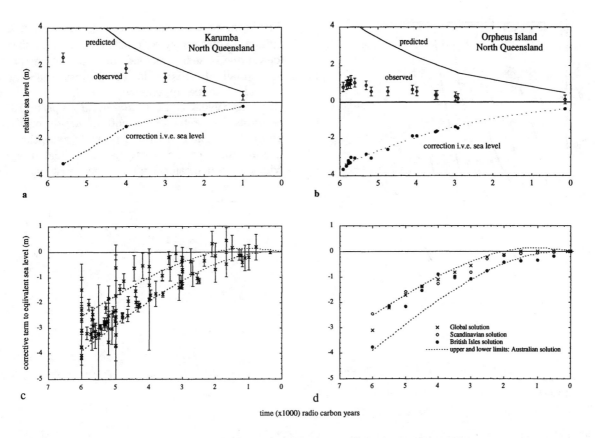

Figure 7. Comparison of predicted and observed late Holocene sea levels based on the earth-model parameters inferred from the differential highstand analysis and on the assumption that no increase in ocean volume occurred after 6000 ^{14}C years BP. The differences correspond to the corrective term $\Delta\zeta_e$ to the ice-volume equivalent (i.v.e) sea level. (a) and (b) illustrate two specific examples. (c) gives the solution for all of the Australian data discussed below. (d) corresponds to similar solutions from other regions. The global solution is from *Fleming et al.* [1998], the Scandinavian solution is from *Lambeck et al.* [1998b] and the British Isles solution is from *Lambeck et al.* [1996].

observed and predicted values. The model parameter space is then searched for the minimum value for this measure of model fit Figure 8. The minimum value found is about 0.5 m, equal to the average accuracy of the observational data, for $H_l = 70\text{-}100$ km and $\eta_{um} = (1.5 - 2.5) \times 10^{20}$. Of note is the correlation that occurs between the two earth-model parameters, thick values for H_l leading to higher values for η_{um}, a correlation previously noted in analyses of sea-level data when the evidence is mainly from the post-glacial period [*Lambeck*, 1993a; *Lambeck et al.*, 1996]. The solution for the corrective term to the ice-volume equivalent sealevel is illustrated in Figure 7c and is similar to that previously found by *Lambeck and Nakada* [1990]. Within the observational uncertainties, this function is not strongly dependent on earth-model parameters in the neighbourhood of the optimum solution.

Figure 9 illustrates the predicted mid-Holocene sea levels around the Australian margin based on the above

rheology and ice-volume equivalent sea level solution. The predicted coastal sea levels are mostly consistent with the observed evidence (c.f. Figure 1c) as should be the case because this observational evidence was used to constrain the solution for model parameters. Levels at 6000 ^{14}C years ago were generally above present with the exception for Tasmania and some other southern margin localities. The ice sheet models used in these predictions contain a substantial component of melt water from Antarctica and Tasmania lies sufficiently close to this former ice load for the glacio-isostatic signal there to be negative and comparable to the magnitude of the hydro-isostatic contribution. The model predicts well the observed gradients in highstand amplitude observed along the deeply indented gulfs of south Australia, as well as across the North Queensland Shelf and along the shores of the Gulf of Carpentaria. In some other localities, such as the southwest corner of Western Australia, quite rapid changes in spatial vari-

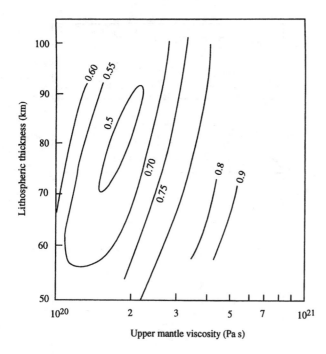

Figure 8. Solution space (H_l, η_{um}) for the Australian late Holocene sea level data for $\eta_{um} = 10^{22}$ Pa s. Contours represent the rms misfit of the model predictions to observed values.

ability of the highstand amplitude are also predicted, reflecting the variability of the hydro-isostatic signal.

4. RECENT SEA LEVEL CHANGE IN THE AUSTRALIAN REGION.

The Australian margin is tectonically quite stable and the observed Holocene sea-level data is compatible with the predictions of glacio-hydro isostatic rebound models to within observational uncertainties. The geological evidence does not require, therefore, the introduction of a tectonic component. In particular, the gradients in the mid-Holocene highstand observed in South Australia do not require a tectonic origin, being wholly explicable in terms of glacio-hydro-isostasy. The apparent absence of the Last Interglacial shoreline in northern Australia may be indicative of some subsidence [*Chappell*, 1987] but it may also reflect a lack of preservation of former shoreline features. Corals from beach rock at about 5 m elevation on Camp Island, North Queensland, give Late Pleistocene ages [*Hopley*, 1971], as do surface outcrops of reefs on Digby and Marble Islands of the central Queensland coast [*Kleypas*, 1996]. In parts of Western Australia the elevations of the older interglacial surfaces point to the possible occurrence of uplift but at rates below 0.1 mm/year [*Lambeck*, 1987].

Furthermore, the Australian margin is characterised by minimal Quaternary sediment loading such that subsidence due to sediment loads or sediment compaction is also small. In Tasmania, in contrast, the Last Interglacial shoreline appears to be well elevated [*Murray-Wallace*, 1990; *Murray-Wallace and Belperio*, 1991] and is indicative of tectonic uplift at a rate of about 0.2 mm/year. Thus, with the possible exception of Tasmania, most of the Australian mainland margin provides a good platform for measurement of recent sea-

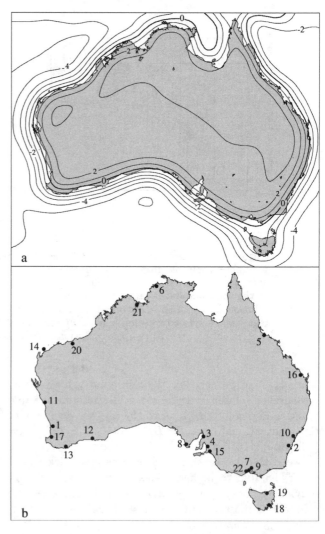

Figure 9. (a) Predicted amplitude of sea-levels around the Australian margin at 6000 ^{14}C years ago based on the glacio-hydro-isostatic rebound parameters and corrective term to the ice-volume equivalent sea-level function derived from the late Holocene sea-level data, Contours are in metres at 1 m intervals. The zero contour is identified by the thicker line. (b) Location of tide-gauge sites with record lengths in excess of 25 years. Numbers refer to the sites listed in Table 1.

Table 1. Summary of observed secular sea-level rates from Australian tide-gauge sites with records longer than 25 years [*Mitchell et al.*, 2000]. Also given are the glacio-hydro-isostatic corrections (Iso. corrn) based on the earth-response function discussed above, and the isostatically corrected (Corr. rate) results. Site locations are shown in Figure 9b.

Site	Tide gauge	Record length (year)	Obs. rate (mm/year)	Iso. corrn (mm/year)	Corr. rate (mm/year)
1	Fremantle WA	90.6	1.38	-0.27	1.65
2	Fort Denison NSW	81.8	0.86	-0.30	1.16
3	Port Pirie SA	63.2	-0.19	-0.46	0.27
4	Port Adelaide SA	55.1	2.07	-0.41	2.48
5	Townsville Qld	38.3	1.12	-0.47	1.59
6	Darwin NT	34.9	-0.02	-0.42	0.40
7	Point Lonsdale Vic	34.4	-0.63	-0.28	-0.35
8	Port Lincoln SA	32.3	0.63	-0.35	0.98
9	Williamstown Vic	31.8	0.26	-0.30	0.56
10	Newcastle NSW	31.6	1.18	-0.30	1.98
11	Geraldton WA	31.5	-0.95	-0.28	-0.67
12	Esperance WA	31.2	-0.45	-0.26	-0.19
13	Albany WA	31.2	-0.86	-0.16	-0.70
14	Thevenard Is WA	31.0	0.02	-0.25	0.27
15	Victor Harbour SA	30.8	0.47	-0.37	0.84
16	Bundaberg Qld	30.2	-0.03	-0.33	0.30
17	Bunbury WA	30.2	0.04	-0.21	0.25
18	Hobart Tas	29.3	0.58	-0.04	0.62
19	George Town Tas	28.8	0.30	-0.14	0.44
20	Port Hedland WA	27.7	-1.32	-0.34	-0.98
21	Wyndham WA	26.4	-0.59	-0.45	-0.14
22	Geelong Vic	25.0	0.97	-0.28	1.25

level change free of tectonic factors in excess of about 0.1 mm/year.

With low background tectonic signals and provided that any tide-gauge site is free of local sediment compaction and subsidence, as is not the case for some of the harbour sites, observations from the Australian margin should provide indicative results of recent sea-level change. Such results become particularly important because there are few long records from the southern hemisphere and estimates of global change tend to be very much dominated by mid- to high-latitude records. Table 1 summarises recent estimates of secular sea-level change from a number of tide gauge sites with records longer than 25 years (see Figure 9b for locations). In two instances, Sydney (Fort Denison) and Fremantle, the records are longer than 80 years [*Mitchell et al.*, 2000]. The Sydney record has been carefully compiled, verified and analysed by numerous investigators over the past two decades (e.g. *Hamon*, 1987). The site itself is on rock, remote from harbour installations, adjacent to relatively deep water and unaffected by fresh-water riverine input. It constitutes the best record of sea-level change from the eastern Australian margin and one of the better southern hemisphere records. The Fremantle record is also from a well maintained and documented tide gauge although the levels are weather affected such that long-term trends are noisier than for most other Australian sites. However, with a 90 year record, this enhanced noise should not impact in a major way on the estimates of long term trends. The rates at these two sites do differ significantly, with the Sydney rate being substantially smaller than the Fremantle rate. Surprisingly, neither of these records are included in recent evaluations of global sea-level change [*Douglas*, 1997; *Peltier*, 2000], the Sydney record being discarded because it was found to be inconsistent with results from the Newcastle tide gauges about 100 km north of Sydney. Records are available from three gauges in Newcastle harbour and the result, of about 30 years duration from Newcastle III - the best of the three - is included in Table 1. This result is about 40% higher than the Sydney result but it is also much less satisfactory because of its shorter record length, its marked perturbation by river discharge and the harbours location on sediments of the river floodplain. Two records from Port Adelaide give consistent results but both sites are potentially perturbed by known sediment subsidence of the harbour environment. The nearby Port Pirie record of 63 year

Table 2. Estimates for earth response parameters (effective lithospheric thickness (H_l, and effective viscosities η_{um}), η_{lm} for the lower and upper mantle) from three different regional solution. The Australian results are based on the analysis of the late Holocene data. The three Scandinavian solutions are from geological evidence since Lateglacial time [*Lambeck et al.*, 1998b] , from tide-gauge analyses [*Lambeck et al.*, 1988a] , and from the Baltic Ice Lake shoreline gradients and the timing of the Baltic lake stages [*Lambeck*, 1999]. The British Isles solution is based on Lateglacial and Postglacial geological data [*Lambeck et al.*, 1996]. Lower mantle estimates in brackets are assumed for these particular solutions.

Solution	H_l (km)	η_{um} (10^{20} Pa s)	η_{lm} (10^{22} Pa s)
Australia	75-90	1.5-2.5	(1)
Scandinavia-1	65-85	3-4	0.6-1.3
Scandinavia-2	80-100	4-5	(1)
Scandinavia-3	40-80	2.5-4.5	0.7-1.5
British Isles	65-70	4-5	0.7-1.3

duration, in contrast, gives a distinctly different result: of an apparent fall in sea level.

Considerable variation occurs in the observed rate of sea-level change, reflecting variability of oceanographic, climatological and geological factors, possibly superimposed on issues related to maintaining tide gauges over very long time intervals. Part of the variability may be a consequence of the glacio-hydro-isostatic contributions. These effects are included in Table 1 for the solution discussed above and in the presence of only these effects the predictions are for falling sea levels at the mainland tide gauges at rates from 0.16 to 0.46 mm/year. (These predictions are based on the assumption of constant ocean volume for the duration of the tide gauge records.) For the Hobart site the predicted rate is much less than for the mainland localities, only 0.04 mm/year, consistent with the previously discussed Late Holocene evidence of an apparent absence of mid-Holocene highstands.

The isostatically corrected values given in Table 1, representing the sea-level change in the absence of the glacio-hydro-isostatic contributions, exhibit a similar spatial variability as the observed values and other, more important, factors must contribute to the observed variability. A number of the Western Australian sites, but not Fremantle, indicate a falling sea level over the past three decades. Tectonic causes, operating over time scales of thousands of years and longer, can be ruled out as an explanation. At Geraldton, for example, this would lead to about 4 m of uplift over 6000 years so that the mid-Holocene highstand would be predicted to occur at elevations of about 5-6 m. Also, the Last Interglacial shoreline would be about 80 m above present

sea level whereas it actually lies at only 3 m elevation, where it is expected to occur in the absence of tectonic movement [*Stirling et al.*, 1998]. If only the two long records from Sydney and Fremantle are accepted then the average corrected sea-level rise is 1.4 mm/year but such averaging may mask information on causes of a spatially variable sea-level response.

5. SOME GLOBAL IMPLICATIONS OF THE AUSTRALIAN SEA LEVEL RESULTS

Upper mantle viscosity is laterally variable. The Australian rheology solution is compared with similar solutions for other regions in Table 2. (All solutions relate to the ^{14}C time scale and the viscosities are in Pa (^{14}C)s.) The Scandinavian results are based on three different types of data and yield concordant estimates for the effective rheological parameters. For both these and the British Isles solutions the parameter search was conducted through a much larger parameter space than for the Australian solution, including two orders of magnitude for the lower-mantle viscosity and some ice-sheet parameters. For the British Isles, mantle models with a greater degree of layering were also explored but only the three-layered models are compared here to facilitate the comparisons between regional solutions. A wholly independent analysis of the European data, independent in the numerical modelling of the earth response and sea-level change as well as in inversion technique, has led to comparable results [*Kaufmann and Lambeck*, 2000]. In all three regional solutions the lithosphere is considered as an unknown parameter and the corrective ice-volume equivalent sea-level term $\Delta\zeta_e$, as well as some ice-sheet scaling parameters in the northern hemisphere solutions, are included.

Of note when comparing the European and Australian results is that the estimates for effective lithospheric thickness are comparable but that the upper-mantle viscosity appears to be lower for the Australian region than for northern Europe. Because of the partial parameter search conducted, and because only the post 6000 year BP data is used (the earlier data contribute to the separation of the H_l and η_{um} parameters), the Australian result must be considered as preliminary but this difference in η_{um} is nevertheless suggestive of the occurrence of lateral variation in mantle viscosity, as already noted by *Nakada and Lambeck* [1991]. Models of upper-mantle structure from inversion of seismic shear wave or attenuation data [e.g. *Romanowicz*, 1998] indicate that considerable lateral variation in structure occurs and that corresponding lateral variations in mantle temperature and hence viscosity can also be expected. Thus the viscosity results are not qualitatively inconsistent with global seismic evidence of higher values beneath shields and stable continents than beneath oceans. In the absence of global solutions for a laterally variable mantle response to surface loading (but see *Kaufmann et al.* [1997]; *Tromp and Mitrovica* [2000]) the strategy adopted here is to estimate viscosities that are representative of the average regional conditions rather than to carry out a global analysis. At a minimum this has the merit of establishing whether lateral variations in mantle viscosity are likely to be significant.

Ocean volume has increased since mid-Holocene time. Solutions for the corrective term $\Delta\zeta_e$ are not limited to far-field locations and the same procedure of solving equation (4) for both rheological and ocean volume parameters (and ice-sheet parameters) has been used in the rebound analyses of formerly glaciated regions. (In these solutions the sea-level variation is effectively separated into a spatially variable part and a spatially invariable but time dependent part and the former constrains the viscosity and ice distribution and the latter constrains $\Delta\zeta_e$.) The corrective term to the ice-volume equivalent sea-level function has been found to be persistently negative immediately after 6000 ^{14}C years BP (Figure 7d). Together, the evidence points to there having been an increase in ocean volume of about 3m equivalent sea-level rise, with the major part of it having occurred in the interval of about 6000-3000 years BP and that any change subsequent to about 2000-3000 years BP and before the recent rise recorded by tide gauges, was at an average rate of only about 0.1-0.2 mm/year

(Figure 7d, see also *Lambeck and Bard* [2000]). The source of this increase cannot be identified from these analyses but a likely major contribution is late melting of the West Antarctic ice sheet as discussed briefly above. The consistency of the results for the various regions indicates that the signal is a global one and not an artifact of analysis procedures; of trade-off between an inadequate modelled regional isostatic response of a nearby ice load and $\Delta\zeta_e$ for example. Also, within observational errors there is no indication of major rapid or abrupt changes in sea level over the last 6000 years by more than about 50 cm at any one time.

Holocene highstands observed at Pacific islands tend to be small in amplitude and to occur later than 6000 ^{14}C years ago [e.g. *Pirazzoli and Montaggioni*, 1988; *Moriwaki et al.*, 2000] (Figure 3). As illustrated schematically in Figure 10, this is a consequence of on-going ocean-volume increase in Late Holocene time. The upper panel illustrates isostatic components for different sites, with the largest values being representative of areas near former ice margins where the glacio-isostatic contribution dominates. The middle panel illustrates the predicted sea levels for the case where all melting ceased 6000 years ago (the ice-volume equivalent sea-level curve is denoted by esl). In this case the predicted mid-Holocene sea-level highstands occur at the time of cessation of melting for all cases. The lower panel is for on-going melting in late-Holocene time. When the isostatic effect is substantial, as in curve 1, this dominates the total sea-level signal and a well-defined highstand develops at the time of cessation of melting. But when the isostatic term is smaller the highstand occurs later and is broader (e.g. curve 3). This predicted behaviour is consistent with the observation that small-island Pacific highstands occur later than at continental margins. It is also consistent with observations from formerly glaciated northern Britain that the timing of the highstand is earliest where the rebound is greatest [*Lambeck*, 1993b]. This characteristic of the highstand cannot be reproduced by modifications of the rheological parameters and the observations are indicative of late-Holocene increases in ocean volume irrespective of the choice of mantle viscosity or lithospheric thickness.

Present-day sea level is spatially variable, even when corrected for glacio-hydro-isostasy. The two long tide-gauge records from Australia indicate a secular rise in relative sea level of 0.9 -1.4 mm/year which, when corrected for the isostatic effect of opposite sign, increases to about 1.2 - 1.6 mm/year respectively. These values are lower than globally averaged estimates of 1.8 - 2.0 mm/year sometimes quoted [e.g. *Peltier and Jiang*,

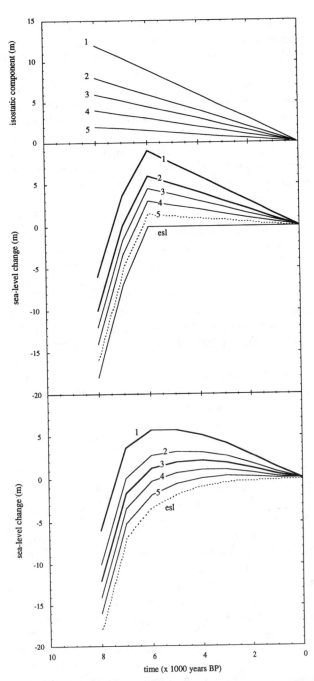

Figure 10. Schematic illustration of the shape of the Late Holocene sea-level curve in the far-field areas in the presence of ocean volume increase. The upper panel illustrates the isostatic components of different magnitudes; the middle panel illustrates the predicted sea levels for the case where all melting ceased at 6000 years ago; the lower panel is for the case of on-going melting in late-Holocene time.

records analysed, or in the corrections applied for tectonic or glacio-hydro-isostatic effects. But it may also be indicative of lateral variation in present-day change due to long term regional changes in climatological and oceanographic factors, including ocean temperatures, salinity, currents and surface-wind stress. Table 3 summarises some results of recent analyses. Of interest is that while the three global solutions are based on very similar isostatic corrections the result based on the far-field analysis by *Davis and Mitrovica* [1996] is less than that resulting from the other two analyses [*Peltier and Jiang*, 1997; *Douglas*, 1997] that incorporate tide gauge data from sites close to or within the former ice margins.

The three global results in Table 3 are all based on the assumption that in calculating the isostatic effects the mantle response is laterally uniform. Furthermore, all three solutions set the lithospheric thickness at an *a-priori* value of 120-125 km and this and the associated viscosity profile may not be appropriate for all locations (c.f. Table 2). This limitation is overcome in the regional solutions included in Table 3 in which the rebound parameters are estimated from the local sea-level evidence and the model becomes an effective interpolator for present-day contributions within the region covered by the analysis. For the Scandinavian solution three approaches have been used to correct the observed tide-gauge for the isostatic effects [*Lambeck et al.*,1998a]. One is to use the geological data and rebound model to estimate the long-term trend and to subtract this from the tide-gauge result (solution SCAN-1, Table 3). In this sense, the rebound model is used as an interpolation device to give a best-fitting sea-level function to the palaeo sea-level indicators and any lack of separation of the model parameters that describe this function (viscosity, lithospheric thickness, the ice-volume equivalent sea-level term, ice model parameters) will not influence greatly the estimation of the present isostatic effects. Another approach is to estimate that part of the Earth's surface where the isostatic effects are zero or very small. Around an ice sheet, for example, there is a zone where the predicted isostatic effect is zero, or nearly so, for a broad range of model parameters and in the case of Scandinavia, a substantial number of quality tide-gauge records lies within this zone (solution SCAN-2). For Scandinavia a third solution is possible, namely to use the observed tilting of some of the large inland lakes [*Ekman*, 1996] to estimate the earth-response parameters and to then use these parameters to correct the marine tide gauges for isostatic effects (solution SCAN-3). All three approaches give consistent results that point to a secular sea level rise in northern Europe of about 1.1 ±0.2 mm/year for the past 100 years.

1997; *Douglas*, 1997] but similar to some of the lower values found from regional analyses (e.g. *Lambeck et al.* [1998a]; *Woodworth et al.* [1999]). Part of the differences may lie in analysis methods used, in lengths of

Table 3. Summary of recent estimates of secular sea level change. All results have been corrected for the glacio-hydro-isostatic contributions. The Australian result is the average of the two longest records from Table 1. The three Scandinavian results are discussed in text. The error bars on the *Woodworth et al.* [1999] solution is from *Shennan and Woodworth* [1992].

Solution/author	Region	Rate mm/year
Australia	regional	1.40 ± 0.25
SCAN-1	regional - Scandinavia	1.1 ±0.2
SCAN-2	regional Scandinavia	1.0 ±0.25
SCAN-3	regional - Scandinavia	1.01 ±0.2
Woodworth et al., 1999	regional British Isles	1.0 ±0.15
Davis and Mitrovica, 1996	regional - N. America E. coast	1.5 ±0.3
Peltier and Jiang, 1977	regional - N. America E. coast	2.0 ±0.6
Mitrovica and Davis, 1995	global (far-field only)	1.4 ±0.4
Peltier and Jiang, 1997	global	1.8 ± 0.6
Douglas, 1997	global	1.8 ±0.1

The solution by *Woodworth et al.* [1999] corrects for the isostatic effects, as well as any other long-term, millennial-scale contributions by differencing the geological and tide gauge records (see also *Shennan and Woodworth* [1992]). This solution is also at the lower end of the estimates. Of the three similar regional solutions for the North American east coast, two give higher values than the third (see Table 3) but the differences are probably statistically insignificant. But the result for all three analyses is significantly higher than that found for the other side of the Atlantic Ocean and this may relate to spatial variability in ocean warming and thermal expansion (e.g. *Levitus et al.* [2000]). The comparisons of estimates for the three regions indicate that the regional variability in secular sea level change, of oceanographic and climatological origin, may be of the order ±0.5 mm/year superimposed on a globally averaged value of about 1.5 mm/year.

Other than thermal expansion, other factors that contribute to spatial variability in long-term sea-level change include shifts in position or in strength of ocean boundary currents. Can variability in the Cape Leeuwin Current or in the East Australian Current, for example, explain the differences in results for the east and west coasts of Australia? Another factor is that when water is added into the oceans from mountain glaciers or from an exchange with ground and surface waters, the redistribution of the added water is not uniformly distributed because of changes in gravitational attraction between land and sea and because of the sea-floor deformation under changing loads. Such variability has been demonstrated for the mountain glacier contribution [*Nakiboglu and Lambeck*, 1991] but will also result from any addition of surface or ground water into the oceans.

What the tide-gauge results in Table 3 suggest, and what our understanding of surface loading problems dictates, is that there may be substantial spatial variability in secular sea-level change, even when the tide-gauge data has been corrected for the glacio-hydro-isostatic effects. Thus the averaging of geographically poorly distributed tide gauges may not lead to a very meaningful estimate of the global average. (This was one of the reasons for the analysis by *Nakiboglu and Lambeck* [1991] in which the tide gauge records were expressed as a low-degree spherical harmonic expansion the zero degree term of which provides the estimate of the globally averaged sea-level rise. Such an analysis has the advantages that it is not necessary to correct for the isostatic effects and that it may provide information on some of the long-wavelength spatial variability of the secular change.)

Irrespective of the detail of this spatial variability, the results do suggest that a globally averaged sea-level rise of about 1.5±0.5 mm/year has occurred for the duration of the tide gauge records, or about 100 years [*Church et al.*, 2001]. Tide gauge records for earlier periods are limited but some of the very long records do indicate an acceleration in mean sea level during the nineteenth century [*Woodworth*, 1999; *Ekman*, 1999; *Maul and Martin*, 1993]. If the secular trend noted for the past century was not a recent development then it would become visible in the geological or archaeological records of sea level change as early as 500 - 1000 years ago with shorelines at 75 - 150 cm above present. High resolution records indicate that little change in sea level has occurred, over and above that which can be attributed to the isostatic factors, during the past 2000 years (Figure 7 and also *Lambeck and Bard* [2000]; *Morhange et al.* [2001]; *Sivan et al.* [2001]) such that the present-day rise must indeed

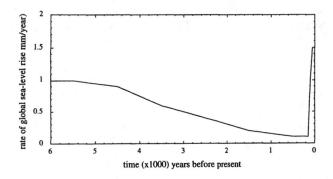

Figure 11. Inferred rates of sea-level change for past 6000 years.

have been a relatively recent phenomenon. The rates of sea-level rise for the past 7000 years are illustrated in Figure 11, based on the average of the results illustrated in Figure 7d and the additional results from the above references. The result for the past century is the global best estimate of *Church et al.* [2001] and this rise is assumed to have been initiated about 100-150 years ago, consistent with the evidence from the longest tide-gauge records.

REFERENCES

Bard, E., M. Arnold, B. Hamelin, N. Tisnerat-Laborde, and G. Cabioch, Radiocarbon calibration by means of mass spectrometric 230Th/234U and 14C ages of corals. An updated data base including samples from Barbados Mururoa and Tahiti, *Radiocarbon, 40*, 1085-1092, 1998.

Bindschadler, R., Future of the West Antarctic ice Sheet, *Science, 282*, 428-429, 1998.

Burne, R.V., Relative fall of Holocene, sea level and coastal progradation, northeastern Spencer Gulf, South Australia, *BMR J. Aust. Geol. Geophys., 7*, 35-45, 1982.

Chappell, J., Late Quaternary sea-level changes in the Australian region, in *Sea-level Changes*, edited by M.H. Tooley, and I. Shennan, pp. 296-331, Basil Blackwell, Oxford, 1987.

Chappell, J., E.G. Rhodes, B.G. Thom, and E. Wallensky, Hydro-isotasy and the sea-level isobase of 5500 BP in north Queensland, Australia, *Marine Geol., 49*, 81-90, 1982.

Chappell, J., A. Chivas, E. Wallensky, H.A. Polach, and P. Aharon, Holocene, palaeo-environmental changes, central to north Great Barrier Reef inner zone, *BMR J. Aust. Geol. Geophys., 8*, 223-235, 1983.

Church, J. A., Gregory, J. M., Huybrechts, P., et al. Changes in sea level. In *Intergovernmental Panel on Climate Change, Third Assessment Report*, Cambridge University Press, Chapter 11, (in press) 2001.

Conway, H., B.L. Hall, G.H. Denton, A.M. Gades, and E.D. Waddington, Past and future grounding-line retreat of the West Antarctic Ice Sheet, *Science, 286*, 280-283, 1999.

Davis, J.L., and J.X. Mitrovica, Glacial isostatic adjustment and the anomalous tide gauge record of eastern North America, *Nature, 379*, 331-333, 1996.

Denton, G.H., and Hughes T.J. (eds), *The Last Great Ice Sheets, Wiley-Interscience*, New York, 484 pp., 1981.

Douglas, B.C., Global sea rise: a redetermination, *Surveys in Geophysics, 18*, 279-292, 1997.

Dyurgerov, M. B. and Meier M. F. Year-to-year fluctuations of global mass balance of small glaciers and their contribution to sea-level changes. *Arctic and Alpine Research, 29*, 392-402, 1997.

Ekman, M., A consistent map of the postglacial uplift of Fennoscandia, *Terra Nova, 8*, 158-165, 1996.

Ekman, M., Climate changes detected through the world's longest sea level series, *Global and Planetary Change, 21*, 215-224, 1999.

Farrell, W.E., and J.A. Clark, On postglacial sea level, *Geophys. J., 46*, 79-116, 1976.

Fleming, K., P. Johnston, D. Zwartz, Y. Yokoyama, K. Lambeck, and J. Chappell, Refining the eustatic sea-level curve since the Last Glacial Maximum using far- and intermediate-field sites, *Earth Planet. Sci. Lett., 163*, 327-342, 1998.

Gregory, J. M. and Oerlemans J. Simulated future sea-level rise due to glacier melt based on regionally and seasonally resolved temperature changes. *Nature, 391*, 474-476, 1998

Hamon, B.V. A century of tide records: Sydney (Fort Denison) 1886-1986, *Tech. Rept. 7* (ISSN 0158-9776), Flinders Institute for Atmospheric and Marine Sciences, 1987.

Hopley, D., The origin and significance of North Queensland island spits, *Z. Geomorph. N.F., 15*, 371-389, 1971.

Hopley, D., *The Geomorphology of the Great Barrier Reef*, Wiley-Interscience, New York, 453 pp., 1982.

Hopley, D., Evidence of 15000 years of sea level change in tropical Queensland, in *Australian Sea Levels in the last 15000 years: A Review*, edited by D. Hopley, pp. 92-104, James Cook University, Dept Geography, 1983.

Huybrechts, P., Formation and disintegration of the Antarctic ice sheet, *Ann. of Glaciol. 20*, 336-340, 1994.

Johnston, P., The effect of spatially non-uniform water loads on the prediction of sea-level change, *Geophys. J. Int., 114*, 615-634, 1993.

Johnston, P., The role of hydro-isostasy for Holocene sea-level changes in the British Isles, *Marine Geology, 124*, 61-70, 1995.

Kaufmann, G., and K. Lambeck, Mantle dynamics, postglacial rebound and the radial viscosity profile, *Phys. Earth Planet. Int., 121*, 301-324, 2000.

Kaufmannn, G., P. Wu, and D. Wolf, Some effects of lateral heterogeneities in the upper mantle on postglacial land uplift close to continental margins, *Geophysical Research Letters, 128*, 175-187, 1997.

Kleypas, J.A., Coral Reef Development under naturally turbid conditions: fringing reefs near Broad Sound Australia, *Coral Reefs, 15*, 153-167, 1996.

Lambeck, K., The Perth Basin: A possible framework for its formation and evolution, *Exploration Geophys., 18*, 124-128, 1987.

Lambeck, K., Glacial rebound and sea-level change: an example of a relationship between mantle and surface processes., *Tectonophysics, 223* 15-37, 1993a.

Lambeck, K., Glacial Rebound of the British Isles. II: A

high resolution, high-precision model, *Geophys. J. Int.,* *115*, 960-990, 1993b.

Lambeck, K., Late Pleistocene and Holocene sea-level change in Greece and southwestern Turkey: A separation of eustatic, isostatic and tectonic contributions, *Geophys. J. Int., 122*, 1022-1044, 1995.

Lambeck, K., Shoreline reconstructions for the Persian Gulf since the Lastglacial Maximum, *Earth Planet. Sci. Lett., 142*, 43-57, 1996.

Lambeck, K., Shoreline displacements in southern-central Sweden and the evolution of the Baltic Sea since the last maximum glaciation, *J. Geol. Soc. London, 156*, 465-486, 1999.

Lambeck, K.,and E. Bard, Sea-level change along the French Mediterranean coast since the time of the Last Glacial Maximum, *Earth Planet. Sci. Lett., 175*, 203-222, 2000.

Lambeck, K., and P. Johnston, The viscosity of the mantle: Evidence from analyses of glacial rebound phenomena, in *The Earth's mantle,* edited by I. Jackson, pp. 461-502, Cambridge University Press, Cambridge, 1998.

Lambeck, K., and M. Nakada, Late Pleistocene and Holocene Sea-Level Change along the Australian Coast, *Palaeogeogr. Palaeoclimatol. Palaeoecol. (Global and Planetary Change Section), 89*, 143-176, 1990.

Lambeck, K., P. Johnston, C. Smither, and M. Nakada, Glacial rebound of the British Isles - III. Constraints on mantle viscosity, *Geophys. J. Int., 125*, 340-354, 1996.

Lambeck, K., C. Smither, and M. Ekman, Tests of glacial rebound models for Fennoscandinavia based on instrumented sea- and lake-level records, *Geophys. J. Int.,* 135, 375-387, 1998a.

Lambeck, K., C. Smither, and P. Johnston, Sea-level change, glacial rebound and mantle viscosity for northern Europe, *Geophys. J. Int., 134*, 102-144, 1998b.

Levitus, S., J.I. Antonov, T.P. Boyer, and C. Stephens, Warming of the World Ocean, *Science, 287*, 2225-2229, 2000.

Maul, G.A., and D.M. Martin, Sea level rise at Key West, florida, 1846-1992: America's longest instrument record?, *Geophysical Research Letters, 20*, 1955-1958, 1993.

Milne, G.A., and J.X. Mitrovica, Postglacial sea-level change on a rotating earth, *Geophys. J. Int., 133*, 1-19, 1998.

Milne, G.A., J. Mitrovica, and J.L. Davis, Near-field hydroisostasy: the implementation of a revised sea-level equation, *Geophys. J. Int., 139*, 464-482, 1999.

Mitchell, W., J. Chittleborough, B. Ronai, and G.W. Lennon, Sea Level Rise in Australia and the Pacific,The South Pacific Sea Level and Climate Change Newsletter, *Quarterly Newsletter, 5*, 10-19, 2000.

Mitrovica, J.X., and W.R. Peltier, On postglacial geoid subsidence over the equatorial oceans, *J. Geophys. Res., 96*, 20053-20071, 1991.

Morhange, C., J. Laborel, and A. Hesnard, Changes of relative sea level during the past 5000 years in the ancient harbor of Marseilles, Southern France, *Palaeogeography, Palaeoclimatology, Palaeoecology, 166*, 319-329, 2001.

Moriwaki, H., M. Chikamori, M. Okuno, and T. Nakamura, Holocene shoreline and sea-level changes in Raratonga, Cook Islands, *Geol. Soc. New Zealand Misc. Publ., 108A*, 104, 2000.

Murray-Wallace, C.V., A. Goede, and K. Picker, Last In-

terglacial coastal sediments at Mary Ann Bay, Tasmania, and their neotectonic significance, *Quat. Australasia,* 8 (2), 26-32, 1990.

Murray-Wallace, C.V., and A.P. Belperio, The last interglacial shoreline in Australia: a review, *Quat. Sci. Rev.* 10, 441-461, 1991.

Nakada, M., and K. Lambeck, Glacial rebound and relative sealevel variations: A new appraisal, *Geophys. J. Roy. astr. Soc., 90*, 171-224, 1987.

Nakada, M., and K. Lambeck, The melting history of the Late Pleistocene Antarctic ice sheet, *Nature, 333*, 36-40, 1988.

Nakada, M., and K. Lambeck, Late Pleistocene and Holocene Sea-Level Change: Evidence for Lateral Mantle Viscosity Structure?, in *Glacial Isotasy, Sea Level and Mantle Rheology*, edited by R. Sabadini, K. Lambeck, and E. Boschi, pp. 79-94, *Kluwer Academic Publ., Dordrecht*, 1991.

Nakiboglu, S.M., K. Lambeck, and P. Aharon, Postglacial sealevels in the Pacific: Implications with respect to deglaciation regime and local tectonics, *Tectonophysics, 91*, 335-358, 1983

Nakiboglu, S.M., and K. Lambeck, Secular Sea-Level Change, in *Glacial Isotasy, Sea Level and Mantle Rheology*, edited by R. Sabadini, K. Lambeck, and E. Boschi, pp. 237-258, Kluwer Academic Publ., 1991.

Peltier, W.R., Global glacial isostatic adjustment, in *Sea level rise: history and consequences*, edited by B. C. Douglas, M. S. Kearney and S. P, Leatherman, *Academic Press* (in press), 2000.

Peltier, W.R., and X. Jiang, Mantle viscosity, glacial isostatic adjustment and the eustatic level of the sea, *Surveys in Geophysics, 18*, 239-277, 1997.

Pirazzoli, P.A., and L.F. Montaggioni, Holocene sea-level changes in French Polynesia, *Palaeogeogr. Palaeoclimatol. Palaeoecol., 68*, 153-175, 1988.

Romanowicz, B. Attenuation tomography of the Earths mantle: a review of current status. *Pure appl. Geophysics, 153*, 257-272, 1998.

Shennan, I., and P.L. Woodworth, A comparison of late Holocoene and twentieth-century sea-level trends from the UK and North Sea region, *Geophysical Journal International, 109*, 96-105, 1992.

Sivan, D., K. Lambeck, E. Galili, and A. Raban, Holocene sea-level changes along the Mediterranean coast of Israel, based on archaeological observations and numerical model, *Palaeogeogr Palaeoclimatol Palaeoecol, 167* 101-117, 2001.

Stirling, C.H., R.M. Esat, K. Lambeck, and M.T. McCulloch, Timing and duration of the Last Interglacial: evidence for a restricted interval of widespread coral reef growth, *Earth and Planetary Science Letters, 160*, 745-762, 1998.

Stuiver, M., P.J. Reimer, E. Bard, J.W. Beck, G.S. Burr, K.A. Hughen, B. Kromer, G. McCormac, J. Van der Plicht, and M. Spurk, INTCAL98 radiocarbon age calibration, 24,000-0 cal BP, *Radiocarbon, 40*, 1041-1083, 1998

Tromp, J., and J. Mitrovica, Surface loading of a viscoelastic planet-III. Aspherical models, *Geophys J Int, 140*, 425-441, 2000.

Woodworth, P.L., M.N. Tsimplis, R.A. Flather, and I. Shennan, A review of the trends observed in British Isles mean sea level data measured by tide gauges, *Geophysical Journal International, 136*, 651-670, 1999.

Zuo, Z. and Oerlemans, J. Contribution of glacier melt to sea level rise since 1865: a regionally differentiated calculation. *Climate Research, 13*, 835-845, 1997.

Kurt Lambeck, Research School of Earth Sciences, Australian National University, Canberra, ACT 0200., Australia. (e-mail: kurt.lambeck@anu.edu.au)

Long-Term Rotation Instabilities of the Earth: A Reanalysis

R. Sabadini

Sezione Geofisica, Dipartimento di Scienze della Terra, Univerità di Milano, Italy

B. L. A. Vermeersen

DEOS, Faculty of Aerospace Engineering, Delft University of Technology, Netherlands

Long-term rotation instabilities are sensitive to surface and internal mass redistribution. The flow of mantle material following the Pleistocene deglaciation cycles and present-day mass imbalance in Antarctica and Greenland are the primary sources of the long-term Earth rotation instability named polar wander, associated with a secular drift of the axis of rotation of about 1 deg/Myr, directed approximately towards Newfoundland. This paper is devoted to a reanalysis of the physics leading to this peculiar Earth rotation component, in terms of the effects of density and viscosity stratification in the transition zone of the mantle. Particular attention is drawn on the effects of density stratification on the amount of polar wander during the ice-age cycles; this displacement arises from either secular or exponential terms in the govering equations. Present-day mass imbalance in Antarctica and Greenland acts in concert with Pleistocene deglaciation to modify previous estimates of lower mantle viscosities, which are based on True Polar Wander (TPW) data and on the assumption that Pleistocene deglaciation is the only forcing mechanism. These new findings support arguments that the lower mantle is more viscous than the upper mantle.

INTRODUCTION

The rotation vector of the Earth changes, in both amplitude and orientation. In the 19th century it was observed that the rotation of the Earth is not regular and we now know that changes in the Earth rotation occur on a broad range of time scales, from shorter than a day to geological periods of hundreds of millions of years.

Ice Sheets, Sea Level and the Dynamic Earth
Geodynamics Series 29
Copyright 2002 by the American Geophysical Union
10.1029/029GD04

The changes in position of the rotation axis can be divided into two main categories: those in which the position of the axis changes with respect to the distant stars but not with respect to the Earth's surface, and vice versa. For the latter category, a hypothetical observer in space will see the Earth shifting underneath its rotation axis as a solid unit while the rotation axis itself remains fixed with respect to the stars. An observer on Earth would see the rotation axis wandering over the Earth's surface. Displacements of the axis of rotation with respect to the 'fixed' stars (changes in which the whole planet is moving rigidly as one unit) are mainly due to external forces, notably the gravitational interactions between the Earth and the Sun, the Moon and the other planets.

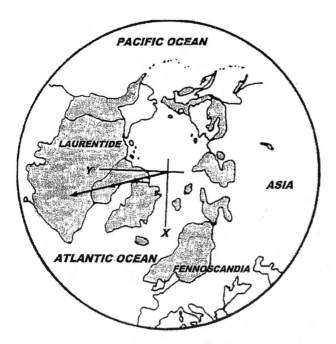

Figure 1. The polar projection shows the geographical coordinates and the location of the major continental ice sheets. The arrow denotes the current direction of polar wander [after *Vermeersen et al,* 1997].

In the present analysis our attention is drawn to the displacements of the axis of rotation with respect to a fixed position on the Earth's surface arising from mass displacements in the interior and in the hydro- and atmosphere. These mass displacements will generally induce changes in the moments and products of inertia. Since the Earth is a deformable body, the rotation axis readjusts to the new situation by shifting over the surface in order that the angular momentum of the Earth is conserved.

There are a number of possible mechanisms responsible for the observed rotational variations, such as growth and decay of ice sheets, variations in sea-level, ocean currents, winds and changes in the pressure distribution of the atmosphere, earthquakes, tectonic plate movements, advection of heterogeneities in the mantle and core, and interactions between the core and mantle. Each of these mechanisms operates on specific time scales and this forcing is reflected in the time scales on which the rotation of the Earth changes.

The observed changes in the position of the rotation axis with respect to the Earth's surface, generally termed True Polar Wander (TPW), consist of two kinds of movements: periodic and linear. The periodic motions consist mainly of two periods. The annual wobble is primarily due to seasonally varying zonal winds. The

cause of the Chandler wobble is still largely unknown, although *Gross* [2000] has argued that the Chandler wobble is due to pressure fluctuations on the bottom of the ocean caused by temperature and salinity variations and wind-driven changes in the circulation of the oceans.

This periodic movement, which is essentially the free precession of the Earth, was theoretically predicted in the 18th century by Euler (therefore it is also called the Eulerian free precession) but not observed until the end of the 19th century.

The contemporary secular drift has been determined by astrometric and space-geodetic observations [for a recent overview: *Gross and Vondrák,* 1999]. Post-glacial rebound resulting from the decay of ice sheets is thought to be the main cause of this secular drift, although present-day ice-mass variations, tectonic processes and mantle convection also contribute. This component of the polar motion is depicted in Figure 1 by the black arrow starting from the North Pole, denoting the direction of this secular drift with respect to the geography.

The secular shift of the rotation axis reflects a true wander with respect to the deep mantle. It is therefore called TPW to be distinguished from the Apparent Polar Wander (APW) that simply reflects the continental drift of the particular plate through which the axis of rotation pierces on the Northern or Southern Hemisphere.

The present study is an overview of the rotational dynamics of a stratified, viscoelastic Earth, that starts from the basic physical principles and deals, in particular, with the polar wander induced by the continuous occurrence of ice ages or by the last Pleistocene deglaciation. Particular attention is devoted to the effects of density stratification and viscosity on the glacially induced TPW on the various terms that enter the governing rotational equations.

ROTATIONAL CHANGES FOR A RIGID EARTH

In this section we assume that the Earth is rigidly rotating and that torques, mass displacements and relative motions can perturb the rotation.

If $\vec{L}$ denotes torque, $\vec{H}$ angular momentum and $\vec{\omega}$ angular velocity, then in a reference frame co-rotating with the Earth, Euler's dynamical equation reads

$$\frac{d\vec{H}}{dt} + \vec{\omega} \times \vec{H} = \vec{L} \qquad (1)$$

with

$$\vec{H} = \int \rho \vec{r} \times (\vec{\omega} \times \vec{r}) dV = \int \rho(r^2\vec{\omega} - (\vec{r}\cdot\vec{\omega})\vec{r})dV \quad (2)$$

or

$$\vec{H} = \mathbf{I} \cdot \vec{\omega} \quad \text{with} \quad I_{ij} = \int \rho(r_k r_k \delta_{ij} - r_i r_j)dV \quad (3)$$

representing the components of the inertia tensor **I**, with δ_{ij} being the Kronecker delta function. This results in

$$\frac{d}{dt}(\mathbf{I}\cdot\vec{\omega}) + \vec{\omega}\times(\mathbf{I}\cdot\vec{\omega}) = \vec{L}. \quad (4)$$

In the general case **I** is time dependent because mass displacements and relative motion, $\vec{u}$, can occur with respect to the axis $\vec{r}$. Therefore $\vec{H}(t)$ can be expressed as

$$\vec{H}(t) = \mathbf{I}(t)\cdot\vec{\omega} + \vec{h}(t) \quad (5)$$

in which

$$\vec{h}(t) = (h_1, h_2, h_3)^T = \int \rho\vec{r}\times\vec{u}dV \quad (6)$$

Substituting (5) and (6) into (4) leads to the so-called Liouville equation

$$\frac{d}{dt}\left(\mathbf{I}(t)\cdot\vec{\omega} + \vec{h}(t)\right) + \vec{\omega}\times\left(\mathbf{I}(t)\cdot\vec{\omega} + \vec{h}(t)\right) = \vec{L} \quad (7)$$

If Ω denotes the mean Earth rotation frequency, then the components of $\vec{\omega}$ can be expressed in the dimensionless quantities m_i as

$$\vec{\omega} = (\omega_1, \omega_2, \omega_3)^T = \Omega(m_1, m_2, 1 + m_3)^T \quad (8)$$

The quantities m_i are small whenever the deviations from the axis of rotation are small. Assuming that $\vec{h}$ and the changes in **I** are small, the inertia tensor can be written as

$$\mathbf{I} = \begin{pmatrix} A + \Delta I_{11}(t) & \Delta I_{12}(t) & \Delta I_{13}(t) \\ \Delta I_{21}(t) & A + \Delta I_{22}(t) & \Delta I_{23}(t) \\ \Delta I_{31}(t) & \Delta I_{32}(t) & C + \Delta I_{33}(t) \end{pmatrix} \quad (9)$$

in which A and C denote the moments of inertia for an equatorial principal axis and the polar principal axis respectively. We assume that the initial mass distribution of the Earth is symmetric with respect to the rotation axis, so that the moments of inertia for the two principal equatorial axes are both equal to A.

With the Eulerian free precession frequency defined as

$$\sigma_r = \frac{C-A}{A}\Omega \quad (10)$$

and assuming that the m_i and ΔI_{ij} are small quantities and with the dot denoting, now on, the time derivative, equation (7) becomes

$$\frac{\dot{m}_1}{\sigma_r} + m_2 = \phi_2 \quad (11)$$

$$\frac{\dot{m}_2}{\sigma_r} - m_1 = -\phi_1 \quad (12)$$

$$\dot{m}_3 = \dot{\phi}_3 \quad (13)$$

in which

$$\phi_1 = \frac{1}{(C-A)\Omega^2}(\Omega^2\Delta I_{13} + \Omega\Delta\dot{I}_{23} + \Omega h_1 + \dot{h}_2 - L_2) \quad (14)$$

$$\phi_2 = \frac{1}{(C-A)\Omega^2}(\Omega^2\Delta I_{23} - \Omega\Delta\dot{I}_{13} + \Omega h_2 - \dot{h}_1 + L_1) \quad (15)$$

$$\phi_3 = \frac{1}{C\Omega^2}(-\Omega^2\Delta I_{33} - \Omega h_3 + \Omega\int_0^t L_3 dt') \quad (16)$$

are the dimensionless excitation functions.

If we only consider mass displacements, neglecting influences of relative motions ($\vec{h}$, $\dot{\vec{h}}$, $\Delta\dot{\mathbf{I}} = 0$), in the absence of external torques ($\vec{L} = 0$), these excitation functions reduce for polar wander to the following excitation function written in complex notation:

$$\Phi_L = \phi_1 + i\phi_2 = \frac{\Delta I_{13}}{C-A} + i\frac{\Delta I_{23}}{C-A} \quad (17)$$

In a deformable Earth there are no stable reference frames. In this case, it is necessary to define such a reference system in a practical way. For short-term polar wander it is convenient to take the geographical frame as reference frame, defined as the mean position of a number of fixed points in stable continental areas. For long-term true polar wander the choice of a reference frame becomes more complicated, as the whole mantle can change its configuration. Usually, the hot spot reference frame is taken as the frame in which the mean mantle material is stable. The relative velocities between the hot spots are generally found to be a factor of one tenth smaller than the relative plate velocities.

In equations (11) and (12), m_1 and m_2 are now the resultant polar shift in radians: m_1 in the x-direction which is chosen to be in the equatorial plane from the center of the Earth towards the Greenwich meridian, and m_2 in the y-direction which is chosen to be in the equatorial plane from the center of the Earth towards 90 degrees East longitude. m_3 gives the change in the length of day.

With this and the polar shift in complex notation $\mathbf{m} = m_1 + im_2$, the linearized Liouville equation for polar wander can be written in complex notation as

$$i\frac{\dot{\mathbf{m}}}{\sigma_r} + \mathbf{m} = \mathbf{\Phi_L} \qquad (18)$$

For loadings that change with much smaller frequencies than σ_r, (18) becomes

$$m_1 = \frac{\Delta I_{xz}}{C - A} \text{ rad} \quad x-\text{component} \qquad (19)$$

$$m_2 = \frac{\Delta I_{yz}}{C - A} \text{ rad} \quad y-\text{component} \qquad (20)$$

Note that the equatorial flattening is of great importance: if C were equal to A, then the Eulerian free precession frequency would be zero, the excitation functions (14) and (15) would be infinite and the polar shift infinitely large.

Equations (14) and (15) or (19) and (20), do not take a shift in the equatorial bulge into account. To put it differently: (14) and (15) or (19) and (20) give the polar wander for a rigid planet. If the rotation axis were to coincide with the axis perpendicular to the plane of the equatorial flattening before a mass change occurs, then after the mass change (19) and (20) would give the new position of the rotation axis that coincides with the axis of maximum moment of inertia. This new position would not in general be perpendicular to the plane of the equatorial flattening. In this case, a wobble is excited with the frequency given by (10). At the end of the 19th century astronomers looked for this frequency in their observations but did not find it. However, they found a strong wobble, the so-called Chandler wobble, that had a period four months greater than the Eulerian free precession period. It was soon realised that the Chandler wobble was in fact the Eulerian free precession and that the four months period extension is due to the fluid and (visco)elastic properties of the Earth. The deformation of the Earth is also responsible for the decay of the wobble amplitudes on time scales of a few decades, implying that the wobbles must be maintained by geophysical forcings. And, as discussed in the intro-

duction, the tidal deceleration of the Earth's rotation also requires that (visco)elastic adjustment of the equatorial bulge be taken into account. Finally, for long-term true polar wander the shift of the equatorial bulge must also be taken into account. Although TPW over geological times amounts to several tens of degrees, the present day equatorial bulge is almost perpendicular to the Earth's rotation axis.

ADJUSTMENT OF THE EQUATORIAL BULGE

If α is the angle between the instantaneous rotation axis $\vec{\omega}$ and the line $\vec{r}$ from the center of mass of the Earth and a perturbation at point P, then

$$\vec{\omega} \cdot \vec{r} = |\vec{\omega}||\vec{r}| \cos \alpha \qquad (21)$$

with $\vec{\omega} \cdot \vec{r} = \omega_1 x_1 + \omega_2 x_2 + \omega_3 x_3$.
The distance l from P perpendicular to the axis of rotation follows now from

$$l^2 = r^2 - r^2 \cos^2 \alpha = r^2 - \frac{(x_1\omega_1 + x_2\omega_2 + x_3\omega_3)^2}{\omega^2} \qquad (22)$$

with $r = |\vec{r}|$ and $\omega = |\vec{\omega}|$.

So with (22) the centrifugal potential $U_c = \frac{1}{2}\omega^2 l^2$ is given by

$$U_c = \frac{1}{2}\omega^2 l^2 = \frac{1}{2}(\omega^2 r^2 - (\omega_1 x_1 + \omega_2 x_2 + \omega_3 x_3)^2) \qquad (23)$$

with

$$U_c = \frac{1}{3}\omega^2 r^2 + \Delta U_c \qquad (24)$$

with $\frac{1}{3}\omega^2 r^2$ the purely radial part of U_c, and ΔU_c the part which describes the equatorial flattening:

$$\begin{aligned}
\Delta U_c = \ & \frac{1}{6}\omega_1^2(x_2^2 + x_3^2 - 2x_1^2) \\
& + \frac{1}{6}\omega_2^2(x_1^2 + x_3^2 - 2x_2^2) \\
& + \frac{1}{6}\omega_3^2(x_1^2 + x_2^2 - 2x_3^2) \\
& - \omega_1\omega_2 x_1 x_2 - \omega_1\omega_3 x_1 x_3 \\
& - \omega_2\omega_3 x_2 x_3 \qquad (25)
\end{aligned}$$

This component of the centrifugal potential can be written in spherical coordinates as

$$\Delta U_c = \frac{1}{6}r^2(\omega_1^2 + \omega_2^2 - 2\omega_3^2)P_{20}(\cos\theta)$$
$$-\frac{1}{3}r^2(\omega_1\omega_3\cos\phi + \omega_2\omega_3\sin\phi)P_{21}(\cos\theta)$$
$$+\frac{1}{12}r^2((\omega_2^2 - \omega_1^2)\cos 2\phi$$
$$-2\omega_1\omega_2\sin 2\phi)P_{22}(\cos\theta) \qquad (26)$$

with the second degree associated Legendre polynomials P_{2m} given by

$$
\begin{aligned}
P_{20}(\cos\theta) &= \frac{3}{2}\cos^2\theta - \frac{1}{2} \\
P_{21}(\cos\theta) &= 3\sin\theta\cos\theta \\
P_{22}(\cos\theta) &= 3\sin^2\theta
\end{aligned}
\qquad (27)
$$

The associated Legendre polynomials $P_{nm}(x)$ of degree n and order m follow from Rodrigues' formula

$$P_{nm}(x) = \frac{(1-x^2)^{m/2}}{2^n n!}\frac{d^{n+m}}{dx^{n+m}}(x^2-1)^n \qquad (28)$$

The potential that deforms the equatorial bulge is given by

$$\Delta U_c' = k_2^T \Delta U_c \quad \text{for } r = R \qquad (29)$$

in which k_2^T is the tidal-effective Love number, also called the centrifugal Love number, of harmonic degree 2. This k_2^T is dependent on the Earth model (stratification, values for density, elasticity, viscosity, etc.). It takes the (visco)elastic deformation of the Earth into account.

The general expression for the degree 2 spherical harmonic gravitational potential field is for $r \geq R$ given as

$$
\begin{aligned}
\Delta U_2 &= \frac{GM}{r}\left(\frac{R}{r}\right)^2 \\
&\cdot \sum_{m=0}^{2}(C_{2m}\cos m\phi \\
&\quad + S_{2m}\sin m\phi)P_{2m}(\cos\theta)
\end{aligned}
\qquad (30)
$$

which becomes for $r = R$ equal to

$$\Delta U_2 = \frac{GM}{R}\sum_{m=0}^{2}(C_{2m}\cos m\phi + S_{2m}\sin m\phi)P_{2m}(\cos\theta) \qquad (31)$$

with C_{nm} and S_{nm} the *Stokes coefficients* of degree n and order m. By comparing (31) with (26) it can easily be derived that

$$C_{20} = \frac{k_2^T R^3}{6GM}(\omega_1^2 + \omega_2^2 - 2\omega_3^2) \qquad (32)$$

$$C_{21} = -\frac{k_2^T R^3}{3GM}\omega_1\omega_3 \qquad S_{21} = -\frac{k_2^T R^3}{3GM}\omega_2\omega_3 \quad (33)$$

$$C_{22} = \frac{k_2^T R^3}{12GM}(\omega_2^2 - \omega_1^2) \qquad S_{22} = -\frac{k_2^T R^3}{6GM}\omega_1\omega_2 \quad (34)$$

The coefficients C_{2m} and S_{2m} can also be expressed as

$$
\begin{aligned}
C_{2m} &= \frac{2-\delta_{0m}}{MR^2}\frac{(2-m)!}{(2+m)!} \\
&\quad \cdot \int_E \rho r^2 P_{2m}(\cos\theta)\cos m\phi\, dV \qquad (35) \\
S_{2m} &= \frac{2-\delta_{0m}}{MR^2}\frac{(2-m)!}{(2+m)!} \\
&\quad \cdot \int_E \rho r^2 P_{2m}(\cos\theta)\sin m\phi\, dV \qquad (36)
\end{aligned}
$$

with ρ the density of a volume element dV of the Earth model E. These Stokes coefficients can in turn be expressed in terms of the moments and products of inertia [*Lambeck*, 1980]

$$C_{20} = \frac{1}{2MR^2}(\Delta I_{xx} + \Delta I_{yy} - 2\Delta I_{zz}) \qquad (37)$$

$$C_{21} = -\frac{\Delta I_{xz}}{MR^2} \qquad S_{21} = -\frac{\Delta I_{yz}}{MR^2} \qquad (38)$$

$$C_{22} = \frac{\Delta I_{xx} - \Delta I_{yy}}{4MR^2} \qquad S_{22} = -\frac{\Delta I_{xy}}{2MR^2} \qquad (39)$$

Comparing (33) and (34) with (38) and (39), the changes in the products of inertia ΔI_{xz} and ΔI_{yz} can be expressed in a linear approximation as

$$
\begin{aligned}
\Delta I_{xz} &= -MR^2 C_{21} = \frac{k_2^T R^5}{3G}\omega_1\omega_3 \\
&\approx \frac{k_2^T R^5 \Omega^2}{3G}m_1(1+m_3) \qquad (40)
\end{aligned}
$$

$$\Delta I_{yz} = -MR^2 S_{21} = \frac{k_2^T R^5}{3G}\omega_2\omega_3$$

$$\approx \frac{k_2^T R^5 \Omega^2}{3G}m_2(1+m_3) \qquad (41)$$

Also, since $m_3 \ll 1$, the forcing function $\mathbf{\Phi_R}$ for rotational deformation can be written in complex notation in the manner of (17) as

$$\mathbf{\Phi_R} = \frac{\Delta I_{xz}}{C-A} + i\,\frac{\Delta I_{yz}}{C-A} \approx \frac{k_2^T}{k_f^T}\mathbf{m} \qquad (42)$$

with (as before) $\mathbf{m} = m_1 + im_2$, and

$$k_f^T = \frac{3G(C-A)}{R^5\Omega^2} \qquad (43)$$

is the so-called fluid Love number.

The linearized Liouville equation for polar wander which includes both loading and centrifugal forcings can be written as

$$i\frac{\dot{\mathbf{m}}}{\sigma_r} + \mathbf{m} = \mathbf{\Phi} \qquad (44)$$

with the forcing function $\mathbf{\Phi}$ consisting of two parts: $\mathbf{\Phi} = \mathbf{\Phi_L} + \mathbf{\Phi_R}$, with $\mathbf{\Phi_L}$ the part describing the direct geodynamic forcing (e.g. an earthquake or changing atmospheric pressure) and $\mathbf{\Phi_R}$ the induced rotational deformation.

The linearized Liouville equation can be expressed as

$$i\frac{\dot{\mathbf{m}}}{\sigma_r} + (1 - \frac{k_2^T}{k_f^T})\mathbf{m} = \mathbf{\Phi_L} \qquad (45)$$

Alternatively, we can write

$$i\frac{\dot{\mathbf{m}}}{\sigma_0} + \mathbf{m} = \mathbf{\Psi_L} \qquad (46)$$

with

$$\sigma_0 = (1 - \frac{k_2^T}{k_f^T})\sigma_r \qquad (47)$$

and

$$\mathbf{\Psi_L} = \frac{k_f^T}{k_f^T - k_2}\mathbf{\Phi_L} \qquad (48)$$

The term σ_0 is the frequency of the Chandler wobble. The frequency σ_0 is smaller than σ_r as a consequence of the term $(1 - \frac{k_2^T}{k_f^T})$ and this agrees with the observation that the Chandler wobble period is 4 months longer than the period of the Eulerian free precession.

In the following we will focus on the solution of Equation (45) for ice loading of a stratified viscoelastic Earth.

LINEARIZED ROTATIONAL THEORY FOR A STRATIFIED, VISCOELASTIC EARTH

Linearized relaxation models constitute the heart of most theoretical models of Earth deformation. Such relaxation models produce the deformation, changes in stress and gravity potential of the solid Earth as a function of time after a load or forcing is applied. The linearized differential equations (conservation of linear momentum, Poisson's equation for the relation between density and gravity potential, and a stress-strain relation) for a viscoelastic body have an equivalent form in the Laplace-transformed time domain as the equations for an elastic body. This was used by *Peltier* [1974] and *Wu* [1978] to develop a normal mode method by which viscoelastic relaxation can be dealt with in much the same way as the easier-to-treat elastic modeling that has been used extensively in studies on free vibrations of the Earth [*e.g., Alterman et al.*, 1959]. For Earth models that contain many layers a numerical propagator method is usually applied, but the propagation can also be performed analytically [*Sabadini et al.*, 1982; *Spada et al.*, 1992; *Vermeersen and Sabadini*, 1997].

In the case of polar wander there are two relaxation mechanisms at work: load relaxation as a consequence of a redistribution of ice and water over the Earth's surface, and tidal-effective (or centrifugal) relaxation as a consequence of the centrifugal force acting on the rotating Earth. This centrifugal force, induced by the displacement of the axis of rotation, causes the equatorial bulge of the Earth to be displaced over the Earth's surface in a manner not unlike a wave traveling over the ocean's surface. This 'polar wander' movement will procede until the Earth's rotation axis coincides with the axis of maximum moment of inertia of the mass distribution.

The theory that is used to study polar wander and the changes in the second degree harmonic of the geoid $\dot{J}_2$ can be found in a number of publications over the past twenty years [*e.g., Nakiboglu and Lambeck*, 1980; *Sabadini and Peltier*, 1981; *Sabadini et al.*, 1982, 1984; *Wu and Peltier*, 1984; *Peltier*, 1985; *Spada et al.*, 1992; *Ricard et al.*, 1992, 1993; *Mitrovica and Peltier*, 1993; *Vermeersen et al.*, 1994, 1997; *Peltier and Jiang*, 1996; *Mitrovica and Milne*, 1998]. The models in all these references employ a viscoelastic Maxwell rheology for a spherical Earth (that is, normal mode theory is first applied to a non-rotating spherical Earth model, than the required rotating ellipsoidal Earth model is obtained by applying the centrifugal potential). Differences in treatments within the above references exist in regard to the number of layers used to discretize the Earth model, the manner in which the differential equations are solved (analytically or numerically), if the Lamé parameter λ is taken as finite or infinite (compressible or incompressible) and whether the models allow only for surface loads or also for internal mantle loads.

The theories developed by *Sabadini et al.* [1982] and by *Wu and Peltier* [1984] have been shown to be equivalent in *Sabadini et al.* [1984], *Vermeersen and Sabadini* [1996] and *Mitrovica and Milne* [1998]. Specifically, *Sabadini et al.* [1984] have shown that the secular polar wander terms are the same, while *Vermeersen and Sabadini* [1996] and *Mitrovica and Milne* [1998] have demonstrated that when the Chandler wobble is filtered from the formulation of *Sabadini et al.* [1982], then the resulting same polar wander curves match those obtained using the theory of *Wu and Peltier* [1984] at all time scales.

In *Spada et al.* [1992], the analytical theory is developed for polar wander predictions for Earth stratifications with five layers at most. *Vermeersen et al.* [1996] and *Vermeersen and Sabadini* [1997] have extended the analytical theory for the relaxation of an Earth model consisting of a general amount of layers. In the following section we concentrate on rotational changes computed by means of analytical models in which the Earth is radially stratified with a general amount of layers.

POLAR WANDER EQUATION IN THE LAPLACE DOMAIN

The Correspondence Principle states that the solution of a viscoelastic problem is equivalent to the elastic one in the Laplace domain. It is thus possible to cast equation (45) for a stratified Earth model in the Laplace domain in the following way

$$\tilde{\mathbf{m}}(s)\left(i\frac{s}{\sigma_r} + 1 - \frac{k_e^T}{k_f^T} - \sum_{i=1}^{M}\frac{k_i^T/k_f^T}{s - s_i}\right) = \tilde{\Phi}(s) \quad (49)$$

as given in *Yuen et al.* [1983]. The tilde denotes Laplace transformed variables and the k_e^T, k_i^T and s_i enter the definition of the tidal-effective Love number obtained from the normal mode theory [*Sabadini et al.*, 1982]. Specifically,

$$k_2^T = k_e^T + \sum_{i=1}^{M}\frac{k_i^T}{s - s_i} \quad (50)$$

with $k_f^T = k_2^T(s = 0)$ being the tidal-effective fluid Love number given by

$$k_f^T = k_e^T - \sum_{i=1}^{M}\frac{k_i^T}{s_i} \quad (51)$$

Using the relation

$$\frac{k_i^T}{s_i} + \frac{k_i^T}{s - s_i} = \frac{k_i^T s}{s_i(s - s_i)} \quad (52)$$

equation (49) becomes

$$s\left(1 + i\frac{\sigma_r}{k_f^T}\sum_{i=1}^{M}\frac{k_i^T}{s_i(s - s_i)}\right)\tilde{\mathbf{m}}(s) = -i\sigma_r\tilde{\Phi}_L(s) \quad (53)$$

We can thus write

$$s\left(1 + \sum_{i=1}^{M}\frac{x_i}{s - s_i}\right)\tilde{\mathbf{m}}(s) = -i\sigma_r\tilde{\Phi}_L(s) \quad (54)$$

with

$$x_i = i\frac{\sigma_r k_i}{k_f s_i} \quad (55)$$

Now, we can also write

$$1 + \sum_{i=1}^{M}\frac{x_i}{s - s_i} = \frac{\prod_{j=1}^{M}(s - s_j)}{\prod_{j=1}^{M}(s - s_j)} + \sum_{i=1}^{M}x_i\frac{\prod_{j\neq i}^{M}(s - s_j)}{\prod_{j=1}^{M}(s - s_j)} \quad (56)$$

where $\prod_{j\neq i}^{M}$ means $\prod_{j=1}^{M}$ without the term $j = i$. The right-hand side of (56) becomes

$$\frac{\prod_{j=1}^{M}(s - s_j) + \sum_{i=1}^{M}x_i\prod_{j\neq i}^{M}(s - s_j)}{\prod_{j=1}^{M}(s - s_j)} \quad (57)$$

can be transformed into

$$\frac{\sum_{i=0}^{M}\alpha_i s^i}{\prod_{j=1}^{M}(s - s_j)} \quad (58)$$

(whereby it immediately follows that $\alpha_M = 1$), and consequently (56) can be rewritten as

$$1 + \sum_{i=1}^{M}\frac{x_i}{s - s_i} = \frac{\prod_{j=1}^{M}(s - a_j)}{\prod_{j=1}^{M}(s - s_j)} \quad (59)$$

where a_i are the M complex roots of the equation

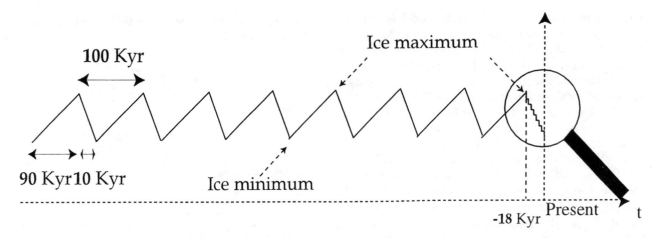

Figure 2. Cartoon showing the ice load forcing used in the models. The ice loads grow linearly over a period of 90,000 years and decline linearly over a period of 10,000 years. Eight ice-age cycles are depicted. The last ice age reaches its maximum ice load at 18,000 years before present [after *Vermeersen et al,* 1997].

$$\sum_{i=0}^{M} \alpha_i s^i = 0 \tag{60}$$

We thus obtain

$$s \frac{\prod_{j=1}^{M}(s - a_j)}{\prod_{j=1}^{M}(s - s_j)} \tilde{m}(s) = -i\sigma_r \tilde{\Phi}_L(s) \tag{61}$$

and so

$$
\begin{aligned}
\tilde{m}(s) &= -i\sigma_r \frac{\prod_{j=1}^{M}(s - s_j)}{s \prod_{j=1}^{M}(s - a_j)} \tilde{\Phi}_L(s) \\
&= -i\sigma_r \left(\frac{A_0}{s} + \sum_{j=1}^{M} \frac{A_j}{s - a_j} \right) \tilde{\Phi}_L(s) \quad (62)
\end{aligned}
$$

$\tilde{m}(s)$ is the Laplace-transformed complex-valued polar wander, defined in such a way that the real-valued component gives the polar wander in the direction of the Greenwich Meridian, and the imaginary-valued component the polar wander in the direction 90 degrees to the east. In this expression, the terms a_j are the inverse relaxation times from the centrifugal problem for the M modes, having the strength given by the residues A_j. The residue A_0 gives the strength of the secular term.

The equations above form the basis of realistic models on Earth rotation that take solid-earth deformation and its consequential shifts of the equatorial bulge self-consistently into account (with one restriction: the Liouville expression has been linearized and therefore polar wander needs to be restricted to about 10 degrees

over the Earth surface, or about 1,000 km, at most). On a time scale of a few million years, the same linearized equations have been used to estimate the effects of mantle convection on polar wander [*Sabadini and Yuen,* 1989; *Ricard and Sabadini,* 1990; *Ricard et al.,* 1992]. The impact of slow varying processes on rotation have been first considered, within a self-consistent fully non-linear approach, by *Ricard et al.* [1993]. Internal mass redistribution could induce a TPW velocity comparable to the observed value of around 1 deg/Myr and could eventually move the Earth's pole by a large amount. For large excursions of the axis of rotation the linearized version of rotation Equation (62) is no longer valid.

The loading term $\tilde{\Phi}_L(s)$ can be expressed in terms of the loading Love number $k_2^L(s)$

$$k_2^L = k_e^L + \sum_{i=1}^{M} \frac{k_i^L}{s - s_i} \tag{63}$$

and the rotation equation (62) becomes, with $\tilde{f}(s)$ denoting the Laplace transform of the time dependent ice load

$$
\begin{aligned}
\tilde{m}(s) = {}&-i\sigma_r \frac{I_{13}^R + iI_{23}^R}{C - A} \left(\frac{A_0}{s} + \sum_{i=1}^{M} \frac{A_i}{s - a_j} \right) \\
&\cdot \left(1 + k_e^L + \sum_{j=1}^{M} \frac{k_j^L}{s - s_j} \right) \tilde{f}(s) \quad (64)
\end{aligned}
$$

It should be noted in comparison with equation (14) in *Yuen et al.* [1983], that the definition of A_0 and

A_i has been changed here, in the sense that $-i\sigma_r$ was included in the older expressions.

The rigid-earth perturbations in the product-of-inertia components I_{13}^R and I_{23}^R are given by equation (28) of *Wu and Peltier* [1984].

$$
\begin{aligned}
I_{13}^R = {} & -M_I a^2 \Big(\frac{1}{6} cos\alpha(1+cos\alpha)P_{21}(cos\theta)cos\phi \\
& -\frac{1}{5}\frac{a_{21}}{a_{00}} \Big)
\end{aligned} \tag{65}
$$

$$
\begin{aligned}
I_{23}^R = {} & -M_I a^2 \Big(\frac{1}{6} cos\alpha(1+cos\alpha)P_{21}(cos\theta)sin\phi \\
& -\frac{1}{5}\frac{b_{21}}{a_{00}} \Big)
\end{aligned} \tag{66}
$$

The quotients a_{21}/a_{00} and b_{21}/a_{00} can be found from the spherical harmonic expansion of the ocean function [*e.g., Lambeck*, 1980]. P_{21} is the second degree, first order Legendre polynomial, given by $P_{21}(cos\theta) = 3\,cos\theta\,sin\theta$ (equation (27)). Note that this is the negative of what is presented in *Wu and Peltier* [1984], resulting in the extra minus sign with respect to the corresponding terms in *Wu and Peltier*'s [1984] equation (28) [see also *Peltier and Jiang*, 1996].

The (negative) inverse relaxation times a_i and their associated complex-valued residues A_i for the M modes i are found by considering the relaxation due to a centrifugal forcing. A_0 is dubbed the secular term [*Munk and MacDonald*, 1960].

Equation (64) can be further simplified to [*e.g., Vermeersen et al.*, 1994]

$$
\begin{aligned}
\tilde{m}(s) = {} & -i\sigma_r \frac{I_{13}^R + iI_{23}^R}{C-A} \\
& \cdot \Big(\frac{A_0^*}{s} + \sum_{i=1}^{M} \frac{\beta_i}{s-s_i} \\
& + \sum_{i=1}^{M} \frac{\gamma_i}{s-a_i} \Big) \tilde{f}(s)
\end{aligned} \tag{67}
$$

in which

$$
\beta_i = A_0 \frac{k_i^L}{s_i} + \sum_{j=1}^{M} \frac{A_j k_i^L}{s_i - a_j} \tag{68}
$$

$$
\gamma_i = A_0(1+k_e^L) - \sum_{j=1}^{M} \frac{A_i k_j^L}{s_j - a_i} \tag{69}
$$

and

$$
A_0^* = A_0(1 + k_f^L) \tag{70}
$$

The fluid limit ($s=0$) of the load Love number is given by

$$
k_f^L = k_e^L - \sum_{i=1}^{M} \frac{k_i^L}{s_i} \tag{71}
$$

After some algebra, it is possible to demonstrate that the coefficients β_i are equal to zero.

LOADING

From oxygen isotope analysis of ocean sediments [*e.g., Shackleton and Opdyke*, 1976] it has been deduced that the great ice sheets had a growth period of about 90,000 years and a decay period of about 10,000 years, so a total period of about 100,000 years for one complete cycle. Figure 2 portrays a cartoon of the ice load model that we will use in the following to study the influence of ice-age cycles on polar wander. As the most recent deglaciation has the largest influence on the present-day secular polar wander we use a more detailed ice (un)loading history for the past 18,000 years, based on the ICE-3G model of *Tushingham and Peltier* [1991].

Apart from the ice load model, which specifies the global distribution of the ice as function of time, an Earth model that specifies the physical properties such as the radial density, the rigidity and the viscosity is also required. For density and rigidity we use values determined from seismological observations, the Preliminary Reference Earth Model, abbreviated PREM, of *Dziewonski and Anderson* [1981]. The viscosity is taken as a free parameter for both upper mantle (extending from the bottom of the lithosphere at 120 km depth till 670 km depth) and lower mantle (extending to the fluid outer core at 2900 km depth).

LONG-TERM BEHAVIOUR OF THE ROTATION EQUATION: CONTRIBUTIONS FROM THE SECULAR AND RELAXATION MODES

The rotation equation (64) is characterized by terms of different nature, the first one with the pole in $s=0$ and the second one consisting in a summation over a series of poles in $s=a_i$. Independently of their strength, the physics carried out by these two contributions is quite different: if $\tilde{f}(s)$ is given by $1/s$ in equation (64) or (67), denoting a constant load in time, then the term

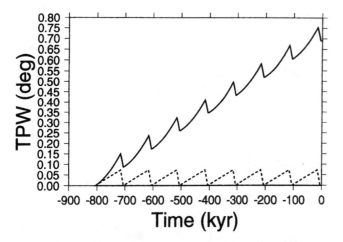

Figure 3. TPW displacement during eight cycles of glaciation-deglaciation, for a 3-layer model consisting of an elastic lithosphere and a uniform viscoelastic mantle of 10^{21} Pa s. All the poles in $s = 0$ and $s = a_i$ are considered for the solid curve, while the dashed one corresponds to $A_0 = 0$.

in A_0 is responsible for a polar wander which grows linearly in time, while the terms in a_i excite exponentially decaying polar shifts.

The impact of the different nature of these terms on polar wander studies has not been emphasized in the literature. Interest has been focussed on the present-day rotational response of the Earth to the last glacial cycle of the Pleistocene, studied in the following section, which has a contribution arising solely from the exponential terms. The following results are devoted to the understanding of the physics underlying these two different classes of s-poles.

It is possible to develop insight into the physics of the $s = 0$ pole, whose strength is given by A_0, in terms of the normal mode theory developed above by studying the long-term behavior of the equation for retrieving $\tilde{m}(s)$. In equation (49) we can take the limit for $|s| \ll |s_i|$ in the left part depending solely on the centrifugal deformation, which means that the Earth rotation is studied for time scales larger than the characteristic relaxation times of the rotational normal modes. In this equation, the terms $k_i/(s - s_i)$ can be expanded to first order

$$\frac{k_i}{(s - s_i)} \approx -\frac{k_i}{s_i}\left(1 + \frac{s}{s_i}\right) \qquad (72)$$

Taking into account the expression of the fluid tidal-effective Love number (equation (51))

$$k_f^T = k_e^T - \sum_{i=1}^{M} \frac{k_i}{s_i} \qquad (73)$$

and the first order approximation of the relaxing terms above, the rotation equation takes the form

$$s\tilde{\mathbf{m}} = \frac{-i\sigma_r \tilde{\mathbf{\Phi}}_{\mathbf{L}}}{\left(1 - \frac{i\sigma_r}{k_f^T}T_1\right)} \qquad (74)$$

If T_1 denotes the time scale of readjustment of the equatorial bulge given by

$$T_1 = \frac{1}{k_f^T}\sum_{i=1}^{N}\frac{k_i^T}{s_i^2} \qquad (75)$$

then equations (62) and (74) imply that the explicit expression of the term A_0 is given by

$$A_0 = \frac{1}{\left(1 - \frac{i\sigma_r}{k_f^T}T_1\right)} \qquad (76)$$

Equation (74) thus becomes in the time domain

$$\frac{d}{dt}\mathbf{m} = \frac{-i\sigma_r \mathbf{\Phi}_{\mathbf{L}}}{\left(1 - \frac{i\sigma_r}{k_f^T}T_1\right)} = -i\sigma_r A_0 \mathbf{\Phi}_{\mathbf{L}} \qquad (77)$$

The basic equations for true polar wander in the $s = 0$ limit can thus be cast in terms of the time scale required for the readjustment of the rotational bulge induced by perturbation in the Earth's rotation [*Sabadini and Yuen*, 1989; *Spada et al.*, 1992]. Through T_1 the long time scale rotation behavior of the Earth depends on the rheology of the mantle.

The following section deals with the displacement of the axis of rotation induced by the continuous occurrence of cycles of glacial loading and unloading. Attention is drawn on the effects of density stratification in the mantle on the relative contributions arising from the $s = 0$ and $s = a_i$ poles in driving a net excursion of the axis of rotation.

ICE-AGE CYCLE AND THE POLAR WANDER PATH

Figure 3 portrays the modeled polar wander over the complete ice-age cycles period of Figure 2, which shows the ice load forcing; the Earth model considered here does not contain the 670 km discontinuity. The mantle is uniform in density and in this sense it is not realistic. This case is, on the other hand, tutorial because it helps to illustrate the effects of density discontinuity at 670 km introduced in Figure 4. The solid curve accounts for the both the $s = 0$ and $s = a_i$ poles, while the dashed one does not contain the contribution from $s = 0$.

The solid curve shows that the axis of rotation oscillates in phase with the loading and unloading phases of the glacial cycles around a mean position which is clearly increasing in time in a linear fashion. Due to this linear growth of the mean position of the pole it can be expected that the $s = 0$ is the major contributor to the net shift of the axis of rotation. That this is the case is shown by the dashed curve at the bottom of Figure 3 corresponding to contributions from the $s = a_i$ poles only. The dashed curve shows that after eight cycles there is no net shift of the axis of rotation. The rotational relaxation due to the $s = a_i$ poles is fast, as indicated by the linear growth and decay of the associated TPW curve at the bottom. The saw-teeth pattern indicates that the axis of rotation is controlled by the pole carrying the Chandler Wobble. The wobbling is averaged out, and the results show the position of the center of the wobble, during the loading and unloading phases. Other $s = a_i$ poles do not contribute substantially, at least for a lower mantle viscosity of 10^{21} Pa s. If the mantle viscosity is increased then an exponential behavior appears within the saw-tooth pattern and the contributon from the $s = 0$ pole is reduced. The results in Figure 3 show that in the absence of chemical density stratification in the mantle, the net shift of the axis of rotation is totally due to the $s = 0$ pole. The velocity of polar wander can be very high, about 1 deg/Myr, for a mantle viscosity of 10^{21} Pa s. If the viscosity is increased, the net velocity of polar wander due to the $s = 0$ pole is reduced. As a matter of fact, the $s = 0$ pole contributes substantially to a net shift of the axis of rotation only if the mantle viscosity is about 10^{21} Pa s. As we have discussed (see equation (76)) the readjustment of the equatorial bulge is completely controlled by the time scale T_1. The terms due to the $s = a_i$ poles do not play any role on the net shift of the pole position, due to the fast relaxation modes a_i. If the mantle is not chemically stratified, the readjustment of the equatorial bulge is fast and large excursions of the rotation pole can be obtained.

The physics underlying the $s = 0$ pole is evident in the structure of equation (74). This equation consists of a rotation part with T_1 in the denominator, multiplied by a term equal to $(1 + k_f^L)$ once the $s = 0$ limit is considered also in the loading part, which quantifies the amount of isostatic disequilibrium after complete relaxation has taken place. It should be emphasized that this term $(1 + k_f^L)$ depends on the elastic properties of the lithosphere, and would be zero if the lithosphere is absent. The term $A_0^* = A_0(1 + k_f^L)$ indicates that the contribution of the $s = 0$ pole depends linearly on

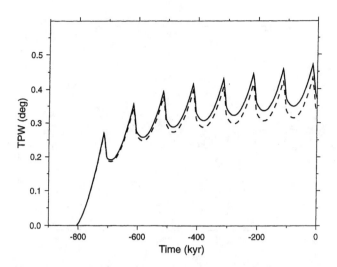

Figure 4. TPW displacement during eight cycles of glaciation-deglaciation, for a 5-layer model consisting of an elastic lithosphere, a stratified mantle with transition zone at 420 and 670 km, and uniform viscosity of 10^{21} Pa s. All the poles in $s = 0$ and $s = a_i$ are considered for the solid curve, while the dashed one corresponds to $A_0 = 0$.

the load which is not isostatically compensated. The contribution from the $s = 0$ pole would be the same if instead of having eight cycle of loading-unloading, we would have a constant load sitting for a span of time corresponding to the eight cycles, with an ice mass which is one half of that at the ice maximum.

Figure 4 deals with a 5-layer, volume-averaged model [*Vermeersen and Sabadini*, 1997], carrying a chemical density contrast at 420 and 670 km depth. In comparison with Figure 3, the noticeable effect of density stratification is a substantial reduction of the net shift of the rotation axis after the eight glacial cycles and the appearance of an exponential behavior in the rotational response of the Earth within each cycle. The most interesting result of Figure 4 is the substantial reduction of the contribution from the $s = 0$ pole on the net shift relative to Figure 3. Comparison between the dashed and solid curves, corresponding to the cases in which the $s = 0$ pole is included (solid) or not included (dashed) shows, in contrast with the previous case of a uniform mantle, that the $s = a_i$ poles now play the major role on the shift of the rotation axis. The density stratification of the mantle is responsible for a rotational stabilization of the Earth, whose rotation behavior is now controlled by the amount of density contrasts in the transition zone. What is the cause of such a drastic modification in the contributions from the $s = 0$ and $s = a_i$ poles to polar wander? Density stratification in the transition zone introduces long relaxation modes,

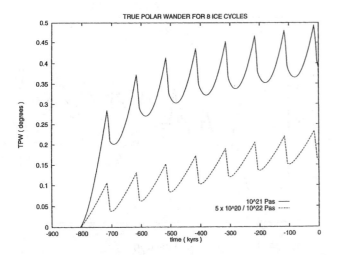

Figure 5. TPW displacement during eight cycles of glaciation-deglaciation for a 31-layer model. The solid curve stands for a uniform viscosity of 10^{21} Pa s, while the dashed one is for a stratified mantle of 5×10^{20} Pa s in the upper mantle and 10^{22} Pa s in the lower mantle [after *Vermeersen and Sabadini*, 1999].

or extremely small s_i, that control the readjustment of the transition zone between the upper and lower mantle during rotational deformation. In fact, from the explicit expression of T_1, it appears that a reduction in the s_i values in the stratified mantle case causes an increase of the characteristic time scale of readjustment of rotational bulge and a corresponding decrease in the velocity of polar wander, since T_1 enters the denominator of the expression for the velocity. In correspondence with this decrease in the contribution from the $s = 0$ pole, there is an increase in the contribution from the a_i modes corresponding to the slowest relaxation rotational mode associated with the density discontinuities of the transition zone. This slowest relaxation is also responsible for the exponential behavior of the net shift of the axis of rotation. It is thus clear that the appearance of the slow relaxation rotational modes in the 5-layer model is responsible for the increase of the strength γ_i/a_i relative to the A_0^* contribution, that depends linearly on A_0, which causes a substantial modification of the pattern of the polar wander curves, from a linear growth to an exponential relaxation. The net shift of the pole is due to the fact that after an ice cycle has been completed the Earth is still out of isostatic equilibrium. The relaxation times of some of the modes by which the Earth relaxes to the changed surface mass distribution are, in fact, longer than the period of the ice cycles. Comparison between Figure 3 and 4 indicates that, since the Earth is chemically stratified, the realistic pattern of polar wander is that portrayed by

the 5-layer model, where the dominant role is played by the $s = a_i$ poles. This has important implications on the net shift of the axis of rotation that the Earth can gain during a finite series of ice ages. The shift that is acquired permanently is that associated with the $s = 0$ pole, while the shift due to relaxation is recovered after the series of cycles, once a sufficiently long span of time has elapsed, governed by the slowest rotational mode. Since the Earth is chemically stratified, ice loading can be responsible for a conspicuous shift of the axis of rotation only during active phases of glaciation.

In Figures 3 and 4 only the magnitude of the polar wander triggered by the Pleistocene ice ages has been plotted, not the direction. Immediately after the model ice begins to grow at 806,000 years before present, the direction of the polar wander is *toward Russia* since a deformable rotating Earth reacts to the largest Laurentide ice sheet through the centrifugal force by moving its rotation axis in such a way that the excess mass will be as far away from the rotation axis as possible. Thus the rotation axis moves in the direction opposite to where the surface load is until the maximum of the ice load has been reached. When the ice starts to melt again the question of how much the ice load has been compensated by solid-earth deformation becomes relevant: if the Earth relaxes quickly to the ice load (deviations from isostatic equilibrium remaining small during ice build-up), then the ice melt will overcompensate the positive anomaly that was formed during times of ice growth. In the situation of Figures 3 and 4 this overcompensation gives rise to the sharp sawteeth during times of ice maximum. During times of ice minimum the overcompensation (from a negative to a positive anomaly) proceeds somewhat slower, giving rise to the smooth reversals at the bottom curves. The direction of polar wander during ice melt is about 180 degrees opposite to the direction during ice growth; that is, during periods of ice melt the rotation axis moves in the direction of Canada, as it is doing at present. The long-term TPW direction is in the same direction as it is during periods of ice growth. The question of whether polar wander during ice ages is mainly caused by the ice ages or by mantle convection / tectonics depends on the time scale being considered: if the present-day (secular) polar wander is to focus, then the direction of polar wander induced by the Pleistocene ice-age cycles is toward Canada. If in contrast polar wander path over the whole Pleistocene is to be considered, then the direction of TPW is toward Russia. The magnitudes of the long-term trend are an order of magnitude smaller than the magnitudes during an ice cycle, thus implying that if both mantle convection and ice-mass variations had a comparable influence on driving polar wander during ice ages on a time scale of one million years, on time scales of 10,000 - 100,000 years ice-mass variations have a far greater influence than mantle convection.

The curves of Figures 4 were compiled by means of a 5-layer model with a uniform viscosity of 10^{21} Pa s. Recent studies by *Vermeersen et al.* [1998] and *Devoti et al.* [2001] show that the viscosity of the upper mantle is more likely to be lower than this value, and the viscosity of the lower mantle is likely to be higher. The effects of viscosity layering can be seen in Figure 5. This figure is taken from *Vermeersen and Sabadini* [1999] and is based on the 31-layer solid-earth model by *Vermeersen et al.* [1997]. The solid and dashed curves depict the case of a uniform viscosity of 10^{21} Pa s and of a stratified mantle carrying a viscosity of 5×10^{20} Pa s in the upper mantle and 10^{22} Pa s in the lower mantle. The large influence that the mantle viscosity has on polar wander is qualitatively the same as that reported in *Sabadini and Yuen* [1989], although their modeling results were based on 4-layer solid-earth models, carrying a single density discontinuity at 670 km. These modeling results indicate that cryospheric forcing is capable of producing TPW because of the basic physics of viscoelastic deformation, and the net shift of the axis of rotation is highly dependent on both the density and viscosity stratification.

PRESENT-DAY POLAR WANDER AND MANTLE VISCOSITY

In this section we explore the effects of viscosity variations in the upper and lower mantle on the present-day TPW rate. The observational datum portrayed in Figure 1 is not related to the net shift of the axis of rotation during the continuous occurrence of ice ages, as shown in the previous figures, but rather to the present-day velocity of polar wander. Figure 6, taken from *Vermeersen et al.* [1997], deals with the velocity of polar wander based on model simulations, after the end of deglaciation. This velocity, evaluated for varying lower mantle viscosity and four fixed values of upper mantle viscosities, can be directly compared with the observational datum of Figure 1. Only the magnitude is portrayed, the direction being in agreement with observations. For low values of the lower mantle viscosity (about 10^{21} Pa s) the polar wander velocity is high and for high values (about 10^{23} Pa s) the velocity is small. For intermediate values of the lower mantle viscosity the behavior of the curves shows deviations from a sim-

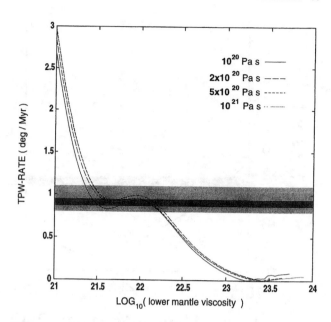

Figure 6. Present-day TPW as a function of lower mantle viscosity based on a 31-layer Earth model. Four cases of upper mantle viscosity ranging from 10^{20} to 10^{21} Pa s are considered [after *Vermeersen et al.*, 1997].

ple monotonic pattern [*e.g., Yuen et al.*, 1986; *Spada et al.*, 1992; *Milne and Mitrovica*, 1996; *Vermeersen et al.*, 1997]. All four curves, for upper mantle viscosities varying from 10^{20} to 10^{21} Pa s, cross the observations over a range of lower mantle viscosities that extends roughly from a few times 10^{21} Pa s to a few times 10^{22} Pa s [see also, *e.g., Mitrovica and Milne*, 1998]. Although the dependence of the predicted TPW rate on the value of the upper mantle viscosity is rather weak, it is not negligible. It is noticeable that the solid line, depicting the case of an upper mantle viscosity of 10^{20} Pa s, starts to cross the observations from the left of the figure at somewhat smaller lower mantle viscosities than the curves with a higher upper mantle viscosity.

TRADE-OFF BETWEEN LOWER MANTLE VISCOSITY AND PRESENT-DAY MASS IMBALANCE IN ANTARCTICA AND GREENLAND

The findings on the sensitivity of TPW data to mantle viscosity opens the question on whether we can use these data to estimate the viscosity profile of the Earth's mantle. The answer to this question is positive, thanks to the growth of satellite geodesy. In the following our aim is to show that TPW and long wavelength geodetic data related to the time dependent gravity field can in principle be used to constrain the mantle viscosity. In

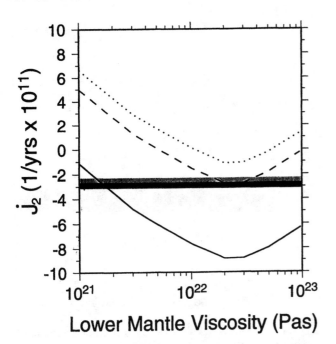

Figure 7. $\dot{J}_2$ as a function of lower mantle viscosity, ranging from 10^{21} to 10^{23} Pa s. Upper mantle viscosity is fixed at 5×10^{20} Pa s. The light and dark greys stand for the *Cheng et al.* [1997] and *Devoti et al.* [2001] solutions, respectively. The solid curve corresponds to Pleistocene deglaciation, the dashed one to Pleistocene plus maximum ice loss in Antarctica and the dotted one includes ice loss in Greenland.

particular, we want to show the impact of present-day mass imbalance in Antarctica and Greenland on viscosity inferences.

Two major mechanisms are responsible for present day changes in the gravitational field: the Pleistocene deglaciation, as described in the previous sections, and the present-day mass instability in Antarctica and Greenland. Comparison between TPW and geodetic data and model results makes it possible to infer the key parameters that control these two major mechanisms.

Figure 7 indicates how the lower mantle viscosity might be retrieved from modern geodetic observations of $\dot{J}_l$, with $l = 2$. This figure is based on two recent analyses of satellite geodetic data from two different research groups [*Cheng et al.*, 1997; *Devoti et al.*, 2001]. The $\dot{J}_2$ data from *Cheng et al.* [1997] and *Devoti et al.* [2001] are $-2.7 \pm 0.4 \times 10^{-11}$ 1/yr and $-2.9 \pm 0.2 \times 10^{-11}$ 1/yr, respectively, with the *Devoti et al.* [2001] solution portrayed by the innermost and darkest grey horizontal stripe. The *Cheng et al.* [1997] solution is represented by the lightest grey stripe. The harmonic J_2 represents the equatorial bulge of the Earth in terms of its moment of inertia, namely

$$J_2 = \frac{C - A}{Ma^2} \qquad (78)$$

where M and a denote the mass and the radius of the Earth. $\dot{J}_2$ is plotted in units of 10^{-11} 1/yr in Figure 7. The fact that the observed quantity is negative indicates that the equatorial bulge of our planet is diminishing. This reduction is due to the mantle flow from the equatorial regions of the mantle towards its polar regions, primarily because of the response of the mantle to the aforementioned process of Pleistocene deglaciation. The mantle is flowing towards the regions where the large ice sheets of the Pleistocene were located, in north America, northern Europe and Antarctica; during the Pleistocene, melting of ice occurred also in Antarctica. The melting (in whole or in part) of these large ice complexes produced a large amount of disequilibrium in the Earth and the mantle material is presently flowing towards the polar regions to compensate the mass deficit left by ice sheet disintegration. In order to infer the viscosity of the mantle it is necessary to compare the $\dot{J}_2$ retrieved from satellite geodesy with the results of our viscoelastic Earth models forced by Pleistocene deglaciation and ice-mass imbalance in Antarctica and Greenland. In Figure 7 the modeled $\dot{J}_2$ is plotted as a function of the lower mantle viscosity in the same fashion as TPW was given in the previous Figure 6. Three curves are provided. The first, denoted by a solid curve, corresponds to the situation in which the disintegration of the Pleistocene ice sheets is the only contributor to the gravity change. The dashed curve corresponds to Pleistocene deglaciation plus ice loss in Antarctica at the maximum rate of melting (-500 Gt/yr) allowed by observations of grounded ice [*Bentley and Giovinetto*, 1991]. The third (dotted curve) corresponds to the situation in which ice-mass loss in Greenland at a rate of -144 Gt/yr is added to the effects considered in the two previous cases. This value agrees with the estimated ice loss of Greenland for 1°C warming [*Oerlemans*, 1991]. The most striking result from this figure is the sensitivity of the $\dot{J}_2$ to lower mantle viscosity variations and the dominant effect of ice loss in Antarctica with respect to Greenland. The time derivative of J_2 shows the tendency to admit two classes of lower mantle viscosity. The lower viscosity solution, the first intersection on the right between the solid curve and the observational data, corresponds to a classical estimate for the mantle viscosity, close to 10^{21} Pa s. The dashed and dotted curves show how ice loss in Antarctica and Greenland impacts the results of the Pleistocene deglaciation [*James and Ivins*, 1997]. With respect to the Pleistocene signal (solid curve) the effects of melting in Antarctica is to displace the peak values in the direction of the observational data. Comparison between the dashed and dotted curves shows that ice loss in Greenland reinforces the effects of Antarctica on the time derivative of $\dot{J}_2$. These results indicate that present-day ice-mass imbalance in the polar regions of the Earth could modify previous estimates of lower mantle viscosity based on the assumption that Pleistocene deglaciation is the only forcing mechanism. It is clear on the basis of the results shown in Figure 7, that there is a trade-off between the lower mantle viscosity and the amount of present-day mass balance in Antarctica and Greenland. In fact when melting in Antarctica is added to Pleistocene deglaciation, the intersection between the model results and observations moves towards higher values of the viscosity relative to the value of 10^{21} Pa s obtained from many post-glacial rebound studies. The intersection between the dashed curve, corresponding to ice loss in Antarctica and the data occurs for a lower mantle viscosity of 2×10^{22} Pa s. This finding suggests that the Earth's mantle is strongly stratified in viscosity, with the lower mantle more viscous than the upper mantle. We also expect that ice-mass imbalance in Antarctica and Greenland affects present-day TPW. That this is the case can be seen from Figure 8, where the longest arrow denotes the observational datum by *McCarthy and Luzum* [1996], while the intermediate length arrow and the shortest one (both scaled with respect to the observational datum) denote the contribution from Greenland and Antarctica ice loss, of -144 and -500 Gt/yr, respectively. Since the observational constrain on TPW is reconciled by Pleistocene deglaciation when a lower mantle viscosity of about 3×10^{21} Pa s is adopted (Figure 6), it is clear from Figure 8 that if present-day melting of Greenland and Antarctica are contributing to TPW, then a viscosity increase in the lower mantle is required to reduce the effects of Pleistocene deglaciation. It should be noted, on the other hand, that the contribution from Greenland does not point in the same direction as the observed TPW as Pleistocene deglaciation does. Therefore, while present-day forcings lead to inferences of a stiffer lower mantle viscosity, a too large contribution from Greenland may degrade the fit to TPW direction. In this respect, our findings agree with those by *Johnston and Lambeck* [1999]. Present-day forcings favour a viscosity increase in the lower mantle, in agreement with the estimates obtained from the long wavelength, static anomalies of the geoid and TPW due to subduction [*Hager*, 1984; *Ricard et al.*, 1992; *Mitrovica and Forte*, 1997] and some recent estimates from post-glacial rebound [*Vermeersen et al.*, 1998].

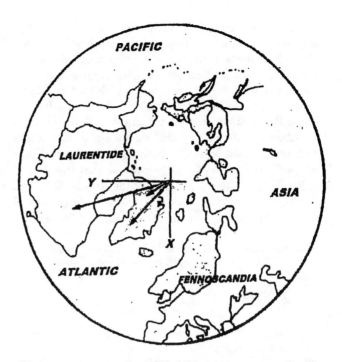

Figure 8. Present-day TPW-rate: the observed present-day TPW-rate from *McCarthy and Luzum* [1996] is given by the long black arrow, while the contribution from present-day melting in the polar region, corresponding to ice loss of -144 and -500 Gt/yr for Greenland and Antarctica, are provided by the intermediate and short arrows, respectively.

CONCLUSIONS

A reanalysis of the polar wander driven by Pleistocene deglaciation has allowed to obtain a deep insight into the physics underlying the poles of different nature ($s = 0$, $s = a_i$) that enter the rotation equations in the Laplace domain. The amount of chemical density stratification in the mantle has a major impact on the relative strength between these two families of poles, resulting in a substantial reduction of the net shift of the axis of rotation for Earth models with chemically stratified transition zones. Since the net shift of the axis of rotation that can be gained permanently after the occurrence of a series of ice ages originates solely from the $s = 0$ pole, it is clear that ice cycles cannot be the major source of TPW on the million year time scale, indirectly supporting the hypothesis that subduction and mountain building are the TPW drivers for times larger than 10^4 yr [*Ricard et al.*, 1992; *Vermeersen et al.*, 1994]. Present-day mass instabilities occurring in the polar regions of the Earth, in Antarctica and Greenland, impact the inference of lower mantle viscosity when simultaneous use is made of the TPW datum referring to the drift of the axis of rotation towards Newfoundland and of the secular drift of the zonal component of the geopotential of harmonic degree two. Ice loss in Antarctica and Greenland have a strong signal on the gravity field when these mass instabilities are chosen in such a way to be coherent with field observations of grounded ice in Antarctica or with climatological models for Greenland. Ice loss in Greenland also impacts the polar motion. These fluctuations may be key to reconcile lower mantle viscosity inferences based on post-glacial rebound with those based on longer time scale geophysical processes. These findings clearly indicate that the secular drift of the zonal components of the gravity field and that of the axis of rotation will remain a fundamental tool to retrieve the dynamical properties of the Earth.

Acknowledgments.   This work has been supported by the GOCE-ITALY project funded by the Italian Space Agency (2000).

REFERENCES

Alterman, Z., H. Jarosch, and C.L. Pekeris, Oscillations of the earth, *Proc. R. Soc.*, A252, 80-95, 1959.

Bentley, C.R. and M.B. Giovinetto, in *Proceeding of the International Conference on the Role of Polar Regions in Global Change*, Univ. of Alaska, Fairbanks, AK, 481-488, 1991.

Cheng, M.K., C.K. Shum, and B.D. Tapley, Determination of long-term changes in the Earth's gravity field from satellite laser ranging observations, *J. Geophys. Res.*, 102, 22,377-22,390, 1997.

Devoti, V. Luceri, C. Sciarretta, G. Bianco, G. Di Donato, L. L. A. Vermeersen, and R. Sabadini, The SLR secular gravity variations and their impact on the inference of mantle rheology and lithospheric thickness, *Geophys. Res. Lett.*, 28, 855-858, 2001.

Dickman, S.R., Secular trend of the Earth's rotation pole: consideration of motion of the latitude observatories, *Geophys. J. R. Astr. Soc.*, 51, 229-244, 1977.

Dziewonski, A.M. and D. Anderson, Preliminary reference earth model, *Phys. Earth Planet. Inter.*, 25, 41-50, 1981.

Farley, K.A., and D.B. Patterson, A 100-kyr periodicity in the flux of extraterrestrial [3]He to the sea floor, *Nature*, 378, 600-603, 1995.

Gross, R.S., The excitation of the Chandler wobble, *Geophys. Res. Lett.*, 27, 2,329-2,332, 2000.

Gross, R.S., and J. Vondrák, Astrometric and space-geodetic observations of polar wander, *Geophys. Res. Lett.*, 26, 2,085-2,088, 1999.

Hager, B.H., Subducted slabs and the geoid: Constraints on mantle rheology and flow, *J. Geophys. Res.*, 89, 6,003-6,015, 1984.

James, T.S. and E.R. Ivins, Global geodetic signatures of the Antarctic ice sheets, *J. Geophys. Res.*, 102, 605-633, 1997.

Johnston, P. and K. Lambeck, Postglacial rebound and sea level contributions to changes in the geoid and the Earth's rotation axis, *Geophys. J. Int.*, 136, 537-558, 1999.

Lambeck, K., *The Earth's Variable Rotation: Geophysical*

Causes and Consequences, Cambridge Univ. Press, New York, 1980.

McCarthy, D.D., and B.J. Luzum, Path of the mean rotational pole from 1899 to 1994, *Geophys. J. Int.*, 125, 623-629, 1996.

Milne, G.A., and J.X. Mitrovica, Postglacial sea-level change on a rotating Earth: first results from a gravitationally self-consistent sea-level equation, *Geophys. J. Int.*, 126, F13-F20, 1996.

Mitrovica, J.X., and A. Forte, Radial profile of mantle viscosity: results from the joint inversion of convection and postglacial rebound observables, Glaciation-induced perturbations in the Earth's rotation: A new appraisal, *J. Geophys. Res.*, 102, 2,751-2,769, 1997.

Mitrovica, J.X., and G.A. Milne, Glaciation-induced perturbations in the Earth's rotation: A new appraisal, *J. Geophys. Res.*, 103, 985-1,005, 1998.

Mitrovica, J.X., and W.R. Peltier, Present-day secular variations in the zonal harmonics of earth's geopotential, *J. Geophys. Res.*, 98, 4,509-4,526, 1993.

Muller, R.A., and G.J.F. MacDonald, Glacial cycles and orbital inclination, *Nature*, 377, 107-108, 1995.

Munk, W.H., and G.F. MacDonald, *The Rotation of the Earth*, Cambridge Univ. Press, New York, 1960.

Nakiboglu, S.M, and K. Lambeck, Deglaciation effects on the rotation of the Earth, *Geophys. J. R. Astr. Soc.*, 62, 49-58, 1980.

Oerlemans, J., The mass balance of the Greenland ice sheet: sensitivity to climate change as revealed by energy balance modelling, *The Holocene*, 1, 40-49, 1991.

Peltier, W.R., The impulse response of a Maxwell earth, *Rev. Geophys. Space Sci.*, 12, 649-669, 1974.

Peltier, W.R., The LAGEOS constraint on deep mantle viscosity: Results from a new normal mode method for the inversion of viscoelastic relaxation spectra, *J. Geophys. Res.*, 90, 9,411-9,421, 1985.

Peltier, W.R., and X. Jiang, Glacial isostatic adjustment and Earth rotation: Refined constraints on the viscosity of the deepest mantle, *J. Geophys. Res.*, 101, 3269-3290, 1996.

Ricard, Y., and R. Sabadini, Rotational instabilities of the earth induced by mantle density anomalies, *Geophys. Res. Lett.*, 17, 627-630, 1990.

Ricard, Y., R. Sabadini, and G. Spada, Isostatic deformations and polar wander induced by redistribution of mass within the earth, *J. Geophys. Res.*, 97, 14,223-14236, 1992.

Ricard, Y., G. Spada, and R. Sabadini, Polar wandering of a dynamic earth, *Geophys. J. Int.*, 113, 284-298, 1993.

Sabadini, R., and W.R. Peltier, Pleistocene deglaciation and the Earth's rotation: implications for mantle viscosity, *Geophys. J. R. Astr. Soc.*, 66, 553-578, 1981.

Sabadini R., and D.A. Yuen, Mantle stratification and long-term polar wander, *Nature*, 339, 373-375, 1989.

Sabadini, R., D.A. Yuen, and E. Boschi, Polar wandering and the forced responses of a rotating, multilayered, viscoelastic planet, *J. Geophys. Res.*, 87, 2,885-2,903, 1982.

Sabadini, R., D.A. Yuen, and E. Boschi, A comparison of the complete and truncated versions of the polar wander equations, *J. Geophys. Res.*, 89, 7,609-7,620, 1984.

Shackleton, N.J., and N.D. Opdyke, Oxygen isotope stratigraphy of Pacific core V28-239, *Mem. Geol. Soc. Am.*, 145, 449-464, 1976.

Spada, G., R. Sabadini, D.A. Yuen, and Y. Ricard, Effects on post-glacial rebound from the hard rheology in the transition zone, *Geophys. J. Int.*, 109, 683-700, 1992.

Tushingham, A.M., and W.R. Peltier, ICE-3G: A new global model of late Pleistocene deglaciation based upon geophysical predications of postglacial relative sea level change, *J. Geophys. Res.*, 96, 4,497-4,523, 1991.

Vermeersen, L.L.A., A. Fournier, and R. Sabadini, Changes in rotation induced by Pleistocene ice masses with stratified analytical Earth models, *J. Geophys. Res.*, 102, 27,689-27,702, 1997.

Vermeersen, L.L.A., and R. Sabadini, Significance of the fundamental mantle rotational relaxation mode in polar wander simulations, *Geophys. J. Int.*, 127, F5-F9, 1996.

Vermeersen, L.L.A., and R. Sabadini, A new class of stratified viscoelastic models by analytical techniques, *Geophys. J. Int.*, 129, 531-570, 1997.

Vermeersen, L.L.A., and R. Sabadini, R., Polar wander, sea-level variations and Ice Age cycles, *Surv. Geophys.*, 20, 415-440, 1999.

Vermeersen, L.L.A., R. Sabadini, R., Devoti, V. Luceri, P. Rutigliano, C. Sciarretta, and G. Bianco, Mantle viscosity inferences from joint inversions of Pleistocene deglaciation-induced changes in geopotential with a new SLR analysis and polar wander, *Geophys. Res. Lett.*, 25, 4,261-4,264, 1998.

Vermeersen, L.L.A., R. Sabadini, and G. Spada, Analytical visco-elastic relaxation models, *Geophys. Res. Lett.*, 23, 697-700, 1996.

Vermeersen, L.L.A., R. Sabadini, G. Spada, and N.J. Vlaar, Mountain building and earth rotation, *Geophys. J. Int.*, 117, 610-624, 1994.

Wu, P., The Response of a Maxwell Earth to Applied Surface Mass Loads: Glacial Isostatic Adjustment, MSc-thesis, University of Toronto, Canada, 171 pp., 1978.

Wu, P., and W.R. Peltier, Pleistocene deglaciation and the Earth's rotation: a new analysis, *Geophys. J. R. Astr. Soc.*, 76, 753-791, 1984.

Yuen, D.A., R. Sabadini and E. Boschi, The dynamical equations of polar wander and the global characteristics of the lithosphere as extracted from rotational data, *Phys. Earth Planet. Inter.*, 33, 226-242, 1983.

Yuen, D.A., R. Sabadini, P. Gasperini, and E. Boschi, On transient rheology and glacial isostasy, *J. Geophys. Res.*, 91, 11,420-11,438, 1986.

R. Sabadini, Sezione Geofisica, Dipartimento di Scienze della Terra, Università di Milano, Via L. Cicognara 7, I-20129 Milano, Italy. (e-mail: roberto.sabadini@unimi.it)

L.L.A. Vermeersen, DEOS, Faculty of Aerospace Engineering, Delft University of Technology, Kluyverweg 1, NL-2629 HS Delft, Netherlands. (e-mail: b.vermeersen@lr.tudelft.nl)

BIFROST: Observing the Three-Dimensional Deformation of Fennoscandia

Hans-Georg Scherneck,[1] Jan M. Johansson,[1,2] Gunnar Elgered,[1] James L. Davis,[3] Bo Jonsson,[4] Gunnar Hedling,[4] Hannu Koivula,[5] Matti Ollikainen,[5] Markku Poutanen,[5] Martin Vermeer,[6] Jerry X. Mitrovica,[7] and Glenn A. Milne[8]

In this paper, we describe the design of the BIFROST GPS network, algorithms for velocity estimation, and statistical assessment of the geodetic time series. This network has been used to make the first detailed three-dimensional map of ongoing crustal deformation due to glacial isostatic rebound in Fennoscandia or elsewhere. We redraw the history of the permanent networks for continuous GPS observations in Fennoscandia as the result of decisions with the aim to obtain the most efficient tools of the time to observe crustal motion to better than parts of millimeters per year. The prerequisite is to measure station positions with sufficiently high repeatability, and to avoid and reduce systematic biases. We show that the data base that this system created achieves the inference of site motion with high confidence. Time-series of site positions provide a large statistical basis to separate deterministic signals in a non-white noise environment. We finally report from the analysis of almost 2500 daily network solutions and present maps with the inferred rates of motion.

[1]Chalmers Centre for Astrophysics and Space Science, Onsala, Sweden

[2]Swedish National Testing and Research Institute, Borås, Sweden

[3]Radio and Geoastronomy Division, Harvard-Smithsonian Center for Astrophysics, Cambridge, Massachusetts

[4]National Land Survey, Gävle, Sweden

[5]Finnish Geodetic Institute, Masala, Finland

[6]Department of Surveying, Helsinki University of Technology, Helsinki, Finland

[7]Department of Physics, University of Toronto, Ontario, Canada

[8]Department of Geological Sciences, University of Durham, Durham, United Kingdom

Ice Sheets, Sea Level and the Dynamic Earth
Geodynamics Series 29
Copyright 2002 by the American Geophysical Union
10.1029/029GD05

1. INTRODUCTION

Geodesy using continous Global Positioning System (GPS) recievers has enabled the determination of subtle patterns of ongoing crustal deformation that even one decade ago was impossible. The low cost of these systems allows a fairly dense network of GPS receivers to be deployed, so that regional problems, including glacial isostatic rebound (GIA), can be addressed. (See also *Wahr and Davis*, this issue.) Using modern space geodetic techniques observe GIA motions is particularly useful, since for the first time the horizontal motions and the radial (vertical) motions can be obtained simultaneously in a consistent reference frame. The BIFROST (Baseline Inferences for Fennoscandian Rebound Observations, Sea level, and Tectonics) GPS networks have already obtained significant results in this regard [*Johansson et al.*, 2001; *Milne et al.*, 2001].

In this paper, we describe the design of the BIFROST GPS network, algorithms for velocity estimation, and statistical assessment of the geodetic time series. We begin in the next section by presenting an account of the planning and design of the BIFROST GPS networks. Some aspects of the design are discussed in *Wahr and Davis* (this issue). However, a detailed discussion of the design philosophy and the practical considerations in implementation have never before been presented. In Section 3 we describe the final network configuration. Data analysis, including reference frame considerations, is discussed in detail in Section 4. This section also has a new statistical analysis of the time series, and present our EOF method, the results of which was used in [*Johansson et al.*] 2001.

2. PLANNING AND DESIGN OF THE NETWORK

In 1991–92 plans were realized by Chalmers and the National Land Survey of Sweden (NLS), creating a permanent station network for GPS receivers that was to become the SWEPOS® network. It was conceived as a multi-purpose network for both geodetic reference and to support fast static and kinematic positioning services.

The emerging investigator group of the BIFROST project was involved already in early stages of the planning and contributed the first generation of instrumentation.

Seen from the perspective of today, a number of important decisions were made at this time that involved a large number of people at different agencies and institutes in different countries. Problems could be solved in a cooperative atmosphere that lead to the continuing success of this project.

A key problem was the demand to furnish the system for obtaining the highest possible stability for the antenna monumentation. This request derived from the origins of the BIFROST initiative dating back to 1990–1991. The SWEPOS network would provide the spatial coverage of the area of study in Sweden, although in some parts the first plans for location of stations would need modification and augmentation. At the instance when the requirements of observing the rebound of Fennoscandia with a GPS network in campaign style where contemplated, that was before SWEPOS was laid out, a sketch emerged that showed a map of the area with about 20 to 30 stations evenly distributed over the land surface. It was designed to provide means for comparison of vertical motion with the classical tide gauge measurements owing to sufficient station proximity, but simultaneously the network would extend into the in-

land. In order to catch the detail of the rebound field at uniform resolution, the network would preferably obtain a uniform mesh width. When the NLS presented their plans for a permanent GPS network in 1991, it coincidentally resembled the BIFROST blueprint at stunning similarity.

We realized the advantage of continuous observations to be able to obtain daily solutions of antenna positions also in quite remote areas difficult to reach in a campaign type of work plan. Continuous observations would provide a rich data base most suitable to study crustal motion together with a careful assessment of perturbing effects. The deployment of such a system would depend on the success to raise considerable funds in Sweden for a park of dedicated receivers. However, regardless of campaign-style or continuous observations, monument stability was a key factor.

In response the NLS devised 20 stations such that they would be based on a standard, heated concrete pillar firmly anchored in crystalline bedrock. More details follow below.

In 1992, the Finnish Geodetic Institute (FGI) decided to build a GPS array of 12 stations in Finland (FinnRef). The network was created between 1994 and 1996. The main objectives of the network were use as a backbone of the national coordinate systems, create a good connection to international reference frames, and crustal deformation studies. Similar antennas and receivers are used as in SWEPOS, but technical solution for antenna platforms differ in some aspects.

The campaign-based collocation of more than 30 GPS antennas and tide gauges was already used in three Baltic Sea Level GPS campaigns in 1990, 1993, and 1997 [*Poutanen and Kakkuri*, 1999]. The sea surface topography of the Baltic Sea was obtained by *Poutanen and Kakkuri* [2000], but these episodic campaigns could not resolve the rebound sufficiently well.

2.1. History of the BIFROST Project

The beginnings of what became the BIFROST Project [*BIFROST Project*, 1996] were initiated in response to the announcement of the Dynamics of the Solid Earth (DOSE) program by NASA. After the Crustal Dynamics Project (CDP) under NASA, which had promoted a decade of studies of crustal motions at interplate scale the focus was shifted to try and detect deformations in plate interiors and off known plate boundaries [*Gordon and Stein*, 1992]. The geodetic systems were developing quickly; in 1990 NASA and cooperators world-wide ran a network of 30 fixed sites with Very-Long Baseline Interferometry (VLBI), 40 fixed sites with Satellite

Laser Ranging (SLR), and 15 fixed GPS sites, of which in particular the latter grew rapidly. NASA provided the coordination framework to utilize the infrastructure, where a core network of space geodetic stations integrating several techniques were forming a Fiducial Laboratory for an International Natural science Network (FLINN in honor of the late Edward A. Flinn) for a wide range of applications [*Flinn,* 1989]. In between FLINN stations, Densely Spaced Geodetic Systems (DSGS) linking to the backbone network would provide the sensor system on a regional scale focusing on particular questions.

After the Crustal Dynamics epoch of the 1980's the focus shifted from the horizontal component of motion as the predominant one in the differential motion between plates (for a comprehensive review of the achievements in global geodynamics of that period we refer to the three volumes in the AGU Geodynamics Series edited by *Smith and Turcotte* [1993a, b, c]). Now the vertical component received increased interest, owing to advances in the modeling of predominantly atmospheric perturbations, which were seen to remove a large fraction of systematic errors. In this light, Fennoscandia was seen to present an important target area for space geodetic activities within DOSE. The area is known to possess up to roughly 10 mm/yr vertical motion, most probably in response to the melting of the large ice sheet that covered the area from about 100 kyr to 9 kyr before present. This motion was known from determination of age versus height of geological markers, from current trends in tide gauge records, and from precise geodetic levelling [*Ekman,* 1991, 1996].

Questions that could be addressed with a regional observation system in Fennoscandia concern the accurate and completely three-dimensional mapping of the rebound motion. The new evidence is then to be compared and combined with, for instance, sea level observations in the area. Another group of problems concern the stability of the lithosphere, its elastic properties, and the rheology of the mantle. A third group of problems concern the role of tectonics, whether motion can be discerned that is unrelated to the rebound, and whether relations with seismic parameters can be established.

Uplift: Space geodesy and tide gauges. Using space geodetic techniques a number of advantages can be exploited. First of all, all three spatial components of crustal deformation can be measured accurately enough on a continental scale. In satellite methods, the vertical component of motion is retrieved as the motion of a station with respect to the centre of the reference frame, in

contrast to surface-bound techniques, where uplift—or more precisely: land emergence—is seen as the relative motion between a station and the sea level. This holds true of tide gauges as this is the parameter they are designed to measure. But also in levelling networks the height systems are linked to realizations of the mean sea level, and larger than regional uplift signatures are not accessible from campaign to campaign.

Conversely, using space geodesy the motion of the crust at tide gauge locations could be inferred or even directly measured. Combining this motion with sea level observations and estimated changes of the gravity potential would provide a regionally representative estimate of sea level change with respect to a global geodetic reference frame.

Geoid. Uplift signatures in the geoid are generated because of the vertical motion of density boundaries (primarily the surface) and because of the net mass flow in the interior of the earth. Model computations show that geoid rates form a regional field with a dome in a similar position as the vertical displacement rates. The dome shape, however, is broader and does not possess a peripheral trough of subsidence. Estimated rates for the geoid in Fennoscandia range from 0.5–0.75 mm/yr [*Peltier,* 1999; *Kakkuri and Poutanen,* 1997; *Ekman and Mäkinen,* 1996; *Mitrovica and Peltier,* 1991]; *Milne, pers. comm.*]. *Mitrovica and Peltier* [1989] showed that an extreme geoid rate compatible with the PREM structure of the earth would be 1.0 mm/yr, assuming a lower mantle viscosity of 2×10^{22} Pa s. PREM is the Preliminary Reference Earth Model obtained from seismic normal mode (free oscillations) analysis by *Dziewonski and Anderson* [1981]. Geoid uplift is needed when inferring sea level change

$$\dot{\zeta} = \dot{w} - \dot{u} \qquad (1)$$
$$\dot{w} = \dot{w}_0 + \dot{N} \qquad (2)$$

where ζ is relative sea level, u vertical displacement of the crust, w absolute sea level and N geoid (w and N are with respect to the same reference surface or frame origin). The term w_0 represents a budget quantity for the global water mass which might carry a mass-independent signature due to steric effects.

Global Positioning System. By 1990 GPS measurements had started to offer competitive precisions with VLBI. Baseline results from campaigns in Southern California presented by *Dixon et al.* [1990] showed repeatabilities of

$$\sigma = \sqrt{16 + (2.2 \times 10^{-8} L)^2} \quad [\text{mm}] \qquad (3)$$

where L is baseline length, i.e. repeatabilities only slightly larger than 0.022 ppm. *Davis et al.* [1989] reported 0.03 to 0.05 ppm. Comparison with *Clark et al.* [1987] suggested that the distance-dependent contribution to the uncertainty was only about three times worse than in VLBI.

The important role that GPS measurements of crustal motion were going to get in world-wide monitoring of sea level was pointed out already by *Carter et al.* [1989], who identified four major areas of application: (1) determining the cause of the sea level change, discerning possible signatures due to global warming; (2) determining variations of geostrophic flow through straits in areas of active crustal motion; (3) calibrating satellite altimeters; and (4) understanding crustal processes and stress response due to plate motion and other tectonics. GPS was seen to reach the required precision in 1993, and its use was boosted from the rapidly developing infrastructure [*Blewitt*, 1994].

Monitoring with continuous GPS. GPS equipment on the market in 1993 offered the opportunity to try and operate stations remotely. Due to the relatively low cost per station a dense network became feasible. Infrastructure for downloading the data and controlling the onsite equipment remotely was available at reasonable expenses using the ground-fixed telephone network.

Possible disadvantages as regards the precision of the technique with respect to VLBI or SLR would be compensated by the large amount of data becoming available from continuous GPS observations. Given stationary error sources, the error levels would efficiently been beaten down. Figure 1 after *Coates et al.* [1985] suggests that daily solutions would provide rate accuracies about 20 times greater than annual campaigns, given all circumstances were the same. The advantage is slightly lower in the case that daily solutions are autocorrelated. Thus, the promises of campaign-based GPS to be able to measure a denser network with a smaller number of receivers were ranked low. Earlier experience with mobile VLBI supported the notion that campaigns contain many difficulties as they require continued logistics and present many risks, technical but not least economical ones.

Experience from permanent GPS networks prior to 1993 stemmed mostly from global tracking networks. In 1987 CIGNET (Cooperative International GPS Network), world-wide GPS tracking sites commenced; in 1993 they became the backbone of the IGS (International GPS Service) infrastructure, see *Blewitt* [1993] and *Segall and Davis* [1997] for a summary. Onsala operated in the CIGNET from 1987 on continuously

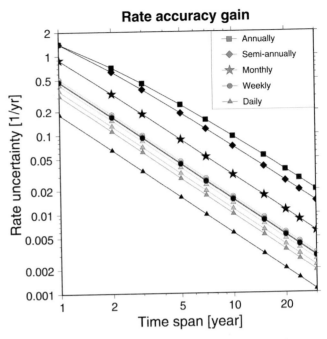

Figure 1. Estimating a constant rate parameter for station motion from repeated position observations. Shown is the uncertainty of the rate assuming a unity uncertainty for an individual position solution. The rate uncertainty is a function of the total duration of the survey (abscissa) and the rate of repetition (curve parameter in the range from daily to annual schemes). Since daily solutions are usually not totally independent we have allowed for an autocorrelation parameter to vary between zero (indicated by black symbols), 0.5 (dark gray), 0.6 (middle gray), and 0.7 (light gray).

[*Johansson et al.* 1992], Metsähovi joined in 1990. A permanent network with a regional focus was developed in Japan. Using 17 months of data from seven stations *Shimada and Bock* [1992] were able to determine 3-dimensional station motion with precisions of 4 to 7 mm/yr, although the satellite system was not completed during most of the period of the study. GEONET, the now gigantic nationwide permanent GPS network of Japan commenced its operation in 1994 [*Miyazaki et al.*, 1998; *Miyazaki and Heki*, 2001]; it currently comprises about 1000 stations.

In Southern California, a permanent network was created (SCIGN). It consisted originally of 40 stations and grew to 225 at present (see the SCIGN home page at http://www.scign.org).

For the part of Onsala Space Observatory, activities in applications to space geodesy date back to the first pilot observations intended to start a long series of projects for the monitoring of plate tectonics and earth rotation [*Scherneck et al.*, 1998].

Thus, the SWEPOS array was among the first continuous observation systems. By establishing links to the global tracking network the geodetic solutions would not only carry relevant information with a regional notion (differential movement). Advances were foreseen in the continuing improvement of the international reference frames and biases between the different techniques were expected to decrease down the road. Existing fundamental geodetic stations in the area and sites where techniques are collocated would contribute to resolving these biases. In 1993 the IGS (*Beutler et al.* [1993]) was recognized by the IAG; the international tracking network developed rapidly. In area several sites were created, like Tromsø and Ny Ålesund (operated by the Norwegian Mapping Authority), Metsähovi in Finland (operated by the Finnish Geodetic Institute), and Kiruna (operated by the European Space Agency). Onsala's and Metsähovi's GPS stations were already operating for the pre-IGS tracking network CIGNET. Collocation between techniques was already supported at Onsala (GPS and VLBI) and Metsähovi (GPS, SLR), and the Norwegian Mapping Authority contributed VLBI at Ny Ålesund. DORIS systems (Doppler Orbitography and Radiopositioning Integrated by Satellite, *Lefebvre et al.* [1996]) were installed and taken into operation in 1990 at Ny Ålesund and Metsähovi. In this paragraph we have not listed infrequent collocation with mobile VLBI and SLR equipment.

2.2. Other Techniques

While the observation efforts in the BIFROST Project would concentrate on GPS we identified a number of relations and application areas that involve other techniques. Continued efforts in these areas and cooperative studies are for instance part of the WEGENER program (Working group of European Geoscientists for the Establishment of Networks for Earth-science Research) in the several areas of geodetic techniques, including satellite laser ranging (SLR), both fixed and mobile, and gravimetry, with an emphasis on absolute gravimetry.

The thematical emphasis of the WEGENER program encompasses seismological studies (source mechanism solutions, inference of strain), regional space geodetic networks that provide densification in areas known or suspected by tectonophysicists to deform at present time, European and global efforts in VLBI to provide global reference constraints, absolute gravimetry (another observable that is sensitive to crustal uplift), and coordination and analysis of the national tide gauge net-

works in the area. With regard to the glacial isostatic adjustment problem, WEGENER envisages continued efforts on tide gauge observations [*Plag et al.*, 1998].

Tide gauge fixing by GPS could be seen as a primary objective of permanent GPS in BIFROST. The task requires tight collocation of receivers and mareographic bench marks. However, *Blewitt* [1994] and others recognized the value of the emerging permanent GPS network in Fennoscandia to support the tide gauge network through a wide catalogue of activities, potentially overcompensating the disadvantage that tide gauge stations were not as tightly collocated as stipulated by the Permanent Service for Mean Sea Level (PSMSL).

3. THE REGIONAL NETWORKS

The present configuration of the GPS network stations is shown in Plate 1. Since the beginning many IGS stations have come online. The result of future evaluation might suggest a new, broader selection of these to be analyzed in BIFROST.

3.1. SWEPOS

The SWEPOS network became complete with 21 stations between summer 1993 and autumn 1995, when the most recent one was sited at the Swedish National Testing and Research Institute (SP) at Borås. This station extended the original plan. SWEPOS was declared as an official real-time positioning and differential navigation facility by the Swedish National Land Survey July 1, 1998. It is coordinated by a group of governmental institutes and companies active in infrastructure. As described above the SWEPOS starting phase is intimately related with the BIFROST project. Apart from maintenance the NLS retrieves and archives the data.

Onsala is both a SWEPOS and an IGS station. At Kiruna, IGS and SWEPOS stations are on different pillars, roughly separated by a five kilometer distance. A comprehensive table of stations, equipment and their refurnishing is contained in *Johansson et al.* [2001]. Figure 2 shows a drawing of the SWEPOS Kiruna station comprising of two pillars and the local survey network with control points for monitoring monument stability. A photo of the Kiruna station is shown in Figure 3. The antenna monuments at the SWEPOS stations are normally of this standard type as shown in these figures. The three meter tall standard concrete pillar is heated by thermostat controlled electrical wire so its temperature is not to drop below 20 °C. Thermal insulation material and an outer cover help to equalize temperature and reduce the effect of unilateral insolation. The

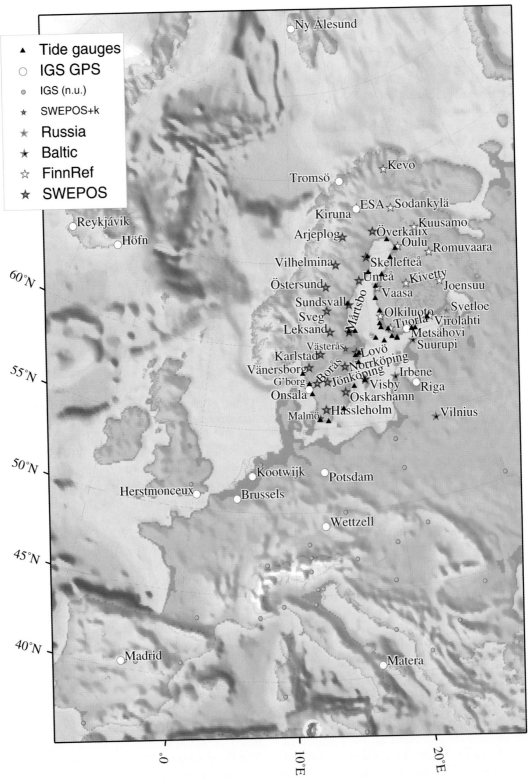

Plate 1. Map of the BIFROST GPS network. Also shown are the stations operating for the International GPS Service (IGS), unlabeled symbols denoting stations not used in BIFROST. The tide gauge stations operating for the Permanent Service for Mean Sea Level (PSMSL) are also included.

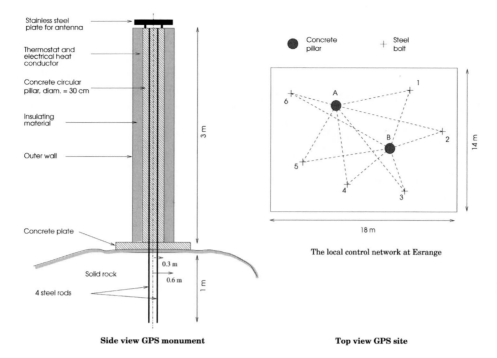

Side view GPS monument

Top view GPS site

Figure 2. Drawing of the monuments and local control network at the SWEPOS station Kiruna. The pillar is of standard SWEPOS type, which implies a steel reinforced concrete structure with a cylindrical shape and a height of three meters, heat insulation and a thermostatic heating system to prevent its temperature to drop below 20° C. The Kiruna site is special in having two pillars.

pillar at Borås contains temperature stabilization that includes also the antenna, the preamplifiers and the cable. Shorter pillars exist at Jönköping, where a taller structure would have interfered with air traffic safety at the end of the local airport's runway, and at Onsala, where monument continuity from the CIGNET epoch was favored. The shorter monuments are not heated. However, the antenna cable at Onsala is thermally stabilized. In each case, Borås and Onsala, the application of GPS to accurate Universal Time services called for controlling the thermal environment.

Antennas in the SWEPOS network are of the Dorne-Margolin type T choke-ring. First equipped in part with TurboRogue SNR 8000 receivers, all SWEPOS stations changed to Ashtech Z12 during 1994–95 as the standard. GLONASS capabilities exist at some stations.

Platforms that enable measurements with absolute gravimeters are installed at the GPS stations Mårtsbo, Onsala and Skellefteå, and at the ESRANGE facility near the Kiruna station.

Selection of stations. The locations of SWEPOS stations were decided on several criteria. First, the terrain and vegetation around a station must guarantee full satellite visibility above 10 degrees of elevation. Sec-

ond, the monumentation must establish a firm mechanical connection with the crystalline bedrock. Third, the preferred station separation was to be 200 km on the average, and as uniform as possible. The requirements for network geometry were raised from the range of applications of the network, mostly regional support of differential and kinematic GPS, and as a national geodetic reference frame. In some of the densely inhabited areas the coverage was increased with new stations at Göteborg, Västerås and Malmö added during 1998 to 2000. These stations are located on buildings; the generallyy poor stability of such sites makes them only potentially useful within the BIFROST project. Future plans aim at a network of about 50 stations in Sweden, where perhaps 10 could be expected to be furnished with concrete monuments matching a high-end stability criterion similar to the first 21 stations.

Site History. The DOSE-93 campaign of 1993 inaugurated the BIFROST project. TurboRogue receivers remained permanently at 16 sites thereafter. Oskarshamn, Norrköping, Lovö and Överkalix were equipped in 1994 with Ashtech receivers. By mid 1995, all stations had been equipped with Ashtech receivers, and a general switch to this receiver type was decided as the

Figure 3. Photograph of the SWEPOS station at Kiruna, showing the pillar and Dorne-Margolin T-antenna, covered with the hemispherical acrylic-glass radome. The instruments and control units are located in the 3 × 4 m wooden hut.

BIFROST standard. The station at Borås became operational in autumn 1995, and an Ashtech receiver was installed in spring 1996.

Conditions providing homogeneous satellite signals were considered important. This argued for not mixing antenna types, but rather keeping a constant design for the antenna and its electrical environment [*Schupler and Clark*, 1991]. This study suggested that in large (more than 1000 km scale) networks distortions in antenna phase diagrams become important since the satellite viewing angles become sufficiently different. The exact origin and nature of phase diagram distortions, e.g. due to radomes covering antennas, still had to be established when the regional, permanent networks were realized.

At first, antennas were covered with radomes provided either by Ashtech or by Delft University of Technology. The disadvantage of these radomes to collect piles of snow and ice during winter was realized by *Jaldehag et al.* [1996b]. The primary effect was seen as a several centimeter large excursion of the vertical position in network solutions. The effect emerged particularly as a consequence of simultaneous troposphere delay estimation. A more pointed radome design was devised, expecting that less snow would stick to steeper surfaces. A Teflon® coating was applied for its well-known non-stick property. Also, near-horizontal surfaces near the antennas, where snow had been piling up before, became completely covered under the new radome. Analyses of antenna sensitivity and lobe pattern signatures by *Jaldehag et al.* [1996a] and by *Ågren*

[1997], however, showed that the parameters for vertical position and for zenith troposphere delay obtained with conical radomes have important and complicated correlations with the lowest satellite elevation at which data are collected. The reason for this behavior is a strongly elevation dependent antenna diagram, being different from reference antenna patterns [*Schupler et al.* 1995]. The distorted antenna diagram interacts with snow. The compromise that the SWEPOS group settled for consisted of a hemispherical acrylic glass radome. It turned out to be sufficiently neutral with respect to satellite elevation; removal of data affected by snow and ice was considered both feasible and to have little effect on the precision of the BIFROST estimates, at least in the long-run. The first radome changes occurred in 1994 and 1995. The hemispherical model was finally installed during 1996. From 1992 to 1999 the Onsala antenna kept a Delft radome as a cover. Instead, due to the different monument design eccosorb was installed around the choke ring assembly in order to avoid signal scattering off the surface of the pillar. For the purpose of tracking the reference position of the Onsala station, long periods were needed when Onsala did not undergo any change while work was proceeding in the remainder of the network and vice versa. This is a consequence of GPS network analysis, where the effect of changes in the antenna patterns is seen not only to affect the position of the very site, but also propagates into the network. (See the "side-effect" parameters in Figures 7–11 below. Finally, the Plexiglas® radome was put on at Onsala in February 1999.

3.2. FinnRef

Finnish Geodetic Institute maintains in Finland 12 permanent GPS stations called FinnRef. All the stations have Ashtech Z-XII3 GPS receivers with choke ring antennas (AOAD_M/B, AOAD_M/T or ASH7009-36A_M). Ashtech choke rings are covered with Ashtech radomes and AOAD antennas with Delft radomes. Metsähovi and Tuorla do not have a radome. At an early stage the FinnRef group made a decision that the antenna construction is not to be touched with changing the radomes etc. in order to guarantee a long uniform data set. Stations are also equipped with weather stations which offer pressure, relative humidity and temperature data with the same 30 s interval that GPS data are stored. Three different antenna mounting platforms have been used as described below. A photo of the Tuorla station is shown in Figure 4.

The base station of FinnRef is Metsähovi Space Geodetic station, which has been collaborating in interna-

Figure 4. Photograph of the FinnRef station at Tuorla, showing the steel grid mast and Dorne-Margolin T-antenna, not radome covered.

tional SLR observation programs for more than two decades. Today it belongs also to the IGS network and together with Joensuu, Vaasa and Sodankylä to the European Permanent Network (EPN). In Metsähovi there is also a GLONASS receiver, a superconducting gravimeter (GWR20) and a Doris beacon.

Selection of stations. The planning of the Finnish part of the Fennoscandian Regional Permanent GPS network was started at FGI during the winter 92/93. Site specifications were similar to the Swedish network, however with additional concern for accommodation of absolute gravity measurements near the stations and a short distance to the points of the precise levelling network. Eleven stations were then chosen for future reconnaissance. The selection of the stations was made according to the following principles: (1) The network should give a good coverage over the whole country in such a way that the maximum land uplift differences in vertical motions between various parts of the country

can be detected; (2) The absolute gravity at, or near the station should be measurable; (3) The site should be located close to the precise levelling network; and (4) The site should be established on the bedrock.

In 1993 steel grid masts of 2.5 m height (Figure 5a) were erected at five sites. The height changes due to the thermal expansion of the steel tower, approximately ±0.8 mm during the course of the year, are small compared to GPS measurement errors (3–5 mm for an individual measurements) and will average out assuming a cyclic variation.

Three sites, Olkiluoto, Kivetty, and Romuvaara, were built in co-operation with the company Posiva Oy. The mount of the antenna is a steel enforced concrete pillar (Figure 5b) which is more stable than the standard mast. The mounts are a part of high precision local GPS control networks, which are periodically remeasured.

Taller steel masts exist at Kevo (5 m), Oulu (8.5 m) and Metsähovi (25 m), and the consequently larger thermal length changes at Oulu and Metsähovi are compensated for, using an antenna suspension structure based on invar-steel (Figure 5c).

Existing buildings were used for the GPS equipment at four stations; Tuorla (the Astronomical Observatory of Turku University), Sodankylä (Sodankylä Geophysical Observatory), Oulu (the Astronomical observatory of Oulu University) and Kevo (the Subarctic Research Institute of Turku University). Separate instrument cabins were built in Virolahti, Joensuu, Vaasa, Kuusamo, Kivetty, Romuvaara and Olkiluoto.

Site History. Metsähovi GPS station has been operational since 1990, first collaborating to the CIGNET network and later after 1992 as a part of the IGS network. Operation in Metsähovi was started with a Rogue SNR-C receiver, which was replaced with a TurboRogue SNR-8100 in 1995 and later in 2000 with the Ashtech Z-

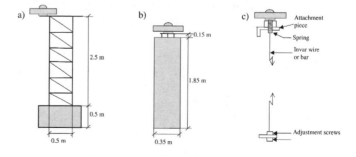

Figure 5. Drawings of three FinnRef monument and antenna assembly types. From left to right: (a) Steel grid mast, 2.5 m high; (b) Concrete pillar; (c) tall steel grid mast, between 5 and 25 m high.

XII3. All the time the same antenna (AOAD_M/B) was used. The first measurements were made at the other 4 stations (Vaasa, Joensuu, Sodankylä, and Virolahti) during the DOSE-93 campaign. At that time only the masts were available and the measurements took place using TurboRogue receivers borrowed from JPL.

Five TurboRogue SNR-8100 GPS receivers were purchased in 1994. The first observations with the new receivers were made during the DOSE-94 Campaign in Aug. 15–26, 1994. The data were collected via telephone line from four sites, Tuorla, Virolahti, Vaasa and Sodankylä. Since the campaign two sites, Virolahti and Sodankylä, have been operating continuously.

Due to frequent malfunctioning of many of the TurboRogues it was decided to change them to Ashtech Z-XII3 receivers in 1995 [*Koivula et al.*, 1998]. The change took place in 1995-1996 except for Metsähovi where the TurboRogue was changed to the Ashtech Z-XII3 receiver in 2000.

Continuous observations were started in 1994 at Virolahti, Tuorla, Sodankylä, and Olkiluoto, and in 1995 at Vaasa, Joensuu, and Oulu. During 1996 four stations, e.g. Romuvaara, Kivetty, Kevo and Kuusamo, were taken into operation.

All stations were equipped with Dorne-Margolin type Choke ring antennas. The radomes used to cover the antennas are made either by the Delft University of Technology or by the Ashtech company. Two of the stations, vis. Tuorla and Metsähovi, have no radome over the antenna. It should be mentioned that no changes of the antennas or the radomes have been made during the whole observation period.

Eight FinnRef stations (Tuorla, Virolahti, Joensuu, Vaasa, Oulu, Sodankylä, Kevo, and Kuusamo) were connected to the precise levelling network by levelling during the years 1995-1998.

Absolute gravity measurements have been made at five FinnRef stations, using both the JILAg-5 of the FGI and various FG5 gravimeters. With the JILAg-5 there are measurements in Metsähovi (more than 80 since 1988), Sodankylä (1988, 1992, and 1998), Vaasa (1995 and 1999), Joensuu (1999), and Virolahti (1999). Five more station will be measured in 2002-2003. With the FG5 there are measurements in Metsähovi by NOAA (FG5-102 in 1993, FG5-111 in 1995) and by BKG (FG5-101 in 2000). Vaasa was measured by NOAA in 1995 (FG5-111).

In the summer of 1998 the FGI made a field calibration for 9 out of 12 antennas of the FinnRef stations [Kylkilahti, 1999] using the method described in [Rothacher et. al., 1995]. In 1998-99 Vaisala PTU 220 type meteosensors were installed for all FinnRef stations. These sensors are connected directly to the receivers and they measure pressure, relative humidity and temperature. The data are downloaded together with the GPS data.

3.3. EUREF/EPN

The permanent network of GPS receivers operating for EUREF (the European Reference Frame) is denoted by EPN. The subset of SWEPOS and FinnRef stations that take part in the EPN, the data that are processed at Onsala Space Observatory consist of observations from Joensuu, Kiruna (both stations), Martsbo, Metsähovi, Onsala, Sodankylä, Vaasa, Vilhelmina, and Visby.

3.4. Campaign Work

Although we established permanent sites in 1993, several campaigns were conducted. Campaigns played the role of creating a first epoch at permanent stations that were planned to become operable later. We also participated in special projects where certain stations, e.g. tide gauges, were to be collocated within the continuous system.

In 1993, as an opening phase of the project, the campaign called DOSE-93 brought TurboRogue receivers to as many permanent monuments as possible in Sweden and Finland. The campaign was repeated in 1994, now bringing in additional Finnish stations. The campaigns were an activity for the NASA Dynamics of the Solid Earth project. A second purpose for these campaigns was to observe the reference points of tide gauge stations. A third and a fourth tide gauge campaign was carried out in 1996 and 1997. The work contributed to vertical datum and reference work within the European Vertical Network (EUVN) and Baltic Sea Level campaigns [*Poutanen and Kakkuri*, 2000].

4. DATA ANALYSIS 1993–1999

4.1. GPS Data Processing

BIFROST GPS observations are processed with GIPSY/OASIS-II software. Alternative solutions are computed with Bernese software (one at Onsala for the European Reference System EUREF from a subset of the Fennoscandian stations, one at the Finnish Geodetic Institute); other GIPSY/OASIS-II solutions are computed (1) for the BALTEX project with focus on the wet troposphere [*Bengtsson*, 2001, *Stoew et al.*, 2001] and (2) a Precise Point Positioning solution [*Zumberge et al.*, 1997]. We will concentrate on the BIFROST-proper solutions. These are described in detail in *Johansson et al.* [2001].

Processing strategies.   Around 1990 the preferred, or more precisely, the desired strategy in GPS projects was to mix a wide range of baseline lengths. The outermost stations would form a network of "fiducial sites". They would frequently be reobserved resulting in tightly constrained coordinates. In a bootstrap procedure, the phase ambiguities would first be solved on the short baselines, taking advantage of the fiducial stations. The coordinate solutions could be refined in a final stage of the solution, with or without ambiguity resolution. But phase ambiguity fixing turned out a difficult process; it turned out viable only when much processing time could be spent on limited data quantities.

In BIFROST one could have expected that the Fennoscandian network of GPS stations would play the role of the regional network within a fiducial network of IGS stations. However, for two reasons development took a different route. First, as a permanent and continuously observed regional network, the two kinds of stations contribute on a quite equal footing. By and large, IGS stations were seen to be replaced, temporarily going out of service, or measurements not reaching satisfying levels of repeatability. Second, in the BIFROST network ambiguity resolution was found to be a time-consuming process beyond practicality.

Such we realized that the stability and continuity of network stations in the Fennoscandian arrays would probably be better than in the IGS network in general. Campaign-style observations of the network would offer little advantage; the solutions would become more vulnerable to changes in conditions of a fiducial network. The processing of less data with greater precision instead would not be a fruitful trade-off.

Occasional tests revealed that 85 to 100 percent of the ambiguities could be resolved despite the long baselines. With increasing computer power routine ambiguity fixing will probably become feasible in 2002.

Going completely nonfiducial.   Regional network observations may be analyzed combined with observations from a set of reference stations in many different ways. In the first years, especially during campaigns, few reference stations at large distances could be used. Their primary role was in the orbit estimation or improvement. Treating them as the ground anchor of the campaign to tie the regional results into a global frame and assuming that the range errors were either negligible or accountable in the regional network analysis, on which they would have an attenuated influence, the final stage of the analysis would treat the positions of the reference stations fixed. Parameters once estimated by one party are taken with that value and with the confidence that the analysis suggested. This is, in short, the concept behind the fiducial approach.

Later on, observation errors at the tracking stations were admitted where they originated in a more balanced error distribution of a nonfiducial reference network. The orbits produced, however, would still be tied closely to the reference stations. The fully nonfiducial concept relaxes also this tie. All variances of estimated positions remain large until the last stage of transformation of estimates into the set of reference stations. As many parameters as possible are represented by their covariances with each other, and a fix is attempted with respect to few, most certain parameters.

Although this method distributes errors more widely between orbit offsets and network offsets, the following relations still have an influence on the observed deviations of position: If an estimated reference station parameter deviates, because, for example, of transient site motion that goes unnoticed, then during the final transformation stage of the network results a part of this deviation will be spread inversely to other station parameters. Studying the station to station correlation in the Empirical Orthogonal Function analysis in Johansson et al. (2001) this is clearly seen as a sharply decreased correlation coefficient between a regional network site and a site used for constraining the transformation into the ITRF, compared to correlation between regional sites only. In Johansson et al. (2001) the case had been demonstrated particularly for the station of Metsähovi, which is located within the region but had to take the role of a constraining site. This reflects the problem that it is difficult to find sufficiently many IGS stations forming a constraining network with a good geometry and simultaneously a long-term continuity. A more ideal situation could be desired, like using a station in western Russia of equally good performance as the Finnish station.

The IGS stations analyzed together with the regional network data, used in the Helmert transformations, are ONSA, METS, MATE, KOSG, NYAL (through 1998), NYA1 (after 1998), MADR (prior to 1997, after which site instabilities occurred), WETB (prior to mid-1996), WTZR (after mid-1996), TROM (through the end of October 1998, with an interruption between January

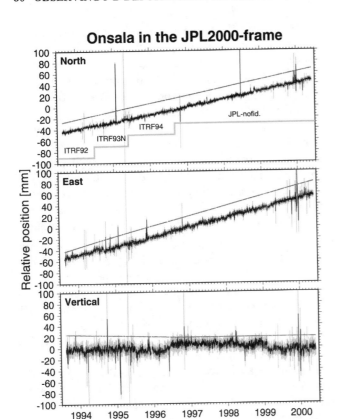

Figure 6. Evolution of site position in the global frame, shown for station Onsala as an example. GPS solutions for vertical, east and north components are shown along the time axis as a black line, standard deviation as gray error bars. Above each series is a reference curve that shows the effect of global frame rotations and translations at Onsala. The top frame shows the periods that relate to different issues of GPS orbit products by their associated terrestrial reference frame. More details are given in the text.

mapping the positions into the ITRF97 frame and using the velocities of the JPL-2000 reference frame. Above each curve of a position component a line is shown that is derived from the joint motion of the mapping sites (rotation, translation, and scale). The curve section after the transition to using JPL nonfiducial orbits in 1996 is due only to a co-rotating frame (no translations, no scale). The earlier curve sections contain a correction term due to the relative rates of rotation and translation of the ITRF frames with respect to the JPL-2000 frame. Since this frame rotation affects the orbits and not the mapping sites (they are already in the latest frame) the effect on the site positions determined with the old orbits is described by the inverse rotation and translation.

The impact of this correction is a 2 mm subsidence of the reference frame between the years 1993.5 and 1996.668. Ignoring this term would bias the vertical rate at Onsala (whole time span average) by almost -0.3 mm/yr. Due to the limited geographic extent of the network, the frame motion looks almost the same at all BIFROST stations.

The choice of mixed reference frames for positions and velocities was based on the observation that, on the average, the ITRF97 shows subsidence at our mapping sites, and especially at Metsähovi the vertical rate is not plausible, whereas problems with consistency appear much less pronounced in the JPL-2000 frame, see Table 1.

4.2. Inference of Regional Motion

In estimating regional motion we first subtract the frame motion from the projected positions (the differ-

1997 and March 1998), and TRO1 (after October 1998). Sites whose data are used but that are not used in the transformation are RIGA, BRUS, POTS, KIRU, HERS, REYK, and HOFN. (For a key to the site abbreviations, see http://igscb.jpl.nasa.gov.)

The stations marked "used for transformation" will, in the discussion below, be designated as "mapping sites." The transformation process contains a pitfall in that the reference positions and velocities changed over the years, and the (fiducial) GPS orbit files used in the GPS analysis are still defined with respect to the old reference frames.

Figure 6 exemplifies the problem. The figure shows the results of the daily analysis of the Onsala site after

Table 1. Comparison of vertical motion of some important stations in the reference frames ITRF97 [*Boucher et al.*, 1999] and JPL-2000 [http://sideshow.jpl.nasa.gov/mbh/series.html].

Station	ITRF97 [mm/yr]	JPL [mm/yr]
Onsala	1.50±0.30	0.35±0.12
Metsähovi	-0.89±0.34	2.94±0.12
Kootwijk	0.27±0.50	-0.57±0.10
Wettzell WTZR	-2.26±0.30	-1.25±0.29
Wettzell WETB		-2.91±0.32
Ny Ålesund NYAL	0.49±0.77	4.96±0.20
Ny Ålesund NYA1		5.17±1.56
Tromsø	-0.24±0.75	1.59±0.15

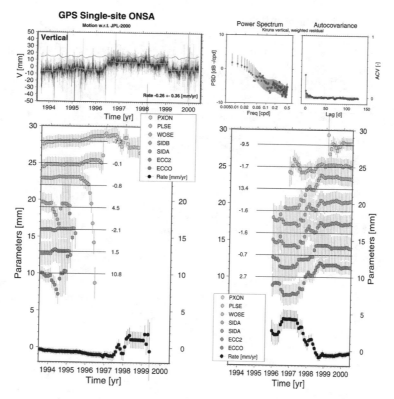

Figure 7. Time series, noise and parameter stability analysis for Onsala. Upper left frame shows the time series of vertical position (black curve with gray error bars corresponding to one standard deviation) together with the fitted model for rate and biases (narrow gray curve near the centre of observations) and seasonal harmonics (wiggly gray curve above). Upper right frames show power spectrum and autocovariance of the residual of the fit. Lower frames show estimated rates and biases versus a moving starting date of the analysis (left frame) or ending date (right frame), respectively. The major purpose of these diagrams is a test of the robustness of the estimated rates. Bias codes designate antenna or radome changes at Onsala or side-effects propagating through the network solutions, see text. The vertical order for the bias parameter curves corresponds to the legend. The average value in millimeter is shown as a number before or after the bias curves.

ence between the black and the gray curves in Figure 6). The remainder is interpreted with different models of motion, and each model is solved using a least-squares fit. In *Johansson et al.* [2001] three different models are fitted, consisting of (1) Standard: constant rates, seasonal harmonics, and bias parameters; (2) Edited: the Standard setup plus predicted atmospheric loading, and outlier editing (5-σ criterion); and (3) EOF: an Empirical Orthogonal Function extension of the Edited solution, including the least-squares residuals of the Edited solution.

The bias parameters are introduced at every event that can be suspected to have an influence on the site position, see *Johansson et al.* [2001] for details. The seasonal harmonics are expected to pick up combined effects due to varying hydrological conditions, e.g. due to loading [*vanDam et al.*, 2001] or due to interaction

with the electrical environment of the antenna caused, for example, by varying ground reflectivity or snow on the radomes [*Jaldehag et al.*, 1996a, b].

To discuss the most important aspects of discrimination of signal in noise we will stay with the standard solution and the vertical component only. Most of our findings from the vertical apply to the horizontal components in a similar way; usually, we find that problems are more profound in the vertical. This concerns for instance the effect of biases and the covariance of the estimated parameters with the rate. Also, the noise color in the horizontal components is generally more white, while the problem signal-to-noise discrimination becomes more accented in brown or red noise color.

In Figures 7–11 we show in five examples (Onsala, Hässleholm, Sveg, Sundsvall, Kiruna) the time series, the fitted rate plus bias model, and the fitted seasonal

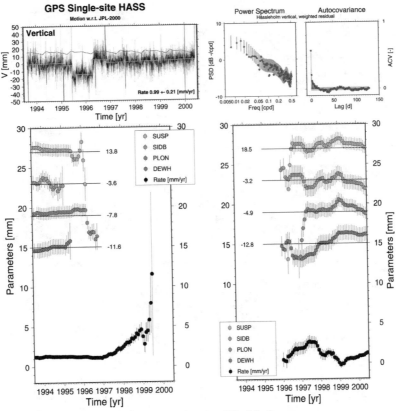

Figure 8. Like Figure 7, but for the site Hässleholm.

model. The figures also show the power spectrum and the autocovariance of the residual of the least-squares fit. These are indicative of the noise color.

We also show the results of a robustness test for the estimated parameters. The test considers the estimated rates and bias parameters as a function of a varying start and end of the observing time span. Each boundary was moved with an increment of 30 days. We notice the large impact of the biases on the rates. When the starting boundary is posterior to the last bias, the rate estimate shows a tendency to lose stability. This is most probably an effect of limited data. The test gives a clue as to the feasibility to estimate the rates from the data obtained after autumn 1996, when most of the radome and antenna conditions remain fixed. In the examples, Kiruna would probably settle at an uplift rate of 7 rather than 8.5 mm/yr and Onsala at 2 rather than 0 mm/yr, while the other three stations are probably less susceptible to sacrificing the early data.

Additional explanations to the robustness plots in Figures 7–11: The biases at Onsala (cf Figure 7) are designated as follows: ECCO and ECC2 - introduction of eccosorb material below the antenna; SIDA, SIDB,

WOSE, and PLSE - side effects due to radome changes at many network stations in four almost simultaneous campaigns; PXON - putting on the (now standard) acrylic-glass radome. When an annual harmonic is estimated and the effective length of the data window is reduced to one year or shorter, the rate becomes highly uncertain. This produces the single outliers in the diagrams.

Power spectrum and autocovariance in Figures 7–11: If the noise process is Gauss-Markov, the autocorrelation coefficient depends on lag t as

$$a(t) = q^t \qquad (4)$$

where $0 < q < 1$ is a constant. The power spectrum has a 3 dB attenuation point at the frequency $-(1/\pi) \log q \, f_\nu$. We typically find Markov-parameters between 0.2 and 0.4.

5. OBSERVED FIELD OF MOTION

In this section we describe a new motion solution based on the same data as in *Johansson et al.* [2001] except that the position solutions of all winter days have

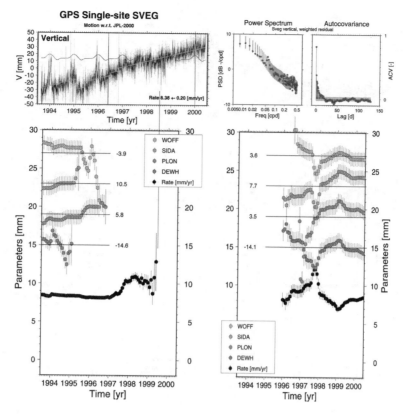

Figure 9. Like Figure 7, but for the site Sveg.

been ignored. The winter editing started at November 1 and ended at March 31 and comprised all sites. The reason for this measure is the uneven bias that is incurred by snow on the antennas. The effect is most profound in the case of the Finnish stations for a number of reasons. First, the station history is shorter, thus irregular winter signatures do not average out sufficiently well. Second, the winters are generally colder, implying more snow pileup. Third, the Finnish stations have a triangular top plate on the monument and an off-center antenna mount (c.f. Figure 5); this causes snow to pile up also at one side, introducing probably a bias also in the horizontal components. In the case of the Swedish stations, the horizontal components are less affected. In order to equalize the winter editing as far as possible, all position solutions have been edited in this manner. This loss of 1057 with respect to the total of 2500 daily solutions is 42 percent.

The position time series are analyzed with a simple least-squares fit of one constant rate per station and component, and antenna biases in the same way as for the "edited solution" in *Johansson et al.* [2001]. Since

the winter editing has reduced most of the annual signature, and since the gaps have an annual repeat pattern, we do not solve for seasonal sinusoids. Although the effect of air pressure loading was found in *Johansson et al.* [2001] to be a minor problem, we include it in this solution. We finally apply an outlier editing recursion based on a criterion of five root-variances of the residual. This weeds out typically one, at maximum ten samples. The normal probability for one sample out of 1400 to exceed five standard deviations is less than one in one thousand.

Time series plots along with the result of the simple motion model are shown in Figures 12–15

The field of motion obtained in this way is compared to the model computations presented in *Milne et al.* [2001]. We use the particular model resolved as best fitting the standard solution of *Johansson et al.* [2001], which implies the Fennoscandian ice load history of *Lambeck et al.* [1998a, 1998b], a 120 km thick lithosphere, an upper mantle viscosity of 8×10^{20} Pa s and a lower mantle viscosity of 1×10^{22} Pa s. Since tectonic plate motion is not included in the model, and since our

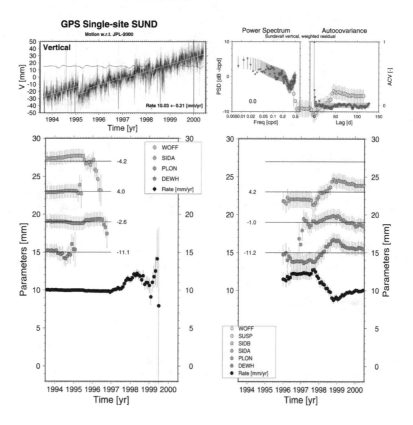

Figure 10. Like Figure 7, but for the site Sundsvall.

reduction for the motion of the Eurasian plate may contain a residual, we allow an adjustment of the horizontal field of motion in the form of a rigid plate rotation. The rates before and after the adjustment, the model rates and the misfit are shown in Table 2. The misfit M is given as follows

$$M = \sqrt{\sum_i \left(\frac{r_i^{(obs)} - r_i^{(mod)} + r_i^{(adj)}}{\sigma_i} \right)^2} \qquad (5)$$

where σ_i is the *a priori* standard deviation and the sum is only over the two horizontal components, since the vertical is practically orthogonal to the residual plate motion.

The *a priori* standard deviation for each station and component along with the factor to obtain the *a posteriori* measure is given in Table 3. Maps of the inferred motion along with the model predictions are shown in Plate 2.

We also show a comparison between terrestrial geodetic inference of vertical motion and BIFROST results.

We use the tide gauge and levelling results of *Ekman* [1996] and add the geoid motion due to our rebound model. In both cases, GPS and terrestrial, we obtain the geometric rate of motion of the earth surface. In the case of GPS the motion is relative to the centre of the satellite frame, which is constrained to the earth's gravity center (how well remains to be determined). In the case of the terrestrial methods the reference is the regional sea surface. The difference between the two data sets interpolated to the point where the inferred land surface motion is zero determines the regional change of sea level with respect to the earth's gravity center.

In the case of Ny Ålesund (NYAL), a careful evaluation of mareograph data was presented by *Breuer and Wolf* [1995], who estimated a relative uplift velocity of 2.6 ± 0.7 mm/yr. The motions given in *Ekman* [1996] and *Breuer and Wolf* [1995] have been included in the computation of the regression line, which is shown in Figure 16. We also show the rates based on the ITRF97 at stations where no terrestrial data is available; they are not included in the fit.

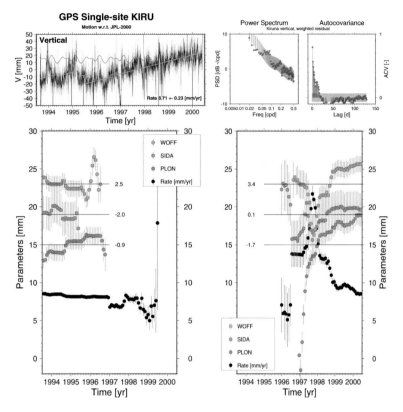

GPS Single-site KIRU

Figure 11. Like Figure 7, but for the site Kiruna.

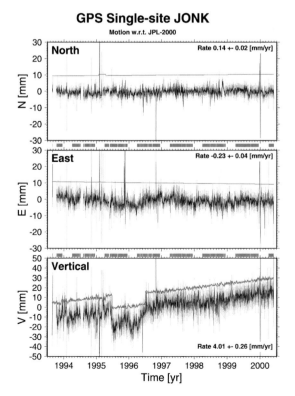

GPS Single-site JONK

FinnRef solution. FGI has made an additional data analysis for the FinnRef part of the network using the Bernese software in differential mode. A different approach to remove the snow-contaminated winter data semi-manually lead to less data rejection with a good and stable solution. It turned out that only the northernmost stations suffered the snow problem at a significant level [*Mäkinen et. al.*, 2000; *Koivula and Poutanen*, 2001].

Another effect was also discovered in FinnRef time series in the form of periodicities in the different time

Figure 12. Evolution of the position of site Jönköping in the regional frame, using Eurasian station motion in the JPL-2000 frame. GPS solutions for vertical, east and north components are shown along the time axis as a black line, standard deviation as gray error bars. The fitted, simple motion model is shown as a thin line, adding an offset for legibility. In the case of the vertical the model included predicted air pressure loading effects. The line above a frame indicates the data that have been subjected to the fit (the north and east components have been treated identically. Radome changes are predominantly seen in the vertical as jumps on the order of ten millimters.

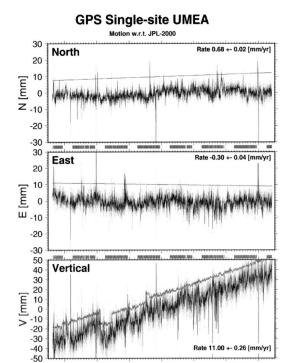

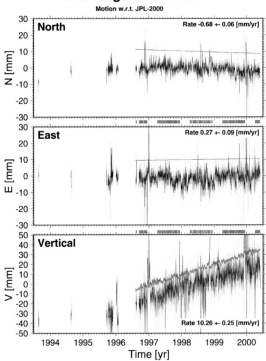

Figure 13. Like Figure 12, but for site Umeå.

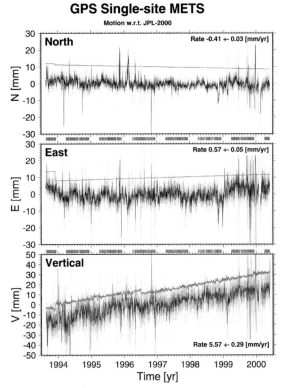

Figure 14. Like Figure 12, but for site Vaasa.

Figure 15. Like Figure 12, but for site Metsähovi.

series. They were found to affect all components including the network scale [*Poutanen et al.*, 2001]. The annual periodic variation in the scale might be attributed to the loading effect capable to cause cm-range vertical motion of the whole network, which is equivalent to a 2–4 ppb variation in the scale parameter. A candidate is hydrological loading, see *vanDam et al.* [2001]. There is danger implied as incomplete annual cycles might bias the rate estimation when annual cycles are not simultaneously solved for. Then again, interannual variations would rather call for simultaneous admittance for predicted loading effects at all stations, including the stations used for constraining the network solution into the reference frames.

5.1. Discussion

Here we discuss the winter-edited solution shortly. Generally, the edited solution of *Johansson et al.* [2001] and the present one are comparable, with larger changes in the case of the east components in Finland. For example, we have Kuusamo (KUUS) before at 2.7 ± 0.4 mm/yr and now 1.6 ± 0.3 mm/yr; Sodankylä (SODA) before at 0.6 ± 0.5 and now -0.26 ± 0.3. The

Table 2. Motions from GPS observations versus glacial isostatic adjustment model. A rigid rotation was estimated using the observed horizontal rates for all the stations except those denoted with a dagger.

Site	lat.	long.	Adjusted obs			Model			Unadjusted obs.			Weighted error	
			V	E	N	V	E	N	V	E	N	after	before
	[°]	[°]	[mm/yr]			[mm/yr]			[mm/yr]			adjustment	
ARJE	18.12	66.32	8.41	-1.64	1.97	9.00	-1.20	1.96	8.41	-1.54	1.62	2.71	2.71
BORA	12.89	57.72	2.85	-1.13	.16	3.00	-.56	-.09	2.85	-1.33	-.10	2.48	3.07
BRUS†	4.36	50.80	-2.10	-.08	-1.07	-.80	-.22	1.20	-2.10	-.52	-1.18	9.84	10.38
HASS	13.72	56.09	1.39	-.38	.57	.80	-.39	.20	1.39	-.65	.29	4.22	2.11
HERS†	0.34	50.87	1.23	1.97	.21	-.77	-.24	1.20	1.23	1.53	.16	8.81	7.63
JOEN	30.10	62.39	5.49	.39	-.60	4.73	.61	.02	5.49	.26	-1.15	2.05	3.61
JONK	14.06	57.75	4.01	-.02	.42	3.23	-.44	-.13	4.01	-.23	.14	6.94	3.53
KARL	13.51	59.44	6.07	-.79	.13	6.17	-.65	-.09	6.07	-.93	-.14	1.46	1.56
KEVO	27.00	69.75	4.60	-.17	1.51	3.47	-.29	2.31	4.60	-.01	1.01	1.65	2.66
KIRU	21.06	67.88	7.12	-1.17	1.99	7.14	-.87	2.33	7.12	-1.03	1.59	1.39	2.21
KIVE	25.70	62.82	7.94	.89	.31	7.70	.54	.07	7.94	.80	-.17	.86	.77
KOSG†	5.81	52.18	-1.10	.12	.01	-1.10	-.27	1.16	-1.10	-.27	-.13	12.77	13.82
KUUS	29.03	65.92	8.09	1.63	.18	8.45	.46	1.02	8.09	1.64	-.35	2.43	3.18
LEKS	14.88	60.72	8.66	-.03	1.16	8.41	-.62	.13	8.66	-.12	.86	4.48	3.26
LOVO	17.83	59.34	6.24	.45	-.07	5.88	-.02	-.23	6.24	.28	-.42	2.18	1.66
MART	17.26	60.60	7.03	-.26	.07	8.26	-.19	-.06	7.03	-.38	-.27	1.66	1.98
METS	24.40	60.22	5.57	.74	.05	4.40	.47	-.12	5.57	.57	-.41	2.58	2.76
NORR	16.25	58.59	5.22	-.43	.07	4.76	-.21	-.22	5.22	-.62	-.25	1.76	1.74
NYAL	11.87	78.93	5.29	-.92	.44	-.61	-1.09	.50	5.29	-.32	.19	8.37	9.52
OLKI	21.47	61.24	7.82	.57	-.31	7.76	.40	-.08	7.82	.46	-.73	.90	2.19
ONSA	11.93	57.40	.46	-.50	.31	2.26	-.63	-.01	.46	-.72	.06	2.86	1.76
OSKA	16.00	57.07	2.39	-.60	.36	2.20	-.23	-.03	2.39	-.84	.04	2.13	2.67
OSTE	14.86	63.44	8.26	-.90	1.22	9.78	-1.25	1.16	8.26	-.89	.92	3.07	3.45
OULU	25.89	65.09	10.46	1.23	.86	9.80	.37	.82	10.46	1.23	.38	1.72	2.02
OVER	22.77	66.32	8.85	-.48	2.24	10.12	-.30	1.68	8.85	-.41	1.81	2.06	1.37
POTS†	13.07	52.38	-1.54	-.21	.39	-.85	-.21	.92	-1.54	-.61	.12	3.36	5.14
RIGA	24.06	56.95	2.54	.74	.20	.45	.16	.24	2.54	.46	-.25	2.82	2.83
ROMU	29.93	64.22	7.25	1.39	.53	7.10	.62	.32	7.25	1.33	-.02	1.24	1.31
SAAR†	20.97	67.86	6.98	-1.13	1.38	7.14	-.87	2.33	6.98	-.99	.98	3.71	5.15
SKEL	21.05	64.88	10.98	-.73	1.17	11.12	-.31	1.13	10.98	-.71	.76	2.40	3.01
SODA	26.39	67.42	9.50	-.27	.65	8.17	.00	1.88	9.50	-.19	.16	2.32	3.15
SUND	17.66	62.23	10.22	-.70	.88	10.32	-.38	.43	10.22	-.76	.53	3.02	1.90
SVEG	14.70	62.02	8.31	-.83	1.00	9.59	-.93	.63	8.31	-.88	.71	3.26	1.53
TROM	18.94	69.66	3.10	-1.58	2.24	2.12	-1.19	2.35	3.10	-1.36	1.87	1.58	1.47
TUOR	22.44	60.42	6.45	.81	-.49	5.80	.45	-.16	6.45	.66	-.92	1.65	2.49
UMEA	19.51	63.58	11.00	-.29	1.05	11.05	-.30	.75	11.00	-.30	.68	1.85	.46
VAAS	21.77	62.96	10.26	.32	-.27	10.00	.24	.31	10.26	.27	-.68	1.78	3.04
VANE	12.04	58.69	4.42	-.87	.21	4.48	-.76	-.09	4.42	-1.03	-.04	2.34	1.64
VILH	16.56	64.70	8.40	-1.74	1.58	10.05	-1.20	1.46	8.40	-1.70	1.25	4.13	3.96
VIRO	27.56	60.54	3.72	.28	-.45	3.38	.47	-.05	3.72	.10	-.96	1.20	2.55
VISB	18.37	57.65	3.59	.05	.39	2.81	.01	-.11	3.59	-.18	.04	4.73	1.96
WTZR	12.88	49.14	-.44	.16	.27	-.54	-.10	1.07	-.44	-.36	.01	4.41	5.79

Table 3. Rates using *a priori* standard deviations. The scale for the uncertainty is calculated to yield a postfit χ^2 residual of unity, accounting for the Gauss-Markov error model.

Site	Vertical			East			North		
	Rate	σ	Scale	Rate	σ	Scale	Rate	σ	Scale
	[mm/yr]	[mm/yr]		[mm/yr]	[mm/yr]		[mm/yr]	[mm/yr]	
ARJE	8.411	0.284	2.59	-1.538	0.040	2.04	1.617	0.027	3.72
BORA	2.852	0.271	1.190	-1.333	0.084	1.50	-0.103	0.053	2.37
BRUS	-2.103	0.255	1.73	-0.517	0.089	1.59	-1.182	0.051	2.29
HASS	1.394	0.262	1.83	-0.645	0.039	1.87	0.289	0.022	2.01
HERS	1.227	0.105	9.48	1.533	0.046	3.11	0.164	0.025	4.77
JOEN	5.487	0.254	1.83	0.259	0.097	2.49	-1.145	0.065	2.61
JONK	4.011	0.259	1.29	-0.230	0.041	1.58	0.136	0.024	1.91
KARL	6.065	0.261	1.48	-0.928	0.063	1.44	-0.143	0.039	2.33
KEVO	4.595	0.298	4.00	-0.005	0.095	2.38	1.014	0.073	3.50
KIRU	7.116	0.282	2.88	-1.027	0.057	2.63	1.588	0.041	4.22
KIVE	7.942	0.297	1.74	0.803	0.108	2.62	-0.169	0.072	2.94
KOSG	-1.096	0.101	1.36	-0.270	0.042	1.36	-0.131	0.023	2.03
KUUS	8.091	0.316	3.15	1.636	0.112	2.86	-0.353	0.079	3.35
LEKS	8.663	0.471	1.99	-0.123	0.088	1.99	0.859	0.049	2.53
LOVO	6.236	0.269	1.15	0.283	0.085	1.40	-0.419	0.054	1.98
MADR	1.625	0.324	2.18	-0.953	0.165	1.90	-0.201	0.074	3.64
MART	7.031	0.260	1.59	-0.377	0.066	2.07	-0.269	0.044	2.13
MATE	-1.262	0.147	1.81	1.882	0.068	1.78	3.504	0.038	2.06
METS	5.569	0.294	1.19	0.572	0.051	1.71	-0.413	0.029	2.38
NORR	5.224	0.266	1.37	-0.616	0.085	1.48	-0.249	0.054	1.98
NYAL	5.291	0.220	1.61	-0.321	0.072	1.19	0.193	0.052	2.51
OLKI	7.819	0.268	1.27	0.455	0.102	1.98	-0.725	0.066	2.24
ONSA	0.460	0.450	1.24	-0.715	0.059	1.36	0.056	0.035	2.03
OSKA	2.390	0.266	1.58	-0.838	0.084	1.37	0.041	0.052	2.71
OSTE	8.263	0.261	1.73	-0.888	0.039	1.80	0.923	0.025	3.08
OULU	10.460	0.277	3.72	1.228	0.098	2.59	0.376	0.068	3.10
OVER	8.846	0.286	1.74	-0.412	0.060	2.67	1.805	0.042	4.38
POTS	-1.536	0.266	0.93	-0.611	0.093	1.21	0.116	0.053	1.64
RIGA	2.536	0.292	1.44	0.455	0.121	1.77	-0.250	0.074	2.78
ROMU	7.247	0.321	2.85	1.326	0.116	2.88	-0.020	0.079	2.81
SAAR	6.981	0.161	2.36	-0.993	0.056	2.57	0.981	0.036	3.66
SKEL	10.982	0.272	1.76	-0.705	0.042	2.11	0.764	0.027	3.36
SODA	9.500	0.328	3.51	-0.185	0.112	2.61	0.164	0.081	3.43
SUND	10.220	0.259	1.65	-0.757	0.038	2.75	0.530	0.024	3.61
SVEG	8.313	0.254	1.88	-0.878	0.038	1.55	0.707	0.024	2.74
TROM	3.095	0.337	1.33	-1.361	0.112	1.56	1.870	0.070	3.97
TUOR	6.446	0.264	1.31	0.657	0.102	1.88	-0.921	0.066	2.57
UMEA	10.996	0.264	1.71	-0.304	0.038	1.97	0.675	0.024	3.42
VAAS	10.260	0.254	2.90	0.273	0.094	2.25	-0.682	0.064	2.55
VANE	4.421	0.270	1.35	-1.029	0.047	1.79	-0.043	0.028	2.37
VILH	8.402	0.265	1.82	-1.698	0.038	1.94	1.250	0.024	3.88
VIRO	3.716	0.261	1.67	0.099	0.102	1.92	-0.955	0.066	2.94
VISB	3.588	0.255	1.81	-0.184	0.039	2.14	0.035	0.023	2.36
WETT	-0.444	0.249	0.84	-0.360	0.105	1.06	0.006	0.059	1.59

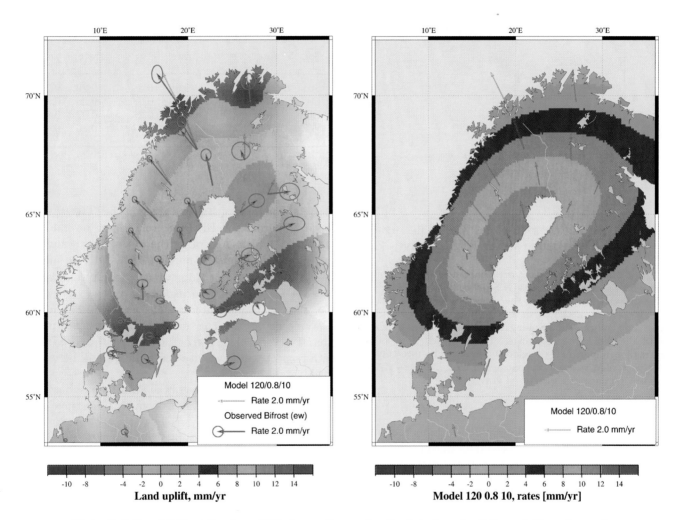

Plate 2. Inferred 3-D motion from GPS observations, using the data from April through November each year only. The observations (left frame) are compared with a Glacial Isostatic Adjustment model (Milne, 2001, right frame), assuming a lithosphere thickness of 120 km, upper mantle viscosity of 0.8×10^{21} Pa s and a lower mantle viscosity of 1×10^{22} Pa s. For ease of comparison, the arrows representing the modeled horizontal motion are shown also in the left frame. The observed horizontal motion is signified by arrows along with the *a posteriori* 63 percent confidence ellipse. The colors for the observed vertical field have been faded out with increasing site distance using a half-width of 200 km.

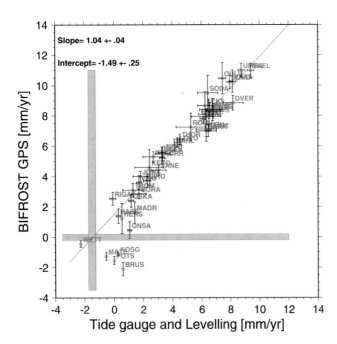

Figure 16. Comparison of terrestrial determinations of relative land surface rise versus GPS-determined, absolute rise. For the absolute estimate the reference frame origin and gravity center of the earth are assumed to be coincident. The tide gauge and levelling rates have been augmented with the geoid rate of change. The expected slope of the regression line should therefore be unity. The intercept indicates a relative sea level change of 1.49 ± 0.25 mm/yr with respect to the earth's center of gravity.

Table 4. Rates determined at Leksand in the different solutions, units are millimeter per year. Solutions are denoted: Std, Standard; Ed, Edited; EOF, Empirical Orthogonal functions; and WEd, Winter-edited. The Std, Ed, and EOF solutions are from *Johansson et al.* [2001];

Sol.	Vertical		East		North	
	rate	σ	rate	σ	rate	σ
Std	8.23	0.9	-0.36	0.3	-0.20	0.2
Ed	8.35	0.9	-0.33	0.3	-0.26	0.2
EOF	9.52	0.5	-0.27	0.2	0.97	0.13
WEd	8.66	0.9	-0.12	0.17	0.86	0.13

scatter. The rates of stations given in the ITRF97 seem to deviate from the BIFROST cluster at several millimeters per year. Accepting the *a posteriori* scaling of the standard deviations implied in the fit, the uncertainty includes a unit slope at 63 percent confidence. The vertical rate for the regional sea level relative to the reference frame center is found at 1.5 ± 0.25 mm/yr. The uncertainty for this rate still only specifies precision. The vertical motion of the origin of the GPS frame with respect to the gravity center might still be rather weakly determined.

An independent solution of the FinnRef part has been compared to the uplift values coming from three precise levellings of Finland. It showed a good agreement with the levelling based values [*Mäkinen et. al.*, 2000] [*Koivula and Poutanen*, 2001].

6. CONCLUSIONS

We have demonstrated the precision and consistency of three-dimensional rate of motion solutions based on 2500 days of continuous BIFROST GPS observations. The continuity enables us to track the evolution of the station position, e.g. by solving offset parameters where observation conditions changed, and estimate rates with support from time series analysis and noise statistics. Still, much more work needs to be done, and we anticipate a more comprehensive analyses of noise and signal estimation, beyond the Empirical Orthogonal Function method employed in this and other BIFROST rate estimations.

A major drawback is the dependence on only a few stations for the representation of the reference system, i.e. for constraining the solutions to the gravity center

decrease of the *a posteriori* confidence limit is mostly due to the avoiding of scattered data during winter.

A larger problem appears to persists at the Leksand site (LEKS) in central Sweden. Before anomalous motion can be concluded we have to keep in mind that the continuity of this station was interrupted at a relatively large number of occasions, eight times during the whole history at which influences of horizontal position can be expected. The reason is that Leksand has been the primary target for monument stability surveys, during which the antenna had to be unmounted. At least during the first surveys exact repositioning did not always succeed.

The unfortunate correlation of estimated offsets with a rate leads us to be cautious in an interpretation. The scatter between the different solutions, as shown in Table 4, is on the order of one millimeter per year, primarily in the north component.

The regression between GPS and terrestrial rates of vertical position in Figure 16 shows remarkably little of the earth. Our solutions differ from ITRF velocity parameters notably in some cases, for instance Wettzell and Metsähovi, where we find 2 to 4 mm/yr greater upward rates.

The present paper has derived a new three-dimensional motion solution by avoiding data taken during the snowy parts of the years. The effect is most probably due to the change of the antenna sensitivity pattern when it is covered by snow. The major bias is found in the vertical, but in some cases the horizontal components suffer owing to a cylindrically asymmetric antenna mount. The effect may amount to the order of one millimeter per year in the vertical. By avoiding the winter months in the data analysis we retain 58 percent of the observations. The *a posteriori* rate uncertainty of this solution is less than in the all-year analysis as the increased variance of position determination from winter data is more severe than the gain from having more data in the fit. The gappy data set, however, causes us to use a simpler motion model.

In the case of the vertical solution, comparison of regional-relative (i.e. differential vertical rates) with precise levelling data may be used in an assessment of the impact of different options in the GPS data processing.

A map of motions emerges that can be reconciled with a model of glacial isostatic adjustment at a high degree of fit. A unique contribution of GPS is in the horizontal components, which previously was observable only in the in this area extremely sparse VLBI network. The uniqueness of GPS lies in the dense spatial sampling of the network.

We need a few years more of observations in order to conclude rates at a safe level of 0.1 mm/yr uncertainty. We can soon sacrifice the data prior to 1996.7, observations which are affected by changes in antenna and monument assemblies. This would imply highly comparable amounts of simultaneous observations at all stations in the network. For the analysis this would also imply a more homogeneous and consistent reference frame, more accurate estimates of satellite parameters, and hopefully less problems at the IGS stations.

An outlook into the future presents opportunities to reprocess the data in order to make it more homogeneous, partly drawing from the experience, partly utilizing increasing computer power. One clue comes from the lower sensitivity to snow conditions when observations also at low satellite elevation angles are taken. Besides several strands of alternative solutions using different software packages we look into the possibility to solve orbits and phase ambiguities.

Acknowledgments. We are grateful for the support by Knut and Alice Wallenberg's Foundation and the Swedish Council for Planning and Coordination of Research (FRN) for equipment. We also like to thank the National Land Survey of Sweden, especially the staff engaged in the establishment, maintenance, and operation of SWEPOS. Some stations in Finland were created with the support of Posiva Oy. Measurement campaigns used equipment made available by the University Navstar Consortium (UNAVCO). Research projects have been funded by the Swedish Natural Science Research Council (NFR, since 2000 renewed by The Swedish Research Council), the Swedish National Space Board (SNSB), and the Swedish Research Council for Engineering Sciences (TFR). NASA has supported the work by grants NAG5-1930 and NAG5-6068, and NSF Geophysics by grant EAR-9526885. NASA has also supported this work under the Dynamics of the Solid Earth initiative (DOSE) and helped to carry out field work. We thank the GIPSY/OASIS group at the Jet Propulsion Laboratory (JPL) for program development, reference frame, and orbit data. Part of this research has been supported by the Smithsonian Institution. We thank the IGS for coordinating international GPS efforts for the benefit of research and the supply with satellite tracking information. Data service resources were used at NASA Crustal Dynamics Data Information System (CDDIS), at the University of California at San Diego (SOPAC), at the Bundesamt für Kartographie und Geodäsie (BKG). Some figures were generated with GMT, the Generic Mapping Tools [*Wessel and Smith*, 1998]. We extend our thanks to the colleagues in the WEGENER project, in part an offspring of NASA-DOSE, which endorses BIFROST. We also like to thank the following individuals for their support, commitment, ideas and interest: Bernt Rönnïg, Irwin I. Shapiro, Biörn I. Nilsson, Martin Ekman, Rüdiger Haas, our PhD students at Onsala Space Observatory, further Rick Bennett, Pedro Elósegui, Ragne Emardson, Kenneth Jaldehag, and Per Jarlemark. Special thanks also to the operators of the permanent stations at Suurupi, Riga, Irbene, and Vilnius, and the many assistants out in the field when we have been conducting campaigns. FinnRef® is a registered trademark of the Finnish Geodetic Institute. Plexiglas® is a registered trademark of Rohm and Haas Co. SWEPOS® is a registered trademark of the National Land Survey of Sweden. Teflon® is a registered trademark of El du Pont de Nemours and Company.

REFERENCES

Ågren, J., Problems Regarding the estimation of tropospheric parameters in connection with the determination of new points in SWEREF 93, in *Report on the Symposium of the IAG Subcommision for the European Reference Frame (EUREF)*, Veröffentlichungen der Bayerischen Kommision für die Internationale Erdmessung, 71–84, 1997.

Bengtsson, L., Numerical modelling of the energy and water cycle of the Baltic Sea, *Meteorol. Atmos. Phys.*, 77, 9–17, 2001.

Beutler, G., P. Morgan, and R. E. Neilan, Geodynamics: Tracking satellites to monitoring global change, *GPS World*, 2, 40–42, 1993.

BIFROST Project (R. Bennett, T. R. Carlsson, T. M. Carlsson, R. Chen, J. L. Davis, M. Ekman, G. Elgered, P. Elósegui, G. Hedling, R. T. K. Jaldehag, P. O. J. Jarlemark, J. M. Johansson, B. Jonsson, J. Kakkuri, H. Koivula, G. A. Milne, J. X. Mitrovica, B. I. Nilsson, M. Ollikainen, M. Paunonen, M. Poutanen, R. N. Pysklywec, B. O. Rönnäng, H.-G. Scherneck, I. I. Shapiro, and M. Vermeer), GPS Measurements to Constrain Geodynamic Processes in Fennoscandia, *EOS Trans. AGU*, 77, 337–341, 1996.

Blewitt, G., The Global Positioning System, in *Report of the Surrey Workshop of the IAPSO Tide Gauge Bench Mark Fixing Committee*, edited by W. Carter, NOAA Tech. Rep. NOSOES0006, pp. 17–26, Silver Spring, Maryland, 1994.

Blewitt, G., Advances in Global Positioning System technology for geodynamics investigations: 1978–1992, in *Contributions of Space Geodesy to Geodynamics: Techniques*, edited by D. E. Smith and D. L. Turcotte, Crustal Dynamics Series, Vol. 25, pp. 195–213, AGU, Washington, 1993.

Boucher, C., Z. Altamimi, and P. Sillard, The 1997 International Terrestrial Reference Frame (ITRF97), *IERS Tech. Note 27*, Observ. de Paris, 191 pp., 1999.

Breuer, D. and D. Wolf, Deglacial land emergence and lateral upper-mantle heterogeneity in the Svalbard archipelago, I. First results for simple load models, *Geophys. J. Int.*, 121, 775–788, 1995.

Carter, W. E., D. G. Aubrey, T. Baker, C. Boucher, C. LeProvost, D. Pugh, W. R. Peltier, M. Zumberge, R. H. Rapp, R. E. Schutz, K. O. Emery, and D. B. Enfield, Geodetic fixing of tide gauge bench marks, *Tech. Rep. WHOI-89-31, CRC-89-5*, Woods Hole Oceanographic Institution, Coastal Research Center, Woods Hole, Massachusetts, 46 pp., 1989.

Clark, T. A., D. Gordon, W. E. Himwich, C. Ma, A. Mallama, and J. W. Ryan, Determination of relative site motions in the Western United States using Mark III Very Long Baseline Interferometry, *J. Geophys. Res.*, 92, 12,741–12,761, 1987.

Coates, R. J., H. Frey, J. Bosworth, and G. Mead, Space-age geodesy: The NASA Crustal Dynamics Project, *IEEE Trans. GARS*, 23, 360–368, 1985.

Davis, J. L., W. H. Prescott, J. L. Svarc, and K. J. Wendt, Assessment of Global Positioning System measurements for studies of crustal deformation. *J. Geophys. Res.*, 94, 13,635–13,650, 1989.

Dixon, T., G. Blewitt, K. Larson, D. Agnew, B. Hager, P. Kroger, L. Krumega, and W. Strange, GPS measurements of regional deformation in Southern California, some constraints on performance, *EOS Trans. AGU*, 71, 1051–1056, 1990.

Dziewonski, A., and D. L. Anderson, Preliminary Reference Earth Model. *Phys. Earth. Planet. Int.*, 25, 297–356, 1981.

Ekman, M., A concise history of postglacial land uplift research (from its beginning to 1950), *Terra Nova*, 3, 358–365, 1991.

Ekman, M., A consistent map of the post glacial uplift of Fennoscandia, *Terra Nova*, 8, 158–165, 1996.

Ekman, M., and J. Mäkinen, Recent postglacial rebound, gravity change and mantle flow in Fennoscandia, *Geophys. J. Int.*, 126, 229–234, 1996.

Flinn, E. A., The role of NASA in geodynamics research in the decade 1991–2000, in *The interdisciplinary role of space geodesy*, edited by I. I. Mueller and S. Zerbini, Lecture Notes in Earth Sciences, Vol. 22, 257–262, Springer, New York, 1989.

Gordon, R. G., and S. Stein, Global tectonics and space geodesy, *Science*, 256, 333–342, 1992.

Jaldehag, R. T. K., J. M. Johansson, B. O. Rönnäng, P. Elósegui, J. L. Davis, I. I. Shapiro, and A. E. Neill, Geodesy using the Swedish Permanent GPS Network: Effects of signal scattering on estimates of relative site positions, *J. Geophys. Res.*, 101, 17,841–17,860, 1996a.

Jaldehag, R. T. K., J. M. Johansson, J. L. Davis, and P. Elósegui, Geodesy using the Swedish Permanent GPS Network: Effects of snow accumulation on estimates of site positions, *Geophys. Res. Lett.*, 23, 1601–1604, 1996b.

Johansson, J. M., G. Elgered, and B. O. Rönnäng, *The space geodetic laboratory at the Onsala Observatory: Site information*, Tech. Rep. 229, Chalmers University of Technology, School of Electrical and Computer Engineering, Göteborg, Sweden, 145 pp., 1992.

Johansson, J. M., J. L. Davis, H.-G. Scherneck, G. A. Milne, M. Vermeer, J. X. Mitrovica, R. A. Bennett, G. Elgered, P. Elósegui, H. Koivula, M. Poutanen, B. O. Rönnäng, and I. I. Shapiro, Contionuous GPS measurements of postglacial adjustment in Fennoscandia: Geodetic results, *J. Geophys Res.*, 106, in press, 2001.

Kakkuri, J., and M. Poutanen, Geodetic determination of the surface topography of the Baltic Sea, *Marine Geodesy*, 20, 307–316, 1997.

Koivula, H., and M. Poutanen, Postglacial rebound from GPS time series of Finnish permanent GPS network FinnRef, *IAG International Symposium on Recent Crustal Movements (SRCM'01), Abstracts*, edited by M. Poutanen, Helsinki, Finland, August 27–31, 2001.

Koivula, H., M. Ollikainen and M. Poutanen, Use of the Finnish permanent GPS betwork (FinnNet), in *Regional GPS Campaigns*, edited by F. K. Brunner, IAG Symposium, vol. 118, 137–142, Springer, New York, 1998.

Kylkilahti, A., *The Antenna Calibration of the Finnish Permanent GPS Network (FinnRef)*, MSc. Thesis, Helsinki Univ. of Tech., Dept. of Surveying, 1999.

Lambeck, K., C. Smither, and P. Johnston, Sea-level change, glacial rebound, and mantle viscosity for northern Europe, *Geophys. J. Int.*, 134, 102–144, 1998a.

Lambeck, K., C. Smither and M. Ekman, Tests of glacial rebound models for Fennoscandinavia based on instrumented sea- and lake-level records, *Geophys. J. Int.*, 135, 375–387, 1998b.

Lefebvre M., A. Cazenave, P. Escudier, R. Biancale, J. F. Cretaux, L. Soudarin, and J. J. Valette, Space tracking system improves accuracy of geodetic measurements, *EOS Trans. AGU*, *77*, 25–29, 1996.

Mäkinen J., H. Koivula, M. Poutanen, and V. Saaranen, Contemporary postglacial rebound in Finland: Comparison of GPS results with repeated precise levelling and tide gauge results, paper presented at the IAG Meeting GGG2000 (Gravity, Geoid and Geodynamics) Banff, Canada, 2000.

Milne, G. A., J. L. Davis, J. X. Mitrovica, H.-G. Scherneck, J. M. Johansson, and M. Vermeer, Space-geodetic constraints on glacial isostatic adjustment in Fennoscandia, *Science*, *291*, 2381–2385, 2001.

Miyazaki, S., and K. Heki, Crustal velocity field of Southwest Japan: subdution and arc-arc collision, *J. Geophys. Res.*, *106*, 4305–4326, 2001.

Miyazaki, S., Y. Hatanaka, T. Tada, and T. Sagiya, The nationwide GPS array as an Earth observation system (on line), *Bull. Geograph. Survey Inst.*, *44*, 1998.

Mitrovica, J. X., and W. R. Peltier, Pleistocene deglaciation and the global gravity field, *J. Geophys. Res.*, *94*, 13,651–13,671, 1989.

Mitrovica, J. X., and W. R. Peltier, On postglacial geoid relaxation over the equatorial oceans, *J. Geophys. Res.*, *96*, 20,053–20,071, 1991.

Peltier, W. R., Global sea level rise and glacial isostatic adjustment, *Global Planet. Change*, 20, 93–123, 1999.

Plag, H.-P., B. Engen, T. A. Clark, J. J. Degnan, and B. Richter, Post-glacial rebound and present-day three-dimensional deformations, *J. Geodynamics*, *25*, 263–301, 1998.

Poutanen, M. and J. Kakkuri (editors), Final results of the Baltic Sea Level 1997 GPS campaign, *Rep. Finn. Geod. Inst.*, *99*, Helsinki, 182 pp., 1999.

Poutanen, M., and J. Kakkuri, The sea surface topography of the Baltic: A result from the Baltic Sea Level Project (IAG SSC 8.1), in *Geodesy beyond 2000: The challenges of the first decade*, edited by K.-P. Schwarz, pp. 289–294, Springer, New York, 2000.

Poutanen M., H. Koivula, and M. Ollikainen, On the periodicity of GPS time series, *Proc. IAG 2001 Scientific Assembly*, in press, 2001.

Rothacher, M., S. Schaer, L. Mervart, and G. Beutler, Determination of antenna phase center variation using GPS data, in *IGS Workshop Proceedings on Special Topics and New Directions*, edited by G. Gendt and G. Dick, 15–18, GeoForschungsZentrum, Potsdam, Germany, 1995.

Scherneck, H.-G., G. Elgered, J.M. Johansson, and B.O. Rönnäng, Space Geodetic Activities at the Onsala Space Observatory: 25 years in the Service of Plate Tectonics, *Phys. Chem. Earth*, *23*, 811–824, 1998.

Schupler, B., and T. A. Clark, How different antennas affect the GPS observable, *GPS World, Nov/Dec*, 32–36, 1991.

Schupler, B., T. A. Clark, and R. L. Allshouse, Characterizations of GPS user antennas: Reanalysis and new results, in *GPS Trends in Precise Terrestrial, Airborne, and Spaceborne Applications*, edited by G. Beutler, G. W. Hein, W. G. Melbourne, and G. Seeber, 338 pp., Springer, New York, pp. 233–242, 1996.

Segall, P., and J. L. Davis, GPS applications for geodynamics and earthquake studies, *Ann. Rev. Earth Planet. Sci.*, *25*, 301–336, 1997.

Shimada, S. and Y. Bock, Crustal deformation measurements in central Japan determined by a Global Positioning System fixed point network, *J. Geophys. Res.*, *97*, 12,437–12,455, 1992.

Smith, D. E., and D. L. Turcotte (editors), *Contributions of Space Geodesy to Geodynamics: Crustal Dynamics*, Geodyn. Series, Vol. 23, 429 pp., AGU, 1993a.

Smith, D. E., and D. L. Turcotte (editors), *Contributions of Space Geodesy to Geodynamics: Earth Dynamics*, Geodyn. Series, Vol. 24, 219 pp., AGU, 1993b.

Smith, D. E., and D. L. Turcotte (editors), *Contributions of Space Geodesy to Geodynamics: Technology*, Geodyn. Series, Vol. 25, 213 pp., AGU, 1993c.

Stoew, B., G. Elgered, and J. M. Johansson, An assessment of estimates of integrated water vapor from ground-based GPS data, *Meteorol. Atmos. Phys.*, *77*, 99–107, 2001.

Tsuji, H., Y. Hatanaka, T. Sagiya and M. Hashimoto, Co-seismic crustal deformation from the 1994 Hokkaido-Toho-Oki earthquake monitored by a nationwide continuous GPS array in Japan, *Geophys. Res. Lett.*, *22*, 1669–1672, 1995.

vanDam, T., J. Wahr, P. C. D. Milly, A. B. Shmakin, G. Blewitt, D. Lavallée, and K. M. Larson, Crustal displacements due to continental water loading, *Geophys. Res. Lett.*, *28*, 651–654, 2001.

Wessel, P. and W. H. F. Smith, New, improved version of the Generic Mapping Tools released, *EOS Trans. AGU*, *79*, 579, 1998.

Zumberge J. F., M. B. Heflin, D. C. Jefferson, M. M. Watkins, and F. H. Webb, Precise point positioning for the efficient and robust analysis of GPS data from large networks, *J. Geophys. Res.*, *102*, 5005–5017, 1997.

J. L. Davis, Harvard-Smithsonian Center for Astrophysics, 60 Garden Street, Cambridge, Massachusetts, MA 02138.

G. Elgered, J. M. Johansson, and H.-G. Scherneck, Chalmers Centre for Astrophysics and Space Science, SE-439 92 Onsala, Sweden. (e-mail: hgs@oso.chalmers.se)

B. Jonsson and G. Hedling, National Land Survey, SE-801 82 Gävle, Sweden.

H. Koivula, M. Ollikainen, and M. Poutanen, Finnish Geodetic Institute, FI-02431 Masala, Finland.

G. A. Milne, Department of Geological Sciences, University of Durham, South Road, Durham, DH1 3LE, United Kingdom.

J. X. Mitrovica, Department of Physics, University of Toronto, 60 St. George Street, Toronto, Ontario M5S 1A7, Canada.

M. Vermeer, Department of Surveying, Helsinki University of Technology, P.O. Box 1200, FI-02015 Helsinki, Finland.

The European Earth Explorer Mission GOCE: Impact for the Geosciences

P.N.A.M. Visser [1], R. Rummel [2], G. Balmino [3], H. Sünkel [4], J. Johannessen [5], M. Aguirre [6], P.L. Woodworth [7], C. Le Provost [8], C.C. Tscherning [9], and R. Sabadini [10]

GOCE will be the first ESA Earth Explorer Mission with a foreseen launch date in 2004. It will also be the first satellite to fly a capacitive gradiometer operating at room temperature. The mission objective is the production of a homogeneous high-resolution, high-accuracy model of the earth's static gravity field, 1 mgal and 1 cm accuracy for gravity anomalies and geoid heights, respectively, at a resolution of 100 km or less. Impact studies have indicated that with such a model significant advances can be made in the fields of solid-earth physics, ocean circulation, geodesy, sea level change monitoring, ice-sheet modeling and positioning.

1. INTRODUCTION

The geodetic and geophysics communities have strived for a dedicated gravity field mission for many years.

[1]Delft University of Technology, Delft Institute for Earth-Oriented Space Research, Delft, The Netherlands.

[2]R. Rummel, Institut für Astronomische und Physikalische Geodäsie, Technische Universität München, Munich, Germany.

[3]Centre National d'Etudes Spatiales, Toulouse, France.

[4]Technical University Graz, Institute of Theoretical Geodesy, Graz, Austria.

[5]Nansen Environmental & Remote Sensing Centre, Marine Monitoring and Remote Sensing Department, Bergen, Norway.

[6]European Space Research and Technolgy Centre, Noordwijk, The Netherlands.

[7]Proudman Oceanographic Laboratory, Bidston Observatory, Prenton, U.K.

[8]Laboratoire d'Etudes en Geophysique et Oceanographie Spatiales LEGOS/GRGS, Toulouse, France.

[9]University of Copenhagen, Department of Geophysics, Copenhagen, Denmark.

[10]University of Milan, Department of Earth Sciences, Milan, Italy.

Ice Sheets, Sea Level and the Dynamic Earth
Geodynamics Series 29
Copyright 2002 by the American Geophysical Union
10.1029/029GD06

Such a gravity field mission is the only way to obtain a homogeneous global high-accuracy and high-resolution model of the earth's gravity field free from parasitic signals such as contained in e.g. altimeter data. For a long period, dating back more than two decades, several workshops and committees indicated the importance of high-accuracy, high-resolution gravity field mapping [*SESAME*, 1986; *Gravity Workshop*,1987; *NRC*, 1997] and several satellite concepts have been proposed and studied, e.g. GRM [*Keating et al.*, 1986; *Wagner and McAdoo*, 1986] and ARISTOTELES [*ESA*,1991; *Rummel and Schrama*,1991; *Lambeck*,1990]. None of these concepts were selected, either due to immaturity of technology or due to budget constraints or insufficient political support. However, the current situation looks very favorable with the launch of CHAMP on 15 July 2000 and the advent of the GRACE and GOCE gravity field missions [*Reigber et al.*,1996; *Watkins et al.*,1995; *Wahr et al.*,1998; *ESA*,1999b]. Two of these missions, GRACE and GOCE, may be regarded as more mature reincarnations of GRM and ARISTOTELES, respectively.

All three missions are linked with each other because each of them will make use of the same accelerometer technology, although for each mission tuned with respect to sensitivity and measurement bandwidth (see also Section 3). CHAMP may be considered as a proof of concept enabling improvement of current long-

Table 1. Required gravity anomaly and geoid height accuracy (static field) for different applications (claimed EGM-96 global RMS accuracy between brackets for relevant resolution interval)

Application	Accuracy		Spatial resolution (km)
	Geoid (cm)	Gravity (mgal)	
SOLID EARTH			
Lithosphere and upper-mantle density		1-2 (4.8)	100
Continental lithosphere			
• sedimentary basins		1-2 (4.8-8.5)	50-100
• rifts		1-2 (4.8->8.5)	20-100
• tectonic motions		1-2 (0.4-4.8)	100-500
Seismic hazards		1 (4.8)	100
Ocean lithosphere and inter-action with asthenosphere		0.5-1 (2.1-4.8)	100-200
OCEANOGRAPHY			
- short scale	1-2 (30)		100
	0.2 (23)		200
- basin scale	≈ 0.1 (4)		1000
ICE SHEETS			
- Rock basement		1-5 (4.8-8.5)	50-100
- Ice vertical movements	2 (4-30)		100-1000
GEODESY			
- Leveling by GPS	1 (4-30)		100-1000
- Unification of datums	1 (0-30)		100-20000
- Inertial Navigation		≈ 1-5 (0.1-4.8)	100-1000
- Orbits[a]	1 (1-10)		100-1000

[a]Radial orbit error

wavelength static gravity field modeling by an order of magnitude. GRACE will enhance the resolution of this modeling in addition to observing time variability of the long- to medium-wavelength part of the gravity field. The foreseen launch date for GRACE is in the fall of 2001 (status January 2001). GOCE aims at very high resolution mapping of the static gravity field, better than 100 km with an accuracy of 1 mgal and 1 cm in terms of gravity anomalies and geoid heights, respectively. GOCE, Gravity field and steady-state Ocean Circulation Explorer, will be the first Earth Explorer Mission in the Living Planet Programme [ESA,1998b; ESA,1999c] of the European Space Agency (ESA). The foreseen launch is in 2004 (status January 2001).

2. MISSION RATIONALE

Although in the last decade significant progress has been made in the field of global gravity field modeling, culminating in e.g. the EGM-96 [Lemoine et al.,1997] and recent GRIM models [Perosanz et al.,1997; Biancale et al,2000], such models are far from homogeneous in terms of accuracy and resolution. This is due to the accumulation of many data sources based on differ-

ent observation techniques, in different reference frames, with different quality and aliasing problems, and different geographical coverage [ESA,1999b]. Although existing models perform very well in precise orbit determinations of e.g. TOPEX/POSEIDON and the ERS satellite with radial orbit error levels in the 2-5 cm range [Tapley et al.,1994; Perosanz et al.,1997; Scharroo and Visser,1998], their accuracy has to be improved by an order of magnitude for several applications in the geosciences. The latter is corroborated by a comparison of the required gravity field model accuracy for many applications with the globally averaged accuracy of the EGM-96 model (Table 1). Locally, the accuracy can be an order of magnitude worse. In addition, the global average is dominated by probably overly optimistic error estimates for the ocean parts, for which the gravity field model may be contaminated by ocean signals due to the incorporation of satellite radar altimeter data.

It will be obvious that existing gravity field models suffer from a lack of accuracy, homogeneity and contamination with non-gravitational phenomena. The only means of overcoming these deficiences in a reasonable time span and at acceptable costs is a dedicated gravity mission. However, the drawback of flying a satellite to this aim is the attenuation of gravity with altitude,

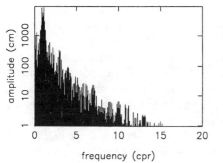 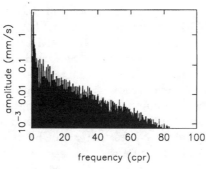 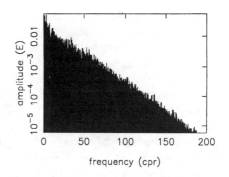

Figure 1. Predicted envelopes for CHAMP (left, altitude perturbations), GRACE (middle, low-low SST) and GOCE (right, Γ_{zz}) gravity field observations based on EGM-96 for a typical orbit altitude of 300 km

which becomes even more pronounced for the higher resolutions.

To ensure the production of a high-accuracy, high-resolution, homogeneous gravity field solution, a gravity mission has to satisfy the following criteria:

- uninterrupted tracking to achieve a more or less homogeneous data distribution and quality;
- the measurement or compensation of non-gravitational forces to prevent contamination;
- an orbit altitude as low as possible to counteract gravity field attenuation with altitude;
- enhance the high resolution part of the gravity field spectrum.

These criteria played a dominant role in the proposed design of the GOCE spacecraft (Section 3).

3. MISSION CONCEPT

The recently launched CHAMP satellite and currently planned GRACE and GOCE missions have different objectives in terms of gravity field sampling. CHAMP will focus on the static long wavelength part (in addition to measuring the geomagnetic field), GRACE on temporal variability and GOCE on the static long to short wavelength gravity field spectrum. This has resulted in three different satellite designs. CHAMP is equipped with a Global Positioning System (GPS) receiver enabling high-precision orbit determination by high-low Satellite-to-Satellite Tracking (SST) to the GPS satellites. Due to its low altitude ($\approx$ 450 km at Begin Of Life, BOL), the CHAMP orbit is perturbed significantly by long wavelength gravity field terms. The GPS receiver will thus provide indirectly the information for modeling the gravity field. However, the CHAMP orbit is perturbed also significantly by atmospheric drag, which can not be modeled with sufficient precision. To overcome this problem, accelerome-

ters are measuring the non-conservative forces allowing a decoupling from gravity field induced perturbations. A typical spectrum of altitude variations that can be derived from the high-low SST observations based on only gravity is displayed in Figure 1 (left). It can be seen that the signal drops by more than three orders of magnitude from 1 cpr (cycles per orbital revolutions, 40,000 km wavelength) to 15 cpr (2700 km wavelength). It can be shown that similar drops occur for the flight and cross-track directions [*Visser*,1992]. CHAMP is also equipped with a Laser Retro-reflector Array (LRA) providing observations that can be combined with the GPS SST observations and/or used for validation purposes. First reports indicate that the GPS receiver and accelerometer are functioning properly, where the resolution of the latter appears to be better than 10^{-9} m/s^2 (priv. comm., G. Balmino, Centre National d'Etudes Spatiales, France). In addition, already a valuable data set of satellite laser tracking observations has been accumulated.

GRACE will basically consist of two CHAMP-type satellites trailing each other at a distance of a few hundreds of kms at about the same altitude ($\approx$ 480 km, BOL). To enhance sensitivity to higher-frequency gravity field perturbations, GRACE will be equipped with a low-low SST device. It can be seen in Figure 1 (middle), that the drop with three orders of magnitude occurs at 80 cpr, or at a wavelength of about 500 km.

The sensitivity to high frequency gravity field perturbations can be further enhanced by adopting a completely new space borne concept, namely Satellite Gravity Gradiometry (SGG). SGG is based on measuring the difference in acceleration of two adjacent proof masses, in this case on board of one and the same satellite. Such a difference measurement is in a very good approximation equal to the second derivative of the local gravity field potential, or the local gravity gradient (denoted by Γ). By taking multiple derivatives, high frequencies are magnified. In this case, the radial SGG component (Γ_{zz}) decays with three orders of magnitude at about 200 cpr (100 km wavelength, Figure 1, right). Conceptually, SGG is thus superior to both high-low and low-low SST when it comes to observing the fine structure of the earth's gravity field. GOCE will be equipped with an electrostatic gradiometer working at room temperature consisting of a triad of three pairs of three-axes accelerometers located on three orthogonal axes with a baseline of about 0.5 m [ESA,1999b]. The gradiometer instrument will be tuned to be particularly sensitive in the 1-100 mHz frequency range, referred to as the Measurement Bandwidth (MB), for which the measurement precision aimed at is 3 mE/$\sqrt{Hz}$ (1 E = 1 Eötvös Unit = 10^{-12} s^{-2}) for the differential accelerometer measurements. This is equivalent to allocating an error budget of 10^{-12} m/s^2 per individual accelerometer for the differential mode where the range must be below 10^{-7} m/s^2 [ESA,1999b]. A trade-off had to be made between sensitivity and dynamic range of the accelerometers, resulting in the specified MB and measurement precision. Outside this MB, the gradiometer will measure with reduced precision. The selected altitude will be 240-250 km, which means that the MB in the frequency domain is equivalent with 80-8000 km in the space domain (half-wavelength). Each accelerometer will have two sensitive and one less-sensitive axes due to the on ground testing in a 1-g environment. The three pairs are in principle able to provide the full SGG tensor. However, the requirement is to provide the three diagonal components only. The off-diagonal components will be used to reconstitute with very high precision the rotational motion of the satellite to eliminate centrifugal and angular acceleration terms from the SGG observations [Aguirre-Martinez,1999].

To take optimal advantage of the information content of the SGG measurements, the position of the instrument has to be known with high precision. For example, a misfit in position of 1 m can lead to an increase of the SGG error budget with about 1 mE due to the central term of the gravity field. Therefore, GOCE will be equipped with a high-quality, dual-frequency GPS receiver. As with CHAMP and GRACE, this will also allow a recovery of the long-wavelength part of the gravity field, i.e. the part for which the gradiometer is less sensitive. The GPS receiver thus has a dual role: provide the SST tracking measurements for a very precise geolocation of the gradiometer instrument (1) and for a precise long-wavelength, complementary, gravity field recovery (2). Both the gradiometer and GPS receiver are able to provide continuous measurements with 1 sec-

ond time interval (1 Hz). Also GOCE will be equipped with a LRA that can be used for validation purposes and partial backup to the GPS receiver.

So far, three of the mission criteria have been met: uninterrupted tracking, low orbit altitude and enhancement of high frequencies. The fourth criterion, elimination or measurement of non-gravitational forces can also be met with the current design. The gradiometer will be able to provide measurements of the non-conservative forces by evaluating the common-mode of the accelerometers. In addition, the implementation of a Drag Free Control (DFC) system is foreseen in the current design in order to prevent saturation of the accelerometers. For example, atmospheric drag will be at a level of 8×10^{-6} m/s^2 compared to a dynamic range of 10^{-7} m/s^2 for the accelerometers along the sensitive axes. The DFC system will consist of ion thrusters to eliminate the large, long-wavelength, components of non-conservative forces (predominantly atmospheric drag), and cold gas proportional thrusters to eliminate the larger part of these perturbations in the gradiometer MB.

The foreseen mission life time for GOCE is 20 months, consisting of a commissioning phase of 3 months and two 6-months measurement periods with a 5-months hibernation period in between during which the satellite experiences relatively large temperature fluctuations due to eclipses. The selected orbit will be a dawn-dusk or dusk-dawn sun-synchronous orbit, with is near-polar. The inclination of the orbit will be 96.6°, i.e. small polar caps will not be covered with observations amounting to less than 1% of the total earth's surface, referred to as the polar gaps. For comparison, CHAMP and GRACE will fly near-polar orbits with an inclination of 89°, filling the gaps left by GOCE for the larger part, although with reduced sensitivity to the high-frequency part of the polar gravity field. Although the satellites will not fly exactly over the poles, the instruments will provide information about the gravity field in the gaps as well, especially the gradiometer due to its 3-dimensional sensing capability [Sünkel,2000]. Finally, it is interesting to note that airborne gravimetry campaigns are planned over the Arctic region that will further complement the GOCE gravity data set (priv. comm., R. Forsberg, Geodynamics Dept., Kort & Matrikelstyrelsen, Denmark).

4. MISSION PERFORMANCE

The proposed GOCE satellite will contain an electrostatic gradiometer, a GPS receiver and a DFC system. An important role in the design was played by

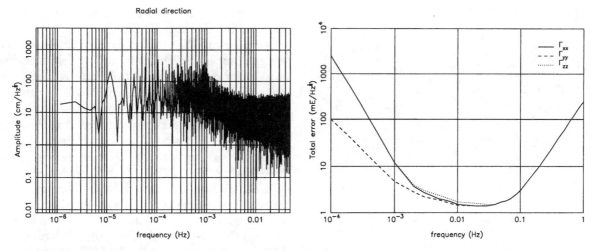

Figure 2. Power Spectral Densities for GOCE radial orbit error (left) and error budget for diagonal SGG components (right). The orbit error integrated over the entire spectrum is 2.5 cm (x, y and z denote the along-track, cross-track and radial direction, respectively)

several mission analysis and error propagation tools to check the mission performance both at the instrument level and in terms of achievable gravity field products [*Alenia*,1998; *ESA*,1998a; *SID*,2000; *CIGAR II*,1990; *CIGAR III*,1993; *CIGAR III*,1995].

The achievable gravity field recovery accuracy and resolution for GOCE depend on the quality of the GPS-based precise orbits and the error budget for the diagonal SGG components. Detailed studies have been conducted to predict the error budgets both for precise orbit determination and SGG measurements [*ESA*,1999b; *SID*,2000; *Visser and van den IJssel*,2000]. For example, estimates of the radial orbit error spectrum based on a kinematic orbit determination approach and the SGG error budget for the diagonal components are displayed in Figure 2. The expected orbit accuracy is at the few cm level, leading to effectively no increase of the SGG error budget due to position uncertainty (with respect to an earth-fixed reference frame) of the gradiometer instrument.

The expected orbit accuracy spectra and SGG error budgets were fed to an error propagation tool to assess the achievable gravity field accuracy. The gravity field is conveniently modeled as a spherical harmonic expansion:

$$U = \frac{\mu}{r} \left\{ 1 + \sum_{l=2}^{\infty} \sum_{m=0}^{l} \left(\frac{a_e}{r} \right)^l \left(\bar{C}_{lm} \cos m\lambda \right. \right.$$
$$\left. \left. + \bar{S}_{lm} \sin m\lambda \right) \bar{P}_{lm}(\sin \phi) \right\} \qquad (1)$$

where μ is the gravity parameter of the earth (the product of the universal gravity constant G and the mass of the earth M), a_e the mean equatorial radius, $\bar{P}_{lm}$ is the fully normalized Legendre polynomial of degree l and order m, $\bar{C}_{lm}$ and $\bar{S}_{lm}$ denote the fully normalized gravity field harmonic or Stokes coefficients. This series is truncated at a certain maximum degree l_{max}. The resolution of such a model is $40,000/l_{max}$ km. With the error propagation tool, the accuracy of the spherical harmonic coefficients can be predicted as a function of the orbit and SGG error spectra [*Colombo*,1984; *Schrama*,1991; *Visser et al.*,1994].

The predicted performance is displayed in Figure 3. The prediction is valid for the area of the earth covered by the ground track of the satellite, *e.g.* for GOCE between -84° and +84° geographical latitude (two polar caps forming less than 1% of the earth's surface are not covered by observations). For comparison, predictions for GRACE and CHAMP are included plus the EGM-96 covariance and the degree variance according to Kaula's rule of thumb [*Kaula*,1966]:

$$\bar{C}_{lm}, \bar{S}_{lm} \div \frac{10^{-5}}{l^2} \qquad (2)$$

It is assumed that the orbit error spectra for all missions look similar, because in all cases use will be made of GPS receivers that are more or less developed in the same time frame for which it is fair to assume similar observation error characteristics. In addition, for all missions orbit determination uncertainty due to nonconservative forces is effectively minimized by the use of accelerometers [*SID*,2000; *Visser and van den IJssel*,2000]. For GRACE, it was also assumed that low-low SST Doppler measurements are available with a

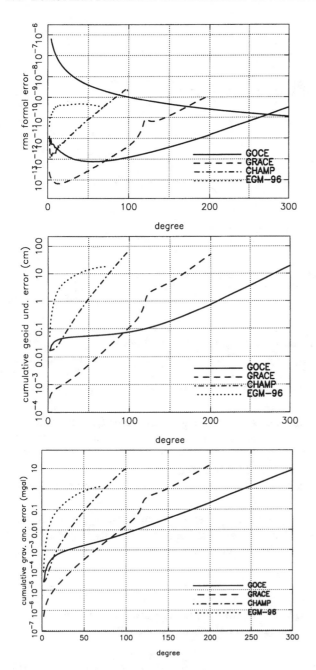

Figure 3. Predicted achievable gravity field recovery as a function of the spherical harmonic degree: degree Root-Mean-Square (RMS ,top), cumulative geoid error (middle) and cumulative gravity anomaly error (bottom). The decreasing line denotes gravity signal variance according to Kaula's rule of thumb

claimed precision of 1 μm/s over the entire MB (no degradation at low and very high frequencies as assumed for GOCE) with an inter-satellite distance of 300 km [*ESA*,1998*a*]. The foreseen mission life times for

CHAMP and GRACE are 5 years compared to 1 year (measurement phase) for GOCE.

According to the predictions, all missions will lead to a significant improvement in gravity field modeling, at least a few orders of magnitude improvement over the state-of-the-art global model EGM-96. It can be seen that the error curve of the latter flattens around degree 20. At higher degrees the error level is reduced significantly by the inclusion of surface gravity and altimeter data [*Lemoine et al.*,1997]. No such data were used in the CHAMP, GRACE and GOCE predictions.

It can also be seen that the GOCE concept is superior at the medium to small wavelengths and GRACE at the long wavelengths: the intersection point is at degree 70 (half-wavelength 285 km). The exceptionally high accuracy predicted for GRACE at the low degrees opens the possibility to generate time series of gravity field solutions, *e.g.* per month with of course lesser accuracy than the 5-year static solution, and study the time variability of the gravity field at the long(er) wavelengths. For GOCE, the signal to noise ratio becomes one around degree 270 or a resolution of about 75 km. For CHAMP, also an improvement of at least an order of magnitude is expected at the long wavelengths. In addition, CHAMP may be considered to be a 'proof of concept', testing accelerometer technology that will also be used for GRACE and GOCE in combination with a high-quality GPS receiver. By comparing Figure 3 with Table 1, it can be seen that most science requirements can be met by GOCE.

5. MISSION PRODUCTS

The scientific instruments on board of GOCE will provide a continuous data stream of GPS SST and SGG observations. In addition, ancillary data are required from the Attitude and Orbit Control System (AOCS) including the DFC and external data from *e.g.* the International GPS Service (IGS) to allow a precise orbit determination. An estimated guess for the total amount of GOCE SST and SGG observations is more than 60 million (for the two 6-months observation intervals). A gravity field model with a resolution of 75 km requires the estimation of more than 270^2 or 72,900 unknowns. Reducing the scientific observations to the gravity field products is a real challenge from a computational point of view. However, sophisticated algorithms have been developed and implemented allowing to conduct this task in a reasonable time frame at acceptable (computing) costs [*CIGAR IV*,1996; *Sünkel*,2000].

It has to be stressed that the primary objective of GOCE is the generation of a high-resolution, high-

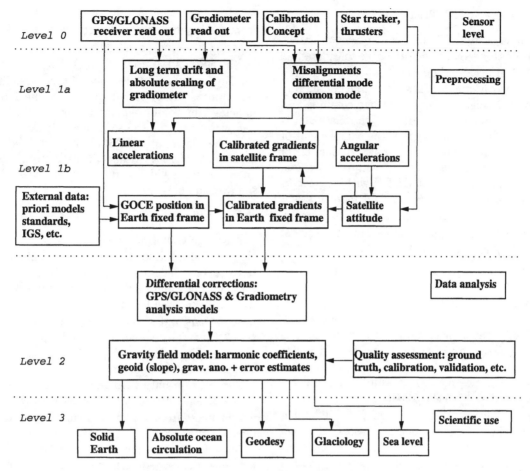

Figure 4. Scheme for GOCE data reduction

accuracy model of the static gravity field. Due to the specific mission scenario, *i.e.* a one-year observation period interrupted by 5 months of hibernation, the time variable gravity field will not average out, especially those signals with a dominant annual cycle. Correcting for time variable gravity field signals will form part of the processing chain [*Sünkel*,2000]. In relation to this issue, it is foreseen that the GOCE data processing will benefit strongly from the GRACE results.

Several stages can be distinguished in the processing of the GOCE observations to the final gravity field products (Figure 4, [*ESA*,1999*b*]). Raw sensor data (*level 0*) of the scientific instruments (GPS receiver and gradiometer) plus ancillary data from the AOCS/DFC (including star tracker and thruster activity data) will be converted to calibrated time series of SST and SGG observations (*level 1a*) and geolocated (*level 1b*) after a preliminary orbit determination making use of external IGS data. Finally, the data will be reduced to a gravity field model together with a quality estimate (*level 2*).

Part of the latter process is the computation of the most precise GOCE orbit possible. The gravity field product may be in the form of a set of spherical harmonic coefficients, a local or global grid of geoid height or gravity anomaly values, a grid of geoid slopes, etc. A quality measure will be attached to the different products. It has to be noted that the calibrated SGG measurements may already be seen as an important geophysical product that can be used directly for certain applications. The production of the *level 0* to *level 1b* products will be taken care of by an assigned Processing and Archiving Facility (PAF). It is foreseen that the reduction to the final gravity field products (*level 2*) will be taken care of by a scientific data consortium that may consist of several participating groups.

The gravity field products will be provided to the scientific community to be used for and incorporated in many applications in *e.g.* the fields of solid-earth research, oceanography, geodesy, glaciology and sea level studies, referred to as *level 3* (Section 6).

6. SCIENTIFIC USE

Gravity plays a dual role in the geosciences. First, in the form of the geoid, which may be considered as the hypothetical ocean surface at rest and serves as a reference for ocean circulation studies and for linking local and global height systems into one common reference frame. Second, as a mirror of the mass structure of the earth's interior, which is the complement of many processes like sea-floor spreading, subduction of oceanic lithosphere, glacial isostatic readjustment, etc. The predicted performance of GOCE with respect to resolution and accuracy in gravity field modeling will allow meaningful applications in many research areas. An inventory was made of the possible impact in several scientific fields, which is highlighted below.

6.1. Solid Earth

In solid-earth research, a high-resolution, high-accuracy gravity field model can serve as an important boundary condition. Gravity field data will enhance images of the density structure of the lithosphere and upper mantle in combination with seismic tomography data, lithospheric magnetic-anomaly measurements and topography data, e.g. [Achache,1994].

Precise knowledge of the density structure will help in improved modeling of e.g. sedimentary basins, rifts, tectonic motions and sea/land vertical motions. Furthermore, high-precision knowledge of the density anomalies in the earth will improve the understanding of the tectonic processes and mechanisms behind earthquakes, e.g. [Negredo et al.,1999; ESA,1999b].

6.2. Ocean Circulation

Also in the case of ocean circulation studies, gravity plays at least a dual role. First and direct in the form of the geoid, the equipotential surface that serves as a reference for ocean dynamics. Second and indirect in the form of orbit perturbations that have to be modeled with high precision to allow the use of satellite radar altimeter data for many applications in the field of oceanography.

A high-precision geoid model will lead to significantly improved and more detailed modeling of ocean currents, leading to reduced uncertainties in volume transports, especially in the upper ocean layers [LeGrand and Minster,1999; Woodworth et al.,1998]. Ocean dynamic modeling plays an important role in modeling the earth's energy/heat budget, transport of nutrients (fishing) and weather prediction.

With CHAMP, GRACE and GOCE the so-called gravity field induced orbit error will become insignificant. Already a valuable altimeter data set consists covering a period of more than two decades, collected by the SEASAT, GEOSAT, TOPEX/POSEIDON and the ERS satellites. The orbits of these satellites can be recomputed using post-flight gravity field models enhancing the point to point accuracy of the altimeter measurements. At the same time, it may be expected that the recomputed orbits will be defined in a more consistent reference frame (Section 2). Also sea level change studies will benefit from improved orbit modeling (Section 6.5).

6.3. Ice Sheets

Although GOCE will cover the larger part of the Arctic and Antarctic ice sheets, the remaining polar gaps have to be filled in to guarantee high-precision gravity field modeling for these areas. However, the GOCE data will be complemented by CHAMP and GRACE observations and airborne gravimetry campaigns (Section 3). It is fair to assume that this complement of data sources will result in high-precision Arctic and Antarctic gravity field modeling.

A precise gravity field over the Arctic and Antarctic ice sheets will, in combination with information on ice thickness from e.g. in-situ surveys, result in better models of the underlying bedrock topography resulting in improved knowledge of ice sheet dynamic behavior, e.g. mass fluxes, especially in the 50-100 km resolution domain. Improved ice dynamics modeling also forms part of the sea level change equation (Section 6.5).

In addition, a precise geoid in the polar areas will enhance geodetic surveying of the ice sheets by e.g. GPS leveling [Roman et al.,1997]. Furthermore, precise gravity field knowledge in the polar areas will practically eliminate the relating orbit error of future altimetric missions facilitating to a larger extent the use of ice topographic data that are collected by such missions, of which two are currently foreseen: Ice, Cloud and land Elevation Satellite (ICESat) and CRYOSAT. ICESat, which will carry the Geoscience Laser Altimeter System (GLAS), forms part of NASA's Earth Science Enterprise (ESE) scheduled for launch in July 2001 [Schutz,1998]. CRYOSAT, which has been selected as the pioneering Earth Explorer opportunity mission in the ESA's new Living Planet program, is scheduled for launch in 2002 [ESA,1999a].

6.4. Geodesy

The field of geodesy encompasses many research and application areas where gravity plays a crucial role. A number of applications will benefit significantly from having a global gravity field model with 1 cm accuracy in terms of geoid heights and 1 mgal accuracy in terms of gravity anomalies at 100 km spatial resolution: leveling by GPS in addition to or replacement of traditional leveling techniques, unification of (local) height systems in order to define one globally consistent datum, orbit determination of satellites and inertial navigation.

Geometric heights determined by GPS can be converted to heights above sea level ('orthometric heights') with the aid of an accurate geoid model [*Rummel*,1992]. This is similar to the extraction of dynamic ocean topography from satellite altimetry in combination with a precise geoid.

In well surveyed areas, predominantly in Europe, North America, Japan and Australia, the GOCE geoid can be combined with very high frequency local terrestrial gravity data resulting in cm-precision local geoids down to a resolution of 5 km. In less surveyed areas, predominantly in developing countries, GPS converted orthometric heights can be obtained free of long-scale biases with sufficient accuracy to satisfy mostly less demanding local needs. Due to missing local gravity information, small scale omission errors of the order of 10-20 cm have to be added [*ESA*,1999*b*]. In general, it may be concluded that height determinations can be conducted faster and at lower cost.

The geoid precision aimed at with GOCE will enable connection of all height systems with cm-precision in one consistent global reference frame, provided that at least one location in each separate system can be positioned with high accuracy using a space based positioning technique such as GPS [*Arabelos and Tscherning*,1999]. Unification of heigh systems allows to bring all sea level recordings into one system (Section 6.5), eliminate height discontinuities between adjacent islands and remove existing biases in terrestrial gravity anomaly data sets.

In precise orbit determination of earth orbiting satellites, gravity field induced orbit errors will become negligible using post-flight GOCE models (the same will be true for post-flight GRACE and to a lesser extent CHAMP models). Precise orbit determinations will not only improve for altimetric satellites (Sections 6.2-6.3), but also for atmospheric profiling missions like the future METOP satellites of which the first will be launched in 2003 (ESA Press Release 06/99, 8 July 1999). The latter missions require (near) real time precise orbits for operational application in *e.g.* numerical wheather prediction models. Improved gravity field knowledge will result in more accurate near real time and predicted orbits to be included in operational applications. Finally, modeling of non gravitational orbit perturbations, such as induced by atmospheric drag or solar radiation, but also modeling of temporal gravity field induced orbit perturbations, *e.g.* caused by tides, will benefit from improved knowledge of the static gravity field.

Inertial navigation is based on single and double integration of measured accelerations to obtain position and velocity changes of a user, *e.g.* land vehicles, aircraft, missiles, submarines, etc. Attitude changes can be derived by making use of gyro's. The accelerometers and gyro's are either mounted on space-stable (or leveled) platforms or fixed to the vehicle to be navigated. The accelerometers measure the sum of the user vehicle and gravity acceleration and precise knowledge of the gravity field will improve overall navigation accuracy and allow an increase in time intervals between velocity and positioning updates [*Schwarz*,1981].

6.5. Sea Level

Sea level change is an aggregate of many different phenomena related to solid-earth dynamics (section 6.1), ocean current systems (section 6.2), ice sheet evolution (section 6.3) and height systems (section 6.4). Different mechanisms may play a role in *e.g.* local sea level change [*Di Donato et al.*,1999]. A proper understanding of the various components of and mechanisms behind sea level change plays a crucial role in climate (change) studies and modeling.

Globally averaged sea level is estimated to have risen by 10-25 cm in the past century and certain predictions indicate an additional rise of approximately another 50 cm in the next century [*Warrick et al.*,1996]. With two thirds of the world's population living in coastal zones, some of which will already have significant elevated risk of flooding with sea level rises of a few decimeters, understanding and being able to predict sea level change is of great importance.

In order to be able to improve and enhance the value of sea level change predictions, it is not sufficient to simply observe total sea level. It is required to understand the various distinct components if accurate predictions are to become available. Moreover, historical tide gauge records and currently available climate models suggest that sea level change has been and will be far from globally uniform [*Warrick et al.*,1996; *Peltier*,1998; *Di Donato et al.*,2000].

Improved gravity field knowledge will aid in improved understanding of (several components of) sea level changes. First, high accuracy geoid modeling will lead to

Table 2. Geographically correlated and anti-correlated radial orbit error for several altimeter satellites (cm)

Satellites	EGM-96		GOCE	
	Correlated	Anti-Corr.	Correlated	Anti-Corr.
ERS-1/2 and ENVISAT	2.24	1.89	0.08	0.08
GEOSAT and GFO	2.51	1.89	0.14	0.13
TOPEX and Jason	0.67	0.58	0.08	0.07
CRYOSAT	7.46	5.52	0.03	0.03

more reliable estimates of ocean and heat fluxes in General Circulation Models (GCMs) that are used in the modeling of sea level change due to thermal expansion. It is expected that thermal expansion is expected to contribute significantly to sea level change in the next century. Second, precise geoid models for the Arctic and Antarctic areas will lead to improved models of ice sheet dynamics. Third, improved gravity field knowledge will allow a better analysis of historical tide gauge records which basically are measures of local sea level with respect to local land level. Different phenomena define changes in local sea level, ranging from *e.g.* changes in ocean circulation to changes in local solid-earth processes. Interpretation of tide gauge records will benefit from improved ocean dynamics and solid-earth modeling. Fourth, unification of height systems will enable comparison of local sea level records in one consistent global reference frame. Fifth and finally, improved gravity field knowledge, especially of the larger wavelengths, will lead to a reduction of the radial orbit error for altimeter satellites. Although already much progress has been made in reducing this error for satellites like TOPEX/POSEIDON and ERS-1/2 [*Tapley et al.,*1994; *Scharroo and Visser,*1998], further reductions are required to improve sea level change estimates based on altimeter measurements and bring the uncertainty significantly below the signal level.

(Re)Computation of the orbits of previous (GEOSAT, ERS-1), current (ERS-2, TOPEX/POSEIDON, GFO) and future (Jason-1, ENVISAT) satellites will result in a multi-decadal record of sea level change estimates of high quality.

The radial orbit error based on the state-of-the-art EGM-96 gravity field model has been assessed for different altimeter missions and serves as an example to indicate the importance of its reduction (Table 2). For GEOSAT, the radial orbit error has an RMS of 2.5 cm. In order to be able to derive sea level change estimates at the mm/year accuracy level (10 cm per century) for a time span of a few decades (GEOSAT flew in the mid eighties), this error has to be reduced to the sub-cm level. With a post-flight GOCE gravity field model this

accuracy level can be achieved. Also note the relatively large radial orbit error for CRYOSAT when using EGM-96. This is due to its almost polar orbit with an inclination of 92°. Gravity field induced orbit perturbations for inclinations close to 90° are very poorly represented in existing models.

As indicated before, sea level changes are expected to be far from globally uniform. This is also true for the gravity field induced radial orbit error. For example, variations in the geographically correlated (average of error on ascending and descending satellite tracks) and anti-correlated (error on ascending minus error on descending track) radial orbit error are of the order of a few cm for the ERS and ENVISAT satellites with EGM-96 (Figure 5). Such errors will result in sea level change estimates that have different errors for different local areas when for example linking altimeter data sets of different altimeter satellites with different radial orbit error spectra that flew in different periods. When comparing *e.g.* GEOSAT and ERS altimeter data, the error can be larger than 5 mm/year.

7. CONCLUSIONS

The first decade of the 21th century, a major step forward will be enabled in the field of geopotential research with the advent of three dedicated satellite gravity missions, GOCE, GRACE and CHAMP, where CHAMP will also measure the geomagnetic field. New technologies, from new generation high-precision GPS receivers to low-low microwave Doppler tracking instruments, ultra-sensitive (arrays of) accelerometers and high-resolution atmospheric drag compensation systems, have been developed that will enable measuring gravity in a space borne environment over a wide wavelength spectrum.

It is expected that CHAMP will provide observations that enable an improvement in gravity field modeling by an order of magnitude over existing models at the long wavelengths (down to ≈ 500 km). The mission can also be seen as a proof of concept of using high-sensitivity accelerometers in combination with GPS (and LRA)

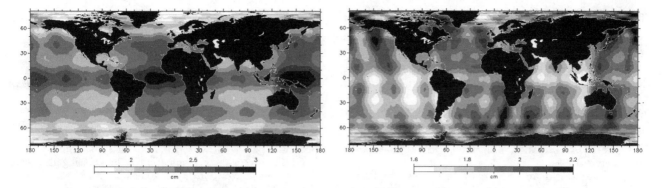

Figure 5. Geographically correlated (left) and anti-correlated (right) radial orbit error for the ERS and ENVISAT satellites based on the EGM-96 calibrated covariance

tracking. The expected performance of GRACE will allow the generation of very precise monthly long to medium wavelength gravity field solutions opening the possibility to study the time variability of the gravity field at these wavelengths. Moreover, GRACE will provide the information for high-precision modeling of the static gravity field as well with unprecedented resolution and accuracy: the gravity signal to noise ratio is expected to reach one at a spherical harmonic degree around 170 (half-wavelength 120 km). The focus of GOCE will be on achieving as high a resolution as possible in modeling the static gravity field. The expected gravity signal to noise ratio is expected to reach one at a degree around 270 (half-wavelength 75 km). The expected accuracy for a gravity field model complete to degree and order 200 (half-wavelength 100 km) is better than the 1 mgal and 1 cm for gravity anomalies and geoid heights aimed at.

The GOCE data products will consist of calibrated and validated gravity field models with associated quality estimates. The models will be provided in several forms: sets of spherical harmonic coefficients with associated (reduced) error/covariance matrices, or local/global grids of gravity anomalies, geoid heights, geoid slopes, etc., with associated error/covariance functions. The GOCE gravity field solutions will be used in a wide field of applications and scientific research.

Significant progress is anticipated in the fields of solid earth research, ocean circulation modeling, ice sheet dynamics, geodesy and the strongly multidisciplinary field of sea level change studies.

Acknowledgments. The authors would like to acknowledge the valuable contributions from a European-wide conglomerate of research institutes, the European Space Agency and an industrial consortium led by Alenia Aerospazio, Turin, Italy.

REFERENCES

Achache, J., Magnetic field, *Report of the ESA Earth Observation User Consultation Meeting, Noordwijk, SP-186*, European Space Agency, 1994.

Aguirre-Martinez, M., Derivation of the satellite gravity gradient observables and recovery of the centrifugal terms in GOCE, ESA-ESTEC working paper EWP-2033, 1999.

Alenia, Gravity Field and Ocean Circulation Explorer (GOCE), End-to-End Simulation Study Summary Report, *ESA Contract 11404/95/NL/CN*, Alenia Aerospazio, November 1998.

Arabelos, D., and C.C. Tscherning, Vertical datum control using GOCE mission error model, *Geophys. Res. Abs.*, *1*(1), 236, 1999.

Biancale, R., G. Balmino, J.-M. Lemoine, J.-C. Marty, and B. Moynot, A new global Earth's gravity field model from satellite orbit perturbations: GRIM5-S1, *Geophys. Res. Lett.*, *27*(22), 3611–3614, 2000.

CIGAR II, Study on precise gravity field determination and mission requirements, *Part 2: Final Report, ESA Contract 8153/88/F/FL*, February 1990.

CIGAR III, Study of the gravity field determination using Gradiometry and GPS, *Phase 1 - Final Report, ESA Contract 9877/92/F/FL*, May 1993.

CIGAR III, Study of the gravity field determination using Gradiometry and GPS, *Phase 2 - Final Report, ESA Contract No. 10713/93/F/FL*, January 1995.

CIGAR IV, Study of advanced reduction methods for spaceborne gravimetry data, and of data combination with geophysical parameters, *CIGAR IV Final Report, ESA Contract No. 152163 - ESTEC/JP/95-4-137/MS/nr*, August 1996.

Colombo, O. L., *The Global Mapping of Gravity With Two Satellites*, vol. 7, no. 3, Netherlands Geodetic Commission, Publications on Geodesy, New Series, 1984.

Di Donato, G., A.M. Negredo, R. Sabadini, and L.L.A. Vermeersen, Multiple Processes causing Sea-Level Rise in the central Mediterranean, *Geophys. Res. Lett.*, *26*(12), 1769–1772, June 1999.

Di Donato, G., L.L.A. Vermeersen, and R. Sabadini, Sea-Level changes, geoid and gravity anomalies due to Pleistocene deglaciation by means of multilayered, analytical Earth models, *Tectonophysics*, *320*, 409–418, 2000.

ESA, The Solid-Earth Mission Aristoteles, Proceedings of an International Workshop, Anacaori, Italy, ESA SP-329, 1991.

ESA, European Views on Dedicated Gravity Field Missions: GRACE and GOCE, *An Earth Sciences Division Consultation Document, ESD-MAG-REP-CON-001*, European Space Agency, May 1998a.

ESA, The Science and Research Elements of ESA's Living Planet Programme, *SP-1227*, European Space Agency, October 1998b.

ESA, Earth Observation Quarterly, ISSN 0256-596X, September 1999a.

ESA, Gravity Field and Steady-State Ocean Circulation Mission, *Reports for Mission Selection, The Four Candidate Earth Explorer Core Missions, SP-1233(1)*, European Space Agency, July 1999b.

ESA, Introducing the Living Planet Programme - The ESA Strategy for Earth Observation, *SP-1234*, European Space Agency, May 1999c.

Gravity Workshop, Geophysical and geodetic requirements for global gravity field measurements 1987-2000, report of a gravity workshop Colorado Springs Geodynamics Branch, Division of Earth science and applications, NASA, February 1987.

Kaula, W. M., *Theory of Satellite Geodesy*, Blaisdell Publishing Co, Waltham, Massachusetts, 1966.

Keating, T., P. Taylor, W. Kahn, and F. Lerch, Geopotential Research Mission science, engineering and program summary, *NASA Tech. Memo. 86240*, 1986.

Lambeck, K., Aristoteles: An ESA Mission to study the Earth's Gravity Field, *ESA Journal, 14*, 1-22, 1990.

LeGrand, P., and J.-F. Minster, Impact of the GOCE gravity mission on ocean circulation estimates, *Geophys. Res. Lett., 26*(13), 1881-1884, July 1999.

Lemoine, F. G., *et al.*, The development of the NASA GSFC and NIMA Joint Geopotential Model, in *International Association of Geodesy Symposia, Gravity, Geoid and Marine Geodesy*, vol. 117, pp. 461-469, Springer-Verlag, Berlin, 1997.

Negredo, A. M., E. Carminati, S. Barba, and R. Sabadini, Dynamic modeling of stress accumulation in central Italy, *Geophys. Res. Lett., 26*(13), 1945-1948, July 1999.

NRC, *Satellite Gravity and the Geosphere: Contributions to the Study of the Solid Earth and Its Fluid Envelopes*, Committee on Earth Gravity from Space, National Research Council, National Academy Press, ISBN 0-309-05792-2, 1997.

Peltier, W.R., Postglacial variations in the level of the sea: implications for climate dynamics and solid-earth geophysics, *Rev. Geophys., 36*(4), 603-689, 1998.

Perosanz, F., J.C. Marty, and G. Balmino, Dynamic orbit determination and gravity field model improvement from GPS, DORIS and laser measurements on TOPEX/POSEIDON satellite, *Journal of Geodesy, 71*(3), 160-170, Feb. 1997.

Reigber, Ch., R. Bock, Ch. Förste, L. Grunwaldt, N. Jakowski, H. Lühr, P. Schwintzer, and C. Tilgner, CHAMP Phase B - Executive Summary, G.F.Z. STR96/13, 1996.

Roman, D.R., B. Csatho, K.C. Jezek, R.H. Thomas, W.B. Krabill, R. von Frese, and R. Forsberg, A comparison of geoid undulations for west central Greenland, *J. Geophys. Res., 102*(B2), 2807-2814, 1997.

Rummel, R., Vertical datum control using GOCE mission error model, *Geodetical Info Mag., 6*(8), 52-56, 1992.

Rummel, R., and E.J.O. Schrama, Two Complementary Systems On-board 'Aristoteles': Gradio and GPS, *ESA Journal, 15*, 135-139, 1991.

Scharroo, R., and P. N. A. M. Visser, Precise orbit determination and gravity field improvement for the ERS satellites, *J. Geophys. Res., 103*(C4), 8113-8127, 1998.

Schrama, E.J.O., Gravity field error analysis: applications of GPS receivers and gradiometers on low orbiting platforms, *J. Geophys. Res., 96*(B12), 20,041-20,051, 1991.

Schutz, B., Spaceborne laser altimetry: 2001 and beyond, in *Book of Extended Abstracts, WEGENER-98, Norwegian Mapping Authority, Honefoss, Norway*, edited by H.P. Plag, 1998.

Schwarz, K.P., Gravity induced position errors in airborne inertial navigation, *Report No. 326*, Department of Geodetic Science and Surveying, Ohio State University, Columbus, 1981.

SESAME, Solid Earth Science & Application Mission for Europe, ESA - Special Workshop, ESA SP-1080, 1986.

SID, GOCE End-to-end Closed Loop Simulation, *Final Report, ESTEC Contract No. 12735/98/NL/GD*, SRON/DEOS/IAPG, January 2000.

Sünkel, H. et al., From Eötvös to mGal, *Final Report, ESA/ESTEC Contract No. 13392/98/NL/GD*, April 2000.

Tapley, B. D., J.C. Ries, G.W. Davis, R.J. Eanes, B.E. Schutz, C.K. Shum, M.M. Watkins, J.A. Marshall, R.S. Nerem, B.H. Putney, S.M. Klosko, S.B. Luthcke, D. Pavlis, R.G. Williamson, and N.P. Zelensky, Precision orbit determination for TOPEX/POSEIDON, *J. Geophys. Res., 99*(C12), 24,383-24,404, 1994.

Visser, P. N. A. M. (1992), *The Use of Satellites in Gravity Field Determination and Model Adjustment*, Ph.D. dissertation, Delft Univ. of Technol., Delft, Netherlands, September 1992.

Visser, P. N. A. M., K. F. Wakker, and B. A. C. Ambrosius, Global gravity field recovery from the ARISTOTELES satellite mission, *J. Geophys. Res., 99*(B2), 2841-2851, 1994.

Visser, P.N.A.M., and J. van den IJssel (2000), GPS-based precise orbit determination of the very low Earth orbiting gravity mission GOCE, *J. Geod., 74*(7/8), 590-602.

Wagner, C.A., and D.C. McAdoo, Time Variations in the Earth's Gravity Field Detectable With Geopotential Research Mission Intersatellite Data, *J. Geophys. Res., 91*(B8), 8373-8386, July 1986.

Wahr, J., M. Molenaar, and F. Bryan, Time-variability of the Earth's gravity field: Hydrological and oceanic effects and their possible detection using GRACE, *J. Geophys. Res., 103*(B12), 30,205-20,229, 1998.

Warrick, R.A., C. LeProvost, M.F. Meier, J. Oerlemans, and P.L. Woodworth Climate Change 1995: The science of climate change, Chapter 7: Changes in sea-level, Cambridge University Press, 1996.

Watkins, M. M., E. S. Davis, W. G. Melbourne, T. P. Yunck, J. Sharma, and B. D. Tapley, GRACE: A New Mission Concept for High Resolution Gravity Field Mapping, European Geophysical Society, Geophysics/Geodesy, Hamburg, Germany, April 3-7, 1995.

Woodworth, P.L., J. Johannessen, P. LeGrand, C. LeProvost, G. Balmino, R. Rummel, R. Sabadini, H. Suenkel, C.C. Tscherning, and P. Visser, Towards the Definitive Space Gravity Mission, *International WOCE Newsletter*, *ISSN 1029-1725*(33), December 1998.

M. Aguirre, European Space Research and Technolgy Centre, Postbus 299, NL-2200 AG, Noordwijk, The Netherlands, (maguirre@estec.esa.nl)

G. Balmino, Centre National d'Etudes Spatiales, GRGS - 18, Av. Edouard Belin, 31401 Toulouse Cedex 4, France, (georges.balmino@cnes.fr)

J. Johannessen, Nansen Environmental & Remote Sensing Centre, Marine Monitoring and Remote Sensing Department, University of Bergen, Norway, (johnny.johannessen@nrsc.no)

C. LeProvost, Laboratoire d'Etudes en Geophysique et Oceanographie Spatiales LEGOS/GRGS, Observatoire de Midi-Pyrenees 14, Avenue Edouard Belin, 31401 Toulouse Cedex 4, France, (Leprovos@pontos.cst.cnes.fr)

R. Rummel, Institut für Astronomische und Physikalische Geodäsie, Technische Universität München, Arcisstrasse 21 D-80290 Munich, Germany, (rummel@step.iapg.verm.tu-muenchen.de)

R. Sabadini, University of Milan, Department of Earth Sciences, Via L. Cicognara 7, I-20129, Milan, Italy, (roberto.sabadini@unimi.it)

H. Sünkel, Technical University Graz, Institute of Theoretical Geodesy, Steyrergasse 30/III, A-8010 Graz, Austria, (suenkel@geomatics.tu-graz.ac.at)

C.C. Tscherning, University of Copenhagen, Department of Geophysics, Juliane Maries Vej 30, Copenhagen, DK-2100, Denmark, (cct@osiris.gfy.ku.dk)

P.N.A.M. Visser, Delft University of Technology, Delft Institute for Earth-Oriented Space Research, Kluyverweg 1, 2629 HS, Delft, The Netherlands, (pieter.visser@lr.tudelft.nl)

P.L. Woodworth, Proudman Oceanographic Laboratory, Bidston Observatory, Bidston Hill, Prenton CH43 7RA, U.K., (plw@mail.nerc-bidston.ac.uk)

Effect of Mantle Structure on Postglacial Induced Horizontal Displacement

Kim O'Keefe and Patrick Wu

Department of Geology and Geophysics, University of Calgary, Calgary, Alberta, Canada

As a result of glacial isostatic adjustment, the Earth's surface is experiencing a slow, three-dimensional deformation. This paper presents a systematic study of the horizontal aspect of this deformation, parallel to the surface of the Earth. The analytical solutions for the horizontal displacement that results from the loading of three simple flat-earth models are derived. The effects of channel flow, elastic lithosphere, discontinuities in density and shear modulus on the excitation strength for horizontal displacement are also investigated. Finally, horizontal displacement in the space-time domain due to the Heaviside loading of a disc load is presented for a channel model and lithospheric earth models that may include asthenosphere, lateral variations in lithospheric thickness or asthenospheric viscosity and nonlinear rheology. It is found that channel flow, nonlinear rheology, and density discontinuities promote positive surface horizontal displacements during loading, but the presence of an elastic lithosphere and discontinuity in shear modulus promote negative surface horizontal displacements during loading.

1. INTRODUCTION

The most commonly studied aspect of the glacial isostatic adjustment process is the change in relative sea-levels associated with the vertical displacement, since the height and age of ancient beaches can be directly measured. However, another important aspect of the motion is the horizontal motion, which, until recently, has been rather difficult to measure. With the advent of new technologies such as GPS (Global Positioning System) and VLBI (Very Long Baseline Interferometry), it is now possible to obtain accurate measurements of the three dimensional motion of the Earth's surface (James and Lambert, 1993; Mitrovica et al., 1994). If measurements are taken in regions that were previously covered by glaciers but are far from the signifi-

cant effects of tectonics, then any motion present may be interpreted as due to glacial isostatic re-adjustment. This gives an additional constraint in determining both ice and earth models.

Gasperini et al. (1990) were the first to compute the horizontal motions induced by deglaciation. They found that horizontal motion is more sensitive to lateral variations in rheology than vertical motion. James and Morgan (1990) investigated horizontal motion for a laterally homogeneous earth and found that horizontal motions are more sensitive to changes in the thickness of the lithosphere than are vertical motions. They suggested that careful analysis of the horizontal motions due to deglaciation could further constrain material properties in the subsurface. James and Lambert (1993) used the more realistic ICE-3G deglaciation chronology (Tushingham and Peltier, 1991) to calculate horizontal velocities and found that glacial rebound induced tangential velocities should be detectable by VLBI. Mitrovica et al. (1994) determined that VLBI could be used to assess the acceptability of ice history and earth model

Ice Sheets, Sea Level and the Dynamic Earth
Geodynamics Series 29

pairs. They also conducted a more detailed analysis of the displacement for various models by using realistic earth models and the ICE-3G loading history to obtain predicted patterns of the horizontal motions for North America and Europe based on specific earth and ice models. Results of Mitrovica et al. (1994) found that the sensitivities of tangential motions to the earth models are a strong function of geographic location and the specific parameter of the earth model. D'Agostino et al. (1997) and Giunchi et al. (1997) have also considered horizontal motions. Both of these papers studied the effect of lateral viscosity variations and deep mantle stratification on glacial rebound. They found that horizontal motions are more susceptible to changes in lateral variations than the corresponding vertical motions.

In the above studies, horizontal surface motion was calculated for multi-layer earth models that contain combined effects of lithosphere, density and viscosity discontinuities and it is not clear how the presence of lithosphere, asthenosphere, discontinuities in density or shear modulus, lateral variations in lithospheric thickness and power-law rheology affect horizontal surface motion individually. Without such understanding, it would be difficult to constrain mantle rheology from the observed horizontal motion.

The purpose of this paper is to analyze the horizontal surface displacement that results from Heaviside loading of simple ice and earth models. First, analytical solutions for horizontal displacement to some simple models are presented. Then, the effects of lithosphere, asthenosphere, channel, discontinuity in density and shear modulus are introduced one at a time. The aim is to study their individual effects on the excitation strength for horizontal displacement. Finally, lithospheric halfspaces with asthenosphere, lateral heterogeneity and power-law rheology will be studies for the horizontal motion in the space-time domain. The solutions are calculated with the spectral technique or finite element method. The results of the two methods have also been demonstrated to agree with each other for horizontal displacements. The details of these methods can be found elsewhere (e.g. Wu & Johnston 1998, O'Keefe & Wu 1998, Wu 1990, Wolf 1985, Cathles 1975) and so will not be discussed here.

In the following, the earth models considered are incompressible, stratified viscoelastic flat-earths with constant properties within each layer. The effect of compressibility, although significant for horizontal displacement (Klemann et al. 2000), will not be considered here. For simplicity, we consider a uniform circular disc load with radius r=2000 km, thickness = 1 km, that is left on the earth's surface (Heaviside loading).

2. RESULTS

2.1 Analytical Solutions

First, we investigate the horizontal displacements for three earth models that are simple enough to give analytical solutions in the wavenumber k domain. These simple examples illustrate that the sign of the displacement is sensitive to the viscosity structure of the earth. These analytical solutions also provide useful checks on the accuracy of the numerical spectral method (see next subsection) and for the finite element method.

The first model to be considered is a uniform viscoelastic halfspace. The Heaviside solution in the k-domain for the horizontal displacement is well known (e.g. Wolf 1998) and is given by:

$$U(k,t) = \frac{\sigma k z e^{kz}}{2\mu k + \rho g}\left\{1 + \frac{2\mu k}{\rho g}\left\{-e^{-\alpha t}\right\}\right\} \quad (1a)$$

where

$$\alpha = \frac{\rho g \mu}{\eta(2\mu k + \rho g)} \quad (1b)$$

σ is the pressure due to the applied load, z is the depth, and ρ, g, μ, η are density, gravity, shear modulus and viscosity respectively. Note that the horizontal displacement at the surface (z=0) is exactly zero for the halfspace.

We have also derived the solution for a viscoelastic channel of thickness H, using a symbolic manipulation program Mathview (Hoffner, 1997) with the rigid boundary condition applied at depth H. The horizontal surface displacement is given by:

$$U(k,t) = \frac{\sigma H^2 k^2}{(2\mu k \gamma + \rho g)\beta}\left[1 + \frac{2\mu k \gamma}{\rho g}\left\{-e^{-\alpha^* t}\right\}\right] \quad (2a)$$

where

$$\alpha^* = \frac{\rho g \mu}{\eta(2\mu k \gamma + \rho g)} \quad (2b)$$

$$\gamma = \frac{\cosh^2 Hk + H^2 k^2}{\cosh Hk \sinh Hk - Hk} \quad (2c)$$

$$\beta = (\cosh Hk \sinh Hk - Hk) \quad (2d)$$

Table 1: The Material Properties in the Channel Model

density ρ (kg m⁻³)	5517
gravitational acceleration g (m s⁻²)	7.365
viscosity η (Pa-s)	1×10^{21}
shear modulus μ (N m⁻²)	1.452×10^{11}
load σ (Pa)	1×10^{7}

Notice that the horizontal displacement is no longer zero except at k=0 and infinity. In fact, U(k,t) is always positive because the factor β is positive for k positive. Physically, when a channel is loaded on the surface, the material underneath the load is forced to flow outwards, thus the sign of the horizontal motion is positive.

Next, we derived, using the symbolic manipulation program Mathview, the analytical solution for an elastic lithosphere overlying an inviscid fluid. For this case, the conditions at the bottom of the lithosphere are that the shear stress is zero and the vertical normal stress is equal to the buoyancy force. Since the lithosphere is elastic and the mantle is inviscid, there is no relaxation. The horizontal surface displacement is given by:

$$U(k) = \frac{\sigma H k \left\{ 2\mu H k^2 - \delta \rho g \right\}}{2\mu k (\rho + \delta \rho) g A + \left\{ 4\mu^2 k^2 + \rho \delta \rho g^2 \right\} B - C} \quad (3)$$

where $A = \cosh Hk \sinh Hk + Hk$,

$B = \sinh^2 Hk$,

$C = 4\mu^2 H^2 k^4$

and H is the thickness of the lithosphere. This result agrees with that derived in section 4.1 of Johnston et al. (1998). Note that all of the above material parameters apply to the properties within the lithosphere and that $\delta \rho$ represents the density difference between the inviscid fluid and the lithosphere (if one exists). Underneath the load where long wavelengths (or small values of wave-number k) dominate, the horizontal displacement is negative. (In the limit that k approaches zero, U becomes proportional to $-\delta\rho/k$.) Thus, within the ice load, an elastic lithosphere gives negative horizontal displacements at the Earth's surface while a viscoelastic channel predicts positive horizontal displacements there. Physically, at the base of the lithosphere underneath the load, the horizontal displacement is directed outwards as in channel flow, and the base of the lithosphere experience tension. However, due to the flexure of the lithosphere, the motion at the top of the lithosphere is opposite to that at the base, thus the surface experiences compression and the horizontal surface motion is inwards (negative).

2.2 Excitation Strengths

Simple analytical solutions only exist for models with "one viscoelastic/elastic layer". In order to investigate the effects of vertical stratification on horizontal displacements, we first look at the excitation strength of some slightly more complex earth models. Excitation strength can be calculated by the method described in Wu & Ni (1996), which involves finding the eigenvalues first and then solving the solution for each eigenvalue using the standard technique of choosing a starting solution that satisfies the lower boundary condition, propagating the solution to the surface of the earth and matching the surface boundary.

We begin with the channel model. The material properties of this model can be found in Table 1. Note that the solutions have been normalized and k is also dimensionless. (To obtain the values of k in terms of km⁻¹, simply divide by 6371 km.) The excitation strength obtained with the propagation technique is found to agree with the second term in the square brackets of equation (2), multiplied by the preceding factor. The result is plotted in Fig.1, which shows that the excitation strength for the horizontal displacement goes to zero for large values of k and decreases linearly for small values in the log-log plot. The results at large k are expected, since short wavelengths "see" the channel as a halfspace, which predicts zero horizontal displacements. As noted above, the excitation strength for horizontal displacement in the channel model is positive.

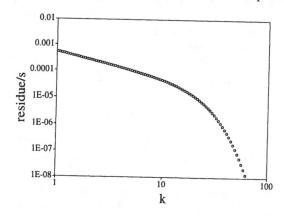

Figure 1. The excitation strength for horizontal displacement in the channel model

Table 2: The Material Properties of Some Two Layer Earth Models

	Lithospheric half-space	Single Density Discontinuity	Single Shear Discontinuity
surface gravity g (m s^{-2})	7.365	7.365	7.365
Thickness of top layer H (km)	150	670	670
density above H (kg m^{-3})	5517	3572	5517
viscosity above H (Pa-s)	infinity	1×10^{21}	1×10^{21}
shear modulus above H (N m^{-2})	1.452×10^{11}	1.452×10^{11}	0.828×10^{11}
density in halfspace below H (kg m^{-3})	5517	6288	5517
viscosity in halfspace below H (Pa-s)	1×10^{21}	1×10^{21}	1×10^{21}
shear modulus in halfspace (N m^{-2})	1.452×10^{11}	1.452×10^{11}	1.715×10^{11}
load σ (Pa)	1×10^{7}	1×10^{7}	1×10^{7}

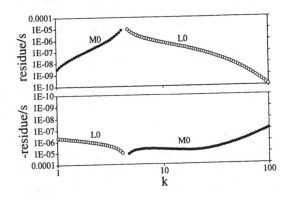

Figure 2. The excitation strength for horizontal displacement in the lithospheric halfspace.

Since the excitation strength represents the viscous portion of the time domain solution, this implies that the subsequent viscoelastic motion will be in the same direction as the initial elastic displacement.

From section 2.1, we saw that an elastic layer overlying an inviscid fluid halfspace predicts negative horizontal displacements. However, that model only has an initial elastic response and no viscoelastic relaxation. In order to study the viscoelastic relaxation of a model with an elastic lithosphere, a uniform viscoelastic halfspace with constant density and shear modulus is added below the lithosphere. This is referred to as the lithospheric halfspace model. The parameters of this model are given in Table 2. It is well known (e.g. Wu & Peltier 1982) that such a model supports two modes, the mantle mode M0 and the lithospheric mode L0. The relaxation diagram and the excitation strength for vertical displacements for this model are very similar to that for the spherical model 9 in Wu & Ni (1996). The excitation strength of the horizontal displacement for the flat-earth case is shown in Figure 2. This is also similar to Fig. 10c in Wu & Ni (1996) or Fig. 15 in p.61 of James (1991) except that the mantle modes M0 for the flat-earth earth have smaller excitation strengths at long wavelengths

implying that the spherical nature of the earth also has a significant effect on the magnitude of horizontal displacement.

For a thinner lithosphere or a weaker shear modulus in the halfspace, the curves in Fig. 2 shift horizontally towards larger k, so that the amplitude of the short wavelength deformation increases. The important point is that the magnitude of the negative excitation strength is generally larger than the positive excitation strength, so that the horizontal displacement turns out to be negative for glacial loading.

Next, we wish to consider the effects of other material discontinuities on the excitation strength. Two models will be considered here. They both have a uniform layer over a uniform halfspace and their parameters can be found in Table 2. The first model has a pure density jump across 670 km depth and the second model has a pure discontinuity in shear modulus. These models are similar to the spherical models 4 and 7 of Wu & Ni (1996) and their relaxation diagrams and excitation strengths for vertical displacements are similar to those shown in Wu & Ni (1996) or Fig. 7 in p.48 of James (1991). Again, the excitation strengths for horizontal displacements of the M0 mode are different between the flat-earth and the spherical earth models, implying that horizontal displacements are also sensitive to the sphericity of the earth.

For the model with a pure density discontinuity, there are 2 mantle modes - M0 is due to a density contrast at the surface and the M1 mode is due to a density contrast at depth. The excitation strength of the horizontal displacement shows that the M0 mode has negative excitation strength while the M1 mode, which is the dominant mode for horizontal displacement, has positive excitation strength (Figure 3). If the depth of the density discontinuity were nearer the surface, then the curves would shift horizontally towards high k values (the amplitude of the short wavelength deformation increases). However, the magnitude of any realistic density jump does not significantly affect the excitation strength. Thus, regardless of the depth of the discon-

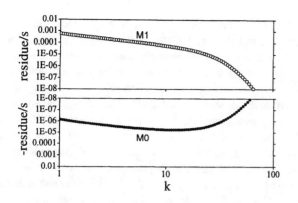

Figure 3. The excitation strength for horizontal displacement in a model with a single density discontinuity

tinuity or its magnitude, the horizontal displacement for a density jump is positive during loading.

For the model with a discontinuity in the shear modulus, three modes can be found (Wu & Ni 1996). They are the mantle mode M0, and the transition modes T1 & T2. The transitional modes exist only if the change in the shear modulus is accompanied by a change in the Maxwell time μ/η (Wu & Ni 1996). The relaxation times of these transition modes are dependent not only on the magnitude of the contrast between the shear moduli of the two layers, but also on the depth at which this discontinuity occurs. As the contrast increases, their relaxation times decrease and as the depth of the discontinuity approaches the surface, their relaxation times are shifted towards shorter wavelengths (larger k). The excitation strength of the horizontal displacement for M0 and T1 modes looks similar to that of the model with a single density discontinuity (c.f. Figures 3 and 4). However, the dominant M0 mode and T1 modes are always negative but the T2 mode is always positive. As the depth of the discontinuity decreases or the shear modulus of the upper layer decreases, the curve for the excitation strength shifts towards large k so that the amplitude of the short wavelength deformation increases. The horizontal displacement for this model is mainly negative during glacial loading.

Thus, in the spectral domain, horizontal surface displacements for the channel model and for the model with a single density discontinuity are dominated by positive excitation strength, while that for lithospheric halfspace and the model with a single discontinuity in shear modulus are dominated by negative contributions.

2.3 Solution in Space-Time Domain

In this section, the horizontal displacement in space-time domain for a channel model and a lithospheric halfspace

will be computed in addition to models with asthenosphere, lateral heterogeneity and nonlinear rheology. The material parameters used for the various models are given in Tables 1-3. For vertically stratified earth models, the horizontal displacement in the space-time domain can be obtained by evaluating the Hankel transform (e.g. Johnston et al. 1998):

$$u(r,t) = \int_0^\infty U(k,t)J_1(kr)J_1(kR)R\,dk \qquad (4)$$

where J_1 is the Bessel function of order one. However, for nonlinear mantle rheology or laterally heterogeneous earths, it is more convenient to use the finite element method. The results of these two methods can be shown to agree with each other (see Fig. 5). Therefore, most of the results in this section were obtained using the finite element package ABAQUS. Several curves are included on each graph to show the variation in time of the horizontal displacement after the load has been emplaced; these times correspond to 0, 1, 5, and 10 thousand years after loading.

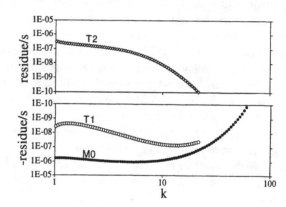

Figure 4. The excitation strength for horizontal displacement in a model with a single discontinuity in shear modulus

Table 3: The Material Properties of a Lithosphere-Asthenosphere over a viscoelastic halfspace

surface gravity g (m s^{-2})	7.365
Thickness of Lithosphere (km)	100
density of Lithosphere (kg m^{-3})	5517
viscosity of Lithosphere (Pa-s)	infinity
shear modulus of Lithosphere (N m^{-2})	1.452×10^{11}
Thickness of Asthenosphere (km)	100
density of Lithosphere (kg m^{-3})	5517
viscosity of Lithosphere (Pa-s)	1×10^{19}
shear modulus of Lithosphere (N m^{-2})	1.452×10^{11}
density of halfspace (kg m^{-3})	5517
viscosity in halfspace (Pa-s)	1×10^{21}
shear modulus in halfspace (N m^{-2})	1.452×10^{11}

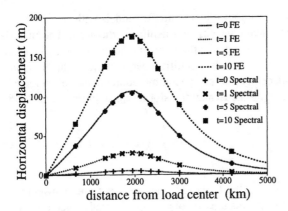

Figure 5. Comparison between the spectral and finite element methods for the computation of horizontal displacement resulting from Heaviside loading of a channel model

Figure 5 shows the Heaviside horizontal displacement of the channel model. Here, the results of the spectral method (symbols) are compared to those for the finite element method (lines). This shows that the results of the two methods compare favorably. The t=0 curve shows the initial elastic displacement and the subsequent curves illustrate the viscoelastic relaxation over time. Note that as discussed in the previous sections, the horizontal displacement that results from the loading of a channel model is positive and the maximum displacement is obtained at the edge of the load. Since the horizontal displacement increases in time during glacial loading, one can easily see that after the load was removed, the horizontal displacement remains positive (i.e. directed outwards from the center of rebound) but the magnitude decreases in time. Thus, the horizontal velocity after deglaciation is negative (i.e. towards the center of rebound).

The next figure (Figure 6) shows the horizontal displacement that results from the loading of a lithospheric halfspace as a function of the distance from the center of the load. One important distinction to be noted immediately is the fact that in contrast to the channel model, the horizontal displacement experienced by this model is mainly negative. This follows from our discussion of the excitation strengths for horizontal displacement in these two models. In both cases the initial elastic displacement is positive, i.e. motion is away from the center of the load, but for the channel model, subsequent motion is also positive while for the lithospheric halfspace, subsequent displacement is mainly negative during loading and the horizontal velocity mainly becomes positive after deglaciation. Also, note that although the horizontal displacement that results from loading a halfspace model is zero, the presence of the lithosphere yields a non-zero horizontal displacement.

A pure channel model is not realistic since the lower mantle is not rigid and a lithosphere always exists near the earth's surface (due to low surface temperature and because rheology is thermally activated). So, in the next model, a low viscosity asthenosphere is inserted between the lithosphere and the uniform halfspace. The lithosphere and low viscosity asthenosphere both have thicknesses of 100 km. The viscosity of the asthenosphere is 1×10^{19} Pa-s that is two orders of magnitude less than that of the underlying halfspace. The horizontal displacement is shown in Figure 7. Comparing this with the results for the lithospheric halfspace (Figure 5), some significant differences are noted. First, the viscoelastic relaxation experienced by the horizontal displacement soon after the emplacement of the load produces a strong positive displacement, which then turns negative after longer time periods. The results of the lithosphere model do not show this initial positive displacement. For times short compared with relaxation of the low viscosity asthenosphere, the horizontal displacements are intermediate between the lithosphere model (Figure 5) and the channel model (Figure 4). At time periods long compared with the relaxation time of the mantle, this model matches more closely with the lithospheric halfspace. In addition, the maximum magnitude in the displacement is slightly greater than that in Fig. 6 since the lithosphere is thinner than the one used in the reference lithosphere models.

Seismic tomography has shown that lateral heterogeneity in seismic properties exist in the subsurface. If these lateral variations are thermally or chemically induced, then lateral variations in viscosity are also likely. It is estimated that upper mantle viscosity can vary by 4 orders of magnitude laterally (e.g. Wu et al., 1998), while lithospheric thickness can vary by a factor of 4 from underneath the oceans to underneath the continents. Thus, it is important to investigate the effects of lateral heterogeneity on the horizontal

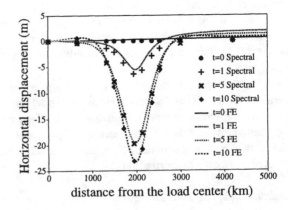

Figure 6. Horizontal displacement due to the loading of a lithospheric halfspace where the lithosphere is 150 km thick.

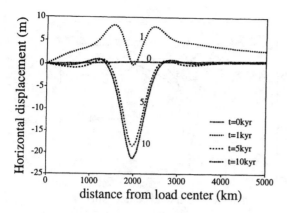

Figure 7. The horizontal displacement for the lithospheric half-space with low viscosity asthenosphere.

displacements that result from surface loading of the earth. The next set of figures is designed to investigate the effects of lateral variations in asthenospheric viscosity and lithospheric thickness with moderate variations. The elastic parameters of the models are the same as the lithospheric halfspace (Table 2). The model with lateral viscosity variation has a 10^{23} Pa-s high viscosity root underneath the lithosphere beneath the load. The high viscosity root extends to a depth of 670 km and the viscosity outside is 1×10^{21} Pa-s (see Figure 8a). Figure 8b shows the effect of this lateral variation in viscosity; the resultant horizontal displacement curve has a very distinct shape. The maximum magnitude in the displacement is no longer obtained at the edge of the load, but rather it is displaced away from the load.

The model in Fig. 9a has a 150 km thick lithosphere in the continental area under the load, while the oceanic lithosphere is 50 km thick. The viscosity of the underlying halfspace is 1×10^{21} Pa-s. The effect of this lateral variation in lithospheric thickness is demonstrated in Figure 9b, which shows that lateral variation in lithospheric thickness results in a larger horizontal displacement near the center of re-

bound but more negative displacement outside the load margin even long after the emplacement of the load. The maximum magnitude in the displacement still occurs at the edge of the load and the magnitude of this displacement is close to the maximum displacement observed for the 100 km thick lithosphere model.

Next we consider lateral viscosity variations in the 100 km thick asthenosphere beneath a uniform 100 km thick lithosphere. For the model shown in Fig.10a, the viscosity directly below the load is 1×10^{20} Pa-s while the viscosity in the rest of the asthenosphere remains at 1×10^{19} Pa-s. The horizontal displacement for this model (Figure 10b) shows a slight difference between the displacement within the region of the load and outside. This can be identified for longer time scales, especially in the peripheral regions.

In all of the models discussed so far it has always been assumed that the rheology of the earth is linear. However, high temperature creep experiments on mantle rocks show that both linear and power-law rheology may operate in the mantle. However, due to the uncertainty of the transition conditions between diffusion and dislocation creep (Ranalli 1987), it is not clear, from the microphysics point of view, which part of the mantle is linear or nonlinear. Recently, Wu (1999) used the sealevel data in and around Laurentia to study the possibility of having nonlinear rheology in different parts of the mantle. The results show that the earth model that best fits the sealevel data is comprised of an elastic lithosphere and a non-linear lower mantle with creep exponent $n = 3$ and creep parameter $A = 3 \times 10^{-35}$ Pa^{-3}s^{-1} where the effective viscosity η_{eff} is defined by:

$$1/\eta_{eff} = 3A\sigma^{n-1} \qquad (5)$$

and σ is the equivalent deviatoric stress. Since only the vertical displacement was investigated in that study, we

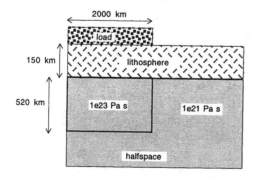

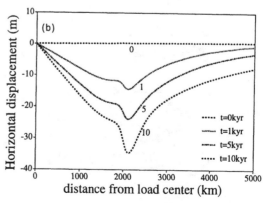

Figure 8. (a) The model with a high viscosity continental root, and (b) its horizontal displacement.

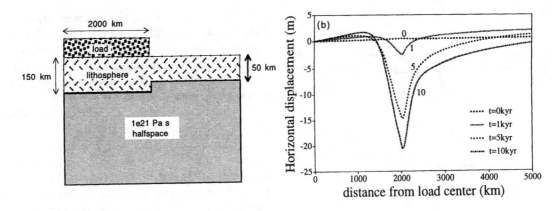

Figure 9. (a) The model with lateral variation in lithospheric thickness and (b) its horizontal displacement.

will investigate the horizontal displacement of this model and another one with nonlinear rheology throughout the mantle.

Figure 11 shows the horizontal displacement for a model with a nonlinear halfspace ($A=3\times10^{-35}$ Pa^{-3}s^{-1}) under an elastic lithosphere. The magnitude of the horizontal displacement curves is predominantly positive just like that for the channel model (Figure 5), however there is a minimum at the edge of the load that decreases with time. As shown in Fig.3 of Wu (1993), the high stress level under the load induces a time dependent low effective viscosity channel underneath the load. This explains why the horizontal displacement of this model is similar to the channel model. However, since the induced channel terminates at the load margin, this also explains why there is a minimum there.

The horizontal displacement for the model with a nonlinear lower mantle, with $A=10^{-36}$ Pa^{-3}s^{-1} under a linear 10^{21} Pa·s mantle and a 150 km thick elastic lithosphere, gives similar results (Figure 12). The maximum displacement is predicted to occur on either side of the edge of the load and this displacement is predicted to be positive after the load

has been emplaced but near the ice margin it would reach a minimum. This implies that the horizontal velocity outside the ice margin after the load is removed would be in a negative direction, towards the center of rebound. But inside the ice margin, the direction of horizontal motion would depend on both space and time.

3. CONCLUSIONS

In this paper, the effects of earth structure on Heaviside horizontal displacement were studied systematically using both spectral and finite element techniques. It is found that positive horizontal surface displacements are predicted during glacial loading if mantle flow is somewhat restricted to lie above a certain depth. This restriction can be due to a rigid lower boundary as in the channel model, a jump in viscosity at the bottom of a low viscosity asthenosphere, a load-induced channel as in power-law rheology or a buoyancy force across a density jump. However, the presence of an elastic lithosphere generally produces negative horizontal surface displacements especially near the edge of the

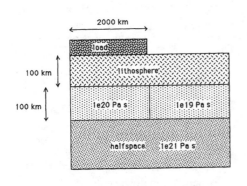

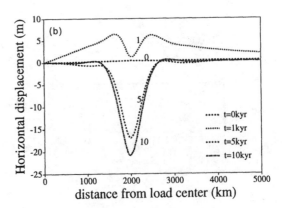

Figure 10. (a) The model with lateral variation in asthenospheric viscosity and (b) its horizontal displacement

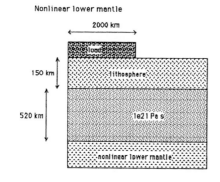

Figure 11. The horizontal displacement for the nonlinear half-space model

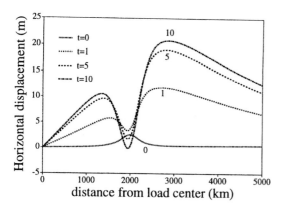

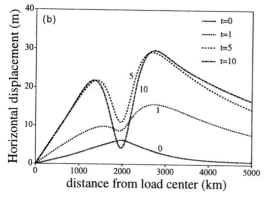

Figure 12. (a) The model with a nonlinear lower mantle and (b) horizontal displacements for the model.

load during the loading phase. The thickness of the lithosphere also has a strong influence on horizontal displacement - the magnitude decreases with a reduction in lithospheric thickness. For models with an asthenosphere beneath the lithosphere, the horizontal displacement will be positive initially, but after the underlying mantle starts to relax, then the horizontal displacement becomes negative. Lateral variation in lithospheric thickness, asthenospheric viscosity or effective viscosity in a power-law medium will

result in asymmetry in the displacement curves. Therefore horizontal motions can be used as an important diagnostic tool to determine the lateral and radial characteristics of the subsurface. In general, most of the curves for the horizontal displacement experience a displacement of approximately 30 meters over ten thousand years. This corresponds to an average speed of 3 mm per year.

In this paper, we have restricted our attention to simple loading histories, but the horizontal motion of more realistic ice models can be found in Wu (2001a) and Kaufmann & Wu (2001) in this volume. The results of this paper also indicate that horizontal motion is sensitive to the spherical shape of the earth. This should be confirmed in future work using spherical earth models. Some progress in this direction has been made in Giunchi & Spada (2000) and Spada & Giunchi (2001), which include sphericity in non-self-gravitating non-Newtonian earth. Recently Wu (2001b) also included self-gravitation in a spherical non-Newtonian earth. Compressibility has been neglected here, but recent work by Klemann et al. (2000) has shown that compressibility has an important effect on horizontal motion. Future investigation of horizontal velocities with spherical and compressible earth models will allow us to better understand mantle rheology and obtain more accurate description of surface motion of the crust. The latter can be used to calibrate shifts in the locations of the stationary GPS base stations and the problem in regular recalibration of the base stations will be reduced.

Acknowledgments. We thank Dr. Thomas James for a constructive review. This research is supported by an NSERC Postgraduate Scholarship to K. O'Keefe and NSERC grant to P. Wu.

REFERENCES

Cathles, L. M.III, *The Viscosity of the Earth's Mantle*, Princeton University Press, Princeton, 1975.

D'Agostino, G., G. Spada and R. Sabadini, Postglacial rebound and lateral viscosity variations: a semi-analytical approach based on a spherical model with Maxwell rheology, *Geophys. J. Int.*, 129, F9-F13, 1997.

Gasperini P., D.A. Yuen & R. Sabadini , Effects of lateral viscosity variations on postglacial rebound: implications for recent sea-level trends, *Geophys. Res. Lett.*, 17, 5-9, 1990.

Giunchi, C. and G. Spada, Postglacial rebound in a non-Newtonian spherical Earth, *Geophys. Res. Lett.*, 14, 2065-2068, 2000.

Giunchi, C., G. Spada and R. Sabadini, Lateral viscosity variations and post-glacial rebound: effects on present-day VLBI baseline deformations, *Geophys. Res. Lett.*, 24, 13-19, 1997.

Hibbitt, Karlsson and Sorensen, Inc., *ABAQUS*, Hibbitt, Karlsson and Sorensen, Inc., 1992.

Hoffner , N.C., *Waterloo Maple Mathview User's Guide*, Waterloo Maple Inc., Waterloo, 1997.

James, T.J., Postglacial Deformation, PhD thesis, Princeton University, Princeton, 1991.

James, T. J. and A. Lambert, A comparison of VLBI data with the ICE-3G glacial rebound model, *Geophys. Res. Lett.*, 20, 871-874, 1993.

James, T. S. and W. J. Morgan, Horizontal motions due to postglacial rebound, *Geophys. Res. Lett.*, 17, 957-960, 1990.

Johnston, P., P. Wu and K. Lambeck, Dependence of horizontal stress magnitude on load dimension in glacial rebound models, *Geophys. J. Int.*, 132, 41-60, 1998.

Kaufmann, G. and P. Wu, Glacial isostatic adjustment on a three-dimensional laterally heterogeneous earth: Examples from Fennoscandia and the Barents Sea, (this volume), 2001.

Kaufmann, G., P. Wu and D. Wolf, Some effects of lateral heterogeneities in the upper mantle on postglacial land uplift close to continental margins, *Geophys. J. Int.*, 128, 175-187, 1997.

Klemann, V., P. Wu and D. Wolf, Compressible viscoelasticity: a comparison of plane-earth solutions, *EOS*, 81, F326, 2000.

Mitrovica, J. X., J. L. Davis and I. I. Shapiro, A spectral formalism for computing three-dimensional deformations due to surface loads 2. Present-day glacial isostatic adjustment, *J. Geophys. Res.*, 99, 7075-7101, 1994.

O'Keefe K. and P. Wu, Viscoelastic Channel Flow, *in Dynamics of the Ice Age Earth: A Modern Perspective*, edited by P.Wu, pp.203-216, Trans Tech Publ., Switzerland, 1998.

Ranalli, G., *Rheology of the Earth: Deformation and flow processes in geophysics and geodynamics*, Allen and Unwin Incorporated, Boston, 1987.

Spada G. and G. Giunchi, Non-Newtonian rebound: new results based on a global Earth model, (this volume), 2001.

Tushingham, A. M. and W. R. Peltier, ICE-3G: A new global model of late Pleistocene deglaciation based upon geophysical predictions of postglacial relative sea level change, *J. Geophys. Res.*, 96, 4497-4523, 1991.

Wolf, D., The normal modes of a layered, incompressible Maxwell halfspace, *Journal of Geophysics*, 57, 106-117, 1985.

Wolf, D., Load-induced viscoelastic relaxation: an elementary example, *in Dynamics of the Ice Age Earth: A Modern Perspective*, edited by P.Wu, pp.87-104, Trans Tech Publ., Switzerland, 1998.

Wu, P., Deformation of internal boundaries in a viscoelastic earth and topographic coupling between the mantle and core, *Geophys. J. Int.*, 101, 213-231, 1990.

Wu, P., Post-glacial Rebound in a Power-law medium with Axial symmetry and the existence of the Transition zone in Relative Sea Level data. *Geophys.J.Int.*, 114, 417-432, 1993.

Wu, P., Modeling postglacial sea levels with power-law rheology and a realistic ice model in the absence of ambient tectonic stress, *Geophys. J. Int.*, 139, 691-702, 1999.

Wu, P., Postglacial Induced Surface Motion, Gravity and Fault Instability in Laurentia: Evidence for Power Law Rheology in the Mantle? (this volume) 2001a.

Wu, P., Effects of nonlinear rheology on degree 2 harmonic deformations in a spherical self-gravitating earth. (submitted) 2001b.

Wu, P. and Z. Ni, Some analytical solutions for the viscoelastic gravitational relaxation of a two-layer non-self-gravitating incompressible spherical earth, *Geophys. J. Int*, 126, 413-436, 1996.

Wu, P. and P. Johnston, Validity of using Flat-Earth Finite Element Models in the study of Postglacial Rebound, *in Dynamics of the Ice Age Earth: A Modern Perspective*, edited by P.Wu, pp.191-202, Trans Tech Publ., Switzerland, 1998.

Wu, P. , Z. Ni and G. Kaufmann, Postglacial Rebound with Lateral Heterogeneities: From 2D to 3D Modelling, *in Dynamics of the Ice Age Earth: A Modern Perspective*, edited by P.Wu, pp.557-582, Trans Tech Publ., Switzerland, 1998.

Wu, P. and W. R. Peltier, Viscous gravitational relaxation, *Geophys. J. Royal Astr. Soc.*, 70, 435-485, 1982.

Kimberley O'Keefe, Department of Geology & Geophysics, University of Calgary, Calgary, Alberta T2N 1N4, Canada.

Patrick Wu, Department of Geology & Geophysics, University of Calgary, Calgary, Alberta T2N 1N4, Canada, (email: ppwu@ucalgary.ca)

A Comparison of Methods of Altimetry and Gravity Inversion to Measure Components of the Global Water Budget

Andrew S. Trupin

Department of Natural Sciences, Oregon Institute of Technology, Klamath Falls, Oregon

C.K. Shum and C.Y. Zhao

Department of Geodetic Sciences, Ohio State University, Columbus, Ohio

We approach the problem of measuring the thickness change of polar ice, and the surface height of the global oceans by interpretation of two independent methods of measurement. These are the inversion of satellite solutions to the low degree zonal coefficients of the Earth's gravitational potential, J_2 through J_8, and the direct measurement of elevation change over the oceans and over Antarctica by satellite altimeters, and over Greenland by airborne surveys. The non-steric changes in sea surface height are estimated by correcting altimeter measurements over the global oceans for a constant rate of thermal expansion, and by calculating the meltwater from melting ice. The seven low degree zonal coefficients provide enough information to simultaneously determine the extent and shape of two distinct regions of thickness change within each ice sheet and the ratio of the lower to upper mantle viscosity. The geopotential is corrected for postglacial rebound and is inverted by means of repeated forward solution to yield a preferred range lower mantle viscosity of 4.5×10^{21} to 10^{22} Pa-s. The predicted sea level rise from Antarctica is 0.12 ± 0.15 mm/yr, and from Greenland, is 0.5 ± 0.5 mm/yr. The total meltwater from both ice sheets is 0.6 ± 0.5 mm/yr. Satellite altimeter measurements predict a sea level contribution from Antarctica of 0.15 ± 0.08 mm/yr and airborne measurements over Greenland show 0.14 ± 0.10 mm/yr sea level rise. There are discrepancies between the observed zonal coefficients and the sum of the components of global gravity calculated from altimeter measurements, including those over the oceans. When thickness change from gridded surface elevation measurements are averaged so that the resolution is comparable to that attainable using gravity inversion, there is general agreement between the thickness changes of both ice sheets using both methods. The discrepancies between observed and altimeter derived zonal coefficients are large enough to make the prediction of lower mantle viscosity uncertain at least until the uncertainties in the gravity coefficients and those of the altimeter measurements are improved.

Ice Sheets, Sea Level and the Dynamic Earth
Geodynamics Series 29
Copyright 2002 by the American Geophysical Union
10.1029/029GD08

INTRODUCTION

In this study, we calculate the thickness change of polar ice by inverting the observed low degree coefficients of the Earth's gravity field and compare the gravity derived thickness change with the thickness change prediced by altimeter measurements of surface elevation corrected for postglacial rebound. While making this comparison, we calculate the the global sea level contribution to gravity, first by including the meltwater from glacial and polar ice, and then by calculating the gravity from satellite altimeter measurements over the oceans. We find that, to within the limited resolution that is currently possible for gravity inversion, the altimeter inferred ice thickness changes agree with those derived from gravity inversion. This agreement is apparent whether the oceanic contribution to gravity is calculated from meltwater or from altimeter measurements.

A second type of comparison is also made between these two independent geodetic methods. The gravity contribution from all the altimeter derived surface mass changes is calculated and compiled into a gravity budget. Some components of this budget, such as mountain glaciers, are not derived from altimeter measurements. The sum of all these gravity contributions is made for three choices of lower mantle viscosity, and and there is less agreement between the time rates of change of the observed coefficients and these sums. This discrepancy may be attributed to a number of factors, such as a latitude gap in the altimeter data for Antarctica, and uncertainties in the ice loading history, lithospheric thickness, and upper mantle viscosity, all of which contribute to the gravity predicted by postglacial rebound, and to the vertical motion from rebound which is used to correct the altimeter measurements.

No attempt is made here to constrain all of the variables that go into postglacial rebound models. We use an upper mantle viscosity of 10^{21} Pa-s, a lithospheric thickness of 120 km, and the ICE-3G loading history [Tushingham and Peltier, 1991] throughout this study. Some of the more recent estimates of the upper mantle viscosity are less than 10^{21} Pa-s, and some models predict greater variation in uplift with upper mantle viscosity than does the model of Han and Wahr [Ivins et al., 2000]. The contributions to gravity from postglacial rebound were provided by John Wahr and were calculated with the Green's functions of the rebound model of Han and Wahr [1995], coupled with ICE-3G loading. These models incorporate the parameters of the Preliminary Reference Earth Model (PREM), [Dziewonski and Andreson 1981]. The ice loading history of the IGE-3G model is derived with the assumption of a lower mantle viscosity of 2.5×10^{21} Pa-s. The range of viscosities in the Han and Wahr model is 10^{21} to 10^{23} Pa-s. By coupling these models, we make the assumption that gravity and uplift predictions of the rebound model depend more strongly on the effect of viscosity on the residues and amplitudes of the Green's functions than they do on the effect of viscosity on the loading history itself. Ideally, each viscosity profile would be associated with its own history if ice loading.

We expect that a full parameterization of these variables could bring the gravity calculated from altimeter data into better agreement with the observed gravity coefficients, but such a study would be difficult on one account: How to correctly fold the huge number of possible ice loading histories, both in time and in terms of the spatial extent of Holocene ice into such a global model. One purpose of this study is to show that more accurate and higher resolution satellite gravity missions can constrain those properties of the Earth that cannot be measured with altimeters alone. The methods presented here can be adapted to solutions of gravity coefficients to high angular order, capable of resolving the surface mass to regional scales.

DATA

All thickness changes are stated in terms of m of water equivalent per year. Thus a thickness change of 0.1 m over 10 sq. m is equivalent to 1000 kg of mass change. Sea level contributions are in mm of water per year.

ERS satellite altimeter measurements of 63 per cent of the Antarctic ice sheet were provided by Justin Mansley of the Department of Space and Climate Physics, University College, London, [Wingham et al., 1998]. These surface elevation data were corrected for post glacial rebound using the model of Han and Wahr [1995]. The average uncertainty of these measurements is ± 5 mm/yr, although this uncertainty may be exceeded locally, where topography is significantly sloped. For the purposes of gravity inversion, a gridded data base of the grounded ice over the Antarctic ice sheet was adapted from the 1° by 1° gridded surface elevation data, with data gaps filled in. The data base, or "template" for gravity inversion does not include the latitude gap or those gaps in the altimeter data resulting from tape recorder limitations. The total area of the grounded ice is 1.22×10^7 km^2, comprising 5645 grid points.

Gridded airborne measurements of surface elevation change for Greenland were provided by Serdar Manizade, NASA Goddard Space Flight Center, Wallops Island, VA, [Krabill et al., 2000]. This ice covered data for Greenland is provided on a 341,585 point grid with squares measuring approximately 2.2 km on a side, giving a total area of 1.71×10^6 km^2. The uncertainty of the airborne measurements, which span a 1000 mm/yr range, is ± 20 mm/yr.

Altimeter data for the global oceans, shown in plate 1 incorporate data from the Geosat (1985-1989), ERS1 (1992-1995), and Topex (1992-2000) missions to achieve coverage from 77° S. Lat to 81° N. Lat. The data are compiled on a 2° latitude by 3° longitude grid, and are average annual rates of change of sea surface height in mm/yr. An area weighted global average of non-steric sea level rise from these data is 2.4 ± 1.5 mm/yr. The effects of thermal expansion vary with location, but the spatial variation is uncertain. We estimated the average

Global Sea Level Trend Observed by GEOSAT, ERS-1 and T/P (1985-1999)

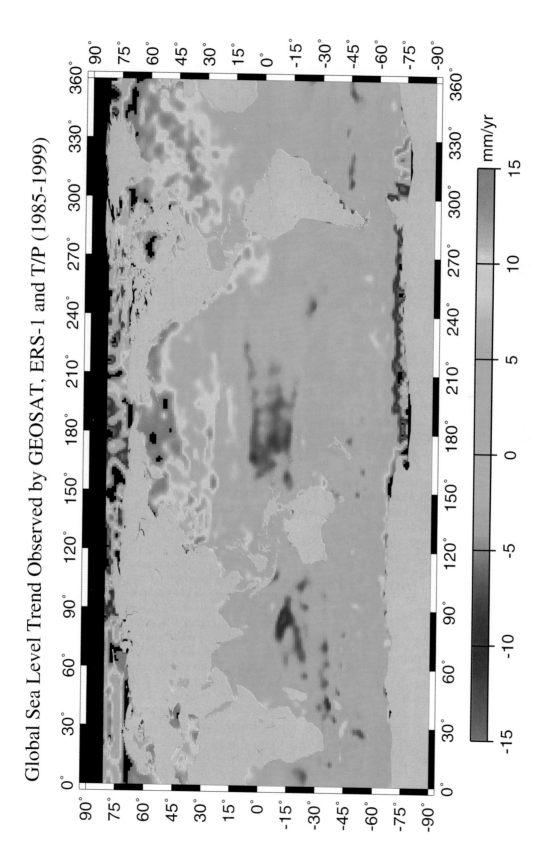

Plate 1. Sea surface height change from satellite altimeters: Geosat (1985-200), ERS1 (1992-1995), and Topex (1992-2000)

secular rate of thermal expansion for the oceans to be 0.55 mm/yr and corrected the altimeter data by this amount before computing the sea level rise.

In the cases where gravity is inverted to obtain ice thickness change, the oceanic altimeter data are used only to estimate the oceanic contribution to gravity when meltwater from polar ice and glaciers is not explicitly included in the gravity budget.

Values for J_2 through J_8 are those presented by [Cheng and Tapley, 2000]. Table 1 summarizes the observed J_l, and the various measured and modeled contributions including those of Antarctica and Greenland.

The contributions of mountain glaciers to gravity can account for roughly half of the magnitude of the uncertainties of the observed low degree zonal coefficients. We estimated the rate of melting that was likely to have occurred during the time period over which satellite solutions generally became available. Measured mass balance data are available for 13 of 31 mountain glacier regions listed in Meier [1984], and they have been used to calculate their contributions to the low degree zonal coefficients, [Trupin, Meier, and Wahr 1992]. Long term volume estimates of glaciers in 4 regions are used in lieu of mass balance data. In these regions the mass balance is estimated at 25 per cent of the long term volume change. For the remaining 14 regions, the mass balance was taken at 5.8 per cent of the amplitude-area product, [Meier, 1984], in order to account for the deceleration of melting during the second half of the Twentieth Century.

The rate of sea level drop from reservoir impoundments since 1950, given in Chao [1996], is 0.26 mm/yr. This drop may be partially or wholly offset by the sea level rise due to underground aquifer depletion, estimated by Meier [1993] to range from 0.06 to 0.3 mm/yr. Since the gravity contributions from ground water depletion are not as yet well known, we treat the anthropogenic effects not as a separate budgetary inputs, but as part of the uncertainty in each of the observed coefficients.

The elastic load Love numbers were those of PREM, and the ocean function coefficients are from Balmino, et al., [1973].

INVERSION OF GRAVITY

This method uses the gravity coefficients and the shape of the ice covered area, alone, to derive thickness change in polar ice. Upper and lower limits of 0.1 and -0.1 m/yr are placed on the thickness change of any large contoured region of the ice sheets Although thickness can vary as much as a meter or more per year over local regions, particularly for coastal Greenland [Krabill et al., 1999], variations this large are not assumed to occur over large areas of either ice sheet. The upper and lower limits of thickness change selected here are comparable to the extremes in the rates of elevation change seen over much of the Antarctic ice sheet from 1992 to 1996.

This study uses a budgetary approach to arrive at mass balance of polar ice, but the gravity budget, not sea level contributions, are the inputs to the model. The variables incorporated into these models are: The size, shape, and magnitude of thickness change in two annular regions of Antarctica and Greenland, and the lower mantle viscosity, which affects the postglacial rebound contribution to observed gravity.

We began by removing, from the 7 observed low degree coefficients, all the gravity contributions not related to the change in thickness of polar ice (table 1). These are late Pleistocene and early Holocene deglaciation, mountain glaciers and small ice systems. The competing effects of impoundment by dams and depletion of groundwater are not explicitly included as inputs to the models even though they are listed in table 1.

To construct realistic boundaries of the accumulation and ablation regions of Greenland, a uniform disk with a radius of 25 grid squares or 510 km and unit thickness Greenland was moved, in a stepwise fashion, throughout the ice covered area. The gravity from the portion of the disk mass that lies within the area of grounded ice is plotted as a function of latitude and longitude of the center of the disk. The result is a map of concentric rings that roughly conform to the coastline near perimeter of the Greenland ice sheet, but are more symmetric toward the interior, as shown in plate 2.

For Antarctica, the radius of the disk is 1000 km, and the contours divide the ice sheet roughly into the East and West Antarctic ice sheets, while conforming to the periphery of the grounded ice.

The contours in plate 2 agree with average isotherms for Greenland, [Radok et al., 1982] and also precipitation contours, [Ohmura, Wild, and Bengtsson, 1996], and ice flow lines in the accumulation regions of the Antarctic ice sheet conform roughly to the outward normals to these contours.

The contours were used to determine the shape of the division between the interior and outer regions. The area of each region and the thickness change for each region in both ice sheets were incremented in steps of .005 m/yr, between -0.1 and 0.1 m/yr for each ice sheet independently. The predicted contributions to gravity for Antarctica and Greenland are added together and then compared to the residuals from the observations to constrain the variables used in the ice models. Only those choices of thickness change which agree well with observed gravity are retained as best fitting solutions. The size of the accumulation regions, and the thickness change for each of these best fitting solutions does not vary widely for a particular choice of lower mantle viscosity.

For the rebound model of Han and Wahr, the low degree zonal coefficients vary with sea level contribution with a slope of the same sign for Greenland, but the even numbered coefficients have positive slope and the odd numbered coefficients have negative slope when plotted

Table 1. Measured and modeled components of the global gravity budget

		X10²¹Pa-S	X10⁻¹¹/Year						
		Lower Mantle Viscosity	$\dot{J_2}$	$\dot{J_3}$	$\dot{J_4}$	$\dot{J_5}$	$\dot{J_6}$	$\dot{J_7}$	$\dot{J_8}$
Observed	Cheng & Tapley, 2000		-2.7 ± 0.5	-1.2 ± 0.5	-1.1 ± 1.0	0.9 ± 1.0	0.4 ± 0.5	-2.4 ± 1.4	1.1 ± 0.8
Mountain Glaciers	Trupin, Meier, & Wahr 1992		.43 ± .2	.23 ± .2	.16 ± .2	.17 ± .2	.03 ± .2	.04 ± .2	.04 ± .2
Reservoirs	Chao, 1996		-.10 ± .3	.01 ± .03	.12 ± .03				
Ground Water			± .4	± .2	± .2	± .2			
Rebound	Han & Wahr 1995	1x	-1.92	0.77	-1.20	1.13	-1.22	1.58	-1.02
		4.5x	-5.31	1.82	-5.60	4.67	-2.33	4.55	-0.96
		10x	-7.31	1.50	-5.92	4.08	-1.61	3.77	-0.60
		20x	-6.99	0.56	-3.10	1.97	-0.82	2.21	-0.28
		50x	-4.89	-0.15	-2.10	1.76	-0.76	1.90	-0.33
		100x	-2.48	0.04	-1.89	1.35	-0.65	1.59	-0.30
Residual Melt-water		10x	4.18	-2.93	4.66	-3.35	1.98	-6.21	1.66
Ocean Altimeters			-2.10	1.89	1.18	-0.38	0.74	0.35	0.63
Residuals Altimeters		4.5x	4.31	-5.16	3.18	-3.61	1.97	-7.34	1.39

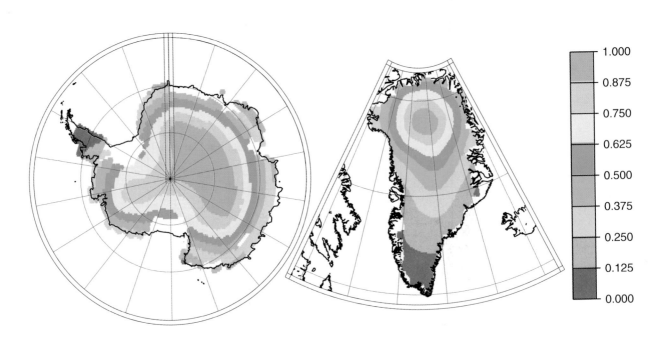

Plate 2. Template that uses low degree zonal coefficients $\dot{J}_2$ through $\dot{J}_8$ to find inverse solutions for thickness change. Seven parameters are adjusted: These are lower mantle viscosity and, for each ice sheet, the location of a contour that divides two regions of differing thickness change, and the amount of thickness change in each region.

Table 2.

(a) Best fitting solutions using melt-water, no altimeter data

I	Thickness change (m/yr) Antarctica	I	Greenland	sea	seg	Sea Level (mm/yr) seatot	fit	visc
5	0.010 -0.045	4	0.000 0.045	-0.03	-0.22	-0.25	2.36	1X
3	0.015 -0.010	3	-0.050 0.050	0.09	0.27	0.36	2.16	4.5X
5	**0.005 -0.045**	**3**	**-0.065 0.050**	**0.11**	**0.52**	**0.64**	**1.83**	**10X**
4	0.005 -0.010	3	-0.050 0.035	0.06	0.44	0.51	2.41	20X
5	0.005 -0.030	2	-0.055 0.015	0.03	0.25	0.27	1.89	50X
3	0.015 -0.005	3	-0.020 0.030	-0.03	-0.01	-0.04	2.03	100X

(b) Best fitting solutions using altimeter data, no melt-water

I	Thickness change (m/yr) Antarctica	I	Greenland	sea	seg	Sea Level (mm/yr) seatot	fit	visc
5	0.005 -0.030	4	-0.005 0.080	0.03	-0.28	-0.25	4.05	1X
4	**0.005 -0.015**	**3**	**-0.065 0.075**	**0.14**	**0.24**	**0.37**	**2.65**	**4.5X**
5	0.000 -0.030	3	-0.075 0.070	0.17	0.46	0.63	3.16	10X
3	0.000 -0.005	2	-0.090 0.025	0.12	0.39	0.51	4.61	20X
5	0.000 -0.015	2	-0.070 0.025	0.08	0.20	0.28	4.01	50X
3	0.010 -0.005	3	-0.030 0.045	0.02	-0.01	0.01	3.53	100X

against sea level contribution for Antarctica, [Trupin and Panfili, 1997]. This is consistent with the finding of James and Ivins [1997], though they used only J_2, J_3, and J_4. If the sea level contribution is large from both ice sheets, the contributions to at least two of the four coefficients will be large. Thus, the modest residual values for all the low degree coefficients, listed in table 1, cannot be obtained simultaneously by differencing large contributions from the two ice sheets. The results of this study show that, for those values of lower mantle viscosity which agree well with observation, the Antarctic and Greenland ice sheets make contributions to overall sea level that are consistent with estimates of global sea level rise derived from tide gauge records.

SEA LEVEL AND LOWER MANTLE VISCOSITY

Table 2 shows two sets of results from gravity inversion, described below. The first set of results includes the meltwater from glacial and polar ice in the gravity signal from the ice. It is also the largest non-steric contribution to sea level rise. Although sea level is not used explicitly as an input to the model for this case, it is calculated separately for each ice sheet, once the thickness change that best fits the observed gravity is determined.

The second set of solutions are calculated without the meltwater contribution to gravity. The oceanic contribution to gravity is computed using altimeter data over the global oceans, listed near the bottom of table 1. Meltwater from glaciers plus polar ice accounts for about 0.7

to 1.6 mm/yr of the global sea level rise versus a 2.4 mm/yr non-steric sea level rise from the altimeter data. By comparison, the IPCC estimate of sea level rise for Antarctica [Warrick et al., 1996] is -1.4 to 1.4 mm/yr.

Prior estimates of the global sea level rise have been determined by averaging global tide gauge data and fitting and removing the effect of postglacial rebound. Some of these estimates are 1.7 mm/yr ± 0.13 mm/yr [Trupin and Wahr, 1990], 1.8 ± 0.1 mm/yr [Douglas, 1997], and 1.8 ± 0.6 mm/yr [Peltier and Jiang, 1997]. In light of the significant spatial variation of altimeter derived ocean height over a 15 year period shown in plate 1, it is reasonable to expect that the concentration of tide gauges in Europe, Japan, and North America might lead to systematic errors in global averages of tide gauge data that exceed the sampling errors quoted in these studies.

The secular gravity signal derived from the oceanic altimeter data is an order of magnitude greater than that calculated from meltwater: -0.12 to +0.1 ×10^{-11}/yr, for meltwater versus -2.1 to + 1.9 ×10^{-11}/yr for the altimeter data. This is due, in part, to the larger altimeter measurement for global sea level rise, but also due to the significant spatial variation in sea surface height shown in the altimeter data that is not present in meltwater. The gravity signals were calculated from the altimeter data, after removing a uniform thermal expansion, by taking each pixel of altimeter data as a point mass load, whereas the gravity signal from meltwater was calculated as a uniform sea level rise (see formulation, below). The spatial variation in the altimeter data may be due to ocean circulation, long period wind forcing, and systematic uncertainties including those encumbered by combining data from different satellite missions. We include these data as a budgetary input for the oceans in table 2(b) to show that the predictions of ice thickness change are not overly sensitive to these inputs.

The lower mantle viscosity enters our analysis as an input when the postglacial rebound is removed from the observed low degree coefficients, leaving a residual from which the mass balance of polar ice is inverted. How many zonal coefficients are sufficient? If the thickness change is assumed constant and uniform for both Antarctica and Greenland, there are just three adjustable parameters, then J_2 through J_5 are sufficient. They include odd coefficients that allow for the separate determination of sea level contributions from each ice sheet, [James and Ivins 1997] or [Trupin and Panfili 1997].

If, however, we choose to obtain some information on differential thickening over each ice sheet, then there are six adjustable parameters for the ice contours and one for viscosity. These are, for each ice sheet, the thickness change inside and outside a contour, and the position of the contour which divides the interior and outer regions. The seven coefficients are sufficient to determine all of the parameters.

Solving for lower mantle viscosity is one objective of this approach. As more accurate altimeter based measurements are made over the entire area of grounded ice, then these can be corrected for ground motion by removing postglacial rebound, which depends on lower mantle viscosity. Ultimately, it will be a combination of good satellite measurements of both gravity and uplift that will help constrain this important variable. Uncertainties in the observed coefficients, differences among postglacial rebound models, and the poorly understood contribution of ground water all contribute to the uncertainty in the predictions for lower mantle viscosity.

FORMULATION

The gravity coefficients calculated here are derived from the gravitational potential, the gradient of which is the total gravitational acceleration minus the acceleration of the Earth's center-of-mass. For this reason, the sum begins at $l=2$ instead of $l=0$. For observations at the surface of the Earth ($r = a$) the gravitational potential is:

$$U(a,\theta,\lambda) = \frac{GM_e}{a} \ \mathrm{Re} \left[\sum_{l=2}^{\infty} \sum_{m=0}^{l} A_{lm} Y_{lm}^*(\theta,\lambda) \right] \quad (1)$$

The spherical harmonics are defined by and normalized according to [Chao and O'Connor, 1988]:

$$Y_{lm}(\theta,\lambda) = [(2-\delta_{m0})(2l+1)(l-m)!/(l+m)!]^{1/2} \times \quad (2)$$
$$P_{lm}(\cos\theta) \ e^{(im\lambda)}$$

$$\int Y_{l'm'}(\theta,\lambda) Y_{lm}^*(\theta,\lambda) \ \sin\theta \, d\theta \, d\lambda = 4\pi(2-\delta_{mo})\delta_{ll'}\delta_{mm'} \quad (3)$$

Re denotes the real part of a complex quantity, * denotes complex conjugation, P_{lm} is the associated Legendre function of degree l and order m, θ is the colatitude, λ is the longitude eastward of the Greenwich meridian, and a is the mean radius of the Earth.

For a realistic Earth model, such as the one used to calculate the rebound rates in Han and Wahr [1995] and Peltier [1985], the time dependence of the harmonic coefficients may be expressed, for each degree l, as an elastic response to present loading, via the load Love number k'_l, and a viscous, or lagged response, as a convolution integral of the loading history over all viscous decay modes, $r_j^l \exp(-s_j^l)$, having characteristic amplitudes or residues r_j^l and inverse decay times s_j^l. For a point mass load, $M_i(t)$, acting at (θ_i,λ_i) and starting at at $t=0$:

$$A_{lm} = \sum_i \frac{Y_{lm}(\theta_i,\lambda_i)}{(2l+1) \ M_e} \times \quad (4)$$
$$\left[M_i(t) \ (1+k'_l) + \sum_j r_j^l \int_0^t d\tau \ M_i(\tau) \ e^{-s_j^l(t-\tau)} \right]$$

The meltwater contribution to the oceans is included in the calculations for deformation via the ocean function, [Chao and O'Connor, 1988]. Six values of Lower mantle Viscosity were used. These are 1, 4.5, 10, 20, 50, and 100 times the upper mantle viscosity of 1×10^{21} Pa-s, with the division between the upper and lower mantle at a depth of 670 km.

The goodness of fit parameter shown in table 2 under the heading "fit", is found by calculating the root mean square of the deviation between observed and predicted harmonic coefficients:

$$\sigma = \left\{ \sum_{l=2}^{8} \left[\dot{j}_l^o - \dot{j}_l^p \right]^2 \right\}^{\frac{1}{2}} \quad (5)$$

where $\dot{J}_l^o$, and the zonal coefficient from the forward solutions are $\dot{J}_l^p$.

RESULTS FOR THR GRAVITY INVERSION MODEL

Gravity inversion results are summarized in table 2 and elsewhere. Table 2(a) shows the best fitting solutions for the thickness change for each ice sheet, and their sea level contributions for each of six values of lower mantle viscosity. The heading. "sea" refers to sea level from Antarctica, "seg" is sea level from Greenland, and "seatot" is the combined sea level contribution. The heading, "fit" is the goodness of fit parameter described above, and "visc" is the ratio of lower mantle to upper mantle viscosity.

The first and third column are integers from 1 to 5, under headings, I, which identify the position of the contour which divides two distinct regions. Both Antarctica and Greenland are divided into 5 contoured regions corresponding to plate 2, and the division between different rates of thickness change is made at the boundary of one of these five divisions. Since the contoured gird in plate 2 varies continuously from 0 to 1.0, an index of 3, for example, means that the division between interior and coastal regions is the contour with a value of 0.6 in plate 2, which is inside the dark green contour. Plate 3 shows the thickness change for the best fitting solutions in table 2(a), which is the one with a lower mantle viscosity that is 10^{22} Pa-s, or 10 times the upper mantle viscosity. The sea level contribution from polar ice is 0.6 mm/yr. If 0.4 mm/yr is attributed to small glaciers, and 0.6 mm/yr is due to thermal expansion, then the predictions of this model lie within the spread of sea level trends derived from tide gauges.

Pursuant to the hypothesis that meltwater, alone, might underestimate the gravity contribution from the oceans, the meltwater was replaced with the oceanic altimeter data. The sea level rise from meltwater is evident in the altimeter data but the altimeter data also include spatial variation of oceanic mass that increases its contribution to the gravity coefficients, while conserving mass in the global water budget. The best fitting results are listed in Table 2(b) and shown in plate 4. The significant changes are a reduction of the sea level contribution from Greenland from 0.52 to 0.24 mm/yr and a reduction of the predicted lower mantle viscosity from 10^{22} Pa-s to 4.5×10^{21} Pa-s. This is surprising, since the gravity signal from oceanic altimeter data is the same order of magnitude as the observed coefficients themselves, and the meltwater contribution from polar and glacial ice is much smaller (see, for example, the two left-hand columns of table 3). Using the oceanic altimeter data in place of the meltwater is equivalent to adding uncertainties to the zonal coefficients that are larger than those quoted by Cheng and Tapley [2000]. We attribute the stability of the solutions to the fact that variation of thickness change over Greenland "soaks up" differences in the residual gravity. The off axis position of the Greenland ice sheet and the 45 per cent reduction in the overall ice volume change between the cases in table 2(a) and table 2(b) closed the gap in gravity residuals created by replacing meltwater with altimeter data.

RESULTS FOR ALTIMETER DATA OVER ICE

The gravity inversion method described above solves for ice thickness that agrees well with observed gravity. We now turn to studies of altimeter data over ice and

reverse the problem: Once the surface elevation of polar ice is well known from altimetric methods, can the study of the global gravity budget be re-directed to predicting the gravity contribution from postglacial rebound and to constraining some of the parameters in the rebound models? Secondly, do the gravity signals derived from altimeter data over polar ice and the oceans agree well with observed gravity? These gravity signals are listed for all the components of the gravity budget in table 3.

The uplift predicted by the model of Han and Wahr is removed from the uncorrected Antarctic altimeter data. Note that figure 2 in Wingham et al., [1998] show these data corrected using both the Greens functions and ice loading history of the IGE-3G model. The sea level contribution from the Antarctic altimeter data varies from 0.12 to 0.16 mm/yr depending on three choices for lower mantle viscosity: 4.5X, 10X and 50X upper mantle viscosity (=10^{21} Pa-s). Whereas vertical uplift is a relatively small correction to the altimetric height map (within 20 per cent of the thickness change over the 63 per cent of the ice sheet that is covered) there is significant variation with viscosity of the rebound contributions to the sums in table 3.

The airborne data over Greenland [Krabill et al., 2000], contributes 0.14 ± 0.10 mm/yr to the sea level. The gridded surface elevation data for polar ice, corrected for uplift, are shown in plate 5. Areas in red are those where ablation is greater than 0.12 m/yr. The extended range for the red color in plate 5 was used so that all of the other colors could match the same color scale for all the plates, so that surface elevation data could be easily compared to the results for gravity inversion.

Discrepancies exist between the sum of the components of gravity including altimeter derived gravity over ice and the oceans (listed under heading "sum1"), and the observed coefficients (listed under the heading, "obs"). These can be quite large for l=2 and l=4. The discrepancies diminish somewhat for the higher value of viscosity. The column, "sum1" includes the column labeled, "oceans", in the sum but not meltwater.

If we replace the gravity signal from oceanic altimeter data with that from meltwater (under heading, "melt") from altimeter data over ice and mass balance of small glaciers, there is slightly better agreement with observation. This sum is under the heading, "sum2". In both cases, many of the summed contributions exceed the uncertainties in the observed coefficients. It is also apparent from table 3 that the uncertainty of the altimeter and airborne data preclude the determination of the viscosity of the lower mantle.

There are two issues that need be mentioned in examining the elevation change data. The first is the effect of the latitude gap in the data for Antarctica. Including artificial data where no altimeter data exists south of 81° Latitude has a small effect on the remaining undetermined parameters, [Trupin and Shum 2000]. The other is the accuracy

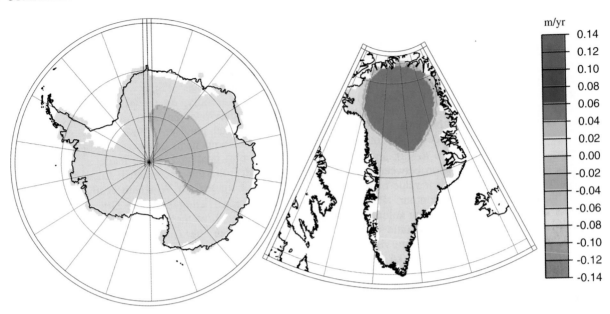

Plate 3. Thickness change of polar ice from gravity inversion with the oceanic zonal coefficients computed from meltwater. Lower mantle viscosity = 1×10^{22} Pa-s; Sea level (mm/yr): Antarctica, 0.11, Greenland, 0.52

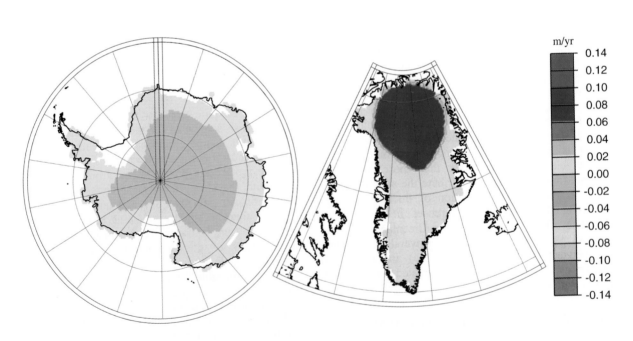

Plate 4. Thickness change of polar ice from gravity inversion with the oceanic zonal coefficients calculated from altimeter data (shown in Plate 1). Lower mantle viscosity = 4.5×10^{21} Pa-s; Sea level (mm/yr): Antarctica, 0.14, Greenland, 0.24

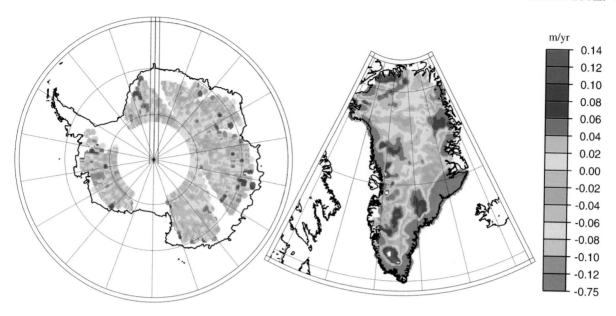

Plate 5. Antarctic elevation change from ERS satellite altimeters [Wingham et. al., 1998] and Greenland elevation change from airborne topographic measurements [Krabill et al., 2000].

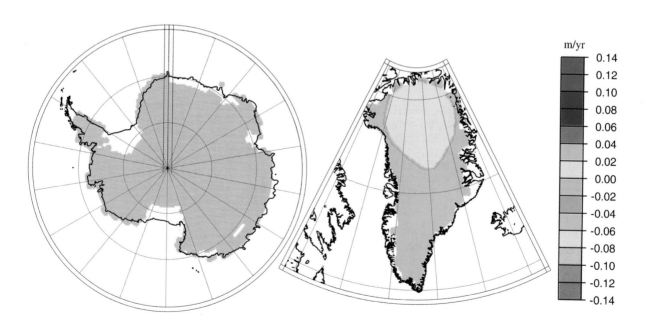

Plate 6. Antarctic altimeter data and Greenland airborne measurements spatially averaged into the regions depicted in Plates 3 and 4. This is what the surface elevation data would look like if those measurements were to have same resolution as the thickness changes predicted by inverting gravity. Lower mantle viscosity = 4.5×10^{21} Pa-s; Sea level (mm/yr): Antarctica, 0.16, Greenland, 0.14

Table 3. Gravity and Sea level Rise from Altimeter and Mass Balance Data

Contributions to Sea Level from Polar Ice

visc	sea	seg	seatot
4.5	0.16	0.14	0.30

Contributions to Zonal Coefficients if the Earth's Gravitational Potential

visc	l	obs	sum1	sum2	ant	green	glaciers	pgr	oceans	melt
4.5	2	-2.70	-5.93	-3.75	0.61	0.47	0.41	-5.31	-2.10	0.08
4.5	3	-1.20	3.74	1.79	-0.63	0.41	0.25	1.82	1.89	-0.06
4.5	4	-1.10	-3.41	-4.56	0.58	0.29	0.14	-5.60	1.18	0.03
4.5	5	0.90	4.16	4.42	-0.49	0.14	0.22	4.67	-0.38	-0.12
4.5	6	0.40	-1.19	-1.91	0.38	0.00	0.03	-2.33	0.74	0.01
4.5	7	-2.40	4.56	4.17	-0.26	-0.12	0.05	4.55	0.35	-0.04
4.5	8	1.10	-0.33	-0.97	0.15	-0.19	0.04	-0.96	0.63	-0.01

Contributions to Sea Level from Polar Ice

visc	sea	seg	seatot
10	0.16	0.15	0.30

Contributions to Zonal Coefficients if the Earth's Gravitational Potential

visc	l	obs	sum1	sum2	ant	green	glaciers	pgr	oceans	melt
10	2	-2.70	-7.93	-5.75	0.58	0.49	0.41	-7.31	-2.10	0.08
10	3	-1.20	3.47	1.52	-0.61	0.43	0.25	1.50	1.89	-0.06
10	4	-1.10	-3.74	-4.89	0.56	0.31	0.14	-5.92	1.18	0.03
10	5	0.90	3.60	3.85	-0.47	0.15	0.22	4.08	-0.38	-0.12
10	6	0.40	-0.48	-1.20	0.36	0.00	0.03	-1.61	0.74	0.01
10	7	-2.40	3.80	3.41	-0.25	-0.12	0.05	3.77	0.35	-0.04
10	8	1.10	0.02	-0.62	0.14	-0.19	0.04	-0.60	0.63	-0.01

Contributions to Sea Level from Polar Ice

visc	sea	seg	seatot
50	0.12	0.14	0.25

Contributions to Zonal Coefficients if the Earth's Gravitational Potential

visc	l	obs	sum1	sum2	ant	green	glaciers	pgr	oceans	melt
50	2	-2.70	-5.69	-3.52	0.45	0.44	0.41	-4.89	-2.10	0.07
50	3	-1.20	1.91	-0.03	-0.47	0.39	0.25	-0.15	1.89	-0.05
50	4	-1.10	-0.08	-1.23	0.43	0.27	0.14	-2.10	1.18	0.03
50	5	0.90	1.37	1.63	-0.35	0.12	0.22	1.76	-0.38	-0.11
50	6	0.40	0.26	-0.47	0.27	-0.01	0.03	-0.76	0.74	0.01
50	7	-2.40	2.00	1.61	-0.18	-0.12	0.05	1.90	0.35	-0.04
50	8	1.10	0.25	-0.39	0.09	-0.20	0.04	-0.33	0.63	-0.01

of the altimeter and airborne data itself (about ± 5 mm/yr for Antarctica and ± 20 mm/yr for Greenland). Plates 3, 4, and 5 all show thinning over the southern half of the Greenland ice sheet. The airborne data shows this thinning confined to the margins of the ice sheet. The satellite solutions to just 7 zonal coefficients does not permit resolution as fine as the airborne data. When plate 5 is averaged over the regions defined by the contours in plates 3 or 4, the result is presented in plate 6. The rough agreement between the measured surface elevation data over both ice sheets, and the thickness change derived from gravity becomes clearer. The gravity differences, listed in table 3, between elevation change data and observed gravity, appear to result from uncertainty of the thickness change of the Greenland ice sheet. This uncertainty may result from interpretation of airborne measurements, or from the various components of the gravity budget.

Some other studies do not indicate thinning of ice over southern Greenland. The spatially averaged Geosat and Geosat-ERM data, [Davis et. al. 1998a and 1998b], including the comments of H. Jay Zwally in the latter reference, shows 0.022±0.009 m/yr for a region on each side of the ice divide extending from 60° to 72° N. latitude, which is predominantly above 2000 m in elevation. Estimates of net mass balance at lower elevations are higher; .06±.03 m/yr.

The problems with gravity inversion that remain are the unknown and unmodeled contributions to gravity from postglacial rebound, continental water storage, and small ice systems, and the limited resolution from small numbers of low degree coefficients. The latter problem will be corrected as GRACE data become available. While it is likely that measurements of surface elevation change will improve upon the gravity inversion approach in determining thickness changes in polar ice, gravity will still be critically important in determining properties of the Earth that altimeters cannot measure alone, such as the viscosity profile of the mantle.

SUMMARY

It is still not possible to constrain the global water budget using the Earth's gravity field alone, but the uncertainties have diminished with the advent of greater numbers of coefficients in the satellite solutions to gravity. The predicted thickness change is sensitive to both the choice of inputs for the contribution from postglacial rebound, and the uncertainties in the time rates of change of the low degree gravity coefficients.

If there are no large unmodeled contributions to the zonal coefficients and large areas of either ice sheet dot not exceed ± 0.1 m/yr in thickness then it is possible that the ice sheets can explain the portion of the sea level rise not attributed to thermal expansion, ground water, impoundments, and melting of small ice systems. The predicted sea level rise from both ice sheets ranges from 0.1 to 1.1 mm/yr. This is close to the predictions of James and Ivins [1997] and Johnston and Lambeck [1999].

Surface elevation measurements, corrected for uplift, over the global oceans and over polar ice have made inroads into our understanding of ice thickness change. They do not agree well with observed gravity, most likely due to the uncertainties in the components of gravity, and the uncertainty of the altimeter measurements, which can be as much as four times as large as the rates of uplift (20 mm/yr versus an average uplift of 5 mm/yr over Greenland, for example). Improvements of altimeter measurements and gravity may allow for the fine tuning of postglacial rebound models, primarily via the constraints placed on mantle viscosity profile and loading history.

Acknowledgments. The authors wish to thank the NASA for their support. We are grateful to John Wahr for providing the contributions of postglacial rebound to gravity and uplift. We are thankful to Bill Krabill for providing the airborne measurements over Greenland and to Duncan Wingham for providing the surface elevation data over Antarctica. Andrew Trupin wishes to thank Ohio State University for its support.

REFERENCES

Balmino, G., K. Lambeck, and W. Kaula, A Spherical Harmonic Analysis of the Earth's Topography, *J. Geophys. Res. 78,* 478-481, 1973.

Chao, B. F., Anthropogenic impact on global geodynamics due to water impoundment in the major reservoirs, I. Geophys. Res. Lett., v. 22 (#24), 3529-3532, 1996.

Chao, B., and W. O'Connor, Effect of a uniform sea-level change on the Earth's rotation and gravitational field, *Geophys. J.R. Astr. Soc., 93,* 191-193, 1988.

Cheng, M.K. and B. D. Tapley, Determination of the long term changes in the Earth's Gravity Field from satellite laser ranging observations, *GGG 2000, Banff, Canada* Jul 31- Aug 4, 2000

Davis, C. H., C. Kluever, B. Haines, Elevation Change of the southern Greenland Ice Sheet, *Science, 279,* March, 1998(a).

Davis, C. H., C. Kluever, B. Haines, Growth of the southern Greenland Ice Sheet, *Science, 281,* August, 1998(b).

Dziewonski, A. and D. Andreson, Preliminary reference Earth model, *Physics of the Earth and Planetary Interiors, 25,* 297-356, 1981.

Han, D. and J. Wahr, The viscoelastic relaxation of a realistically stratified Earth, and a further analysis of postglacial rebound, *Geophys. J. Int. 120,* 287-311, 1995.

Ivins, E., Wu, X., Raymond, C., Yoder, C., and James, T., Temporal Geoid of a Rebounding Antarctica and Potential Measurement by GRACE and GOCE Satellites, in *IAG Symposium Series 120,* (ed. M. Sideris), Springer Verlag, N.Y., 2001.

James, T. and E. Ivins, Global Geodetic Signatures of the Antarctic Ice Sheet, *J. Geophys. Res., Vol 102, no. B1* 605-633, 1997.

Johnston, P. and K., Lambeck, Postglacial rebound and sea level contributions to changes in the geoid and the Earth's Rotation Axis *Geophys. J Int., 136,* 537-558, 1999.

Krabill, W., E Frederick, S. Manizade, C. Martin, J. Sonntag, R. Swift ,R. Thomas, W. Wright, and J. Yungel, Rapid thinning of parts of the southern Greenland Ice Sheet, *Science, 283,* 1522-1529, 1999.

Krabill, W., W. Abdalati, E. Frederick, S. Manizade, C. Martin, J. Sonntag, R. Swift, R. Thomas, W. Wright, and J. Yungel, Greenland Ice Sheet: High-Elevation Balance and Peripheral Thinning, *Science, 289,* 428-430, 2000.

Meier, M.F., Contribution of small glaciers to global sea level, *Science, 226,* 1418-1421, 1984.

Meier, M.F., Ice, climate and sea level: Do we know what is happening?, in *Ice in the Climate System,* ed. W.R. Peltier, Springer-Verlag, Berlin, 1993.

Ohmura, A., M. Wild, and L. Bengtsson, A possible change in mass balance of Greenland and Antarctic Ice Sheets in the coming century, *J. of Climate, Vol. 9,* 2124-2135, 1996.

Peltier, W.R. The LAGEOS constraint on deep mantle viscosity; results from a new normal mode method for the inversion of viscoelastic relaxation spectra, *J. Geophys. Res. 90,* 9411-9421, 1985.

Radok, U., R. Barry, D. Jenssen, R. Keen, G. Kiladis, and B. McInnes, Climatic and physical characteristics of the Greenland Ice Sheet, parts I-IV, *Cooperative Institute for Research in the Environmental Sciences,* University of Colorado, Boulder, CO 80309 USA, 1982.

Tushingham, A.M., and W.R. Peltier, ICE-3G: A new global

model of late Pleistocene de-glaciation based upon geophysical predictions of post glacial relative sea level change, *J. Geophys. Res.,* vol 96, No. B3, 4497-4523, 1991.

Trupin A., Effects of polar ice on the Earth's rotation and gravitational potential *Geophys. J. Int, 113,* 273-283, 1993.

Trupin, A. and R. Panfili, Gravity and rotation changes from mass balance of polar ice, *Surveys of Geophysics,* 18, 313-326, 1997.

Trupin, A. and C.K. Shum, Determination of mass balance of polar ice from gravity, in *IAG Symposium Series, 120, (ed. M Sideris), Springer Verlag, N.Y., 2001.* Jul 31 - Aug 4, Banff Canada, 2000.

Wingham, D., A. Ridout, R. Scharroo, R. Arthern, C.K. Shum, Antarctic elevation change, *Science,* Vol 282, 456-458, 1998.

Andrew Trupin, Department of Natural Sciences, Oregon Institute of Technology, 3210 Campus Dr., Klamath Falls, OR 97601 (trupina@oit.edu)

Late-Pleistocene, Holocene and Present-day Ice Load Evolution in the Antarctic Peninsula: Models and Predicted Vertical Crustal Motion

Erik R. Ivins and Carol A. Raymond

Jet Propulsion Laboratory, California Institute of Technology, Pasadena, California, USA

Thomas S. James

Geological Survey of Canada, Sidney, British Columbia, Canada

New computations of the present-day postglacial vertical isostatic motion in the Antarctic Peninsula are presented. Part of the present-day isostatic response must be driven by a fading memory of the late-Pleistocene to early-Holocene Antarctic Peninsula Ice Sheet (APIS). The geometry and collapse history of the APIS is somewhat constrained by numerical models and marine sedimentary data. Mid to late-Holocene ice mass change within the Peninsula region may also be significant. Climate and oceanographic studies indicate that periodic ice mass imbalance of the region may be of larger magnitude than elsewhere in Antarctica. Consequently, the bedrock response to a series of more recent ice fluctuation scenarios are computed using a simple gravitating incompressible Earth model consisting of an elastic lithosphere and a viscoelastic mantle half-space. Isostatic adjustment of the solid Earth to glacial mass changes integrated over the last several thousand years could drive present-day changes in topography at geodetically detectable rates ($\geq$ 4 mm/yr) over a broad range of mantle viscosity values. In the presence of oscillatory mass change, and mantle viscosity below $2 - 3 \times 10^{20}$ Pa s, the concomitant present-day isostatic motion may be large in magnitude, $\mathcal{O}(10$ mm/yr), significantly phase-lagged with respect to surface load change, and have a complex pattern of bulge migration. The key parameters which control phase-lagged responses are elucidated by using a simplified circular disk load and 3-phase history. The complexity of this younger glacioisostatic adjustment process is exacerbated by the overlapping of load history and wavenumber-dependent viscoelastic relaxation time scales. In contrast, isostasy driven by continuous drawdown of ice mass during the mid-to late-Holocene does not exhibit this delicate, and cancelling, competition between memory of load accumulation and memory of ablation. Consequently, in a continuous ice mass drawdown mode, vertical uplift may be quite large at the present-day: $\mathcal{O}(20$ mm/yr).

Ice Sheets, Sea Level and the Dynamic Earth
Geodynamics Series 29
This paper not subject to U.S. copyright
Published in 2002 by the American Geophysical Union
10.1029/029GD09

1. INTRODUCTION

Understanding the hydrologic mass exchanged between oceans and ice once grounded on the Antarctic continent during the past 20,000 years is of basic importance to paleoclimatology and paleoceanography. This mass exchange causes a gravitational disequilibrium in the deep Earth. Highly accurate geodetic measurements of the present-day isostatic recovery from disequilibrium could provide data that bound the mass, timing and spatial detail involved in the continent-ocean exchange. Glacial rebound in Antarctica, however, is complicated by the likelihood of an inquiet mass balance state during the past 5 thousand years [Bindschadler and Vornberger, 1998; Goodwin, 1998]. Here we address several fundamental questions regarding recent mass variations, as these may occur as oscillations, or, alternatively, as continuous (monotonic) variations. The Antarctic Peninsula region is well-suited to such a study, since a variety of new constraints have recently been placed on the early glacial extent and timing of the retreat from Last Glacial Maximum (LGM) [Ingólfsson et al., 1998; Bentley and Anderson, 1998]. We incorporate the present-day ice mass exchange into models of both the regional Neoglacial advances/retreats and the Peninsula LGM [Payne et al., 1989] in order to predict the present-day vertical isostatic response. These crustal motions could be measured using Global Positioning System (GPS) observations, if taken over a sufficient duration of time [Scherneck et al., 2001].

Climatic conditions in this region are extremely dynamic [Clapperton, 1990; Cullather et al., 1998] and ice core evidence indicates that interannual variability in oceanic-atmospheric temperature and precipitation are more variable in the Peninsula than elsewhere on the Antarctic continent [Peel, 1995]. The fact that ice mass imbalance may be larger in this region than anywhere else on Earth, and that a regional GPS network has been collecting data since January of 1995 [Dietrich et al., 2001], make the predictions of regional crustal motions due to ice loading events of interest to both the glaciological and geodetic communities.

Emerging from our study of the Antarctic Peninsula is the question of how to better describe the operative physics in the mantle-lithosphere system when driven by disequilibrium loads. The level of complexity of the geodetic responses associated with the finite glacial evolution phase is intertwined with the poorly constrained mantle viscosity. If the ice change and the solid Earth isostatic response time scales are similar, then phase-lagged (nonequilibrium) effects become important to the prediction of present-day crustal motion. *Mitrovica and Davis* [1995] demonstrated that the combination of a relatively high mantle viscosity ($\eta \geq 10^{21}$ Pa s) and a variable finite growth phase (before deglaciation from Würm-Wisconsin LGM during 21-6 kyr BP) is important to the calculation of present-day vertical crustal motion at the 0.1 - 0.5 mm/yr level. The role of viscoelastic phase-lags in the mantle/lithosphere beneath presently evolving glacial systems with lower mantle viscosity (i.e., $\eta \leq 5 \times 10^{20}$ Pa s) has seen relatively little treatment, although the response may be an order-of-magnitude larger. Recent work by *Wolf et al.* [1997] for the Vatnajökull ice cap in Iceland, and by *Ivins and James* [1999] for the Patagonian ice fields, has offered some elucidation of the physics involved. One of the goals of this paper is to refine our understanding of this physics.

In the first part of this paper we discuss the isostatic response to late-Pleistocene and early-Holocene glacial change. Our new model for the Antarctic Peninsula region is primarily based upon the numerical model results of *Payne et al.* [1989]. In this model deglaciating ice in the Peninsula region accounts for slightly more than 3 meters of equivalent eustatic sea-level rise (e.s.l.r) during the period 21 to 5 kyr BP. The model represents an improvement, in both space and time, over the models of *Tushingham and Peltier* [1991] (ICE-3G) and of *Denton et al.* [1991] (termed *D91* by *James and Ivins* [1998]). However, the total mass involved in the ice model depends upon poorly understood basal flow conditions and ice stream fluxes, along with other poorly constrained ice sheet, ice shelf, atmospheric and oceanic parameters. Furthermore, the total volume of ice that is accumulated and/or ablated from the region during the mid-to late-Holocene is potentially significant and may imply contributions to global sea-level change at the level of several tenths of a meter, or more, during the last 4000 years [Goodwin, 1998].

In the second part of this paper we address the differing styles of isostatic response due to this younger, though considerably smaller, continent-ocean hydrological exchange. As the Neogene tectonics of the Antarctic Peninsula beg the question of what influence a weaker upper mantle rheology (compared to Fennoscandia or Laurentia) might have, we consider fluctuations in mass that may occur during the current millennium, up to, and including, the present-day. Indeed, it may be

that the most recent mass load evolution (6.5-0 kyr BP, or mid-to late-Holocene) in the Peninsula is of central importance to understanding regional uplift data.

A systematic study of the phase-lagged vertical motion is also presented in which the influence of load dimension, lithospheric thickness, viscosity and growth phase duration are all accounted for. The Earth model employed here is extremely simple, consisting of an elastic incompressible lithosphere over a Maxwell viscoelastic half-space. Some basic cases of mid-to late-Holocene ice loading include: (i) a series of 4 Neoglaciations during the last 4 kyr BP, including mass discharge that initiates at AD 1850 and continues at a constant rate to the present day; (ii) a series of 3 Neoglaciations during the last 4 kyr BP, with the last phase reaching its peak 1300 years ago, then decaying until AD 1850, after which mass accumulation is maintained to the present-day; (iii) a case with minimal LGM loading (e.s.l.r. amounting to less than 0.75 meters), but with a continuous drawdown of mass to the present-day; (iv) the same continuous drawdown history, but with complete termination of ice mass loading/unloading at 2 kyr BP. These cases represent realistic, end-member, scenarios and we explore the influence that each might have on present-day crustal motion predictions as a function of the assumed model viscosity.

2. LGM AND COLLAPSE HISTORY

It is now widely appreciated that the past 10,000 years of Antarctic ice sheet change has the potential for driving a substantially larger geodetic signal than those driven by the Earth's instantaneous elastic isostatic response to present-day ice mass change [*Greischar and Bentley*, 1980; *Lingle and Clark*, 1985; *James and Ivins*, 1995; 1998; *Wahr et al.*, 1995; 2000]. Prediction of a robust Antarctic postglacial signal emerges from two physical conditions: (1) The collapse of ice sheet mass is relatively substantial after the Antarctic Cold Reversal (ACR) (13.0 - 11.5 kyr B.P.) and the northern hemispheric Younger Dryas (YD) event (11.5 - 10.2 kyr B.P.) has terminated [*Jouzel et al.*, 1996]. This viewpoint is supported by global paleosealevel reconstructions [*Nakada and Lambeck*, 1989; *Tushingham and Peltier*, 1991; *Peltier*, 1994]: (2) Viscous isostasy of the mantle preserves a 'memory' of post-ACR Antarctic glacial unloading that may be manifest at the present-day in the form of observable time-evolving crustal displacement and gravity [*James and Ivins*, 1998]. However, the 'memory' of deglaciation from LGM may be complicated by glacial changes that are mid-to late-Holocene in age (6.5 - 0 kyr B.P.). These younger

ice load changes occur during periods of readvancement and retreat in the Antarctic Peninsula region [*Clapperton and Sugden*, 1982; 1988; *Drewry and Morris*, 1992; *Pudsey et al.*, 1994; *Goodwin*, 1998]. Late Holocene ice mass variability in the region, although of a smaller scale than that of continent-wide post-ACR unloading, could drive a potent isostatic motion due to its relative youth.

A primary goal of this paper is to quantitatively examine possible scenarios of mid-to late-Holocene (6.5-0 kyr B.P.) ice mass change and to compute the corresponding vertical crustal motion rates that could be detected using GPS. A parallel goal is then to examine the possible responses caused by late-Pleistocene to mid-Holocene deglaciation (21-5 kyr BP). These two sets of predictions are extremely sensitive to both mantle viscosity and lithospheric thickness.

2.1. Constraints on Paleohydrology

Late-Pleistocene glacial evolution of the Peninsula region is recorded in both offshore sedimentary deposits and in a variety of glacial landforms that indicate a more expansive former ice sheet [*John and Sugden*, 1971; *Elverhoi*, 1981; *Kennedy and Anderson*, 1989; *Dasilva et al.*, 1997; *Canals et al.*, 2000]. The intensity of glacial erosion and the relative scarcity of datable organic material in the southernmost portion of the Peninsula region make reconstructions there more problematic. In contrast, 15 degrees north of George IV Sound, in the southernmost Andes, land-based moraine deposits provide a somewhat more detailed glacial history than in the Antarctic Peninsula. From this record three major phases of mid-to late-Holocene glaciation have been inferred [*Mercer*, 1982; *Porter*, 1981; *Rabassa and Clapperton*, 1990; *Aniya*, 1995]. Evidence suggests that similar glacial fluctuations occurred at 0 - 5 kyr BP in the Antarctic Peninsula and environs [*Clapperton and Sugden*, 1988; *Mäusbacher et al.*, 1989; *Clapperton*, 1990; *Björck et al.*, 1991; 1996; *Yoon et al.*, 1997]. *Porter and Denton* [1967] termed the Holocene advances "Neoglacials" and they are found to have occurred along the entire high latitude Cordilleran mountains that border the eastern Pacific rim. Significant changes in regional climate may also occur on hundred year time scales, as evidenced by diatomaceous sediment core data south of Brabent Is. (Figure 1) which reveal a 200-300 year cyclicity linked to the presence of fresh melt water supply [*Leventer et al.*, 1996].

Evidence in the Antarctic Peninsula region for mid-to late-Holocene fluctuation in grounded ice cover includes: (1) Glaciomarine sequences in the South Shet-

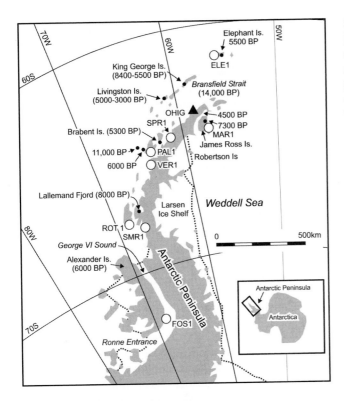

Figure 1. Map of the Antarctic Peninsula region. Locations of radiocarbon dates that provide minimum ages for glacial retreat from LGM [*Ingólfsson et al.*, 1998] are indicated by the small dots. Eight of the 17 SCAR GPS sites [*Dietrich et al.*, 2001]) and the continuous tracking station O'Higgins (OHIG) are shown as large solid dots and solid triangle, respectively. The SCAR network is also described at `http://www.tu-dresden.de/ipg/tpgsc98.html`. (From *Ivins et al.*, [2000]).

land Is. that, for example, contain a variety of distinctive basal till deposits [*Yoon et al.*, 1997]. Although the sedimentary evidence indicates that glacial retreat from LGM was well underway by 8 kyr BP [*Hjort et al.*, 1992], and probably was complete by 6 - 5 kyr BP [*Clapperton and Sugden*, 1988; *Mäusbacher et al.*, 1989], several readvances of the ice margin probably have occurred circa 1 - 3 kyr BP [*Mäusbacher et al.*, 1989; *Clapperton and Sugden*, 1988; *Yoon et al.*, 1997]. (2) The disappearance of a late-Pleistocene to early-Holocene ice shelf in George VI Sound at 6.5 kyr BP was followed by the advance of several valley outlet glaciers whose terminal moraines overlie an earlier sequence of ice shelf-derived moraines [*Clapperton and Sugden*, 1982].

2.2. An LGM Maximum Model

A reconstruction of the Antarctic Peninsula ice sheet (APIS) by *Payne et al.* [1989] employed numerical techniques described by *Mahaffy* [1976] and *Budd and Smith* [1982]. Both the regional motion of maritime air masses and ice sheet/bed rock topography were accounted for and mass wasting at the margins was simulated by calving and basal melting. Exposure to the warm Pacific westerlies and to influences of the polar water of the Weddell Sea gyre to the east were included. In the simulation, 'glacial' conditions were imposed from -40,000 to -18,000 yr, with the maximum extent of the ice sheet achieved at -14,000 yr.

The modeling of *Payne et al.* [1989] produced three important results: (1) Model predictions were strongly influenced by assumed sea-level, surface accumulation rate, marine basal melt and calving rates; (2) A 3.5 km thick ice dome is predicted that straddles the southern extent of the George VI Sound and the Ronne Entrance; (3) The timing of the growth and collapse of the APIS tends to be dominated by changes in oceanic conditions. Reasonable boundary conditions yielded simulations consistent with known constraints on the timing of the grounding line retreat at the coastlines of Alexander Island and Palmer Land at 6.5 kyr BP [*Clapperton and Sugden*, 1982]. A series of snapshots of one probable scenario for the evolution of ice sheet pattern and thickness are provided at 14, 10, 7 and 4 kyr BP.

2.3. The APIS Model Wedded to a Continental-Scale Reconstruction

The APIS model provides a basis for ice sheet evolution between 68°S and 78°S. The changing ice load outside the region of simulation must also be accounted for since rebound has long wavelength components that, for example, are influenced by ice mass evolution in the Ellsworth Mountains and the post-ACR ungrounding of ice to the south and east of Palmer Land [*James and Ivins*, 1998]. In order to overcome this problem the APIS model is merged with a model of *Denton et al.* [1991] (*D91*) that was discussed in detail by *James and Ivins* [1998]. The 1 kyr-interval digitization of the *D91* model was modified to include additional square-edged disks (also see *Tushingham and Peltier* [1991]) to give sufficient spatial detail for the APIS model.

It was also necessary to extend the APIS model north of 68°S. Here ice masses are assumed to exist out to the 500 meter bathymetric contour, with thicker ice at higher bedrock elevations. For disks located outside of the region simulated in the APIS model, the timing of retreat was assumed to follow the same trend as the APIS model with marine-based collapse underway by approximately 14 kyr BP. Compelling evidence that an active ice stream occupied the Bransfield Basin during LGM has been presented by *Canals et al.* [2000]. This

stream was grounded very locally to a depth of 1000 meters below the present-day sea-level and likely represents the maximum seaward bathymetric penetration of the LGM ice sheet in the Peninsula region.

Pallàs et al. [1998] examined raised beaches on the South Shetland Is. (Figure 1) and compared dated paleoshorelines to predictions of the ICE-3G model of *Tushingham and Peltier* [1991]. Predictions using this model are significantly larger than the relative sea-level observations. One reason for the ICE-3G over-prediction is the presence at LGM of grounded ice cover in unrealistically deep offshore waters (≥ 750 m). This deficiency originates, in part, from a very coarse disk-grid structure employed for this region in the ICE-3G model. In the present revision no LGM ice load is assumed where present-day water depths are in excess of 500 - 700 meters (also see *Bentley and Anderson* [1998]). In addition, grounded ice having no net contribution to change in ocean volume was not treated as an effective ice load. The thickest LGM load disk (2155 meters in height) covers the southern extent of George VI Sound and includes the sites on Alexander Island, including Ablation Point (50 km north of FOS1, see Figure 1), that were described by *Clapperton and Sugden* [1983].

The *D91* model, now merged with the Peninsula model, was modified in an additional way. The total melt water loss in the *D91* model of *James and Ivins* [1998] was equivalent to 24.5 meters e.s.l.r. Here the modeled Antarctic LGM collapse has a total volumetric loss equivalent to 20 meters e.s.l.r. All loading initiates at 108 kyr and increases linearly to 18 kyr BP. All disks of the previous *D91* model assumed a constant LGM height between 18 and 12 kyr BP, followed by a linear collapse that terminates at 5 kyr BP [*James and Ivins*, 1998]. All non-Peninsula disk loads retain this timing. The timing for the new continent-wide model (henceforth, *D91'*) in the Antarctic Peninsula region follows primarily from *Payne et al.* [1991]. Slow deglaciation initiates at 18 kyr BP on marine-based disks, with a broad-scale collapse (on all 38 new Peninsula disks) initiated at 14 kyr BP and fully completed at 5 kyr BP. In light of the revised timing and a 19% reduction in total LGM mass, predictions of smaller rebound signal, for a given Earth rheological structure, are expected.

2.4. Uplift Rate Predictions from D91'

Computations of the vertical isostatic motion ($\dot{w}$) for the Antarctic Peninsula are shown in Plate 1 assuming the *D91'* model and three pairs of mantle viscosity, η, and lithospheric thickness, h, values. The uplift rate predictions for a mantle viscosity of $\eta = 10^{21}$ Pa s and a lithospheric thickness of $h = 120$ km are shown

in Plate 1a. These values were assumed for the upper mantle and lithospheric thickness in the global ICE-3G model of *Tushingham and Peltier* [1991]. The predictions shown in Plate 1a might be confidently accepted if the values of h and η that are derivable from the global models, in fact, apply to the Peninsula region. More recent global modeling of the geoid, postglacial relative sea-level, mantle convection, plume migration speeds and polar wander, however, suggest that the bulk of the upper mantle has a viscosity in the range $1.0 - 7.0 \times 10^{20}$ Pa s [e.g., *Mitrovica*, 1996; *Peltier*, 1996; *Steinberger and O'Connell*, 1998; *Lambeck et al.*, 1998a; *Mitrovica and Forte*, 1997; *Vermeersen et al.*, 1998]. Consideration of lateral heterogeneity within the upper mantle may further broaden the range of possible viscosities and lithospheric thicknesses [e.g., *Sigmundsson*, 1991; *Ivins and Sammis*, 1995; *Kaufmann and Wolf*, 1996; *Hirth and Kohlstedt*, 1996; *Ivins and James*, 1999]. Weaker mechanical structure has a strong affinity with relatively youthful (≤ 15 Ma) volcanism and extensional tectonism [*Liu and Furlong*, 1994]. The northernmost Antarctic Peninsula region near the Bransfield Strait [*Hole and Larter*, 1993] may have exactly such an affinity. We return to the question of tectonic history in Section 5.1.

Plate 1 also shows the uplift rate predictions computed assuming the *D91'* load history and weaker mantle-lithosphere mechanical structure; $h = 70$ km, $\eta = 5.0 \times 10^{20}$ Pa s (b) and $h = 50$ km, $\eta = 1.5 \times 10^{20}$ Pa s (c), respectively. Clearly, these two cases reveal a diminished signature that is caused by the assumption of weaker Earth rheological structure. More than 90% of the reduction seen in Plates 1b and 1c, with respect to the uplift signature of Plate 1a, is due to the decrease in viscosity, as opposed to thinning of the lithosphere. Below we investigate the role of recent ice mass changes over a broader range of possibilities for mantle viscosity. Plate 1d gives the general character of the history of *D91'* disks. This character, with evolution terminating at 5 kyr BP, is to be contrasted to the styles of deglaciation that are considered below.

3. MODELS FOR MID TO LATE-HOLOCENE GLACIER CHANGES

3.1. Present-day Mass Balance and its Backward Extrapolation

The Antarctic Peninsula region receives a relatively large annual precipitation. Mean annual accumulation maps produced by numerical weather forecasting indicate that this region sustains the highest long-term coastal accumulation rate by at least a factor of 2 and,

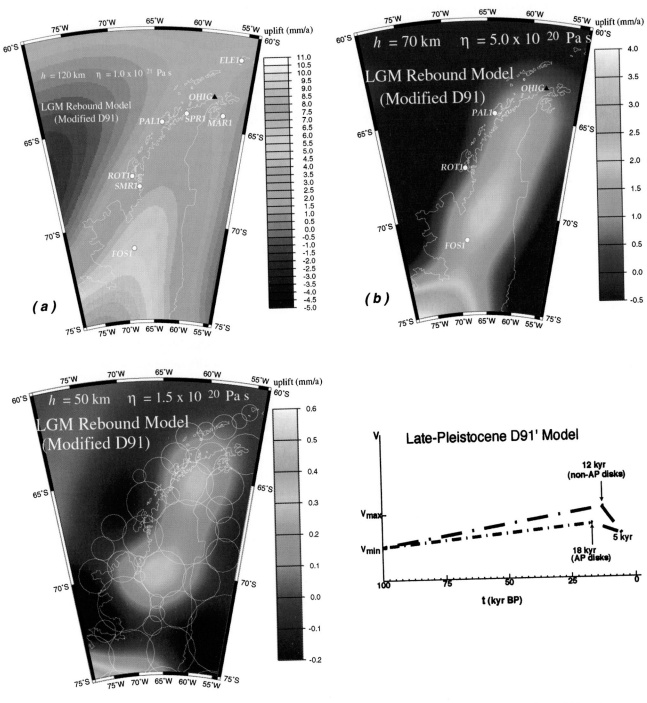

Plate 1. Present-day vertical crustal motion from LGM ice disintegration (in mm/yr) for three different viscosity structures. The disks for late-Pleistocene and early-Holocene glacial load for the *D91'* model in the Peninsula region are shown in frame (c). Rates above 3-4 mm/yr are likely to be detected from half a decade of GPS observations. Ice density and Earth parameters are given in Table 1. Frame (d) shows the general character of the load history assumed for the LGM results given in frames (a-c).

more generally, by a factor 10, over other regions of Antarctica [*Genthon and Krinner*, 1998]. The Peninsula receives nearly one-tenth of the total continental annual snow fall, despite grounded ice cover there being, in area, a mere 2.5% of the Antarctic continent [*Doake*, 1985]. Nevertheless, if substantial basal melting occurs beneath ice shelves and if the recent breakup of the Larsen Ice Shelf [*Vaughan and Doake*, 1996; *Doake et al.*, 1998] is an indicator of more widespread processes underway, then it is conceivable that the region is in a state of substantial negative mass balance. A rise in mean annual temperature and net glacial ablation near Fossil Bluff (FOS1 in Figure 1) support this interpretation [*Morris*, 1999]. In sum, however, reliable observations are too sparse to come to any firm conclusions regarding the regional 20th Century net mass balance state [*Giovinetto et al.*, 1990].

Bentley and Giovinetto [1991] estimated the surface mass balance for an ice catchment area that drains into the Bellingshausen Sea and ocean waters of the George VI Sound at +48 Gt/yr (-1 mm e.s.l.r. = 357 Gt continental ice gain). However, a large basal melt rate could more than cancel this positive surface rate [*Jacobs et al.*, 1992]. If the surface balance estimate roughly corresponds to the net imbalance rate over the entire Antarctic Peninsula region, then it would exceed negative balance estimates of *Aniya et al.* [1997] for the Patagonian icefields by about a factor of five. Such a regional imbalance amplitude (± 48 Gt/yr) sustained for 1500 years would cause a change in sea-level of 0.2 meters, an amount far too small to violate far-field observations of paleoshorelines [*Nydick et al.*, 1995]. Ablation-atmospheric temperature relations [*Morris*, 1999] suggest that this might correspond to a regional increase of mean summer air temperature of 2 °C, which is a factor of three above estimates for the average southern hemispheric warming over the past 150 years [*Jones et al.*, 1986]. The main point is that while we do not know the mass imbalance for the Peninsula region, a value of 48 Gt/yr is not inconsistent with the annual hydrological throughput estimates and will be used in our current study of isostasy.

3.2. Sawtooth Approximation for 0 - 5 kyr BP Glacial Fluctuations

Data capable of constraining the past 5000 years of glacial loading conditions in the Antarctic Peninsula region are scarce. It is necessary, therefore, to employ a proxy model of glacial fluctuations in the region. The proxy model is realistic in that numerous observations suggest such fluctuations and that it is based upon the

timing of glacial advances and retreats known to occur in the southernmost Andes of South America [*Mercer*, 1982; *Rabassa and Clapperton*, 1990; *Aniya*, 1995].

The climates of southernmost South America and the Antarctic Peninsula are dominated by maritime conditions. The glacial complexes north of the Drake Passage in southern Patagonia are in the pathway of the same eastward moving storm systems of the Antarctic circumpolar current that carry humid air masses and feed Antarctic Peninsula glaciers. The moisture content and temperatures of these storms control precipitation and glacier sustenance [*Heusser*, 1989; *Clapperton*, 1990; *Rabassa and Clapperton*, 1990; *Warren*, 1993; *Hulton et al.*, 1994]. In both the Antarctic Peninsula and in Patagonia, the glacial reaction time to increased (or decreased) moisture flux from the Pacific southern westerlies is about 200-300 years [*Drewry and Morris*, 1992; *Oerlemans*, 1997; *Goodwin*, 1998]. This equilibrium reaction time is considerably shorter than that of the vast Antarctic ice sheet south of 75°S.

Figure 2 shows two possible Neoglacial histories that are assumed for computation of mid-to late-Holocene glacially forced uplift/subsidence rates. The Neoglacials (I, II, III and IV) are somewhat ad-hoc since the evidence for retreat and advancement, and for climate

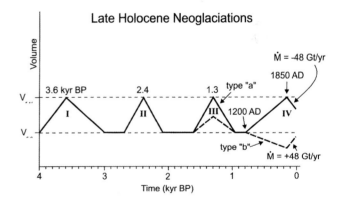

Figure 2. Changes in glaciation assumed for the Antarctic Peninsula region since 4 kyr BP. The location of model load disks are shown in Figure 3. The total variation in volume $\Delta V = 5.65 \times 10^4$ km^3 (e.s.l.r. $= \Delta \xi = 0.157$ m) is identical for each of the 4 Neoglacials (except for phases III and IV which are halved for type "b" loads, as discussed in the text). Post-1850 AD change is an equivalent sea-level rate change of ± 0.134 mm/yr. The present-day mass loss rate is scaled to the surface mass balance of a subregion within the Antarctic Peninsula as given by *Bentley and Giovinetto* [1991] and *Vaughan et al.* [1999]. In all the sawtooth Holocene cases a minor LGM load is included of total volume ΔV, initialized at 108 kyr BP, reaching a maximum at 14 kyr BP, and linearly (and completely) deglaciating until 9 kyr BP.

change in general, are limited in both space and time [e.g., *Aristarain et al.*, 1990; *Björck et al.*, 1991; 1996; *Peel*, 1995]. There is some scant evidence that a transition to arid conditions took place shortly after 3 kyr BP, based upon biogenic lake deposits recovered from James Ross Is. [*Björck et al.*, 1996]. Such a change could, for example, correspond to the sawtooth with Neoglacial I as shown in Figure 2. However, core data from these same lake deposits would imply the absence of Neoglacial II and a glacial advance and return to humid conditions at 1.2 kyr, roughly corresponding to Neoglacial III of Figure 2. Somewhat ambiguous evidence [*Aristarain et al.*, 1990; *Peel*, 1995; *Fabrés et al.*, 2000] can be presented regarding Neoglacial IV, a phase that is, as it will be shown, quite significant to the prediction of the present-day isostatic motion of the crust. The existence of a glacial advancement at this time (see Figure 2) would correspond to a southern hemispheric 'Little Ice Age' (LIA). Analyses of lake sedimentary core records by *Mäusbacher et al.* [1989] indicate that southern Patagonia and the Antarctic Peninsula have synchronous fluctuation in both humidity and glacial advance between 3 and 1 kyr BP.

Two different cases are computed in this section in which a present-day ice imbalance rate of $|\dot{M}_{p.d.}| = 48$ Gt/yr is assumed (Figure 2). The first (type "a") has a negative present-day balance state. The second assumes a positive balance state at the present-day (type "b"), having a sawtooth function for phases I and II that is identical to that assumed for type "a". However, a near-mirror image of phase IV of type "a" is assumed from AD 1200 to the present-day in type "b" loading. The two cases have the opposite sign for present-day Peninsula net mass balance. The type "b" glacial history of effectively deglaciates from AD 700 (with a brief pause from AD 1000 to 1200) to 1850, with continuous accumulation of ice mass after AD 1850. However, both backward extrapolations from the present-day use the regional mass balance magnitude suggested by *Bentley and Giovinetto* [1991], who favored a positive mass balance at the present-day. The sawtooth types "a" and "b" shown in Figure 2 represent the first two cases of mid to late-Holocene history that are considered in this paper.

A refined disk load is assembled to study the response to mid-to late-Holocene glacial change in the Peninsula region. None of the effects from larger scale continent-wide ice mass changes are included.

3.3. Predicted Isostatic Response to Sawtooth 0 - 5 kyr BP Glacial Fluctuation

3.3.1. Neoglacial Type "a" Predictions. Figure 3 shows contours of the predicted present-day uplift rate (in mm/yr) assuming the type "a" Neoglacial mass changes shown in Figure 2. It features loading from AD 1200 to 1850 with subsequent continuous mass loss at a rate of 48 Gt/yr to the present-day. Figure 3a shows the disk-grid structure employed. A relatively thick lithosphere ($h = 120$ km) and high value for the mantle viscosity ($\eta = 10^{21}$ Pa s) is assumed for the computation in Figure 3a. The remaining cases assume thinner lithosphere ($h = 50$ km) and smaller mantle (half-space) viscosity, $\eta = 1.5 \times 10^{20}$ (Figure 3b) and 1.5×10^{19} Pa s (Figure 3c). Ongoing subsidence is predicted throughout the region for the two stronger mechanical structure assumptions in frames (a) and (b) of Figure 3, with the rate of subsidence increasing as the viscosity decreases by one order of magnitude below 1.0×10^{21} Pa s. This subsidence is predicted despite 150 years of substantial mass loss, an unloading which would cause uplift if the earlier history were ignored. This predicted subsidence demonstrates the dominance of the steadily increasing load that was assumed from the year AD 1200 to AD 1850. The last thousand years of glacial history has relatively greater significance to the present-day deformation than do earlier Holocene Neoglacials of equivalent total mass fluctuation.

The surface crustal response is significantly phase-lagged with respect to the glacial forcing. This general phenomenon is encountered in coupled viscous and elastic mechanical systems subjected to periodic forcing. The phase-lag in crustal motion can be especially dramatic if the amplitude of the forcing reaches a maximum (or minimum) during the past 150 - 200 years with a more prolonged phase of glacial change during the preceding centuries [*Wolf et al.*, 1997; *Ivins and James*, 1999]. Predictions of phase-lagged crustal motion are quite sensitive to the assumed viscosity. For very weak mechanical structure, as assumed in Figure 3c, the rate at which the mantle isostatically equilibrates increases, so that the present-day response remains in-phase with the last 150 years of ice load forcing. This is analogous to the expected elastic-gravitational deformation response, but here (Figure 3c) it is a viscous response and the rate enhancement over the elastic case exceeds one order of magnitude (see figures 5 and 11 of *Ivins and James* [1999]). For the type "a" load a purely elastic effect drives vertical uplift at the present-day with maximum rates (not shown in Figure 3) of only $\approx +$ 1.5 mm/yr [*Ivins et al.*, 2000]. The glacial load model (Figure 2) uses the disk structure shown in Figure 3a. The present-day uplift pattern shown in Figure 3a displays a prominent short wavelength deformation component comparable to the load disk diameters that is absent in computations having lower viscosity (Figures 3b and 3c). The fine-scale in Figure 3a is caused by the

Mass loss rate = 48 Gt/a begins at 1850 AD with type "a" history

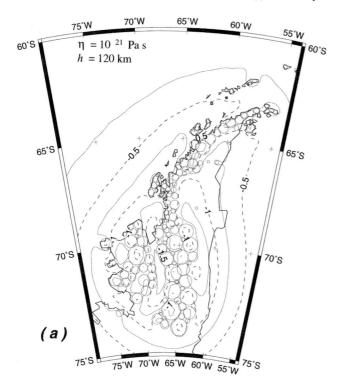

(a) $\eta = 10^{21}$ Pa s $h = 120$ km

Mass loss rate = 48 Gt/a begins at 1850 AD with type "a" history

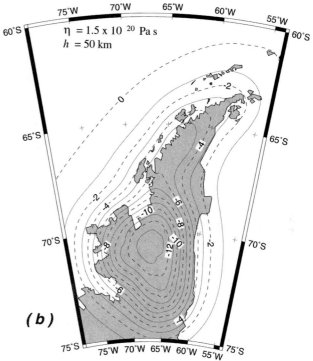

(b) $\eta = 1.5 \times 10^{20}$ Pa s $h = 50$ km

Mass loss rate = 48 Gt/a begins at 1850 AD with type "a" history

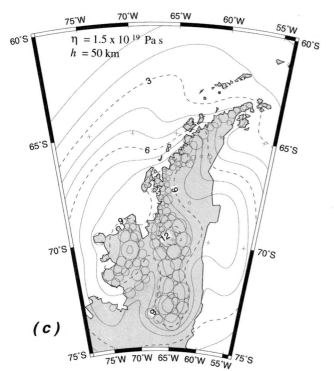

(c) $\eta = 1.5 \times 10^{19}$ Pa s $h = 50$ km

Figure 3. Present-day vertical crustal motion (in mm/yr) from type "a" history and present-day ice mass change (see Figure 2). The viscosity and lithospheric thickness are indicated on each frame. Other density and elasticity parameters are as in Plate 1. Disk locations are indicated in frames (a) and (c).

strong influence of elasticity in the deformation field in this higher viscosity case. The instantaneous motion caused by elasticity in Figures 3a and b oppose the viscous-gravitational motion in the half-space. The latter motions, and their relative strength, are relicts of the 650 year-long loading phase between AD 1200 and 1850 (see Figure 2). The subsidence in Figures 3a and 3b is caused by the viscous memory of the mantle.

3.3.2. Neoglacial Type "b" Predictions.

The type "b" loading retains the phase I and II portions of the history shown in Figure 2, but has half the volume change for phases III and IV ($\Delta V = (V_{\max} - V_{\min})/2$). This maintains the same total volume differential between 1200 and 1850 AD (in type "a") and between 700 and 1850 AD (in type "b"). The rate of ice volume change since 1850 AD for type "b" is identical in magnitude, but of opposite sign, to those of type "a". For type "b" the load growth begins in phase III at 1.5 kyr BP, is followed by ablation from 1.3 to 1.0, a pause from 1.0 to 0.8 kyr BP, and continued ablation from AD 1200 to 1850. Thereafter, mass accumulates to the present-day (see Figure 2). The loading assumes that the total amplitude of ice mass loss from 1.3 to 0.15 kyr BP is identical to the total changes assumed in each of phases I and II. The present-day uplift predictions are shown in Figure 4 for a series of differing viscosity assumptions. In this case the highest viscosity (Figure 4a) is similar to that appropriate to Fennoscandian rebound [Mitrovica, 1996; Lambeck et al., 1998a,b; Wieczerkowski et al., 1999; Scherneck et al., 2001].

Might we anticipate that type "b" loading drives an enhanced phase-lag in comparison to those appearing in the type "a" load study of Figure 3 since a longer pre-1850 phase is assumed? The answer to this question is no. Figure 4b, in fact, clearly reveals a diminished phase-lagged response, as the present-day $|\dot{w}|$ values are smaller than in Figure 3b (which offers a direct comparison). Obviously, a more subtle set of relationships among phase-lag components is operating. At viscosity values of $\eta = 3 \times 10^{19}$ and 1.5×10^{19} Pa s (Figures 4c and 4d, respectively) the direct, and in-phase, viscous motion caused by loading is dramatically enhanced. Note that Figures 3c and 4d are directly comparable, only differing in the assumed history type. The in-phase motions in Figure 4d have peak values $|\dot{w}| \approx 20$ mm/yr, as opposed to Figure 3c in which peak $|\dot{w}| \approx 12$ mm/yr. So it seems that a longer pre-1850 AD loading phase diminishes the viscous phase-lag, probably due to the establishment of near isostatic equilibrium during the earlier phase. In the next section we study the phase-lagged crustal motion behavior associated with a single

square-edged circular disk and a 3-phase load history. The simplicity of this model helps isolate the important role played by the rate of the loading during the pre-1850 AD history.

3.4. Single Disk with 3-Phase Load History

Depressions are created more rapidly during loading of a weaker, less viscous, mantle. Owing to density and gravity, the isostatic flow eventually approaches a new state of equilibrium. A series of computations, of the type described by Figures 3 and 4, reveal that as the mantle viscosity is lowered below about 3.0×10^{19} Pa s this deeper depression and the shorter memory relaxation time both act to produce senses of motion at present-day that reflect the last 150 years of monotonic load evolution. In other words, the exponential time constants of the mechanical-gravitational system are then sufficiently reduced that the memory of the previous (650 to 1500 year-long) load phase is, effectively, lost and the post-1850 viscoelastic gravitational flow field is in-phase with the present-day forcing.

A combined, wavelength-dependent, relaxation time for surface loading establishes the time scale for such memory loss. There may be a complicated tradeoff among the relative amplitudes of each decay mode [e.g., Wolf, 1985]. In particular, it is possible for the load to continue to force the crust to subside (emerge) at present, even when the net ice mass is negative (positive), as shown in the preceding section. It is important to establish criteria for the occurrence of phase-lagged behavior and to isolate the pertinent physics involved. Toward this end, some further simplifications are useful. The load history is parameterized as in Figure 5 for a single disk of constant radius α. Only three phases of linear temporal evolution of the load are included. The first phase allows ice mass to grow or decay and has a duration Δt_1. This initial phase is followed a period of quiescence of duration Δt_q. The final phase is of duration Δt_2 and at time $t \equiv t_1 + \Delta t_2$ ($t_1 \equiv \Delta t_1 + \Delta t_q$) the modeled "present-day" observation of vertical motion, $\dot{w}$, is recovered. Under these restrictive assumptions, at time t, the surface motion has an analytical form:

$$\dot{w} = -\frac{g}{\pi \mu_2^e \alpha} \cdot \langle \Gamma'(k') \{ \frac{dM_2}{dt} [a_2'(k') + \frac{\mu_2^e}{4k'\mu_1^e}]$$
$$+ \nu_p'(k') \{ \frac{dM_2}{dt} [1 - e^{-\gamma_p(k')(t-t_1)}] \quad (1)$$
$$- \frac{M_{\max}}{\Delta t_1} (1 - e^{\gamma_p(k')\Delta t_1}) \cdot e^{-\gamma_p(k')t} \} \} \rangle,$$

with the assumption of a hydrostatically pre-stressed, two-layered gravitational half-space with the deepest layer having Maxwell rheology of instantaneous shear

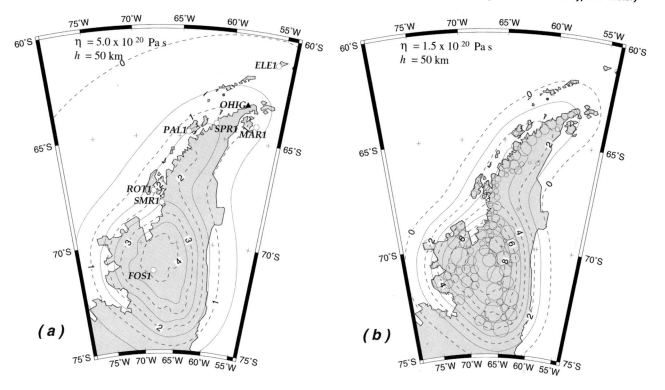

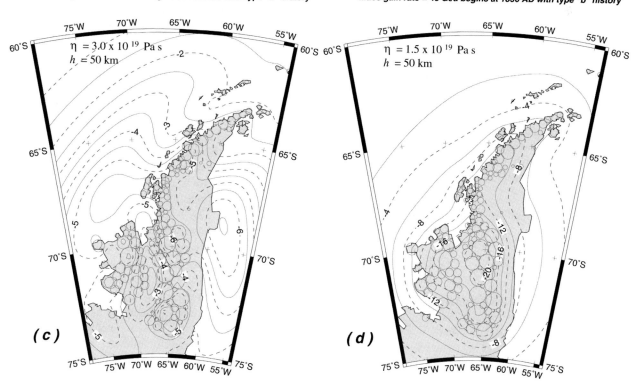

Figure 4. Present-day vertical crustal motion (in mm/yr) from type "b" Holocene and present-day ice mass change (see Figure 2). Locations of SCAR campaign and permanent GPS sites are indicated in frame (a), see Figure 1.

3-Phase, Single Disk History Scheme

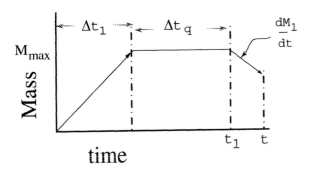

Figure 5. Load history with three phases as assumed in Section 3.4 and Figure 6. (Also see definitions in Table 2).

modulus μ_2^e, density ρ_2 and viscosity η. The top layer (lithosphere) has thickness h, shear modulus μ_1^e and density ρ_1. The solution has a dependence on the wavenumber k', indicated explicitly in the ratio of amplitude to inverse decay times, $\nu_p'(k')$, and exponential decay factors (inverse decay times) $\gamma_p(k')$ (see *Wolf* [1985] and *Ivins and James* [1999] for further details and analytic expressions). In (1) a sum is implied over the two decay modes (p), $\gamma_p(k') \geq 0$, and the convolution from wavenumber (k') to radial position r' away from the disk center is

$$\langle \cdots \rangle \equiv \int_0^\infty \cdots J_0(k'r') J_1(k'\alpha') \ dk',$$

where the n-th order Bessel function is J_n. A common factor in (1) is $\Gamma'(k') \equiv 4\mu_1^e k'/(2\mu_1^e k' + gh\rho_1)$.

The simplicity of the incompressible model allows for an analytical verification that the 1st term on the RHS of (1) associated with $\dot{M}$ is entirely elastic. The expression for $a_2'(k')$ contains no viscosity-related terms:

$$a_2'(k') = -2e^{2k'} \mu_2^e k'[\mu_2^e k'^2(gh\Delta\rho + 2\Delta\mu k')$$
$$+ \ \mu_1^e(gh\Delta\rho + 2\Delta\mu k'^2)] \ /$$

$$\Big(gh\rho_1\Big\{-\Delta\mu(gh\Delta\rho + 2\Delta\mu k') + e^{2k'}\Big[2e^{2k'}\mu_1^{e2}k'+$$

$$\mu_2^e(e^{2k'} - 4k')(gh\Delta\rho + 2\mu_2^e k') \ +$$

$$\mu_1^e\Big[(-2 + e^{2k'})gh\Delta\rho + 4e^{2k'}\mu_2^e k'\Big]\Big]\Big\}$$

$$+2\mu_1^e k'\Big\{\Delta\mu(gh\Delta\rho + 2\Delta\mu k') \ +$$

$$e^{2k'}\Big\{2(-2 + e^{2k'})\mu_1^{e2}k' \ +$$

$$\mu_1^e\Big[e^{2k'}gh\Delta\rho + 4k'(e^{2k'}\mu_2^e + gh\rho_2 + 2\Delta\mu k'^2)\Big]$$

$$+ \ \mu_2^e\Big[(2 + e^{2k'})(gh\Delta\rho + 2k'\mu_2^e)$$

$$+ \ 4k'^2(-gh\Delta\rho + 2\Delta\mu k')\Big]\Big\}\Big\}\Big) \quad (2)$$

All terms that contain the factors $1 - e^{\cdots \gamma_p(k') \cdots}$ in (1) are related to viscous deformations. In fact, Equation (1) may also be written as:

$$\dot{w} \ = \ \text{Present} - \text{day Mass Balance Rate} \ \times$$
$$\Big(\text{Elastic Deformation} + \qquad (3)$$
$$\text{Viscous Memory of Most Recent Evolution} \Big) +$$
$$\Big(\text{Most Recently Terminated Mass Balance Rate} \ \times$$
$$\text{Viscous Memory of Most Recently Terminated Change} \Big)$$

Figure 6 shows three model experiments used to compare predictions of $\dot{w}$ beneath the center of a single ice disk for differing load histories. Equation (1), or, equivalently, (3), provides the requisite theoretical architecture. The three experiments have solid Earth parameters as given in Table 1 and ice load parameters as given in Table 2. The experiments are designed to visit, in a simple way, a broad spectrum of realistic phase-lagged behavior, similar to that found for the Antarctic Peninsula uplift maps of Figure 3 and 4. Experiment 1 (Figure 6a) characterizes, quite generally, the final sawtooth of type "a" loading used in computing the uplift maps of Figure 3. In Experiment 2 there is a long period of quiescence during which the load mass does not change (Figure 6b). There is also the realistic situation of a continuous, long-term, load change that continues into the present-day. The latter is simulated in Experiment 3 (Figure 6c) with an extremely short ablation period, followed by a long-term growth phase. The total mass involved in each experiment does not vary. Later in this paper we shall conduct numerical experiments relevant to the Peninsula region that contain this "continuous mode" change featured in Experiment 3.

3.4.1. Experiment 1: Results. Experiment 1 has a long period of ice growth, followed by a short epoch of net ablation. It is assumed that the mass rate of growth (dM_1/dt) is equal and of opposite sign to the rate of ablation (dM_2/dt). The growth period is 1350 years and the ablation period is 150 years. Figure 6a

Experiment 1
(no period of quiescence)

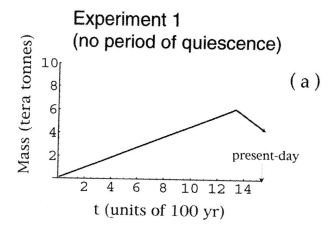

(a)

present-day

Experiment 2
(long period of quiescence)

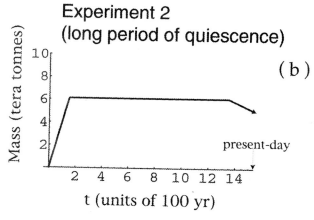

(b)

present-day

Experiment 3
(long period of mass change
at constant rate)

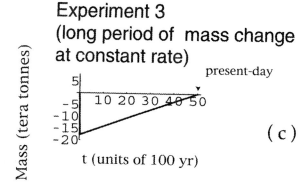

(c)

present-day

Figure 6. Three experiments for the single disk, 2 and 3-phase load histories. Frames (a-c) correspond to the model results shown in Plate 2(a-c), respectively. Two experiments, (a) and (c), assume that there is no quiescent phase (i.e., $\Delta t_q = 0$).

shows the mass history that is assumed. Plate 2 (frame a) shows the predicted central uplift (positive $\dot{w}$) over a range of half-space viscosities and disk radii for two values of the lithospheric thickness h.

For Experiment 1 the main competition among memory sources is between the 2nd and 3rd terms on the

RHS of (1). At low viscosity values ($2 - 5 \times 10^{19}$ Pa s) the $e^{-\gamma_p(k')t}$ term in the 3rd term on the RHS of (1) diminishes the phase-lagged behavior, and consequently, uplift is predicted. Note in Plate 2a that the uplift predictions (in-phase with present-day load forcing) are more robust than are the phase-lagged subsidence rates. Except for loads of small radius (α), lithospheric thickness plays a minor role, as the two different surfaces in Plate 2a primarily reflect the difference in predicted amplitude.

3.4.2. Experiment 2: Results. Experiments 1 and 2 are identical in both the total mass change (or equivalent $\Delta\xi$) and in evolution details after 0.15 kyr BP. However, the long period of quiescence ($\Delta t_q = 1200$ yr) assumed in Experiment 2 has a dramatic influence on the predicted surface crustal motion rate $\dot{w}$, for, in fact, no phase-lagged behavior is found. Here a long period having no surface ice imbalance allows gravitational (isostatic) equilibrium to be achieved (at least across the Earth parameter spectrum investigated). As the 0.15 kyr BP mass loss is initiated, the solid Earth rapidly responds, in-phase (via elastic *and* viscoelastic deformation modes) in order to stay in gravitational equilibrium with the changing surface load. (See the 1st and 2nd terms on the RHS of (1), respectively.)

3.4.3. Experiment 3: Results. The results of Experiment 2 provide impetus for Experiment 3 in which a long period of change is maintained and continues to the present-day. No phase-lags are induced since the initial phase, Δt_1, is of very short duration. (Note both exponential terms in the 3rd term on the RHS of Eq. (1) suggest severe inhibition of any phase-lagged memory effects). Also note, by comparing Plate 2c to 2a and 2b, that the amplitudes of crustal motions for a long-sustained history are the largest predicted in any of the simple tutorial experiments. Minimizing phase-lags, then, also tends to maximize the combined amplitude of direct elastic plus viscous responses.

Table 1. Model Solid Earth Mechanical Parameters

symbol	definition	model value
ρ_{ice}	ice density	$917.0\ kg\ m^{-3}$
ρ_1	lithospheric density	$3380\ kg\ m^{-3}$
ρ_2	mantle (half-space) density	$3590\ kg\ m^{-3}$
g	uniform gravity	$9.83218\ m\ s^{-2}$
μ_1^e	lithospheric rigidity	$67\ GPa$
μ_2^e	mantle rigidity	$145\ GPa$
η	mantle (half-space) viscosity	variable
h	lithospheric thickness	variable

Table 2. Single-disk 3-Phase Load History Parameters

symbol	definition
$M_{\max}$	total mass of loading density
Δt_1	time duration of initial glacial loading phase
Δt_q	time duration of a quiescent phase with load in place
Δt_2	time duration of final load phase (includes present-day)
dM_1/dt	net mass gain (loss) rate during phase 1 (initial)
dM_2/dt	net mass loss (gain) rate during phase 2 (final)
$\dot{M}_{\mathrm{p.d.}}$	mass rate at present-day
α	square-edged circular disk radius of the ice load
$\Delta\xi$	total sea-level rise (lowering) during phase 1 (initial)
$\dot{\xi}$	equivalent rate of sea-level rise during phase 2 (final)
$\dot{w}$	predicted present-day (end of phase 2) vertical crustal rate

In the section that follows we present a numerical simulation using the same disk load structure as in Figures 3 and 4. Here we investigate how long-term mass drawdown during the Holocene influences our predictions of vertical motion rates in the Peninsula region.

4. CONTINUOUS MODE HOLOCENE GLACIAL CHANGE

4.1. Additional Ice History Scenarios

The combined results in Plate 2 indicate that non-oscillatory loading events in the Antarctic Peninsula region are of interest. Thus far in this paper we have considered two general cases of loading: (1) rebound from a robust LGM load ($\Delta\xi_{\mathrm{LGM}} = 3.12$ m) in which all mass exchange with the oceans terminates after 5.0 kyr BP (Plate 1), and (2); rebound from an extremely minimalistic LGM ($\Delta\xi_{\mathrm{LGM}} = 0.157$ m), but with an oscillatory mid-to late-Holocene series of exchanges (Figures 2, 3 and 4). Two additional cases are now treated: (3); a minimalistic LGM ($\Delta\xi_{\mathrm{LGM}} = 0.72$ m) but which experiences a small load increase at 5.5 kyr BP (amounting to 0.016 m of e.s.l.r.), and subsequent decrease until AD 1850, thereafter sustaining mass loss at the rate $\dot{M}_{\mathrm{pd}} = -48$ Gt/yr, as in the cases for Figure 3. Mass loss from 5.0 to 0.15 kyr BP is 0.016 m of e.s.l.r. (see Figure 7). Finally, we consider case (4), which is identical to that in Figure 7, except that the hydrological load exchange with the oceans terminates at 2 kyr BP, with the final phase of unloading (5.0 to 2.0 kyr BP) amounting, also, to a total of 0.016 m of e.s.l.r. These additional cases aid in quantifying the differences between continuous load change to the present (case 3) and those featuring a Holocene history that terminates at 2 kyr BP (case 4).

4.2. Drawdown to the Present-day

Figure 8 shows three predictions of present-day $\dot{w}$ for lithospheric thickness $h = 50$ km in loading case (3) described above. The case of highest viscosity is likely to correspond to that of the Fennoscandian upper mantle (though the lithospheric thickness is near the lower limits [e.g., *Wieczerkowski et al.*, 1999]). The predictions are surprisingly robust in amplitude, considering the minimalistic LGM that is assumed. This case shows the important influence that a continuous drawdown mode has on the prediction of crustal motion rates in the presence of a present-day ice sheet, ice cap or ice field mass imbalance. Some aspects of this type of prediction were reported by *Le Meur and Huybrechts* [1996] and also by *Wahr et al.* [1995] who computed rebound from a continuously evolving ice sheet model. Previous studies, however, have not attempted a systematic quantification of the combined sensitivity to load history, mantle viscosity and present-day load change.

As the viscosity is lowered from 5×10^{20} ("Fennoscandian" (Figure 8a) to 1.5×10^{20} Pa s (Figure 8b), fairly small changes in the map view plots of $\dot{w}$ are predicted, with the lower viscosity case apparently capable of exciting stronger long wavelength deformation. But even the differences at long wavelength are small. Note, in contrast, how large the differences are between the LGM cases shown in Plate 1b and 1c, even though an identical viscosity differential is involved (cf. Figures 8a and 8b). In the drawdown history case, for very weak viscosity ($\eta = 1.5 \times 10^{19}$ Pa s), the predictions are dramatically larger, peaking at 30 mm/yr (see Figure 8c). In this very weak viscosity case the direct and recent viscoelastic isostatic motions combine with the continuous millennial time-scale changes, driving a vigorous rebound at the present-day.

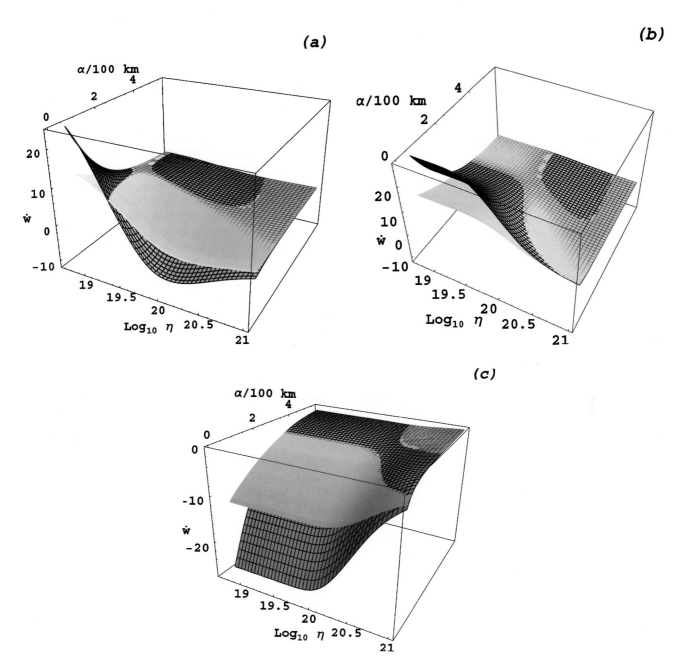

Plate 2. Uplift rates $\dot{w}$ (mm/yr) plotted in perspective (z-axis) vs. half-space viscosity ($\log_{10}$ Pa s) and single disk radius α (10^2 km). Surfaces with mesh assume lithospheric thickness $h = 35$ km and those without mesh, $h = 65$ km. Frames (a), (b) and (c) correspond to Experiments 1, 2 and 3 (see Figure 6), respectively. In all cases $dM_2/dt = \dot{M}_{p.d.}$. For frame (a): $dM_2/dt = -dM_1/dt$, $\dot{M}_{p.d.} = -4.5$ Gt/yr, $\Delta\xi = -1.7$ cm, $\Delta t_1 = 1350$ yr, $\Delta t_q = 0$ and $\Delta t_2 = 150$ yr. For frame (b): $dM_2/dt = -dM_1/dt$, $\dot{M}_{p.d.} = -4.5$ Gt/yr, $\Delta\xi = -1.7$ cm, $\Delta t_1 = 150$ yr, $\Delta t_q = 1200$ yr and $\Delta t_2 = 150$ yr. For frame (c): $|dM_1/dt| \gg |dM_2/dt|$, $\dot{M}_{p.d.} = +4.5$ Gt/yr, $\Delta\xi = +6.25$ cm, $\Delta t_1 = 10$ yr, $\Delta t_q = 0$ and $\Delta t_2 = 5000$ yr.

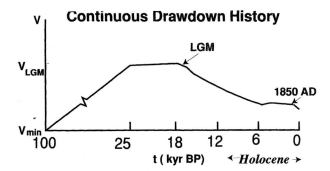

Figure 7. Continuous drawdown load history for the spatial configuration of disks as shown in Figure 3. Ordinate axis (V) is not to scale after t = 25 kyr BP. For the case in which the load terminates at 2 kyr BP (see Section 4.3) the load volume drops to V_{min} and is unchanging thereafter. (From *Ivins et al.*, [2000]).

4.3. Drawdown Truncated at 2 kyr BP

Figure 9 shows the predicted uplift for the load history shown in Figure 7, except that all loading/unloading ceases after 2 kyr BP. This truncation of the load means that the volume drops to V_{min} and is unchanging after 2 kyr BP. A comparison to the case $\eta = 5 \times 10^{20}$ Pa s, $h = 50$ km (not shown in any map view figure here) reveals that there is remarkably little difference (for this viscosity/lithospheric thickness pair) between the truncated and untruncated drawdown predictions. The map view of $\dot{w}$ shown in Figure 9a is computed with a thicker lithosphere, but slightly lower viscosity than in Figure 8a. The amplitudes in the case of drawdown truncated at 2 kyr BP are slightly smaller than in the untruncated drawdown case (Figure 8a) due to the thicker lithosphere that is assumed.

Horizontal bedrock motions may be an important diagnostic for constraining large glacial mass changes [*James and Ivins*, 1998; *Sauber et al.*, 2000]. In Figure 9b the horizontal motions predicted by this loading are also plotted, here computed on an incompressible spherical Earth model with a fluid core and a homogeneous sub-lithospheric Maxwell viscoelastic shell [*James and Morgan*, 1990]. Figure 9b reveals a scaling between peak horizontal motions and uplift of roughly one to ten, or more, similar to that discussed by *James and Lambert* [1993] and *Mitrovica et al.* [1994]. The spherical model has a truncation at harmonic degree $\ell = 650$ and the vertical uplift rate (contours in Figure 9b) compares extremely well with the cylindrical geometry calculations of Figure 9a. As anticipated, the reduction of viscosity by one order-of-magnitude below 4×10^{20} Pa s has a dramatic effect on the solution: with a viscosity of order 4×10^{19} Pa s, the memory of pre-0 AD ice

load change is altogether lost and peak vertical uplift predictions are at the level of 0.02 mm/yr (Figure 9c), far below the level of detectability.

5. SUMMARY AND CONCLUSIONS

5.1. Phase-lagged Isostasy

In this paper we have presented a series of models for the glacial history of the Antarctic Peninsula region and examined the interplay of load history, mantle viscosity and lithospheric thickness. We also developed a 3-phase single-disk model to elucidate the physics that controls a glacioisostatic phase-lagged response for viscosity values less than 5×10^{20} Pa s. The single-disk model makes clear the different role played by a long continuous drawdown of ice mass versus those having oscillatory mass change during the last few thousand years. Both the existence and character of this phase-lagged isostasy is strongly viscosity-dependent. Our summary, consequently, includes a brief discussion of the mantle-lithosphere kinematics of the Peninsula region, as the late-Cenozoic tectonics may favor a relatively weak upper mantle viscosity.

5.2. Tectonic Framework

Inversion of near-field relative sea level data for the upper mantle viscosity in Scandinavia yield values that are roughly bounded by $4.0 \times 10^{20} \leq \eta \leq 1.1 \times 10^{21}$ [*Mitrovica*, 1996; *Lambeck et al.*, 1998b; *Wieczerkowski et al.*, 1999]. However, far from the former great continental ice sheet centers of the northern hemisphere, there exists evidence that Iceland, the Svalbard Archipelago and the northern Cascadia subduction zone have a substantially weaker upper mantle viscosity [*Sigmundsson*, 1991; *Kaufmann and Wolf*, 1996; *James et al.*, 2000]. It is not fortuitous that the first two of these three regions are characterized by ocean ridge tectonics. Furthermore, the northern Cascadia subduction zone absorbs relatively youthful oceanic lithosphere [*Thorkelson and Taylor*, 1989]. The upper mantle beneath the northern Antarctic Peninsula has been the site of ridge subduction and formation of a slab-free window [*Hole and Larter*, 1993]. Consequently, it is quite plausible that the upper mantle viscosity might be an order-of-magnitude lower than those estimated for Fennoscandia, possibly near 5×10^{19} Pa s, or lower. *Ivins and James* [1999] argued that evidence for a weak mantle viscosity beneath Patagonia comes from the Plio-Pleistocene Austral volcanics of the southernmost Andean Cordillera which preserve evidence of a significant and geologically youthful slab dewatering [*Peacock et al.*, 1994].

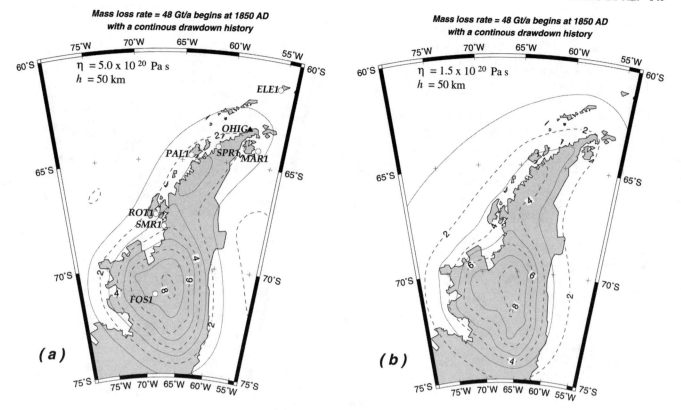

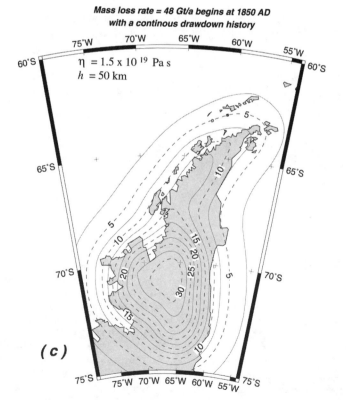

Figure 8. Present-day vertical crustal motion (in mm/yr) from continuous drawdown model of Figure 7. (Additional cases of lithospheric thickness and mantle viscosity are reported for the same loading model in *Ivins et al.* [2000].) Frames (a) through (c) show the effect of reducing the assumed half-space (mantle) viscosity.

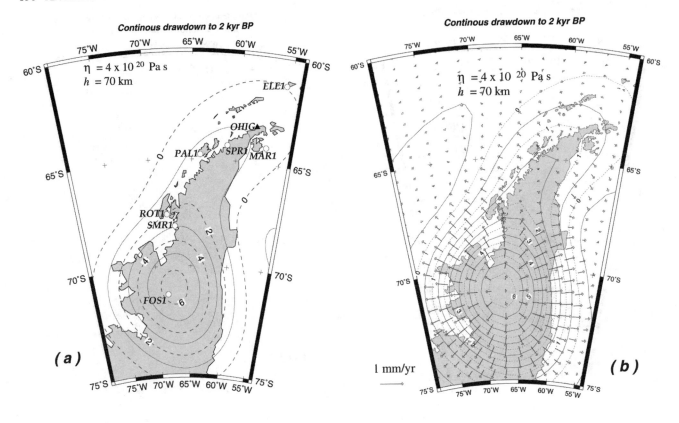

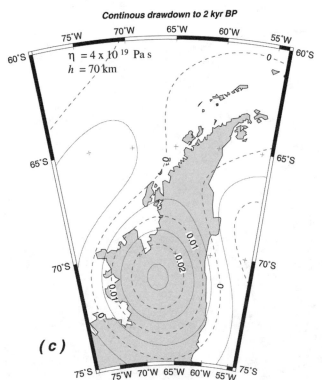

Figure 9. Present-day vertical crustal motion (in mm/yr) from continuous drawdown model, but truncated at 2 kyr BP. The contrast of frames (a) and (c) show the influence of the assumed half-space (mantle) viscosity. Frame (b) is computed with a spherical Earth model truncated at degree 650 (see text for other characteristics). The horizontal velocity is also shown in frame (b). Frames (a) and (b) have all parameters identical and, hence, provides benchmarking and checking on the numerical solutions. The differences are below 1.5%.

Although subduction of ocean ridge segments characterizes much of the late-Cenozoic tectonics of the Antarctic Peninsula, it is clear that ridge subduction and slab window formation are older tectonic events than in southern Patagonia [*Larter and Barker,* 1991]. Rifting is presently active within the Bransfield Strait north of Brabent Is. However, late-Cenozoic ridge-trench collisions are, generally, older than 4 Ma [*Hole and Larter,* 1993] and the last active rifting in the George IV Sound dates to at least 30 Ma [*Bell and King,* 1998]. Mafic intrusions associated with the last ridge-trench collision near the George VI Sound region predate 50 Ma [*Scarrow et al.,* 1998]. The antiquity of these tectonic events suggests that mantle rheology beneath the Antarctic Peninsula region may not be as weak, for example, as beneath the arc environment of Japan nor beneath the San Andreas shear zone in western North America (e.g., *Maeda et al.* [1992]; *Pollitz et al.* [2000]). Thermomechanical models of mafic underplating by *Liu and Furlong* [1994], in fact, suggest that the post-orogenic lithosphere and mantle becomes rheologically stiffened after the associated initial thermal pulse is diffused and advected away from the underplated region. We suggest that the regional tectonic history favors an upper mantle viscosity in which $\eta > 2 \times 10^{19}$ Pa s. If the viscosity of the mantle beneath the Antarctic Peninsula south of Brabent Is. is near such a lower bound, then vertical crustal motion in the region is, most likely, driven by the past 5 thousand years of intergrated glacial load history. It is likely, however, that the 1500 km-long spine of the Antartcic Peninsula is laterally heterogeneous in lithospheric thickness and upper mantle viscosity. This feature might be included in future finite element modeling, but other features, such as ocean loading, viscous stratification and compressibility should also be considered. Although we have not included these features in the models presented here, they are unlikely to significantly alter our main conclusions regarding differing load history models. If viscosity stratification involves an order-of-magnitude change, however, the mantle relaxation spectrum is substantially affected. Such a feature should be revisited with respect to the load model variations that were explored in this paper. An example might be a low viscosity basal crust, as this implies a more complex load-response wavenumber-dependency [e.g., *Di Donato et al.,* 2000] than found in the present study.

5.3. Geodetic Predictions

The models presented for Antarctic Peninsula ice loading here are probably all equally viable interpreta-

tions. The mid-to late-Holocene models are more speculative since they involve less ice and, therefore, evidence about their volume and evolution is sparse. The geodetic predictions are summarized in terms of the probable end-members of the rheological-mechanical spectrum appropriate to the Peninsula region. Table 3 gives vertical rate predictions at four SCAR sites, forming a south (FOS1) to north (OHIG) profile (see Figure 1). In Table 3 the numbered history types, (1) through (4), correspond to those enumerated in Section 4.1. As a reasonably "soft" rheological end-member, the parameter pair $h = 35$ km and $\eta = 3.0 \times 10^{19}$ Pa s is selected. A reasonably "stiff" end-member has a "Fennoscandian" character, with $h = 70$ km and $\eta = 5.0 \times 10^{20}$ Pa s, or a factor of 2 increase in lithospheric thickness and a factor of 17 increase in viscosity. These are reasonable, but not extreme, end-members since higher or lower viscosity values than those given in Table 3 cannot be ruled out.

Table 3 shows that if the D91' model, with termination of deglaciation at 5 kyr BP, is the "best" among rival load models, then mantle viscosity like that of Fennoscandia must exist for a detectable vertical uplift signal to be manifest at the present-day. This feature was also demonstrated in Plate 1. On the other hand, if the "soft" end of the viscosity spectrum is preferred, then termination of unloading, even as late as 2 kyr BP, produces virtually undetectable crustal motions at present-day. In summary, Table 3 prominently reveals that those ice models having finite mass exchange with the ocean up to, and including, the present-day produce large uplift signals. These signals are large enough to be detected using multiple years of continuous GPS tracking data [*Zumberge et al.,* 1997; *Milne et al.,* 2001]. A continuous mode of mass drawdown may, indeed, operate during the mid-to late-Holocene across the Antarctic continent. This appears to be a viable concept from the perspective of both regional evidence [e.g., *Kellogg and Kellogg,* 1987; *Goodwin,* 1998; *Bindschadler and Vornberger,* 1998] and from far-field sea-level records [e.g., *Lambeck and Bard,* 2000].

Although the earth models assumed here are quite simple, we are able to establish some generalizations from our extensive study of mid-to late-Holocene glacial loading styles. Ice load changes during this period are capable of exciting significant present-day crustal motion. Oscillatory histories tend to be self-cancelling, while, in contrast, mid-to late-Holocene continuous drawdown excites a relatively larger present-day glacioisostatic response. The modeled responses are, however, highly dependent on mantle viscosity. For sufficiently low viscosity, present-day uplift rates may be marginally different from zero, but only if the load changes ter-

Table 3. Predicted Uplift Rates $\dot{w}$ for "Stiff" and "Soft" Rheologies at Selected SCAR Sites

Load type¶	"Fennoscandian" structure		$\eta = 5 \times 10^{20}$ Pa s	$h = 70$ km
	FOS1	ROT1	PAL1	OHIG
(1)‡	+2.2 mm/yr	+1.0 mm/yr	+0.8 mm/yr	+0.7 mm/yr
(2)†	−4.6	−2.5	−2.2	−0.9
(3)†	+7.7	+2.7	+2.5	+2.1
(4)‡	+6.5	+2.2	+1.3	+0.9
	"Arc-related" structure		$\eta = 3 \times 10^{19}$ Pa s	$h = 35$ km
(1)‡	0.1 mm/yr	0.0 mm/yr	0.0 mm/yr	0.0 mm/yr
(2)†	−13.2	+2.1	−2.2	−1.0
(3)†	+21.2	+11.9	+8.0	+6.8
(4)‡	0.0	0.0	0.0	0.0

†Net present-day mass balance = - 48 Gt/yr.
‡No present-day mass imbalance.
¶(1) ⟹ max LGM; (2) ⟹ Holocene oscillatory; (3) ⟹ Holocene continuous drawdown; (4) ⟹ like (3) but terminating at 2 kyr BP (see text §4.1).

minate at sometime in the past. If the loading continues into the 20th and 21st Centuries, however, isostatic responses may be phase-lagged and geodetically detectable across a broad spectrum of mantle viscosity values.

Acknowledgments. We are grateful to Bert Vermerseen, Jeanne Sauber and Masao Nakada for very helpful reviews and comments. This work was performed at the Jet Propulsion Laboratory, California Institute of Technology, under contract with NASA and at the Pacific Geoscience Centre, an office of the Geological Survey of Canada. The JPL Supercomputing Office has provided support of its facilities. A grant from the Solid Earth and Natural Hazards Program of NASA's Earth Science Office to E.R.I. and NSF grant OPP-9527606 to C.A.R. has provided support for this research. Geological Survey of Canada Contribution 2000244.

REFERENCES

Aniya, M., Holocene glacial chronology in Patagonia: Tyndall and Upsala glaciers, *Arctic and Alpine Res., 27,* 311-322, 1995.

Aniya, M., H. Sato, R. Naruse, P. Svarca, and G. Casassa, Recent glacier variations in the Southern Patagonian Icefield, South America, *Arctic and Alpine Res., 29,* 1-12, 1997.

Aristarain, A.J., J. Jouzel and C. Lorius, A 400 year isotope record of the Antarctic Peninsula climate, *Geophys. Res. Lett., 17,* 2369-2372, 1990.

Bell, A.C. and E.C. King, New seismic data support Cenozoic rifting in George VI Sound, Antarctic Peninsula, *Geophys. J. Int., 134,* 889-902, 1998.

Bentley, M.J. and J.B. Anderson, Glacial and marine geological evidence for the ice sheet configuration in th Weddell Sea-Antarctic Peninsula region during the Last Glacial Maximum, *Antarctic Sci., 10,* 309–325, 1998.

Bentley, C.R. and M.B. Giovinetto, Mass balance of Antarctica and sea level change, in *Proc. Int. Conf. on the Role of the Polar Regions in Global Change,* edited by G. Weller, C.L. Wilson and B.A.B. Severin, pp. 481-488, Univ. Alaska, Fairbanks, 1991.

Bindschadler, R. and P. Vornberger, Changes in the West Antarctic ice sheet since 1963 from declassified satellite photography, *Science, 279,* 689-692, 1998.

Björck, S., N. Malmer, C. Hjort, P. Sandgren, Ó. Ingólfsson, B. Wallén, R.I.L. Smith and B.L. Jónsson, Stratigraphic and paleoclimatic studies of a 5500-year-old moss bank on Elephant Island, Antarctica, *Arctic and Alpine Res., 23,* 361-374, 1991.

Björck, S., S. Olsson, C. Ellis-Evans, H. Håkansson, O. Humlum and J.M. de Lirio, Late Holocene paleoclimatic records from lake sediments on James Ross Island, Antarctica, *Paleogeog. Paleoclimatology, Paleoecology, 121,* 195-220, 1996.

Budd, W.F. and I.N. Smith, Large scale numerical modeling of the Antarctic ice sheet, *Ann. Glaciology, 3,* 42-49, 1982.

Canals, M., R. Urgeles and A.M. Calafat, Deep sea-floor evidence of past ice streams off the Antarctic Peninsula, *Geology, 28,* 31-34, 2000.

Clapperton, C.M., Quaternary glaciations in the Southern Ocean and Antarctic Peninsula area, *Quaternary Sci. Rev., 9,* 229-252, 1990.

Clapperton, C.M. and D.E. Sugden, Late Quaternary glacial history of George VI Sound area, West Antarctica, *Quaternary Res., 18,* 243-267, 1982.

Clapperton, C.M. and D.E. Sugden, Geomorphology of the Ablation Point massif, Alexander Is., Antarctica, *Boreas, 12,* 125-135, 1983.

Clapperton, C.M. and D.E. Sugden, Holocene glacier fluctuations in South America and Antarctica, *Quat. Sci. Rev., 7,* 185-198, 1988.

Cullather, R.I., D.H. Bromwich and M.L. Van Woert, Spatial and temporal variability of Antarctic precipitation from atmospheric methods, *J. Climate, 11,* 334-367, 1998.

Dasilva, J.L., J.B. Anderson and J. Stravers, Seismic facies changes along a nearly continuous 24-degrees latitudinal transect- the fjords of Chile and the northern Antarctic Peninsula, *Marine Geol., 143*, 103-123, 1997.

Denton, G.H., M.L. Prentice and L.H. Burckle, Cainozoic history of the Antarctic ice sheet, in *Geology of Antarctica,* edited by R.J. Tingey, pp. 365-433, Oxford Univ. Press, Clarendon, 1991.

Di Donato, G., J.X. Mitrovica, R. Sabadini and L.L.A. Vermeersen, The influence of a ductile crustal zone on glacial isostatic adjustment: geodetic observables along the U.S. East Coast, *Geophys. Res. Lett., 27*, 3017–3020, 2000.

Dietrich, R., R. Dach, G. Engelhardt, J. Ihde, W. Korth, H.-J. Kutterer, K. Lindner, M. Mayer, F. Menge, H. Miller, W. Niemeier, J. Perlt, M. Pohl, H. Salbach, H.-W. Schenke, T. Schöne, G. Seeber, A. Veit and C. Völksen, ITRF coordinates and plate velocities from repeated GPS campaigns in Antarctica - an analysis based on different individual solutions, *J. Geodesy, 74*, 756–766, 2001.

Drewry, D.J. and E.M. Morris, The response of large ice sheets to climate change, *Phil. Trans. R. Soc. London, Ser. B, 338*, 235-242, 1992.

Doake, C.S.M., Antarctic mass balance: glaciological evidence from Antarctic Peninsula and Weddell Sea sector, *Glaciers, Ice Sheets and Sea-level: Effect of a CO_2-induced Climate Change,* Report to Workshop held in Seattle, WA Sept. 13-15, 1984, DOE/EV/60235-1, pp. 197-209, 1985.

Doake, C.S.M., H.F.J. Corr, H. Rott, P. Skvarca and N.W. Young, Breakup and conditions for stability of the northern Larsen Ice Shelf, Antarctica, *Nature, 391*, 778-780, 1998

Elverhoi, A., Evidence for a late-Wisconsin glaciation of the Weddell Sea, *Nature, 293*, 641-642, 1981.

Fabrés, J., A. Calafat, M. Canals, M. A. Bárcena and J. A. Flores, Bransfield Basin fine-grained sediments: late-Holocene sedimentary processes and Antarctic oceanographic conditions, *The Holocene, 10*, 703-718, 2000.

Genthon, C., and G. Krinner, Convergence and disposal of energy and moisture on the Antarctic polar cap from ECMWF reanalysis and forecasts, *J. Climate, 11*, 1703-1716, 1998.

Giovinetto, M.B., N.M. Waters, and C.R. Bentley, Dependence of Antarctic surface mass balance on temperature, elevation, and distance to open ocean, *J. Geophys. Res., 95*, 3517-3531, 1990.

Goodwin, I.D., Did changes in Antarctic ice volume influence late-Holocene sea-level lowering? *Quaternary. Sci. Rev., 17*, 319-332 1998.

Greischar, L. L. and C. R. Bentley, Isostatic equilibrium grounding line between the west Antarctic inland ice sheet and the Ross Ice Shelf, *Nature, 283*, 651-654, 1980.

Heusser, C.J., A polar perspective of late-Quaternary climates in the southern hemisphere, *Quaternary Res., 31*, 423-425, 1989.

Hirth, G. and D.L. Kohlstedt, Water in the oceanic upper mantle: implications for rheology, melt extraction and the evolution of the lithosphere, *Earth Planet. Sci. Lett., 144*, 93-108, 1996.

Hjort, C., Ó. Ingólfsson and S. Björck, The last major deglaciation in the Antarctic Peninsula region - A review of recent Swedish Quaternary research, in *Recent Progress in Antarctic Science,* edited by Y. Yoshida, K. Maminuma

and K. Shiroishi, pp. 741-743, Terra Scientific, Tokyo, 1992.

Hole, M.J. and R.D. Larter, Trench-proximal volcanism following ridge-crest-trench collision along the Antarctic Peninsula, *Tectonics, 12*, 897-910, 1993.

Hulton, N., D. Sugden, A. Payne, and C. Clapperton, Glacier modeling and the climate of Patagonia during the last glacial maximum *Quaternary Res., 42*, 1-19, 1994.

Ingólfsson, Ó., C. Hjort, P.A. Berkman, S. Björck, E. Calhoun, I. Goodwin, B. Hall, K. Hirakawa, M. Melles and M.L. Prentice, Antarctic glacial history since the Last Glacial Maximum: an overview of the record on land, *Antarctic. Sci, 10*, 326-344, 1998.

Ivins, E.R. and C.G. Sammis, On lateral viscosity contrast in the mantle and the rheology of low-frequency geodynamics, *Geophys. J. Int., 123*, 305–322, 1995.

Ivins, E.R. and T.S. James, Simple models for late-Holocene and present-day Patagonian glacier fluctuation and predictions of a geodetically detectable isostatic response, *Geophys. J. Int., 138*, 601–624, 1999.

Ivins, E.R., C.A. Raymond and T.S. James, The influence of 5000 year-old and younger glacial mass variability on present-day rebound in the Antarctic Peninsula, *Earth, Planets and Space, 52*, 1023–1029, 2000.

Jacobs, S. S., H. H. Helmer, C. S. M. Doake, A. Jenkins and R. M. Frolich, Melting of ice shelves and the mass balance of Antarctica, *J. Glaciology, 38*, 375–387, 1992.

James, T.S. and E.R. Ivins, Present-day Antarctic ice mass change and crustal motion, *Geophys. Res. Lett., 22*, 973–976, 1995.

James, T.S. and E.R. Ivins, Predictions of Antarctic crustal motions driven by present-day ice sheet evolution and by isostatic memory of the Last Glacial Maximum, *J. Geophys. Res., 103*, 4993–5017, 1998.

James, T.S. and A. Lambert, A comparison of VLBI data with the ICE-3G glacial rebound model, *Geophys. Res. Lett., 20*, 871–874, 1993.

James, T.S. and J.P. Morgan, Horizontal motions due to postglacial rebound, *Geophys. Res. Lett., 17*, 957–960, 1990.

James, T.S., J.J. Clague, K. Wang and I. Hutchinson, Postglacial rebound at the northern Cascadia subduction zone, *Quaternary Sci. Rev., 19*, 1527-1541, 2000.

John, B.S. and D.E. Sugden, Raised marine features and phases of glaciation in the South Shetland Islands, *Br. Antarctic Surv. Bull., 24*, 45-111, 1971.

Jones, P.D., S.C.B. Raper and T.M.L. Wigley, Southern Hemisphere surface air temperature variations: 1851-1984, *J. Climate and Appl. Meterology, 25*, 1213-1230, 1986.

Jouzel, J., C. Waelbroeck, B. Malaize, M. Bender, J. R. Petit, M. Stievenard, N. I. Barkov, J. M. Barnola, T. King, V. M. Kotlyakov, V. Lipenkov, C. Lorius, D. Raynaud, C. Ritz, and T. Sowers, Climatic interpretation of the recently extended Vostok ice records, *Climate Dynamics, 12*, 513-521, 1996.

Kaufmann, G. and D. Wolf, Deglacial land emergence and lateral upper-mantle heterogeneity in the Svalbard Archipelago: 2. Extended results for high-resolution load models, *Geophys. J. Int., 127*, 125-140, 1996.

Kennedy, D.S. and J.B. Anderson, Glacial-marine sedimentation and Quaternary glacial history of Marguerite Bay, Antarctic Peninsula, *Quaternary. Res., 31*, 255-276, 1989.

Kellogg, D.E. and T.B. Kellogg, Microfossil distributions in modern Amundsen Sea sediments, *Marine Micropaleontology*, *12*, 203-222, 1987.

Lambeck, K., C. Smither and M. Ekman, Tests of glacial rebound models for Fennoscandinavia based on instrumented sea- and lake-level records, *Geophys. J. Int.*, *135*, 375-387, 1998a.

Lambeck, K., C. Smither and P. Johnston, Sea-level change, glacial rebound and mantle viscosity for northern Europe, *Geophys. J. Int.*, *134*, 102-144, 1998b.

Lambeck, K. and E. Bard, Sea-level change along the French Mediterranean coast for the past 30,000 years, *Earth Planet. Sci. Lett.*, *175*, 200-222, 2000.

Larter R.D. and P.F. Barker, Effects of ridge crest-trench interaction on Antarctic-Phoenix spreading: forces on a young subducting plate, *J. Geophys. Res.*, *96*, 19,583-19,607, 1991.

Le Meur, E. and P. Huybrechts, A comparison of different ways of dealing with isostasy: examples from modelling the Antarctic ice sheet during the last glacial cycle, *Ann. Glaciol.*, *23*, 309-317, 1996.

Leventer, A., E.W. Domack, S.E. Ishman, S. Brachfeld, C.E. McClennen and P. Manley, Productivity cycles of 200-300 years in the Antarctic Peninsula region: understanding linkages among the sun, atmosphere, oceans, sea ice, and biota, *Geol. Soc. Amer. Bull.*, *108*, 1626-1644, 1996.

Lingle, C.S. and J.A. Clark, A numerical model of interactions between a marine ice sheet and the solid earth: application to a West Antarctic ice stream, *J. Geophys. Res.*, *90*, 1100-1114, 1985.

Liu, M. and K.P. Furlong, Intrusion and underplating of mafic magmas: thermal-rheological effects and implications for Tertiary tectonomagmatism in the North American Cordillera, *Tectonophysics*, *237*, 175-187, 1994.

Maeda, Y., M. Nakada, E. Matsumoto, and I. Matsuda, Crustal tilting from Holocene sea-level observations along the east coast of Hokkaido in Japan and upper mantle rheology, *Geophys. Res. Lett.*, *16*, 857-860, 1992.

Mahaffy, M.A.W., A numerical three-dimensional ice flow model, *J. Geophys. Res.*, *81*, 1059-1066, 1976.

Mercer, J.H., Holocene glacier variations in southern South America, *Striae*, *18*, 35-40, 1982.

Mäusbacher, R., J. Müller and R. Schmidt, Evolution of postglacial sedimentation in Antarctic lakes (King George Island), *Zeitschrift für Geomorphologie*, *33*, 219-234, 1989.

Milne, G.A., J.L. Davis, J.X. Mitrovica, Scherneck, H-G., J.M. Johansson, M. Vermeer and H. Koivula, Space-geodetic constraints on glacial isostatic adjustment in Fennoscandia, *Science*, *291*, 2381-2385, 2001.

Mitrovica, J.X., Haskell [1935] revisited, *J. Geophys. Res.*, *101*, 555-569, 1996.

Mitrovica, J.X. and J.L. Davis, The influence of a finite glaciation phase on predictions of post-glacial isostatic adjustment, *Earth Planet. Sci. Lett.*, *136*, 343-361, 1995.

Mitrovica, J.X., J.L. Davis and I.I. Shapiro, A spectral formalism for computing three-dimensional deformations due to surface loads 2. Present-day glacial isostatic adjustment, *J. Geophys. Res.*, *99*, 7075-7101, 1994.

Mitrovica, J.X., and A.M. Forte, Radial profile of mantle viscosity: results from the joint inversion of convection and postglacial rebound observables, *J. Geophys. Res.*, *102*, 2751-2769, 1997.

Morris, E.M., Surface ablation rates on Moraine Corrie Glacier, Antarctica, *Global and Planetary Change*, *22*, 221-231, 1999.

Nakada, M. and K. Lambeck, Late Pleistocene and Holocene sea-level change in the Australian region and mantle rheology, *Geophys. J.*, *96*, 497-517, 1989.

Nydick, K.R., A.B. Bidwell, E. Thomas and J.C. Varekamp, A sea-level rise curve from Guiliford, Connecticut, USA, *Marine Geol.*, *124*, 137-159, 1995.

Oerlemans, J., Climate sensitivity of Franz Joseph Glacier, New Zealand, as revealed by numerical modeling, *Arctic Alpine Res.*, *29*, 233-239, 1997.

Pallàs, R., T.S. James, F. Sábat, J.M. Vilaplana and D.R. Grant, Holocene uplift in the South Shetland Islands: Evaluation of tectonics and glacio-isostasy, in *The Antarctic Region: Geological Evolution and Processes*, edited by C.A. Ricci, pp. 861-868, Terra Antarctica Pub., Siena, Italy, 1998.

Payne, A.J., D.E. Sugden and C.M. Clapperton, Modeling the growth and decay of the Antarctic Peninsula Ice Sheet, *Quaternary Res.*, *31*, 119-134, 1989.

Peacock S.M., T. Rushmer, A.B. and Thompson, Partial melting of subducting oceanic crust, *Earth Planet. Sci. Lett.*, *121*, 227-244, 1994.

Peel, D.A. Ice core evidence from the Antarctic Peninsula region, in *Climate Since A.D. 1500*, edited by R.S. Bradley and P.D. Jones, pp. 446-462, Routledge, New York, N.Y., 1995.

Peltier, W.R., Ice age paleotopography, *Science*, *265*, 195-201, 1994.

Peltier, W.R., Mantle viscosity and ice-age ice sheet topography, *Science*, *273*, 1359–1364, 1996.

Pollitz, F.F., G. Peltzer and R. Bürgmann, Mobility of continental mantle: evidence from postseismic geodetic observations following the 1992 Landers earthquake, *J. Geophys. Res.*, *105*, 8035-8054, 2000.

Porter, S.C., Pleistocene glaciation in the southern Lake District of Chile, *Quaternary Res.*, *16*, 263-292, 1981.

Porter, S.C. and G.H. Denton, Chronology of Neoglaciation in the North American Cordillera, *American J. Sci.*, *265*, 177-210, 1967.

Pudsey, C.J., P.F. Barker and R.D. Larter, Ice sheet retreat from the Antarctic Peninsula shelf, *Continental Shelf Res.*, *14*, 1647-1675, 1994.

Rabassa, J. and C.M. Clapperton, Quaternary glaciations of the southern Andes, *Quatern. Sci. Rev.*, *9*, 153-174, 1990.

Sauber, J., G. Plafker, B.F. Molina and M.A. Bryant, Crustal deformation associated with glacial fluctuations in eastern Chugach Mountains, Alaska, *J. Geophys. Res.*, *105*, 8055–8077, 2000.

Scarrow, J.H., P.T. Leat, C.D. Wareham and I.L. Millar, Geochemistry of mafic dykes in the Antarctic Peninsula continental-margin batholith: a record of arc evolution, *Contrib. Mineral. Petrol.*, *131*, 289-305, 1998.

Scherneck, H-G., J.M. Johansson, M. Vermeer, J.L. Davis, G.A. Milne and J.X. Mitrovica, BIFROST Project: 3-D crustal deformation rates derived from GPS confirm postglacial rebound in Fennoscandia, *Earth, Planets and Space*, *53*, 703-708, 2001.

Sigmundsson, F. Postglacial rebound and asthenospheric

viscosity in Iceland, *Geophys. Res. Lett.*, *18*, 1131-1134, 1991.

Steinberger, B. and R.J. O'Connell, Advection of plumes in mantle flow: implications for hotspot motion, mantle viscosity and plume distribution, *Geophys. J. Int.*, *132*, 412-434, 1998.

Thorkelson, D.J. and R.P. Taylor, Cordilleran slab windows, *Geology*, *17*, 833-836, 1989.

Tushingham, A.M. and W.R. Peltier, ICE-3G: A new global model of late-Pleistocene deglaciation based upon geophysical predictions of postglacial relative sea-level, *J. Geophys. Res.*, *96*, 4497-4523, 1991.

Vaughan, D.G. and C.S.M. Doake, Recent atmospheric warming and retreat of ice shelves on the Antarctic Peninsula, *Nature*, *379*, 328-331, 1996.

Vaughan, D.G., J.L. Bamber, M. Giovinetto, J. Russell and A.P.R. Cooper, Reassessment of net surface mass balance in Antarctica, *J. Climate*, *12*, 328-331, 1999.

Vermeersen, L.L.A., R. Sabadini, R. Devoti, V. Luceri, P. Rutigliano, C. Sciarretta and G. Bianco, Mantle viscosity inferences from joint inversions of Pleistocene deglaciation-induced changes in geopotential and with new SLR analysis and polar wander, *Geophys. Res. Lett.*, *25*, 4261-4264, 1998.

Wahr, J., D. Han and A.S. Trupin, Predictions of vertical uplift caused by changing polar ice volumes on a viscoelastic Earth, *Geophys. Res. Lett.*, *22*, 977-980, 1995.

Wahr, J., D. Wingham and C. Bentley, A method of combining ICESAT and GRACE satellite data to constrain Antarctic mass balance, *J. Geophys. Res.*, *105*, 16,279-16,294, 2000.

Warren, C.R., Rapid recent fluctuations of the calving San-Rafael glacier, Chilean Patagonia: climatic or non-climatic? *Geografiska Annal.*, *75 A*, 111-125, 1993.

Wieczerkowski, K., J.X. Mitrovica and D. Wolf, A revised relaxation-time spectrum for Fennoscandia, *Geophys. J. Int.*, *139*, 69-86, 1999.

Wolf, D., The normal modes of a layered, incompressible Maxwell half-space, *J. Geophysics*, *57*, 106-117, 1985.

Wolf, D., F. Barthelmes and F. Sigmundsson, Predictions of deformation and gravity caused by recent change of Vatnajökull ice cap, Iceland, *Comp. Rend. J. Luxemb. Geodyn.*, *82*, 36-42, 1997.

Yoon, H.I., M.W. Han, B-K. Park, J-K. Oh, and S-K. Chang, Glaciomarine sedimentation and palaeo-glacial setting of Maxwell Bay and its tributary embayment, Marian Cove, South Shetland Islands, West Antarctica, *Marine Geol.*, *140*, 265-282, 1997.

Zumberge, J.F., M.B. Heflin, D.C. Jefferson, M.M. Watkins, and F.H. Webb, Precise point positioning for efficient and robust analysis of GPS data from large networks, *J. Geophys. Res.*, *102*, 5005-5017, 1997.

E. R. Ivins and C. A. Raymond, Division of Earth and Space Sciences, MS 300-233, Jet Propulsion Laboratory, California Institute of Technology, 4800 Oak Grove Dr., Pasadena, CA. 91109-8099, USA. (e-mail: eri@fryxell.jpl.nasa.gov; car@orion.jpl.nasa.gov)

T. S. James, Geological Survey of Canada, 9860 W. Saanich Road, Sidney, British Columbia, V8L 4B2 Canada. (e-mail: james@pgc.nrcan.gc.ca)

Recent Advances in Predicting Glaciation-Induced Sea-Level Changes and Their Impact on Model Applications

Glenn A. Milne

Department of Geological Sciences, University of Durham, Durham, UK

In the past decade, the theory most commonly adopted to predict glaciation-induced sea-level changes has been extended to include: time-dependent coastline migration, a more accurate treatment of the ocean-load increment in areas characterized by marine-based ice, and a rotating Earth model. The changes required to incorporate each of these extensions is reviewed in detail. Predictions of relative sea level (RSL) in both near- and far-field regions are generated to review and extend previous analyses that considered the impact of the above model extensions and their significance for model parameter estimation. The results, based on a single Earth/ice model pair, show that the water influx mechanism and time-dependent coastline migration can significantly impact near-field RSL predictions in Fennoscandia and the Barents Sea. A suite of RSL predictions at four far-field sites where long time-series data have been obtained show that each of the three model extensions can also significantly perturb the total signal. Therefore, future analyses that employ sea-level observations to obtain inferences of model parameters relating to either mantle rheology or ice-ocean surface mass redistribution should consider the impact of each of the above model extensions.

1. INTRODUCTION

The mass transfer between ocean water and continental ice during periods of Earth glaciation and the corresponding deformation of the solid Earth represents a unique interaction between the Earth's oceans, ice sheets and solid interior. Glacial isostatic adjustment (GIA) is the study of this interaction through the comparison of surface observables related to the glaciation-deglaciation process with predictions generated from quantitative physical models. By making this comparison, it is possible to infer values for model parameters that relate to the rheology of the solid Earth [e.g., *Haskell*, 1935; *Vening Meinesz*, 1937; *McConnell*, 1968; *Peltier and Andrews*, 1976; *Wu and Peltier*, 1983; *Sabadini et al.*, 1985; *Nakada and Lambeck*, 1989; *Ivins et al.*, 1993; *Lambeck*, 1993; *Davis and Mitrovica*, 1996; *Kaufmann and Wolf*, 1996; *Mitrovica*, 1996; *Wu*, 1999; *Milne et al.*, 2001a] as well as past and present ice sheet evolution and climate change (e.g., *Wu and Peltier*, 1983; *Nakada and Lambeck*, 1989; *Peltier and Tushingham*, 1991; *Tushingham and Peltier*, 1991; *Lambeck*, 1993; *James and Ivins*, 1997; *Yokoyama et al.*, 2000; *Mitrovica et al.*, 2001a].

There are four primary GIA data types that can be employed in the parameter inference procedure: changes in relative sea level (RSL), 3-D motions of the solid surface, the shape of the geopotential and the time derivative of this field, and changes in the Earth's rotation

Ice Sheets, Sea Level and the Dynamic Earth
Geodynamics Series 29
10.1029/029GD10

vector. The first of these data types (RSL) provides a quantitative measure of the GIA process extending back to the last glacial maximum (LGM) or earlier. The latter three data types, in contrast, provide only a present-day 'snap-shot' of the current GIA process, which is a combination of the ongoing isostatic response to past deglaciation events as well as the response to ice-ocean mass flux occurring at present. Of these four GIA data types, RSL observations have proven to be the most useful for constraining both Earth rheology and changes in ice mass distribution, largely due to the extended time period covered by these data.

A typical GIA forward model comprises three key elements: a model of glacial growth and melting, a model of the contemporaneous sea-level change, and a model of Earth rheology. The first two components represent the surface loading function, or 'input forcing', and the third component defines the Earth 'filter' that maps the input forcing into a response that can be compared with the observations. Changes in sea level are, therefore, not only an important observable, but also a component of the surface loading that influences each of the predictable data types listed above.

The theory presented by *Farrell and Clark* [1976] is widely regarded as the reference work on glaciation-induced sea-level change. Their theory is embodied in the so-called 'sea-level equation', which has been adopted in a number of subsequent analyses. Solutions of the sea-level equation are termed 'gravitationally self-consistent' since the predicted sea-level change complies with gravitational perturbations associated with mass flux on and within the Earth model (see Section 2). In their analysis, *Farrell and Clark* [1976] explicitly adopted two assumptions when applying their theory: that the geometry of their Earth model's coastlines was time invariant and that their Earth model was non-rotating. During the past decade, algorithms for solving the sea-level equation have been extended to incorporate these two mechanisms.

The geometry of the Earth's coastlines changed considerably during previous glaciations. In glaciated regions, the advance and retreat of ice sheets with a significant marine-based component resulted in the closure and opening of inland seas such as Hudson Bay in northeastern Canada and the Gulf of Bothnia in northwestern Europe. In coastal areas that were not glaciated, the magnitude and sign of the relative sea-level change and the slope of the topography govern the extent of the marine transgression or regression. That is, for a given relative sea-level change, the shallower the gradient of the land, the greater the coastline migration.

Coastline migration caused by the advance and retreat of marine-based ice sheets has rarely been discussed in the literature and so it is not clear which previous studies included this aspect of shoreline migration. Coastline migration driven by GIA-induced sea-level change has been addressed by a number of authors to varying degrees of accuracy (e.g., *Lambeck and Nakada*, 1990; *Johnston*, 1993; *Peltier*, 1994; *Milne and Mitrovica*, 1998a]. The studies by *Johnston* [1993] and *Peltier* [1994] were the first to solve the sea-level equation in a gravitationally self-consistent manner while adopting a time-varying coastline geometry. This extension of the model has a significant impact on RSL predictions in both near- and far-field regions [e.g., *Johnston*, 1993; *Milne and Mitrovica*, 1998a].

While considering the retreat of marine-based ice sheets and the consequent change in near-field coastline geometry, *Milne* [1998] discovered a significant limitation of the *Farrell and Clark* [1976] sea-level theory. At the time of ice retreat, the sub-geoidal volume once occupied by ice is replaced by ocean water. The *Farrell and Clark* [1976] theory does not predict this water influx and so a discrepancy of up to several hundred meters in the ocean load is produced in these locations at this model time step (see Section 2.3). Predictions based on a revised theory [*Milne*, 1998; *Milne et al.*, 1999] show that near-field water influx can significantly affect predictions of RSL in both near- and far-field regions.

The glaciation-induced surface ice-ocean mass redistribution is the dominant Earth forcing that drives the GIA process. A secondary forcing is induced by this surface mass flux and the resulting solid Earth deformation. These processes both act to perturb the Earth's rotation vector [e.g., *Sabadini et al.*, 1982; *Wu and Peltier*, 1984; *Mitrovica and Milne*, 1998], and the corresponding change in the Earth's rotational potential contributes to the net GIA-induced sea-level change. A number of studies have considered the impact of the GIA-induced perturbation to the rotational potential on sea-level change [*Han and Wahr*, 1989; *Bills and James*, 1996; *Milne and Mitrovica*, 1996; *Peltier*, 1998a]. *Milne and Mitrovica* [1998b] pointed out that approximations adopted in the analyses by *Han and Wahr* [1989] and *Bills and James* [1996] led to exaggerated predictions of the rotation-induced signal. In the context of predicting late Pleistocene sea-level histories, Milne and Mitrovica (1998b) concluded that the rotation-induced signal was of too small a magnitude to be of consequence in near-field regions but could be significant when considering the applications of far-field RSL data (see Section 2.4).

The goal of the current study is to review the theoretical advances introduced above and their significance for GIA model parameter estimation. The discussion will consider only the prediction of RSL observations and the inferences that result from these. It is important to note, however, that these model extensions also impact predictions of other GIA observables [e.g., *Mitrovica et al., 2001b*].

In the next section a brief description is given of the sea-level equation and the general procedure commonly adopted for its solution. Following this, a short account of the model extensions – time-dependent ocean margin, near-field water influx, and rotation – is provided. The third section includes a discussion of the influence of each of these model extensions on predictions of RSL and the consequent impact on model parameter estimation. The main results are summarized in Section 4.

2. SEA LEVEL MODEL

2.1. Introduction

In the following discussion, the term 'sea level' represents the vertical height between the geoid (or equilibrium ocean surface) and the ocean floor. Relative vertical motion of these surfaces therefore results in a change in sea level. *Farrell and Clark* [1976] described the principal physical mechanisms associated with Earth glaciation that produce sea-level changes on a viscoelastic, self-gravitating, non-rotating Earth model. In their eqn (8), they sum the contributions of these mechanisms to give the so-called 'sea-level equation',

$$S_G(\theta,\psi,t) = \frac{1}{g}\Phi^L(\theta,\psi,t) - R^L(\theta,\psi,t) + G^M(t), \quad (1)$$

where the parameters θ, ψ and t are, respectively, co-latitude, east longitude, and time relative to the onset of surface loading. The functions Φ^L and R^L represent the load-induced perturbations (hence the superscript L) to the geopotential and the vertical position of the solid surface. The Φ_L term is normalized by g, the surface gravitational acceleration, to convert the perturbation of the geopotential into a height shift of the equipotential that defines the ocean surface at the onset of surface loading. The term G^M describes a spatially uniform height shift of the geoid. This term is added to ensure that the system conserves surface ice/water mass. The change in volume bounded by the geoid and ocean floor must be consistent with the volume of ice

lost or gained by the adopted ice model. This mass conservation term is defined as [*Farrell and Clark, 1976*],

$$G^M(t) = \frac{-M_I(t)}{\rho_W A_O} - \frac{1}{A_O}\left\langle \frac{1}{g}\Phi^L(\theta,\psi,t) - R^L(\theta,\psi,t) \right\rangle, \quad (2)$$

in which $M_I(t)$ is the mass increase in grounded ice since the beginning of the loading episode, ρ_W is the density of water and A_O is the area of the ocean basins. (*Farrell and Clark* [1976] assumed this value to be constant and equal to the area of the ocean at present.) The angled brackets denote integration over the ocean basin area.

The load-induced perturbations to both the geopotential, Φ^L, and the vertical position of the solid surface, R^L, are computed by convolving, in space and time, the appropriate Green's functions with the GIA surface load [e.g., *Farrell and Clark, 1976; Peltier and Andrews, 1976; Clark et al., 1978*],

$$\Phi^L(\theta,\psi,t) = \int_{-\infty}^{t}\iint_{\Theta} a^2 L(\theta',\psi',t')\Upsilon^L(\gamma,t-t')d\Theta'\, dt',$$

$$(3)$$

and

$$R^L(\theta,\psi,t) = \int_{-\infty}^{t}\iint_{\Theta} a^2 L(\theta',\psi',t')\Gamma^L(\gamma,t-t')d\Theta'\, dt'.$$

$$(4)$$

In the above, Θ signifies spatial integration over the unit sphere, and a is the radius of the Earth model. The parameter, γ, is the great circle angle between a load point (θ',ϕ') and the 'observation' point (θ,ϕ), and $\Upsilon^L(\gamma,t)$ and $\Gamma^L(\gamma,t)$ are the impulse response Green's functions for the load-induced perturbation to the geopotential at the undeformed solid surface and the radial displacement of the solid surface, respectively [*Peltier, 1974*]. The total surface load, $L(\theta,\psi,t)$, is given by,

$$L(\theta,\psi,t) = \rho_I I(\theta,\psi,t) + \rho_W S(\theta,\psi,t), \quad (5)$$

where ρ_I is the density of ice and I gives the thickness of ice distributed over the globe as a function of time. The temporal convolution in (4) and (5) is necessary due to the viscous 'memory' of the Earth model to past surface mass flux.

The above equations summarize the key analytical relationships that describe glaciation-induced sea-level change on a viscoelastic Earth model. The remainder of the discussion within this sub-section will focus on the numerical solution of (1).

In order to compute the convolutions (3) and (4), it is necessary to prescribe a temporal form to the ice and sea-level change loading history. It is conventional to define the time evolution of these required inputs as a series of instantaneous changes at a set of discrete times [e.g., *Farrell and Clark*, 1976; *Peltier and Andrews*, 1976],

$$L(\theta, \psi, t) = \sum_{n=1}^{N} \left[\rho_I \delta I^n(\theta, \psi) + \rho_W \delta S^n(\theta, \psi) \right] H(t - t_n).$$

$$\tag{6}$$

The spatial extent of the ice and ocean load increments are given by the functions δI^n and δS^n, respectively, whereas the time evolution is specified by the Heaviside step function, H, and the loading times, t_n. The δI^n and the t_n are prescribed *a priori* based on various types of independent constraints [e.g., *Andrews*, 1982]. The remaining task, once the viscoelastic Earth structure is defined and the Green's functions are computed, is to determine the δS^n.

Computation of the sea-level increments must, of course, be based on (1). If t_0 defines the onset of surface loading, then the global sea-level increment at t_1 is given by,

$$\delta S_G^1(\theta, \psi) = S_G(\theta, \psi, t_1) - S_G(\theta, \psi, t_0), \tag{7}$$

which leads to the more general relation,

$$\delta S_G^j(\theta, \psi) = S_G(\theta, \psi, t_j) - S_G(\theta, \psi, t_{j-1}). \tag{8}$$

The subscript G denotes a global change in sea level. Of course, the sea-level loading is limited to ocean regions and so is determined by multiplying (8) by the so-called 'ocean function', $O(\theta, \psi)$ [*Munk and MacDonald*, 1960],

$$\delta S^j(\theta, \psi) = O(\theta, \psi) \delta S_G^j(\theta, \psi), \tag{9}$$

where $O(\theta, \psi)$ takes the value unity in ocean regions and zero elsewhere.

A general procedure that can be adopted to solve (1) for the δS^j is illustrated by considering the solution for the jth sea-level increment. Since the unknown sea-level change appears on the right-hand side of (1) as a component of the surface load (see eqns 3-5), the sea-level equation is an integral equation and so is commonly solved using an iterative algorithm [e.g., *Farrell and Clark*, 1976; *Mitrovica and Peltier*, 1991]. The zeroth order iterate for the jth sea-level increment is commonly taken to be the eustatic component of the prediction,

$$\left[\delta S^j \right]^{i=0} = O(\theta, \psi) \left[\frac{M_I(t_{j-1}) - M_I(t_j)}{\rho_W A_O} \right], \tag{10}$$

in which i is an iteration counter. The zeroth iterate loading increment (10) is then employed along with the

$\delta I^n(\theta, \psi)$ and the previously determined $\delta S^n(\theta, \psi)$ for $n = 1, j - 1$ in (1) to determine the next iterate of $S_G(\theta, \psi, t_j)$. The associated iterate of the sea-load increment is then obtained via eqns (8) and (9). This procedure is repeated until the difference between successive iterate solutions is less than a specified tolerance (three to five iterations are normally adequate for convergence). As stated in the Introduction, solutions of the sea-level equation are termed 'gravitationally self-consistent' since the perturbation to the geoid, which is an integral aspect of the sea-level change, is consistent with both the surface ice and water mass redistribution and the load-induced internal mass redistribution.

A variety of algorithms have been applied to solve the sea-level equation including both space domain [e.g., *Farrell and Clark*, 1976; *Peltier and Andrews*, 1978] and spectral domain [e.g., *Mitrovica and Peltier*, 1991; *Johnston*, 1993] approaches. In recent years, an iterative algorithm that involves calculations in both the spatial and spectral domains [*Mitrovica and Peltier*, 1991] has been adopted by a number of research groups due to its high computational efficiency.

Note, from the above, that the solution of (1) leads to the determination of the $\delta S^n(\theta, \psi)$, $n = 1, N$ and the global set of predictions, $S_G(\theta, \psi, t_n)$, $n = 1, N$. The $\delta S^n(\theta, \psi)$ can be subsequently applied to predict a number of other GIA observables such as 3-D crustal deformations [e.g., *Mitrovica et al.*, 1994; *Milne et al.*, 2001a] and time variations in the Earth's gravitational potential [e.g., *Mitrovica and Peltier*, 1993a]. The $S_G(\theta, \psi, t_n)$ $(n = 1, N)$ are applied to calculate changes in RSL,

$$RSL(\theta, \psi, t_j) = S_G(\theta, \psi, t_j) - S_G(\theta, \psi, t_p), \tag{11}$$

where t_p is the present time.

2.2. Time-Dependent Ocean Margin

As described in the Introduction, changes in the geometry of the ocean margin can be produced either by the growth/ablation of marine-based ice or marine transgression/regression due to relative sea-level change. In this section, the extensions of the sea-level algorithm necessary to incorporate a time-dependent ocean margin (TDOM) are described.

Extending the sea-level algorithm to incorporate coastline migration in a gravitationally self-consistent manner is not straightforward. In ice-free regions, the coastlines are located where the geoid and the solid surface intersect (i.e., where the topography is zero). Thus, in order to accurately predict the position of the Earth's coastlines in the past, the evolution of the global topography field, $T(\theta, \psi, t)$, must be known. The time evolution of $T(\theta, \psi, t)$ due to GIA can be determined from the relation,

$$T(\theta, \psi, t_j) = T(\theta, \psi, t_p) - RSL(\theta, \psi, t_j), \qquad (12)$$

where t_j is the time of a general loading increment and t_p is the present time. From (12) it is clear that in order to accurately compute the coastline evolution due to GIA, the relative sea-level change must first be determined. Due to the integral nature of the sea-level equation, any changes in coastline geometry will affect the predicted sea-level change and vice versa. Therefore, an accurate determination of the coastline evolution requires the implementation of a second iterative loop over an entire glacial cycle, rather than a single time step [e.g., *Peltier*, 1994].

The sea-level equation is first solved using a set of N ocean functions computed from a present-day topography data set and a chosen glaciation history. In ice-covered regions, the ocean function at a general loading time, t_j, is given the value unity where the topography is negative (i.e., the solid surface lies below the geoid) unless the model ice sheet is thick enough to be grounded. That is, unless,

$$I(\theta, \psi, t_j) > |T(\theta, \psi, t_p)| \frac{\rho_W}{\rho_I}, \qquad (13)$$

where $I(\theta, \psi, t_j)$ is the ice thickness. This set of zeroth iterate ocean functions defines a coastline geometry that is fixed except in glaciated regions.

The predicted sea-level change, described by the set of $S_G(\theta, \psi, t_n)$, $n = 1, N$ can then be adopted to calculate, *a posteriori*, the paleotopography at the distinct loading times, t_n, via eqns (11) and (12). A second set of N ocean functions can then be computed based on this time-varying topography field (for example, $T(\theta, \psi, t_p)$ is replaced by $T(\theta, \psi, t_j)$ in eqn 13). This procedure is repeated until successive iterates converge to within a specified tolerance (three iteration loops are usually adequate).

Some minor modifications to the theory outlined in the previous sub-section are required to accommodate a set of time-dependent ocean functions. Specifically, eqn (9) is modified to,

$$\delta S^j(\theta, \psi) = O(\theta, \psi, t_j) \delta S_G^j(\theta, \psi). \qquad (14)$$

By employing (14), it is assumed that the coastline is cliff-like and defined by $O(\theta, \psi, t_j)$ between the times t_{j-1} and t_j. The loss in accuracy introduced by this approximation can be minimized by increasing the temporal resolution of the calculation. It is also possible to introduce a revised ocean function at t_j that places the coastline midway between the predicted locations at t_{j-1} and t_j [e.g., *Johnston*, 1993].

Note that the relation,

$$\delta S^j(\theta, \psi) = O(\theta, \psi, t_j) S_G(\theta, \psi, t_j) - O(\theta, \psi, t_{j-1}) S_G(\theta, \psi, t_{j-1}), \qquad (15)$$

would appear to remove the approximation explicit in (14). However, the application of (15) leads to an incorrect prediction of the jth sea-level increment in the region defined by the ocean function migration between the two time steps [*Milne*, 1998; *Milne et al.*, 1999]. For example, consider the case of marine regression and the corresponding seaward migration of the coastline. Within the region traversed by the coastline, the ocean function $O(\theta, \psi, t_{j-1})$ is defined to be unity, whereas the ocean function $O(\theta, \psi, t_j)$ is defined to be zero. Thus, (15) reduces to,

$$\delta S(\theta, \psi, t_j) = -S_G(\theta, \psi, t_{j-1}), \qquad (16)$$

which equates the jth sea-level increment to the negative of the *total* sea-level change predicted from the onset of loading to the time t_{j-1}. Applying a similar rationale leads to the equally erroneous result for the case of a local marine transgression: the incremental sea-level change within the region defined by landward migration of the shoreline is equal to the total change in sea level from the onset of surface loading until the time t_j.

A second modification of the time-invariant coastline theory involves the calculation of the spatially uniform geoid shift required to conserve surface ice-water mass (eqn 2). When a discrete set of different ocean functions are adopted, the area of the oceans, A, becomes time-dependent. The *incremental* height shift of the geoid at the time step t_j is given by,

$$\delta G^M(t_j) = \frac{M_I(t_{j-1}) - M_I(t_j)}{\rho_W A_O(t_j)} - \frac{1}{A_O(t_j)}$$
$$\times \left\langle \frac{1}{g} \left[\Phi^L(\theta, \psi, t_j) - \Phi^L(\theta, \psi, t_{j-1}) \right] \right.$$
$$\left. - R^L(\theta, \psi, t_j) + R^L(\theta, \psi, t_{j-1}) \right\rangle_j, \qquad (17)$$

where the subscript j after the angled brackets denotes that the integration is over the area of the model ocean

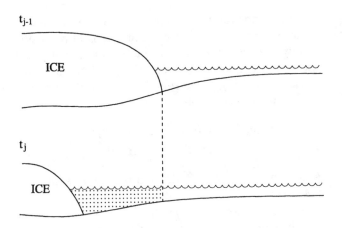

Figure 1. Schematic diagram illustrating the water load applied at the jth loading increment in regions characterized by the retreat of marine-based ice. (Taken from *Milne* [1998].)

at $t = t_j$. The net height shift of the geoid, since the onset of surface loading, is related to the δG^M via,

$$G^M(t_j) = \sum_{n=1}^{j} \delta G^M(t_n). \qquad (18)$$

2.3. Near-Field Water Influx

Milne [1998] first showed that the sea-level theory proposed by *Farrell and Clark* [1976] produces incorrect results in regions covered by marine-based ice at the times of ice retreat. Figure 1 is a schematic illustration of shoreline migration and sea-level change driven by the retreat of a marine-based ice sheet between the times t_{j-1} and t_j. The sea-level increment within the hatched region of the figure is given by the *total* height between the geoid and the solid surface and not the relative change between these two surfaces during this time increment. Therefore, the incremental sea-level change defined by (14) is not correct for this special case.

A significant revision of the standard sea-level theory is required [*Milne*, 1998] to correct this shortcoming. A revised theory, which ensures that sub-geoidal basins exposed by a retreating ice sheet are flooded with water and subsequently subjected to sea-level changes governed by the standard sea-level equation, is based on replacing the relation (14) with [*Milne*, 1998],

$$\delta S(\theta, \psi, t_j) = O(\theta, \psi, t_j)\Big[\zeta(\theta, \phi, t_j)\delta S_G^j(\theta, \psi)$$

$$+ \Big\{\zeta(\theta, \psi, t_j) - 1\Big\}T(\theta, \psi, t_j)\Big], \qquad (19)$$

where ζ takes the value zero in areas where ice has retreated and the value unity elsewhere. (This function is, therefore, completely specified by the input ice model.) The first term in the large square brackets describes the same field as in eqn (14), except in regions of ice retreat where ζ is zero. In these regions, the second term becomes non-zero and gives the negative of the paleotopography at this time (i.e., the absolute height difference between the geoid and the solid surface).

The remainder of the theory described in Section 2.2 is valid. The general procedure adopted to calculate the sea-level change based on (1) also remains the same, except that now (14) is replaced by (19) and the incremental height shift of the geoid is given by,

$$\delta G^M(t_j) = \frac{M_I(t_{j-1}) - M_I(t_j)}{\rho_W A_O(t_j)} - \frac{1}{A_O(t_j)}\Big\langle \zeta(\theta, \psi, t_j)$$

$$\times \Big[\frac{1}{g}\Big\{\Phi^L(\theta, \psi, t_j) - \Phi^L(\theta, \psi, t_{j-1})\Big\} - R^L(\theta, \psi, t_j)$$

$$+ R^L(\theta, \psi, t_{j-1})\Big] + \Big[\zeta(\theta, \psi, t_j) - 1\Big]T(\theta, \psi, t_j)\Big\rangle_j. \qquad (20)$$

As described above, two iterative loops are employed to ensure that the functions $\delta S^n(\theta, \psi)$ and $T(\theta, \psi, t_n)$ are computed in a gravitationally self-consistent manner.

Peltier [1998b] also recognized the limitation of the *Farrell and Clark* [1976] theory described above. However, the methodology adopted by *Peltier* [1998b] to solve the problem is different to that outlined in this sub-section.

2.4. Earth Rotation

In the following discussion, a right-handed Cartesian co-ordinate system is adopted which has its origin at the center of mass of the Earth model prior to surface loading. The x_1 axis is aligned with Greenwich longitude and the x_2 axis is 90 degrees east of x_1. Before surface loading begins, the model rotation vector, $\vec{\omega}(t)$, is $(0, 0, \Omega)$, where Ω is assumed to be the present-day Earth rotation rate. Subsequent to GIA surface mass flux and the associated Earth deformation, the inertia tensor of the system is perturbed and the components ω_i become non-zero (in general). The ω_i are conventionally written as [e.g., *Munk and MacDonald*, 1960],

$$\omega_i(t) = \Omega\big(\delta_{i3} + m_i(t)\big), \qquad (21)$$

where the $m_i(t)$ represent perturbations from the equilibrium state. The perturbation in the rotation vector defined by the m_i produces a perturbation in the rotational potential, which is most conveniently expressed as a sum of spherical harmonic functions [e.g., *Lambeck*, 1980; *Milne and Mitrovica*, 1998b],

$$\Lambda(\theta,\psi,t) = \Lambda_{0,0}(t)Y_{0,0}(\theta,\psi) + \sum_{m=-2}^{2} \Lambda_{2,m}(t)Y_{2,m}(\theta,\psi),$$
(22)

where

$$\Lambda_{0,0}(t) = \frac{a^2\Omega^2}{3}\left[m^2(t) + 2m_3(t)\right]$$

$$\Lambda_{2,0}(t) = \frac{a^2\Omega^2}{6\sqrt{5}}\left[m_1^2(t) + m_2^2(t) - 2m_3^2(t) - 4m_3(t)\right]$$

$$\Lambda_{2,1}(t) = \frac{a^2\Omega^2}{\sqrt{30}}\left[m_1(t)(1 + m_3(t)) - im_2(t)(1 + m_3(t))\right]$$

$$\Lambda_{2,2}(t) = \frac{a^2\Omega^2}{\sqrt{5}\sqrt{24}}\left[(m_2^2(t) - m_1^2(t)) + i2m_1(t)m_2(t)\right],$$
(23)

with

$$\Lambda_{2,-m} = (-1)^m \Lambda_{2,m}^\dagger.$$
(24)

and where the $Y_{\ell,m}(\theta,\psi)$ are spherical harmonic functions normalized such that,

$$\iint_\Omega Y_{\ell',m'}^\dagger(\theta,\psi)Y_{\ell,m}(\theta,\psi)\,\sin\theta\,d\theta d\psi = 4\pi\delta_{\ell',\ell}\delta_{m',m}.$$
(25)

In the above, i represents the complex number $\sqrt{-1}$, $\dagger$ denotes the complex conjugate and $\delta_{i,j}$ is the Kroneker delta symbol (which equals unity when $i = j$ and zero otherwise). The GIA-induced perturbation to the rotational potential, Λ, is completely described by degree zero and degree two harmonics. However, the $\Lambda_{2,1}$ coefficient is dominant since the m_3 perturbation is orders of magnitude smaller than m_1 and m_2, and $\Lambda_{2,1}$ is the only coefficient that contains first order terms in either m_1 or m_2.

Perturbations in ω_1 and ω_2 correspond to a change in the orientation of the rotation vector relative to its equilibrium position (in contrast to ω_3 which relates to a change in vector magnitude only). This re-orientation is termed true polar wander (TPW) as it refers to motion of the rotation vector with respect to the surface geography. Thus, the GIA rotational feedback is dominated by TPW which excites, predominantly, a degree 2 order 1 response of the planet [e.g., *Han and Wahr*, 1989; *Mitrovica et al.*, 2001b]. The geometric form of

the $Y_{2,1}$ spherical harmonic is schematically illustrated in Figure 2.

Incorporating the GIA rotational forcing into the revised sea-level equation (19) in a gravitationally self-consistent manner is non-trivial. An outline of the methodology is given here, but the reader is referred to *Milne and Mitrovica* [1998b] for more detail. The predicted change in global sea level, $S_G(\theta,\psi,t_j)$, on a rotating Earth model is given by,

$$S_G(\theta,\psi,t_j) = \frac{1}{g}\left[\Phi^L(\theta,\psi,t_j) + \Phi^R(\theta,\psi,t_j)\right]$$
$$+ G^M(t_j) - R^L(\theta,\psi,t_j) - R^R(\theta,\psi,t_j),$$
(26)

where

$$\Phi^R(\theta,\psi,t_j) = \int_{-\infty}^{t_j} \Lambda(\theta,\psi,t')\Upsilon^T(t_j - t')\,dt', \quad (27)$$

$$R^R(\theta,\psi,t_j) = \frac{1}{g}\int_{-\infty}^{t_j} \Lambda(\theta,\psi,t')\Gamma^T(t_j - t')\,dt', \quad (28)$$

and

$$\delta G^M(t_j) = \frac{M_I(t_{j-1}) - M_I(t_j)}{\rho_W A_O(t_j)} - \frac{1}{A_O(t_j)}\Bigg\langle \zeta(\theta,\psi,t_j)$$
$$\times \Bigg[\frac{1}{g}\Big\{\Phi^L(\theta,\psi,t_j) - \Phi^L(\theta,\psi,t_{j-1}) + \Phi^R(\theta,\psi,t_j)$$
$$- \Phi^R(\theta,\psi,t_{j-1})\Big\} - R^L(\theta,\psi,t_j) + R^L(\theta,\psi,t_{j-1})$$
$$- R^R(\theta,\psi,t_j) + R^R(\theta,\psi,t_{j-1})\Bigg] + \Big[\zeta(\theta,\psi,t_j) - 1\Big]$$
$$\times T(\theta,\psi,t_j)\Bigg\rangle$$
(29)

In the above, Φ^R and R^R are the rotation-induced perturbations to the geopotential and the radial position of the solid surface, respectively. $\Upsilon^T(t)$ and $\Gamma^T(t)$ are the viscoelastic Green's functions that govern, respectively, the predicted perturbation to the geopotential on the undeformed surface and the vertical displacement of the solid surface driven by a potential (tidal) forcing. For consistency, the temporal evolution of the GIA-induced perturbation to the rotational potential is described in the same manner as the surface loading: as a series of instantaneous changes at the times t_n,

$$\Lambda(\theta,\psi,t) = \sum_{n=1}^{N} \delta\Lambda^n(\theta,\psi)\,H(t - t_n).$$
(30)

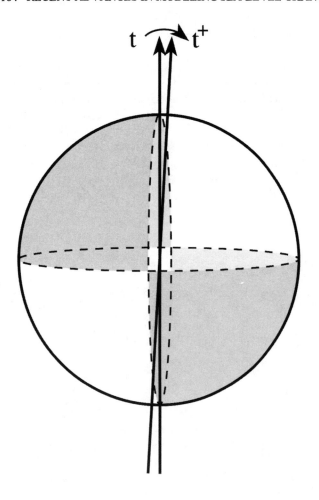

Figure 2. Schematic diagram illustrating the spherical harmonic degree 2 order 1 geometry that dominates the perturbation to the rotational potential driven by clockwise wander of the rotation pole between the times t and t^+. Shaded areas indicate a positive perturbation whereas non-shaded areas indicate a negative perturbation. (Adapted from *Mound and Mitrovica* [1998].)

As discussed above, the change in sea level at each time step is determined in an iterative fashion. The GIA-induced perturbation to the ω_i can be calculated if the surface loading history and the Earth model rheology are specified [e.g., *Wu and Peltier*, 1984; *Milne and Mitrovica*, 1998b]. During a specific time step iteration, the $\delta\Lambda^j(\theta,\psi)$ is determined via eqns (22) to (25) and is used to compute the rotation-induced perturbations to the geoid and the solid surface which are then employed in eqn (26) to calculate the global sea-level change. The remainder of the calculation is based on eqns (19) and (29) and is executed as outlined in the previous subsection. Incorporating the rotational component of the calculation within the iteration loop at each time step

does not significantly add to the total computation time and ensures that the sea-level calculation is performed in a gravitationally self-consistent manner.

3. IMPACT ON MODELING APPLICATIONS

The predictions shown in the following are all based on the same Earth and ice model pair. The Earth model is spherically symmetric, compressible and self-gravitating, and has a Maxwell viscoelastic rheology. The elastic structure has a depth resolution of 10km in the crust and 25 km in the mantle and is taken from PREM [*Dziewonski and Anderson*, 1981]. The viscosity structure is more crudely parameterized into three layers: a 100 km elastic lithosphere, an upper mantle region extending from the base of the lithosphere to the 660 km seismic discontinuity with a viscosity of 5×10^{20} Pa s, and a lower mantle region extending to the core-mantle boundary with a viscosity of 10^{22} Pa s. This viscosity profile is broadly compatible with the inferences obtained in a number of recent analyses [e.g., *Nakada and Lambeck*, 1989; *Mitrovica and Forte*, 1997; *Lambeck et al.*, 1998; *Peltier*, 1996; *Wieczerkowski et al.*, 1998; *Milne et al.*, 2001a].

The adopted ice model is a hybrid of two published models. To illustrate the consequence of implementing a time-dependent ocean function and the near-field water influx mechanism, it is important to adopt a high-resolution regional ice model. For this purpose, the Fennoscandian and Barents sea ice models proposed by *Lambeck* [1995, 1996] and *Lambeck et al.* [1998] were adopted. Ice histories in all other regions are taken from the ICE-3G deglaciation model [*Tushingham and Peltier*, 1991]. In this case, the timing of the melt increments for the ICE-3G model were altered to match those for the Fennoscandian and Barents Sea regional model. A glaciation phase for the ICE-3G component of the model was constructed by reversing the time evolution of the deglaciation history and extending the increment between loading episodes from 1 cal. kyr to 7 cal. kyr.

Solutions of the sea-level equation were obtained using the pseudospectral algorithm [*Mitrovica and Peltier*, 1991] with harmonic expansions truncated at degree and order 256.

3.1. Near-Field Relative Sea-Level Data

The magnitude of GIA-induced sea-level change is largest in near-field regions due to the influence of ice loading. As a consequence, data from these regions have the greatest sensitivity to both Earth rheology and the space-time history of the local ice sheets. A large number of GIA modeling analyses have been based on near-field data, particularly those from Fennoscandia [e.g., *Haskell*, 1935; *Vening Meinesz*, 1937; *McConnell*, 1968; *Wolf*, 1987; *Tushingham and Peltier*, 1991; *Fjeldskaar*, 1994; *Mitrovica*, 1996; *Lambeck et al.*, 1998; *Wieczerkowski et al.*, 1999] and northeastern Canada [e.g., *Peltier and Andrews*, 1976; *Wu and Peltier*, 1983; *Tushingham and Peltier*, 1991; *Mitrovica and Peltier*, 1995; *Mitrovica and Forte*, 1997]. In the following we consider the influence of the GIA model extensions described in Section 2 on predictions of near-field relative sea-level change and the consequences of this influence on traditional parameter estimation procedures.

Milne et al. [1999] showed that the water influx mechanism significantly impacts RSL predictions in northeastern Canada, especially in the vicinity of Hudson Bay where a large component of the Laurentide ice sheet was marine-based. To complement these results, the following discussion will focus on the impact of both a TDOM and water influx in northwestern Europe. In regions of major glaciation activity, such as northeastern Canada and northwestern Europe, the magnitude of the load-induced RSL signal is several hundreds of meters immediately preceding and during the early Holocene. In contrast, the rotation-induced RSL signal is around 1 m in magnitude at this time [e.g., *Milne and Mitrovica*, 1998b]. Thus, the rotation-induced signal is too small to be of consequence and so will not be discussed further in this sub-section.

Figure 3 shows predictions of the sea-load increments in the Fennoscandian region during the end of the deglaciation phase of the adopted high-resolution regional ice model (left-hand column). The predictions shown in the columns labeled 'A', 'B' and 'C' are based on progressively more complex sea-level models and illustrate well the extensions described in Section 2. The sea-loading increments in column A are computed by solving the original sea-level equation as defined by *Farrell and Clark* [1976], assuming that the ocean function remains fixed except where the marine-based components of ice sheets advance or retreat (hereafter referred to as 'model A'). These predictions show that the sea-level change is negative in the immediate vicinity of the waning ice sheet but becomes positive beyond a clearly defined periphery. The negative values are a result of two dominant mechanisms: the rebound of the solid Earth in consequence of the reduced surface load as well as the diminishing direct gravitational attraction of the ice sheet. Both of these effects reduce in magnitude with distance from the glaciation center, with the

crustal motion changing from uplift to subsidence in the so-called 'peripheral bulge' region. Note that, for this set of predictions, any changes in the coastline geometry are a consequence of the shifting ice sheet margin only.

The predictions shown in column B were computed in the same manner as those in column A except that the sea-level algorithm was extended to incorporate a time-dependent ocean function in a gravitationally self-consistent manner (hereafter denoted as 'model B'). The geometry of the coastline changes dramatically in regions characterized by large amplitude sea-level variations and low-lying topography (e.g., much of Finland and parts of southern Sweden). The magnitude of the sea-load increments is, in contrast, comparable to those shown in column A. Note that in columns A and B, the sign of the predicted sea-load increments is consistently negative, illustrating the limitation of the original sea-level theory in regions covered by marine-based ice at times of ice retreat (see Fig. 1). Application of the revised sea-level theory (see eqn 19) that incorporates both near-field water influx and a gravitationally self-consistent and time-dependent ocean function, gives the result shown in column C (this model is hereafter referred to as 'model C').

The water influx mechanism is most pronounced at the 10.2 cal. kyr BP time step, when the marine-based component of the model ice sheet retreats from the south to the north of the Gulf of Bothnia. As discussed above, this episode of ice retreat is accompanied by an influx of water to the basin vacated by the ice sheet. The sea load increment at this time is positive with a magnitude of up to several hundreds of meters, compared to the negative load of a few tens of meters in columns A and B (compare these results to those shown in Figs 4 and 5 of *Milne et al.* [1999]). A comparison of the predictions in columns B and C indicates that this non-perturbative extension of the theory has a dramatic effect on the predicted signal and results in a significantly more accurate simulation of the sea load in the vicinity of a retreating marine-based ice sheet.

There are several interesting features to note in column C of Fig. 3. The predictions suggest that Lake Vänern in southern Sweden may once have been connected to both the Baltic Sea and the North Sea. Also note that the island Saaremaa (approx. 22.5°E, 58.5°N) is submerged at 11.6 cal. kyr BP and then begins to emerge at 10.2 cal. kyr BP as a result of the sea-level fall in this region. The increasing surface area of the island Gotland (approx. 18.5°E, 57.0°N) during the three time steps considered also illustrates this effect.

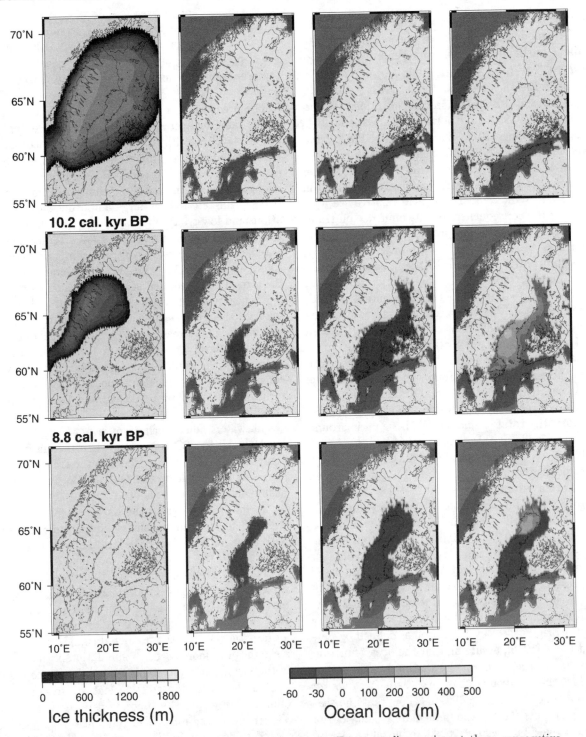

10.2 cal. kyr BP

8.8 cal. kyr BP

0   600   1200   1800
Ice thickness (m)

-60 -30   0   100 200 300 400 500
Ocean load (m)

Figure 3. Predictions of the water-load increments in the Fennoscandian region at three consecutive time steps in the deglaciation of the adopted high resolution ice model (left-hand frames) [*Lambeck et al.*, 1998]. The light-grey shade in all frames corresponds to areas that are ice-free and above the geoid. The column labeled 'A' shows the ocean load increments at the times indicated in the left-hand column predicted using the traditional sea-level theory (time-dependence of the ocean function due to ice migration is included). The results in columns B and C show equivalent predictions to those in column A except that they are based on versions of the sea-level theory extended to include gravitationally self-consistent coastline migration (B), or both gravitationally self-consistent coastline migration and near-field water influx (C).

The predicted timing and rate of these 'events' is, of course, dependent on the specific choice of input glaciation history and Earth model.

The predictions in column C also provide evidence of a limitation of the sea-level theory [*Milne*, 1998; *Milne et al.*, 1999]. As described in the previous section, the ocean function is defined to be unity where the topography is negative (i.e., where the solid surface lies below the geoid) and where there is no grounded ice cover. As a result, some inland lakes that lie below contemporaneous sea level are automatically defined as ocean areas by the model even though they are not connected to the open ocean. For example, the water load increment predicted within Lake Vänern at 8.8 cal. kyr BP illustrates the error introduced by this limitation of the sea-level algorithm.

The model extensions illustrated in Fig. 3, columns B and C, impact predictions of RSL in near-field regions, especially those deglaciated by ice sheets with marine-based components [e.g., *Milne et al.*, 1999]. The impact on RSL predictions due to both a time-dependent ocean function and near-field water influx are considered for Fennoscandia and the Barents Sea in Figure 4.

The top map in Fig. 4 shows the difference between predictions based on model B and those based on model A at 10.2 cal. kyr BP. As one would expect, the largest discrepancies are located where the sea-level driven coastline change and the magnitude of the ocean load are large. Predictions of RSL in much of Finland and parts of southern Sweden are most significantly affected, with model B predicting higher relative sea levels by up to ~15 m. The sign of the discrepancy in these areas is a consequence of the larger area of ocean being subject to the (negative) sea-loading increments, which leads to a larger sea-level fall at these locations. The time dependence of the difference between these two sets of model predictions is illustrated in Figure 5, which shows predictions of RSL at six locations where sea-level data have been obtained [see *Lambeck et al.*, 1998 and references therein]. The RSL predictions based on model B (dashed line) lie above those based on model A at all of the chosen locations. Of the sites chosen, the maximum discrepancy is ~9 m at site S3.

The lower map in Fig. 4 shows the predicted discrepancy between model C and model B. This discrepancy is due to water influx only and can therefore be compared with Figure 5 in *Milne et al.* [1999]. The magnitude of the signal is considerably larger than that shown in the top map (peak to peak amplitude of ~50 m compared to ~15 m) indicating that this extension of the model is the more significant of the two considered in this sub-

section. The geometry and sign of the signal shown in the lower map of Fig. 4 can be understood by considering the difference between the sea-level loads predicted by models B and C. The differential load (model C minus model B) is, to a first approximation, positive and located in the Gulf of Bothnia and the northwestern part of the Barents Sea. Areas immediately under this load will thus experience a 'differential' sea-level rise due to the load-induced solid-surface subsidence and the geoid rise associated with the direct gravitational attraction. Therefore, a reduced sea-level fall is predicted for model C compared to model B. In contrast, the positive sign of the discrepancy in the region peripheral to the differential water load is largely a result of crustal subsidence.

RSL predictions based on model C at the six sites indicated in Fig. 4 are shown by the solid lines in Fig. 5. The predicted sea-level fall is considerably lower for model C compared to either B or A. The maximum and minimum discrepancies between model C and model B predictions are 36 m at site S3 and 20 m at site S6, respectively, at 10.2 cal. kyr BP.

The results shown in Figs 3 and 5 are consistent with those presented by *Milne et al.* [1999] for northeastern Canada. The difference in RSL predictions based on model C, compared to either models B or A, is generally negative in the Gulf of Bothnia and the Barents Sea. Thus, application of this more accurate model will result in estimates of ice thickness in these regions that are considerably thicker. For example, employing data from sites S2 to S6 to infer ice thickness within the southern part of the Gulf of Bothnia would result in an increase in thickness, compared to the same analysis based on either model B or model A, by about 20% (i.e., the average of the discrepancy at these five sites).

Inferences of mantle viscosity or lithospheric thickness based on near-field data from this region will also be significantly biased by the adoption of an inaccurate sea-level theory (the level of bias is not straightforward to determine since the predicted RSL signal is not linearly dependent on these parameters). Note, however, that inferences of viscosity based on the decay-time parameterization of near-field data [e.g., *Mitrovica and Peltier*, 1993b; *Mitrovica and Forte*, 1997] are less affected than the 'raw' RSL data by this extension to the sea-level theory. This results from the relative insensitivity of the decay-time data to variations in the loading history [e.g., *Mitrovica and Peltier*, 1993b]. For example, data from sites S2 and S6 were adopted to infer viscosity using the decay-time analysis [*Mitrovica and Forte*, 1997]. The values obtained for the decay

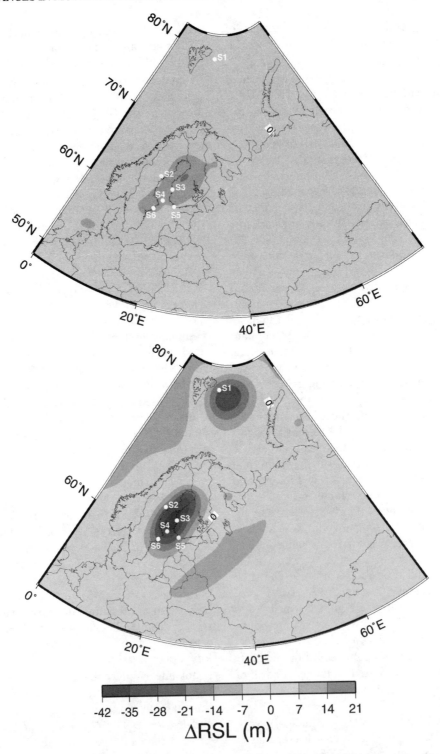

Figure 4. The top frame shows the differential RSL signal at 10.2 cal. kyr BP determined by subtracting predictions based on the model illustrated in column A (Fig. 3) from predictions based on the model illustrated in column B (Fig. 3). The bottom frame shows the differential RSL signal at 10.2 cal. kyr BP determined by subtracting predictions based on the model illustrated in column B (Fig. 3) from predictions based on the model illustrated in column C (Fig. 3). The labeled dots mark selected locations where data have been obtained.

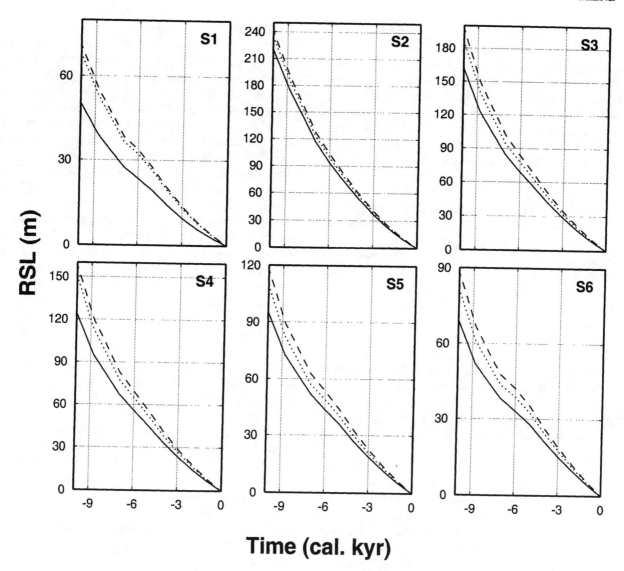

Figure 5. RSL predictions at the six sites indicated in Fig. 4 based on: the traditional sea-level theory (with an ice-induced time-dependent ocean margin) (dotted lines); the extended theory that includes a gravitationally self-consistent ocean margin (dashed line); or the extended theory that includes both a gravitationally self-consistent ocean margin and near-field water influx (solid line).

times based on models A, B and C are, respectively, 11.60, 11.76 and 12.23 kyr for site S2 and 15.67, 14.79 and 14.54 kyr for site S6. This variation in the predicted decay-time values is smaller than the calculated 1-σ uncertainties of 1.35 kyr (S2) and 2.50 kyr (S6).

3.2. Far-Field Relative Sea-Level Data

At a considerable distance from the major glaciation centers, the influence of ice loading is significantly reduced and the sea-level signal is dominated by the meltwater rise. Far-field data thus provide important con-

straints on the rate and magnitude of meltwater influx to the oceans during deglaciation which are, in turn, key signatures of paleoclimatic change on the ice-age Earth [e.g., *Clark and Mix*, 2000]. The meltwater, or eustatic, signal can be readily converted into an estimate of grounded continental ice volume at a past time compared to the present, and thus serves as an important integral constraint on ice sheet reconstructions [e.g., *Flemming et al.*, 1998; *Lambeck et al.*, 2000; *Milne et al.*, 2001b].

High-stands in RSL are a common feature of the far-field sea-level record during the mid-Holocene. These data mark the time when the globally uniform fall of the geoid due to ocean syphoning [*Mitrovica and Peltier*, 1991] surpassed the magnitude of geoid rise due to meltwater addition to the oceans. These data are sensitive to both local and far-field rheology, as well as the occurrence of glacial melt during the mid- to late Holocene [e.g., *Nakada and Lambeck*, 1989]. *Nakada and Lambeck* [1989] considered differences between high-stands observed at various locations around the Australian continent to constrain local rheological parameters (the differencing procedure acts to remove data sensitivity to syphoning and late Holocene melting) and then adopted the best-fitting rheology model to predict and thus correct the high-stand data for ocean load induced deformation and syphoning. They then applied these data to infer a late Holocene melt signal. This is an important application of far-field data that can provide relatively robust constraints on both mantle rheology and climate parameters.

In the following, the impact of the model extensions described in Section 2 on predictions of far-field sea-level change is considered. In particular, the discussion will focus on the significance of these model extensions for the applications outlined above.

Figure 6 shows predictions at four far-field sites where data have been employed to constrain the meltwater signal [e.g., *Flemming et al.*, 1998; *Milne et al.*, 2001b; *Peltier*, 1994; *Yokoyama et al.*, 2000]. The solid line in the left-hand frames is based on a sea-level model that includes all of the extensions discussed in Section 2 (hereafter referred to as the 'reference' model). The remaining curves in these left-hand frames are based on the reference model with one of the following three mechanisms removed: TDOM, near-field water influx or rotation (see caption for more information). The curves in the right-hand frames show the difference between the reference model predictions and those of the three simpler models and so they show, directly, the influence of the three model extensions described in Section 2.

Note that the ice model adopted in the present analysis contains an equivalent melt water rise of $\sim$108 m. This value is likely a $15 - 20\%$ underestimate of the actual eustatic sea-level rise since the LGM [*Milne et al.*, 2001b]. As a result, the influence of a TDOM on RSL predictions is underestimated by a similar amount, while the impact of near-field water influx is reduced by a relatively smaller value.

The impact of a gravitationally self-consistent TDOM on the predictions is shown by the dashed lines in Fig. 6 (right-hand frames). At three of the four sites (Barbados, Huon Peninsula and Tahiti), the inclusion of a

TDOM results in a monotonic sea-level rise of about 4 m from the LGM until the early Holocene. At Bonaparte Gulf, however, the TDOM produces a nearly monotonic fall in sea level of magnitude $\sim$4.5 m from 20 cal. kyr BP to the present.

Incorporating a TDOM produces a sea-level rise at Barbados, Huon Peninsula and Tahiti through two different mechanisms. The eustatic rise in sea level due to meltwater addition to the oceans is given by the first term on the right-hand side of eqn (29). When a TDOM algorithm is adopted, the area of the ocean (A_O) is less than that for a fixed (to present-day) continental margin at each time step, and so a larger eustatic sea-level rise is predicted. The second term (enclosed in angular brackets) in eqn (29) describes the contribution to the uniform geoid shift resulting from GIA-induced deformations of the geoid and the solid surface (as well as any contribution from near-field water influx). The effect of the ocean loading on this term is most relevant here. When a TDOM is adopted, the area traversed by the model coastlines during deglaciation experiences a negative load (relative to the fixed margin case) in both near- and far-field regions, and so a sea-level fall is predicted (compared to the fixed margin model; see *Milne and Mitrovica* [1998a] for a schematic illustration of this point). To compensate for this reduction in volume between the geoid and the ocean floor, the geoid height is uniformly increased (via eqn 29) to ensure mass conservation and this contributes to the predicted sea-level rise at the three sites.

The TDOM-induced sea-level signal at Bonaparte Gulf is caused by the dramatic coastline migration at this location [see, for example, Fig. 1 of *Milne and Mitrovica*, 1998a] (Bonaparte Gulf is located at approximately 12°S and 128°E). The shelf area in this region was exposed from the LGM until the Lateglacial and so the net ocean loading was significantly less than in other regions where the coastline remained relatively stationary. This differential negative load leads to a local sea-level fall of around 8 m and thus explains the anomalous signal at this site compared to the other three considered.

At the four sites considered in the present analysis, the error induced by not considering a TDOM is at the 5% level, which is approximately the magnitude of the observational precision. However, the discrepancy is more significant at other locations. *Milne and Mitrovica* [1998a] considered the effect of a TDOM on RSL predictions in Australia and New Zealand and the impact of this effect on constraining mantle viscosity and melt history. Their Fig. 3 shows that the discrepancy

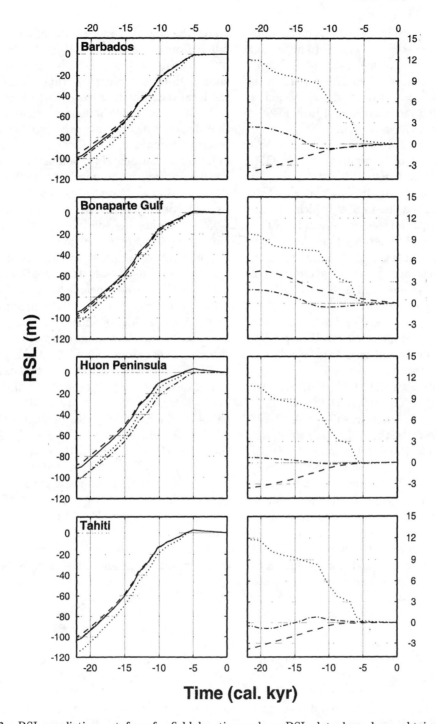

Figure 6. RSL predictions at four far-field locations where RSL data have been obtained. In the left-hand column, predictions based on the full theory that incorporates a gravitationally self-consistent TDOM, near-field water influx and GIA-induced perturbations to the Earth model's rotation vector are shown by the solid line. The dashed, dotted and dashed-dotted lines show predictions based on the full theory minus predictions that do not incorporate, respectively, a gravitationally self-consistent TDOM, near-field water influx, and rotational effects. The right-hand column shows the difference between the predictions based on the full theory (solid line in left-hand frames) to those based on the three less sophisticated sea-level models (shown by the dashed, dotted and dashed-dotted lines in the left-hand frames).

introduced by not incorporating a TDOM is highly site-dependent and ranges between ∼-4 m to ∼16 m around the time of the LGM, and between ∼-1 and ∼3 m during the mid-Holocene. These results led *Milne and Mitrovica* [1998a] to conclude that this model extension is necessary to obtain accurate inferences of melt history and mantle viscosity from far-field data.

The recent acquisition of data from Indonesia also highlights the importance of incorporating a TDOM. *Hanebuth et al.* [2000] obtained RSL data with ages that extend from the LGM to the Lateglacial. This data was obtained from the Sunda Shelf which was exposed for much of this period [see, for example, Fig. 6 of *Peltier,* 1994]. Figure 7 illustrates that the impact of a TDOM is significant in this region and, in particular, in areas of data collection; thus, the application of these data to infer the glacial meltwater signal should be based on a sea-level algorithm that includes a TDOM.

Of the three model extensions considered, near-field water influx has the largest effect on the RSL predictions at the four sites shown in Fig. 6 (e.g., see dotted line in right-hand frames), with a predicted sea-level fall of between 10-12 m since the LGM. This sea-level fall is primarily due to the influence of the water influx mechanism on the spatially uniform shift of the geoid (see eqn 29, the last term on the right-hand side). As described by *Milne et al.* [1999], the retreat of marine-based components of the major ice sheets (e.g., Laurentide, Fennoscandia, Barents Sea, Inuition, Antarctic) produced large basins that flooded with water leading to a lowering of the geoid height in order to ensure that the volume of water bounded between the geoid and the ocean floor is conserved. The 'step-like' form of the dotted line in Fig. 6 (right-hand frames) indicates three main episodes of marine-based ice retreat in the adopted ice model: 20 − 18 cal. kyr BP, 12 − 8 cal. kyr BP, and 7 − 6 cal. kyr BP. The effect is most pronounced between 12 and 6 cal. kyr BP, during which ∼8 m of sea-level fall is predicted.

The effect of near-field water influx on the uniform geoid shift does not completely explain the results shown in Fig. 6. The volume of ocean floor exposed by the retreating model ice sheets produces a geoid drop of just over 9 m. The total signal, which can reach ∼12 m, also contains an ocean loading component. The average, far-field ocean load is approximately 10% less due to the pooling of water in near-field basins, relative to a model that does not include this effect, and so the contribution of the local ocean loading to the sea-level rise is reduced by about the same amount. This effect increases the predicted discrepancy by 1 − 2 m and ex-

plains the slightly lower water-influx induced sea-level fall at Bonaparte Gulf and Huon Peninsula. The effect of the ocean load at these two sites is reduced due to either the time-varying coastline (Bonaparte Gulf) or the local coastline geometry (Huon Peninsula).

Applying RSL predictions calculated from a theory that does not incorporate near-field water influx would result in estimates of meltwater flux (or, equivalently, differential continental ice volume) that are biased low. By not including this mechanism, the predicted sea-level rise is exaggerated. The relative error introduced by neglecting near-field water influx can be as large as ∼25% (compare the solid and dotted lines in the left-hand frames in Fig. 6). Note that the sign of this bias is consistent with the error introduced when estimating ice sheet thicknesses from near-field data using a model that does not include water influx (see Section 3.1). Also, perturbations to the predicted differential (between sites) high-stands at 5 cal. kyr BP, arising from this model extension, can be ∼0.5 m and so viscosity inferences based on differential high-stands might also be significantly impacted by this extension.

The dashed-dotted lines in the right-hand frames of Fig. 6 show the influence of the GIA-induced perturbation to the Earth model's rotation vector on the sea-level predictions. The spatial variation of this signal between the four sites reflects the degree 2 order 1 spherical harmonic geometry illustrated in Fig. 2. Both Bonaparte Gulf and Huon Peninsula are located in the same quadrant in the southern hemisphere of the rotational driving potential [see Fig. 3 of *Milne and Mitrovica,* 1998b]. Barbados is located in the northern hemisphere quadrant that is antipodal to the quadrant occupied by Bonaparte Gulf and Huon Peninsula and so experiences a similar form of rotation-induced forcing (see Fig. 2). Tahiti, in contrast, is located in the southern hemisphere quadrant of opposite polarity and so exhibits a signal that is reversed in sign relative to the other three sites.

Of the three model extensions considered, the rotation-induced sea-level signal has the least significant influence on the predictions shown in Fig. 6. The maximum error introduced at a given site by not incorporating this signal is ∼3 m at the LGM. However, none of the four sites considered are located where the signal is a maximum within each quadrant: that is, none are located at mid-latitudes, ∼90 degrees from the lines of latitude that define the longitudinal boundaries of each quadrant. At these locations the predicted signal can reach an amplitude of ∼7 m at the LGM and ∼1 m during the mid-Holocene [*Milne and Mitrovica,* 1998b]. The non-

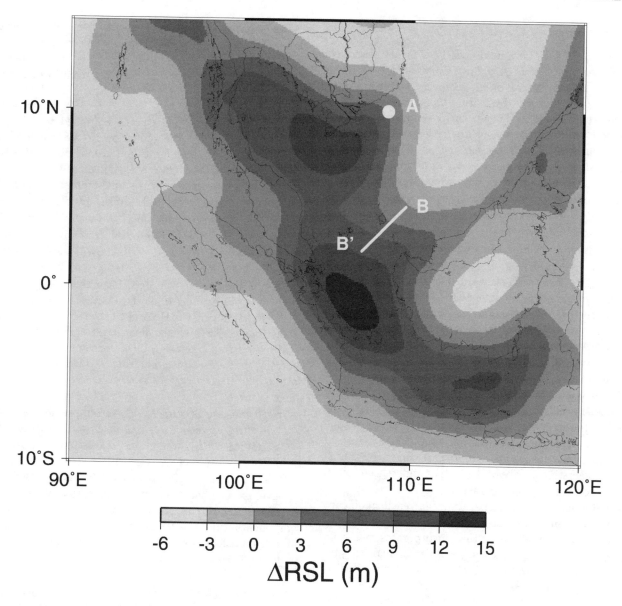

Figure 7. Differential RSL in the Indonesian region calculated by subtracting predictions based on the traditional sea-level theory to those that incorporate a gravitationally self-consistent TDOM. The locations marked by A and the transect B – B' indicate where recent RSL data covering the period between the LGM and the Lateglacial have been obtained [*Hanebuth et al., 2000*].

monotonic form of the rotation-induced component of sea-level change is a consequence of the time evolution of the relative magnitudes of solid Earth response and the geoid response to the rotational potential [see *Milne and Mitrovica*, 1998b for an in-depth discussion].

Due to the long wavelength of the rotation-induced signal, spatial differencing between sites in relatively close proximity (up to ~1000 km) effectively removes this signal from the predictions. As a consequence, in-

ferences of mantle viscosity based on differences of far-field high-stands predicted with a non-rotating Earth model are not significantly biased [*Milne and Mitrovica*, 1998b]. In contrast, the rotation-induced signal can have a significant magnitude during the mid-Holocene (~1 m) and therefore the application of GIA-corrected high-stand data to infer mid- to late Holocene melting events should be performed using a rotating-Earth sea-level theory.

4. SUMMARY

During the past decade, the sea-level theory described by *Farrell and Clark* [1976] has been extended to incorporate: a time-varying ocean geometry, water influx to regions once occupied by marine-based ice, and the influence of GIA-induced perturbations in Earth rotation. These theoretical extensions are reviewed in Section 2. The consequence of these extensions for RSL prediction and GIA model parameter estimation, considered in Section 3, is summarized in the following.

Incorporating a gravitationally self-consistent TDOM produces a significant impact on RSL predictions in both near- and far-field regions [*Johnston*, 1993; *Milne and Mitrovica*, 1998a]. In areas where the coastal topography is free of large gradients, a regional sea-level fall (near field) leads to a seaward migration of the model coastlines, whereas a regional sea-level rise (far field) produces a landward migration of the coastlines. The influence of a TDOM can be separated into a globally uniform and a local sea-level component. The globally uniform signal is caused by the reduced ocean area predicted by the TDOM theory (due to the dominance of marine transgression during deglaciation), relative to an ocean function fixed at the present-day geometry, as well as the integrated effect of sea-level fall around the ocean margin caused by the differential load. Both of these mechanisms lead to a uniform rise in sea level during deglaciation of about 4 m. The differential ocean load produced by incorporating a TDOM results in a spatially varying RSL signal that can reach significant magnitudes in both near- and far-field regions.

Application of a theory that incorporates near-field water influx results in a considerably more realistic simulation of the ocean loading in regions characterized by marine-based ice. At times of ice retreat the differential ocean load (compared to the theory that does not include this extension) can be several hundreds of meters in the Gulf of Bothnia and Hudson Bay [*Milne et al.*, 1999]. The RSL signal due to near-field water influx comprises spatially uniform and local components that can have considerable magnitude. The local signal is confined to regions where ice sheets with significant marine-based components existed. In these regions, the predicted sea-level fall is considerably less than that obtained from a theory that does not incorporate this extension. For example, the water influx produces a sea-level fall that is reduced by ~40 m in some parts of Fennoscandia. This level of perturbation is large enough to significantly bias estimates of either regional ice thickness or mantle rheology.

The spatially uniform component of the water influx signal is a consequence of the volume of ocean basin exposed by the retreating marine-based ice. This volume is flooded by ocean water, producing a uniform geoid fall of ~9 m during the period of ice model deglaciation (~21 cal. kyr BP to ~5 cal. kyr BP). In far-field regions, this sea-level fall can be amplified to ~12 m at some sites due to the reduction in the ocean load produced by the water influx extension. Estimates of the glacial meltwater signal (or, equivalently, grounded continental ice volume relative to the present volume) can, consequently, be biased low if the adopted sea-level algorithm does not include near-field water influx.

The predicted rotation-induced component of sea-level change is accurately described by a degree 2 order 1 spherical harmonic geometry [*Han and Wahr*, 1989] and a non-monotonic temporal form that can reach a magnitude of up to ~7 m at the LGM. The magnitude of this signal during the Holocene is too small to impact predictions of RSL in near-field regions. However, the rotation-induced signal can be of consequence when estimating the meltwater signal from far-field data if these data are located in regions where the rotation-induced signal is large.

The above results confirm those from previous analyses: the three extensions of the sea-level model considered can impact RSL predictions in a significant manner and thus affect inferences of mantle viscosity, ice sheet histories, and the deglacial meltwater signal based on sea-level data. Therefore, future analyses that attempt to refine previous inferences of these key GIA model components using the RSL observational record should carefully consider the impact of each of these model extensions.

Acknowledgments. I'd like to thank Jerry Mitrovica for his support and inspiration during my PhD when most of the research presented above was carried out. All figures were created using the GMT software package.

REFERENCES

Andrews, J. T., 1982. On the reconstruction of Pleistocene ice sheets: A review, *Quat. Sci. Rev.*, *1*, 1–30.
Bills, B. G., and T. S. James, 1996. Late Quaternary variations in relative sea level due to glacial cycle polar wander, *Geophys. Res. Lett.*, *23*, 3023–3026.
Clark, J. A., W. E. Farrell, and W. R. Peltier, 1978. Global changes in postglacial sea level: a numerical calculation, *Quat. Res.*, *9*, 265–287.
Clark, P. U. and A. C. Mix, 2000. Ice sheets by volume, *Nature*, *406*, 689–90.
Davis, J. L., and J. X. Mitrovica, 1996. Glacial isostatic ad-

justment and the anomalous tide gauge record of eastern North America, *Nature, 379*, 331–333.

Dziewonski, A. M., and D. L. Anderson, 1981. Preliminary Reference Earth Model (PREM), *Phys. Earth Planet. Inter., 25*, 297–356.

Fairbanks, R. G., 1989. A 17,000-year glacio-eustatic sea-level record: influence of glacial melting rates on the Younger Dryas event and deep ocean circulation, *Nature, 342*, 637–642.

Farrell, W. E., and J. A. Clark, 1976. On postglacial sea-level, *Geophys. J. R. astr. Soc., 46*, 647–667.

Fjeldskaar, W., 1994. Viscosity and thickness of the asthenosphere detected from the Fennoscandian uplift, *Earth Planet. Sci. Lett., 126*, 399–410.

Han, D. and J. Wahr, 1989. Post-glacial rebound analysis for a rotating Earth, in *Slow Deformations and Transmission of Stress in the Earth*, edited by S. Cohen and P. Vanicek, pp. 1–6, *AGU Mono. Series 49*.

Hanebuth, T., K. Stattegger, and P. M. Grootes, 2000. Rapid flooding of the Sunda Shelf: A Late Glacial sea-level record, *Science, 288*, 1033–1035.

Haskell, N. A., 1935. The motion of a viscous fluid under a surface load, *Physics, 6*, 56–61.

Ivins, E. R., C. G. Sammis, and C. F. Yoder, 1993. Deep mantle viscous structure with prior estimate and satellite constraint, *J. Geophys. Res., 98*, 4579–4609.

James, T. S., and E. R. Ivins, 1997. Global geodetic signatures of the Antarctic ice sheet, *J. Geophys. Res., 102*, 605–633.

Johnston, P., 1993. The effect of spatially non-uniform water loads on predictions of sea-level change, *Geophys. J. Int., 114*, 615–634.

Kaufmann G., and D. Wolf, 1996. Deglacial land emergence and lateral upper-mantle heterogeneity in the Svalbard Archipelago – II. Extended results for high-resolution load models, *Geophys. J. Int., 127*, 125–140.

Lambeck, K., 1993. Glacial rebound of the British Isles–II. A high resolution, high-precision model, *Geophys. J. Int., 115*, 960–990.

Lambeck, K., 1995. Constraints on the Late Weichselian ice sheet over the Barents Sea from observations of raised shorelines, *Quat. Sci. Rev., 14*, 1–16.

Lambeck, K., 1996. Limits on the areal extent of the Barents Sea ice sheet in Late Weichselian time, *Palaeogeogr. Palaeoclimatol. Palaeoecol., 12*, 41–51.

Lambeck, K. and M. Nakada, 1990. Late Pleistocene and Holocene sea-level change along the Australian coast, *Palaeogeog. Palaeoclimat. Palaeoecol., 89*, 143–176.

Lambeck, K., C. Smither, and P. Johnston, 1998. Sea-level change, glacial rebound and mantle viscosity for northern Europe, *Geophys. J. Int., 134*, 102–144.

Lambeck, K., Y. Yokoyama, P. Johnston, and A. Purcell, 2000. Global ice volumes at the Last Glacial Maximum and early Lateglacial, *Earth Planet. Sci. Lett., 181*, 513–527.

McConnell, R. K., 1968. Viscosity of the mantle from relaxation time spectra of isostatic adjustment, *J. Geophys. Res., 73*, 7089–7105.

Milne, G. A., Jan. 1998. Refining models of the glacial isostatic adjustment process, Ph.D. thesis, 124 pp., Univ. of Toronto, Toronto, Canada.

Milne, G. A., J. .L. Davis, J. X. Mitrovica, H.-G. Scherneck, J. M. Johansson, M. Vermeer, and H. Koivula, 2001a. Space geodetic constraints on glacial isostatic adjustment in Fennoscandia, *Science, 291*, 2381–2385.

Milne, G. A., and J. X. Mitrovica, 1996. Postglacial sea-level change on a rotating Earth: first results from a gravitationally self-consistent sea-level equation, *Geophys. J. Int., 126*, F13–F20.

Milne, G. A., and J. X. Mitrovica, 1998a. The influence of a time-dependent ocean-continent geometry on predictions of post-glacial sea level change in Australia and New Zealand, *Geophys. Res. Lett., 25*, 793–796.

Milne, G. A., and J. X. Mitrovica, 1998b. Postglacial sea-level change on a rotating Earth, *Geophys. J. Int., 133*, 1–19.

Milne, G. A., J. X. Mitrovica, and J. L. Davis, 1999. Near-field hydro-isostasy: the implementation of a revised sea-level equation, *Geophys. J. Int, 139*, 464–482.

Milne, G. A., J. X. Mitrovica, and D. P. Schrag, 2001b. Estimating past continental ice volume from sea-level data, *Quat. Sci. Rev.*, in press.

Mitrovica, J. X., 1996. Haskell [1935] revisited, *J. Geophys. Res., 101*, 555–569.

Mitrovica, J. X., J. L. Davis, and I. I. Shapiro, 1994a. A spectral formalism for computing three dimensional deformations due to surface loads, 2. Present-day glacial isostatic adjustment, *J. Geophys. Res., 99*, 7075–7101.

Mitrovica, J. X., and A. M. Forte, 1997. The radial profile of mantle viscosity: Results from the joint inversion of convection and post-glacial rebound observables, *J. Geophys. Res., 102*, 2751–2769.

Mitrovica, J. X., and G. A. Milne, 1998. Glaciation-induced perturbations in the Earth's rotation: a new appraisal, *J. Geophys. Res., 103*, 985–1005.

Mitrovica, J.X., G.A. Milne, and J.L. Davis, 2001a. Glacial isostatic adjustment on a rotating Earth, *Geophys. J. Int.*, in press.

Mitrovica, J. X., and W. R. Peltier, 1991. On post-glacial geoid subsidence over the equatorial oceans, *J. Geophys. Res., 96*, 20,053–20,071.

Mitrovica, J. X., and W. R. Peltier, 1993a. Present-day secular variations in the zonal harmonics of the Earth's geopotential, *J. Geophys. Res., 98*, 4509–4526.

Mitrovica, J. X., and W. R. Peltier, 1993b. A new formalism for inferring mantle viscosity based on estimates of post glacial decay times: Application to RSL variations in N.E. Hudson Bay, *Geophys. Res. Lett., 20*, 2183–2186.

Mitrovica, J. X., and W. R. Peltier, 1995. Constraints on mantle viscosity based upon the inversion of post-glacial uplift data from the Hudson Bay region, *Geophys. J. Int., 122*, 353–377.

Mitrovica, J. X., M. E. Tamisiea, J. L. Davis, and G. A. Milne, 2001b. Recent mass balance of polar ice sheets inferred from patterns of global sea-level change, *Nature, 409*, 1026–1029.

Munk, W. H., and G. J. F. MacDonald, 1960. *The Rotation of the Earth*, 329 pp., Cambridge University Press, Cambridge.

Mound, J. E., and J. X. Mitrovica, 1998. True polar wander as a mechanism for second-order sea-level variations, *Science, 279*, 534–537.

Nakada, M., and K. Lambeck, 1989. Late Pleistocene and Holocene sea-level change in the Australian region and mantle rheology, *Geophys. J. Int.*, *96*, 497–517.

Peltier, W. R., 1974. The impulse response of a Maxwell Earth, *Rev. Geophys.*, *12*, 649–669.

Peltier, W. R., 1994. Ice age paleotopography, *Science*, *265*, 195–201.

Peltier, W. R., 1996. Mantle viscosity and ice-age ice sheet topography, *Science*, *273*, 1359–1364.

Peltier, W. R., 1998a. 'Implicit ice' in the global theory of glacial isostatic adjustment, *Geophys. Res. Lett.*, *25*, 3955–3958.

Peltier, W. R., 1998b. Postglacial variations in the level of the sea: implications for climate dynamics and solid-earth Geophysics, *Rev. Geophys.*, *36*, 603–689.

Peltier, W. R., and J. T. Andrews, 1976. Glacial isostatic adjustment–I. The forward problem, *Geophys. J. R. astr. Soc.*, *46*, 605–646.

Peltier, W. R., and A. M. Tushingham, 1991. Influence of glacial isostatic adjustment on tide gauge measurements of secular sea level change, *J. Geophys. Res.*, *96*, 6779–6796.

Sabadini, R., D. A. Yuen, and E. Boschi, 1982. Polar wander and the forced responses of a rotating, multilayered, viscoelastic planet, *J. Geophys. Res.*, *87*, 2885–2903.

Sabadini, R., D. A. Yuen, and P. Gasperini, 1985. The effects of transient rheology on the interpretation of lower mantle viscosity, *Geophys. Res. Lett.*, *12*, 361–364.

Tushingham, A. M., and W. R. Peltier, 1991. ICE-3G: A new global model of late Pleistocene deglaciation based on geophysical predictions of post-glacial relative sea level change, *J. Geophys. Res.*, *96*, 4497–4523.

Vening Meinesz, F. A., 1937. The determination of the Earth's plasticity from the post-glacial uplift of Scandinavia: isostatic adjustment, *Koninklijke Akademie van Weten-schapen*, *40*, 654–62.

Wahr, J. M., M. Molenaar, and F. Bryan, 1998. Time variability of the Earth's gravity field: Hydrological and oceanic effects and their possible detection using GRACE, *J. Geophys. Res.*, *103*, 30,205–30,229.

Wieczerkowski, K., J. X. Mitrovica, and D. Wolf, 1999. A revised relaxation-time spectrum for Fennoscandia, *Geophys. J. Int.*, *139*, 69–86.

Wolf, D., 1987. An upper bound on lithosphere thickness from glacio-isostatic adjustment in Fennoscandia, *J. Geophys.*, *61*, 141–149.

Wu, P., 1999. Modelling postglacial sea levels with power-law rheology and a realistic ice model in the absence of ambient tectonic stress, *Geophys. J. Int*, *139*, 691–702.

Wu, P., and W. R. Peltier, 1983. Glacial isostatic adjustment and the free air gravity anomaly as a constraint on deep mantle viscosity, *Geophys. J. R. Astr. Soc.*, *74*, 377–449.

Wu, P., and W. R. Peltier, 1984. Pleistocene deglaciation and the Earth's rotation: A new analysis, *Geophys. J. R. astr. Soc.*, *76*, 753–792.

Yokoyama, Y., K. Lambeck, P. De Deckker, P. Johnston, and L. K. Fifield, 2000. Timing of the Last Glacial Maximum from observed sea-level minima, *Nature*, *406* 713–716.

G. A. Milne, Department of Geological Sciences, University of Durham, Science Labs, South Road, Durham, DH1 3LE, UK. (e-mail: g.a.milne@durham.ac.uk)

Contributions of Ineffective Ice Loads on Sea-Level and Free-Air Gravity

Jun'ichi Okuno

Earthquake Research Institute, The University of Tokyo, Bunkyo, Japan

Masao Nakada

Department of Earth and Planetary Sciences, Faculty of Sciences, Kyushu University, Fukuoka, Japan

We investigate the effects of ice that does not act as an unloading ice load and that contributes to the global water (ocean) load, on predictions of the sea-level changes and free-air gravity anomalies associated with glacial isostatic adjustment (GIA). The space occupied by ice below the geoid in the process of GIA is simultaneously replaced by sea-water when the ice melts. This ice load is here referred to as an ineffective ice load. The thickness of the ineffective ice load in Hudson Bay estimated by considering the palaeotopography is greater than 300 m for the ARC3 (ICE-1) ice model, and the improper treatment of this 'load' will affect predictions of various observables. In particular, significant local negative free-air gravity anomalies less than -30 mGal are predicted for the areas with ineffective ice mass even for a relatively uniform viscosity model. On the other hand, the magnitude of the relative sea-level prediction in glaciated regions including the effects of ineffective ice mass and water influx is smaller than that without these loads. An accurate evaluation of ineffective ice mass as well as water influx is, therefore, required when we examine the melted ice thickness by using sea-level variations.

1. INTRODUCTION

The surface mass redistribution associated with the Late Pleistocene glacial cycles induces changes of various geophysical observables such as sea-level variations, gravity anomalies and radial and horizontal deformation rates. A theory of postglacial sea-level change caused by surface loads, the so-called sea-level equation, was first described by *Farrell and Clark* [1976]. This theory has been used to predict many features of the global postglacial sea-level variation [e.g., *Farrell and Clark*, 1976; *Peltier and Andrews*, 1976; *Wu and Peltier*, 1983; *Nakada and Lambeck*, 1989; *Mitrovica and Peltier*, 1991; *Lambeck*, 1993].

To predict relative sea-level (RSL) changes and other geophysical signals, it is very important to accurately describe the surface loads. *Nakada and Lambeck* [1987] indicated that high-degree spatial resolution of coastline geometry is required to accurately predict the Holocene sea-levels at the far-field sites (see also *Nakada*[1986]). *Lambeck and Nakada*[1990] also pointed out the importance of time-dependent ocean geometry on RSL predictions. On the other hand, the meltwater load has often been simplified. Some previous models have assumed a spatially uniform meltwater load equal to equivalen-

Ice Sheets, Sea Level and the Dynamic Earth
Geodynamics Series 29

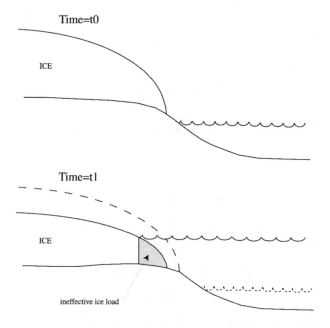

Figure 1. A schematic diagram illustrating the ineffective ice load. At time t0 years BP, the entire ice sheet acts as a load. The shaded part of ice below sea-level at time t1 years BP (t0 ≥ t1), however, does not act as an unloading ice load. This part of ice, referred to as ineffective ice load here, transforms to a global water (ocean) load at this stage because the space occupied by this ice is simultaneously replaced by sea-water when this ice melts.

·t sea-level (ESL) change [*Nakada and Lambeck*, 1987], where ESL is the volume of meltwater as a function of time divided by the area of the oceans. If the meltwater load is assumed to be spatially uniform, the error in sea-level predictions will be largest close to the former ice sheet, where the sea-level change is drastically different from the ESL. Thus, an evaluation of the spatially non-uniform water load is very important for sea-level predictions at sites near or within ice sheets [*Mitrovica and Peltier*, 1991; *Johnston*, 1993]. Furthermore, *Milne et al.* [1999] examined the effects of the meltwater influx in previously ice-covered subgeoidal geographic regions. They showed a discrepancy of ∼ 40 per cent in RSL predictions around the Hudson Bay region compared with solutions without the effects of the meltwater influx.

An accurate evaluation of the Earth's deformation due to surface loads is, therefore, required in order to examine the mantle viscosity from relative sea-level changes, surface deformation rates and gravity anomalies. In particular, the time-dependent ocean function considering the effects of both the Earth's deformation and ice distribution has to be precisely included in the modelling, because the crustal movement associated with the glacio- hydro- isostatic adjustment is significant in glaciated regions.

In Fig.1, we show a schematic figure of ice load distribution at time t0 and t1 years BP (t0 ≥ t1). At time t0, the whole part of ice sheet works as an ice load. At time t1, however, the ice below the sea surface (shaded region) does not work as an unloading ice load and transforms to global water (ocean) load at this stage. That is, the space occupied by this ice is simultaneously replaced by sea-water when this portion of the ice melts. This means that the shaded region of ice does not work as unloading ice load. This ice load is, therefore, referred to as ineffective ice load here. To consider this problem, we have to evaluate the palaeotopography based on the predicted sea-level changes. The proper treatment of the ineffective ice load was incorporated into previous treatments of water dumping [e.g., *Milne et al.*, 1999]. Specifically, a simple check was applied to the ice load during the application of the water dumping algorithm: if any column of a marine-based ice sheets were less massive than a water column would be for the same location, the ice in this location would be removed from the model. In this paper we focus on the error that is introduced if, instead, one permits the ineffective ice to load the model.

2. EVALUATION OF INEFFECTIVE ICE LOAD

A spatio- and time-dependent ocean function is required to evaluate the water loads precisely. Particularly, the evaluation of an ocean function is very important in calculating the sea-level changes for the glaciated regions characterized by both large crustal deformation due to glacial rebound and the existence of ice sheets [e.g., *Milne et al.*, 1999]. We use a formulation of the water load component introduced by *Milne et al.* [1999]. *Milne et al.* [1999] used an ocean function based on palaeotopography (including the height of ice sheet) in which they considered the water loads due to the influx of meltwater to subgeoidal solid surface regions once covered with the marine-based Late Pleistocene ice sheets. In fact, the water influx in this region significantly contributes to the surface load.

In the calculations of sea-level variations and other geophysical signals, we use an Earth model characterized by an elastic structure given by seismically determined Preliminary Reference Earth model (PREM) [*Dziewonski and Anderson*, 1981]. We also adopt two viscosity models, A and B. The upper mantle viscosity is 10^{21} Pa s for these models, and the lower mantle viscosity below 670 km depth is 2×10^{21} Pa s for model A and 10^{22} Pa s for model B. The thickness of the elastic lithosphere is 100 km for both models.

In this study, we used two ice models, i.e., AR-C3+ANT4b [*Nakada and Lambeck*, 1988; 1989] and

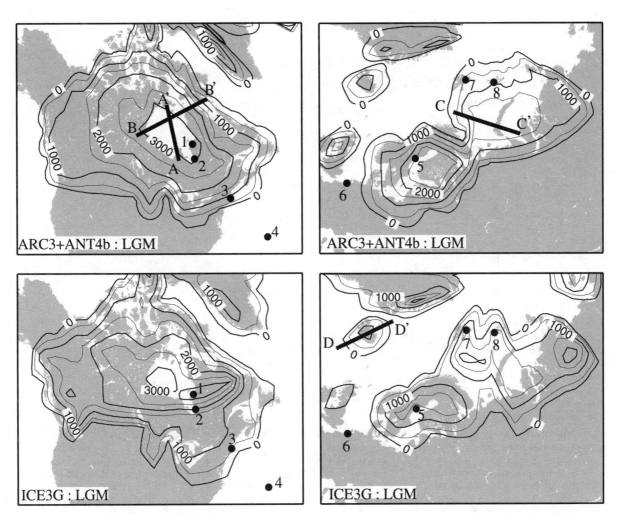

Figure 2. Thicknesses of Arctic ice removed since the LGM for the ice models ARC3 and ICE-3G. Contour interval is 500 m. The circles represent locations at which relative sea-levels are calculated (1: Richmond Gulf, 2: James Bay, 3: Boston, 4: Bermuda, 5: Angerman River, 6: Zuid Holland, 7: Gustav Adolf Land, 8: Hooker Island). The ineffective ice loads along profiles A-A', B-B', C-C' and D-D' are discussed in detail (see Fig.3).

ICE-3G [*Tushingham and Peltier,* 1991] to clearly indicate the effects of the meltwater influx and the ineffective ice load. The Arctic ice model ARC3 includes the Laurentide, Fennoscandia and Barents-Kara ice sheets in which the ice model for Laurentide and Fennoscandia ice sheets corresponds to ICE-1 of *Peltier and Andrews* [1976]. The Barents-Kara ice sheet model in the ARC3 corresponds to the maximum model of *Denton and Hughes* [1991]. The ICE-1 model is based on geological and geomorphological observations. ICE-3G is, however, constructed by comparing sea-level observations and predictions for a relatively uniform viscosity model. Thicknesses of Arctic ice removed since the Last Glacial Maximum (LGM) are shown in Fig.2 for ice models of ARC3 and ICE-3G. The Antarctic ice model ANT4b,

with a minor Holocene melting, was originally generated by maximum reconstruction in Antarctica by *Denton and Hughes* [1981].The minor Holocene melting ($\sim$ 3 m for ESL) for the ANT4b is supported by sea-level observations in the far-fields [*Nakada and Lambeck,* 1988; 1989; *Okuno and Nakada,* 1998] and those from British Isles [*Lambeck et al.,* 1996].

In evaluating the spatially non-uniform water load [*Mitrovica and Peltier,* 1991; *Johnston,* 1993], we use an iterative pseudospectral method formulated by *Mitrovica and Peltier* [1991]. Ocean function and water loads are based on the formulation by*Milne et al.* [1999]. The meltwater influx obtained from this formulation without a check for ineffective ice is referred to as model WD. The correction for ineffective ice (FI) is then

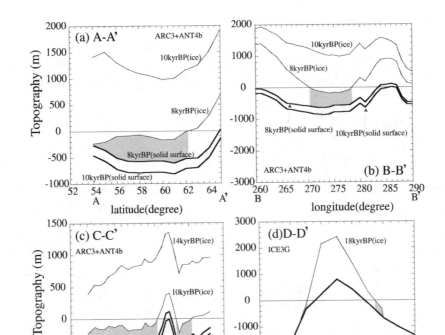

Figure 3. The profiles of the solid surface of the Earth (palaeotopography) and the ice height relative to the palaeotopography (PT) for four sections (A-A',B-B',C-C' and D-D') shown in Fig.2. The adopted ice model is ARC3+ANT4b for (a), (b) and (c) and ICE3G for (d). The viscosity model is A with the upper and lower mantle viscosities of 10^{21} Pa s and 2×10^{21} Pa s. The shaded regions represent the ineffective ice loads derived from the PT and the ice height relative to the PT.

added to the WD model, and this model is referred to as WD+FI.

Fig.3 depicts the profiles of ice height and solid surface of the Earth (palaeotopography) for earth model A. These are evaluated at four sections (A-A', B-B', C-C' and D-D') shown in Fig.2. 'Solid surface' in Fig.3 indicates the palaeotopography (PT) derived from the predicted relative sea-levels and present topography. 'Ice' in Fig.3 represents the ice height of ice model (ARC3 or ICE-3G) relative to the PT. Figs.3a, b and c illustrate the profiles of the palaeotopography and ice height relative to the PT for the ARC3 ice model. Those at 10 and 8 kyr BP for section A-A' are shown in Fig.3a. This section traverses the central part of the Laurentide ice sheet. The ineffective ice mass (shaded region) develops at 8 kyr BP and its thickness is about 500 m. The same magnitude of the ineffective ice load is also predicted at 8 kyr BP for the section of B-B'. These results indicate that the ineffective ice loads with the thickness of about 500 m develops at 8 kyr BP in the central part of Hudson Bay. Fig.3c shows the profiles in

Barents Sea region (section C-C'). In this area, the ineffective ice mass develops at 10 kyr BP and its thickness is about 200 - 300 m. The profiles across the Iceland are shown in Fig.3d for ice model of ICE-3G. The ineffective ice mass for this section exists at the LGM, and its thickness is about 200 m.

Fig.4 shows the spatial distribution of water loads (WD) and ineffective ice loads (FI) during the past 18 kyr for ice models ARC3+ANT4b and ICE-3G. We only show the results for the earth model A, because the general tendency for the earth model B is similar to that for model A. In these figures showing the water influx denoted by WD, a negative value indicates the unloading caused by crustal uplift and positive value indicates loading. The ineffective ice mass in Figs.4b, 4d, 4f and 4h does not work as ice unloads. Significant ineffective ice loads are predicted for Hudson Bay and Barents-Kara Sea regions for the ice model ARC3+ANT4b, and for Greenland and Iceland regions for the ICE-3G. These regional differences are attributed to the difference of the ice thickness in these two ice mod-

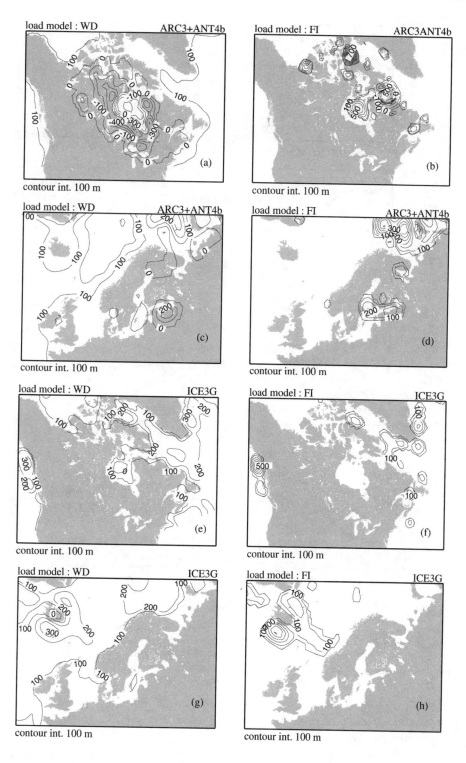

Figure 4. Spatial variations of the water load (a,c,e,g) and ineffective ice (b,d,f,h) during the past 18 kyr for the ice models of ARC3+ANT4b and ICE-3G. The viscosity model is A with the upper and lower mantle viscosities of 10^{21} Pa s and 2×10^{21} Pa s. Contour unit is m. In figures (a), (c), (e) and (g), the negative value indicates the unloading and the positive value indicates the loading.

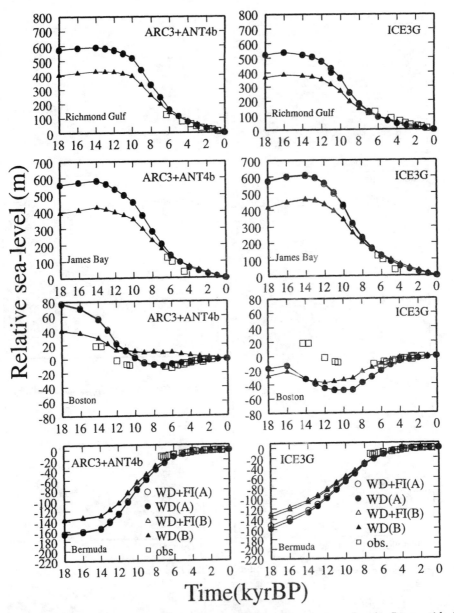

Figure 5. Relative sea-level curves in glaciated and intermediate regions for the Laurentide ice sheet. Ice models are ARC3+ANT4b and ICE-3G, and viscosity models are A and B. In these figures, we show the results for two models, i.e., WD and WD+FI. The WD model is based on the formulation of water load introduced by *Milne et al.* [1999] without their a priori correction for the ineffective ice load. The effect of ineffective ice (FI) is then added to the WD model, and this model is referred to as WD+FI.

els. For example, the ice sheets for Hudson Bay and Barents-Kara Sea regions of ARC3+ANT4b are thicker than those for the ICE-3G (see Fig.2). If the glaciation started on land at about 100 kyr BP (we adopt this assumption in this study), the ice loads transformed to ineffective ice loads will provide a significant contribution to the present-day free-air gravity anomaly as shown below. Thus, the condition of the start of glaciation at about 100 kyr BP has to be included in the glacial re-

bound modelling as suggested by *Kaufmann and Wolf* [1996] and *Milne et al.* [1999].

3. PREDICTIONS OF SEA-LEVEL CHANGES AND FREE-AIR GRAVITY ANOMALIES

ICE-1 (ARC3) has a maximum melted ice thickness of about 3500 m in the central part of the Laurentide ice sheet. The thickness of ICE-3G [*Tushingham and Pelti-*

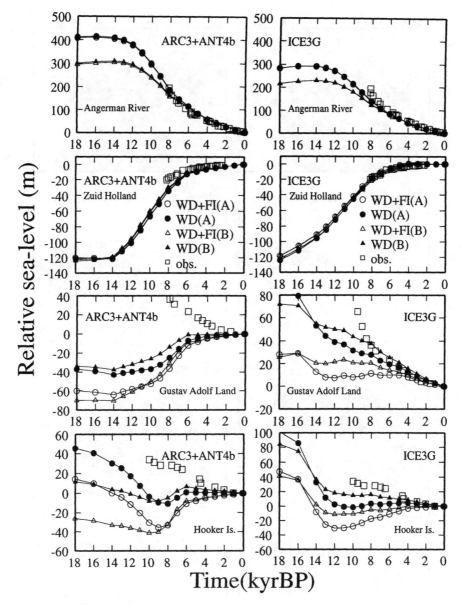

Figure 6. Same as Fig.5 but sea-level curves for Fennoscandia.

er, 1991] is reduced to about 3000 m in order to yield better sea-level predictions. The difference of thicknesses removed since the LGM between these ice models is similar to the magnitude of the ice load reduction related to water influx and ineffective ice loads, indicating that an accurate evaluation of water loads may be important in the construction of the ice model.

We calculate the relative sea-level changes at sites shown in Fig.2, in which observations for these sites have been compiled by *Walcott* [1972] and *Lambeck et al.* [1998]. Figs.5 and 6 show the relative sea-level curves in glaciated and intermediate regions for the Laurentide and Fennoscandia ice sheets, respectively.

Sea-level predictions in the central parts of these ice sheets are insensitive to the distribution of ineffective ice mass, but those of Iceland and Barents Sea regions (Gustav Adolf Land and Hooker Island in Fig.2) are very sensitive to its distribution. That is, careful correction for the ineffective ice mass is required in predicting the relative sea-level changes around the marine-based ice sheets. In the central part of the ice sheets, these loads reduce the magnitude of the relative sea-level predictions in the deglaciation phase as was indicated for water influx by *Milne et al.* [1999]. In fact, comparisons of sea-level variations between observations and predictions for models ARC3+ANT4b and ICE-3G sug-

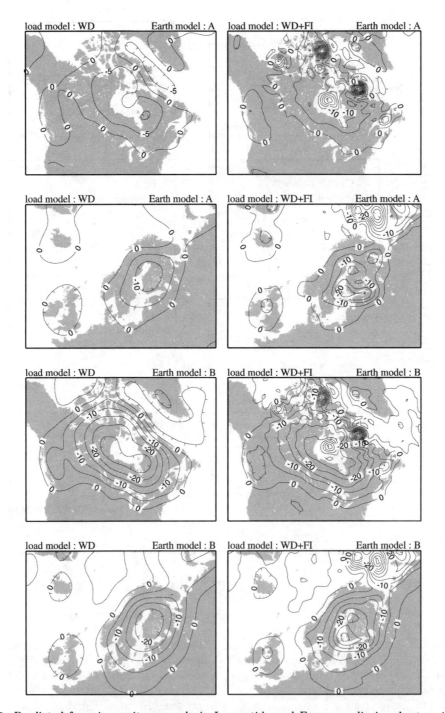

Figure 7. Predicted free air gravity anomaly in Laurentide and Fennoscandia ice sheet regions. The ice model is ARC3+ANT4b, and viscosity models are A and B. We show the results for the model WD and WD+FI. Contour interval is 5 mGal. The predictions for the WD+FI show significant local peaks of negative anomalies for the regions with ineffective ice (see Fig.3).

gest that the ice thickness adopted for the Hudson Bay region in ARC3 (ICE-1) seems to be reasonable (see the predictions for Richmond Gulf and Angerman River). In the calculations of free-air gravity anomaly, we

therefore show the results for the ice model of AR-C3+ANT4b.

The predicted present-day free air gravity anomalies are illustrated in Fig.7. In the WD+FI model, significant local negative free-air gravity anomalies are predicted in the Hudson Bay and Barents Sea regions for ARC3+ANT4b. The difference between the free-air gravity anomaly for models WD+FI and WD is attributed to the correction for ineffective ice during the past 18 kyr (see Fig.4). That is, the deformation of the Earth associated with the ice load transformed to ineffective ice mass still remains. The negative local peak is predicted in these regions for both earth models. The predicted solid surface gravity change, and vertical and horizontal deformation rates for WD+FI are insignificantly different from those obtained for WD, although we do not show those predictions here.

4. CONCLUDING REMARKS

In this study, we indicated that the evaluation of surface loads associated with the water influx corrected for the ineffective ice mass may be very important in predicting the sea-level variations and free-air gravity anomalies associated with the glacial isostatic adjustment. Effects on observables are significant in glaciated regions, particularly for the regions covered by marine-based ice sheets. The ice unloading in these regions is effectively reduced by the ineffective ice mass, and the deformation of the Earth due to the ice load transformed to ineffective ice mass still remains. Thus, significant negative peaks of free-air gravity anomaly are predicted for the regions with ineffective ice mass. For example, the predicted free-air gravity anomaly for the ice model ARC3+ANT4b is about -30 mGal in Hudson Bay even for the viscosity model with a lower mantle viscosity of 2×10^{21} Pa s. Sea-level predictions for the regions covered by marine-based ice sheets such as Barents Sea are also sensitive to the correction for the ineffective ice mass. These results indicate that an accurate ocean function and evaluation of ineffective loads when treating water influx are required when ice thickness are inferred using sea-level variations.

REFERENCES

Denton, G. H. and T. J. Hughes, *The Last Great Ice Sheets*, Wiley, New York, 1981.

Dziewonski, A. M. and D. L. Anderson, Preliminary reference Earth model, *Phys. Earth Planet. Int.*, *25*, 297-356, 1981.

Farrell, W. E. and J. A. Clark, On postglacial sealevel, *Geophys. J. R. astr. Soc.*, *46*, 637-667, 1976.

Johnston, P., The effect of spatially non-uniform water loads on the prediction of sea-level change, *Geophys. J. Int.*, *114*, 615-634, 1993.

Kaufmann, G. and D. Wolf, Deglacial land emergence and lateral upper-mantle heterogeneity in the Svalbard Archipelago -II. Extended results for high-resolution lad models, *Geophys. J. Int.*, *127*, 125-140, 1996.

Lambeck, K., Glacial rebound of the British Isles-II. A high resolution, high precision model, *Geophys. J. Int.*, *115*, 960-990, 1993.

Lambeck, K. and M. Nakada, Late Pleistocene and Holocene sea-level change along the Australian coast, *Palaeogeogr. Palaeoclimatol. Palaeoecol.*, *89*, 143-176, 1990.

Lambeck, K., P. Johnston, C. Smither, and M. Nakada, Glacial rebound of the British Isles - III. Constraints on mantle viscosity, *Geophys. J. Int.*, *125*, 340-354, 1996.

Lambeck, K., C. Smither, and P. Johnston, Sea-level change, glacial rebound and mantle viscosity for northern Europe, *Geophys. J. Int.*, *134*, 102-144, 1998.

Milne, G. A., J. X. Mitrovica and J. L. Davis, Near-field hydro-isostasy: the implementation of a revised sea-level equation, *Geophys. J. Int.*, *139*, 464-482, 1999.

Mitrovica, J. X. and W. R. Peltier, On post-glacial geoid subsidence over the equatorial oceans, *J. Geophys. Res.*, *96*, 20053-20071, 1991.

Nakada, M., Holocene sea levels in oceanic island: implications for the rheological structure of the Earth's mantle, *Tectonophysics*, *121*, 263-276, 1986.

Nakada, M. and K. Lambeck, Glacial rebound and relative sea-level variations: a new appraisal, *Geophys. J. R. astr. Soc.*, *90*, 171-224, 1987.

Nakada, M. and K. Lambeck, The melting history of the late Pleistocene Antarctic ice sheet, *Nature*, *333*, 36-40, 1988.

Nakada, M. and K. Lambeck, Late Pleistocene and Holocene sea-level changes in Australian region and mantle rheology, *Geophys. J.*, *96*, 497-517, 1989.

Okuno, J. and M. Nakada, Rheological structure of the upper mantle inferred from the Holocene sea-level change along the west coast of Kyushu, Japan, *Dynamics of the Ice Age Earth: A Modern Perspective*, Trans Tech Publications Ltd, Brandrain, Switzerland, 443-458, 1998.

Peltier, W. R. and J. T. Andrews, Glacial isostatic adjustment - I. The forward problem, *Geophys. J. R. astr. Soc.*, *46*, 605-646, 1976.

Tushingham, A. M. and W. R. Peltier, ICE-3G: a new global model of late Pleistocene deglaciation based upon geophysical predictions of postglacial relative sea level change, *J. Geophys. Res.*, *96*, 4497-4523, 1991.

Walcott, R. I., Late Quaternary vertical movements in eastern North America : Quantitative evidence of glacial-isostatic rebound, *Rev. Geophys.*, *10*, 849-884, 1972.

Wu, P. and W. R. Peltier, Glacial isostatic adjustment and the free-air gravity anomaly as a constraint on deep mantle viscosity, *Geophys. J. R. astr. Soc.*, *74*, 377-450, 1983.

J. Okuno, Earthquake Research Institute, The University of Tokyo, Bunkyo 113-0032, Japan. (e-mail: okuno@eri.u-tokyo.ac.jp)

M. Nakada, Department of Earth and Planetary Sciences, Faculty of Sciences, Kyushu University, Fukuoka 812-8581, Japan. (e-mail: mnakada@geo.kyushu-u.ac.jp)

On the Radial Profile of Mantle Viscosity

Jerry X. Mitrovica

Department of Physics, University of Toronto, Toronto, Canada

Alessandro M. Forte

Department of Earth Sciences, University of Western Ontario, London, Canada

We present radial viscosity profiles determined from non-linear, iterative, Occam-style inversions of an extensive set of surface geophysical observables related to glacial isostatic adjustment (henceforth GIA) and mantle convection. The GIA data include decay times determined from the post-glacial uplift of sites in Fennoscandia and Hudson Bay, as well as the newly revised relaxation spectrum for Fennoscandia. The convection observables include long-wavelength free-air gravity anomalies, plate motions, dynamic surface topography, and the excess ellipticity of the core-mantle-boundary. The principal distinction between our previous joint inversions of GIA and convection data and those reported here, is the inclusion of the relaxation spectrum and plate-motion data. Of all the convection data cite above, the plate motions provide the only constraints on the absolute value of mantle viscosity which are independent of those provided by GIA data. The GIA and convection data sets can be simultaneously reconciled with a radial viscosity profile that is characterized by an increase, from the base of the lithosphere to ~2000 km depth, that exceeds two orders of magnitude. These inversions support our recent inference of a high-viscosity peak at 2000 km depth based on convection data alone. Furthermore, the existence of a solution that satisfies both the GIA and convection data sets suggests that transient viscosity need not be invoked to achieve a reconciliation; this result supports our previous conclusion based on joint inversions of a considerably more limited data set.

1. INTRODUCTION

The inference of the radial profile of mantle viscosity is a long-standing problem, and a source of continuing contention, in geophysics. Traditionally, these inferences have been derived from the analysis of data related to the glacial isostatic adjustment (henceforth GIA) of the Earth and surface observables connected to mantle convective flow. Indeed, the intensity of the ongoing debate reflects the breadth of geodynamic applications in which mantle viscosity exerts a fundamental control. As an example, GIA studies that once focused on Holocene sea-level variations and anomalies in the Earth's gravitational and rotational state, have broadened to include, for example, the correction of tide gauge records to estimate present-day sea-level change, the connection between post-glacial lithospheric stress

Ice Sheets, Sea Level and the Dynamic Earth
Geodynamics Series 29

10.1029/029GD12

regimes and seismicity, and perturbations in the Earth's orbital parameters. Furthermore, viscosity plays a central role in convective flow modelling of the long-term evolution of the Earth's thermochemical and rotational state, the Earth's present-day gravitational field, plate tectonic motions, dynamic topography and the geological record.

Haskell's [1935] study of the post-glacial uplift of Fennoscandia, and his inference of an average viscosity of 10^{21} Pa s in the top 1000-1500 km of the mantle, is widely acknowledged as the first seminal contribution to the problem. Subsequent work, leading to *McConnell's* [1968] analysis of his so-called Fennoscandian relaxation spectrum, led to the prevailing view that the mantle viscosity increased significantly with depth while still satisfying the Haskell average. This view was overturned in the early 1970's, when the analysis of sea-level and uplift records by *Cathles* [1971, 1975] and *Peltier and Andrews* [1976], based on a new generation of global GIA models, suggested that the viscosity increased only moderately, if at all, from the base of the lithosphere to the core-mantle-boundary. Peltier and colleagues added support to this argument in a series of analyses involving progressively larger data sets [e.g., *Wu and Peltier*, 1983, 1984; *Tushingham and Peltier*, 1991].

Efforts to constrain viscosity using surface observables associated with mantle convection can be traced to *Hager* [1984], *Richards and Hager* [1984] and *Ricard et al.* [1984]. These earliest studies were based on the Earth's gravity field and they suggested a significant (factor of ∼30) increase of viscosity with depth in the sub-lithospheric mantle. This conclusion was supported by *Forte and Peltier* [1987], who extended the viscous flow modelling to include plate velocities, and by numerous subsequent analyses of similar type [e.g., *Ricard et al.*, 1989; *Ricard and Vigny*, 1989; *Hager and Clayton*, 1989; *Forte and Peltier*, 1991; *Forte et al.*, 1991, 1993, 1994; *King and Masters*, 1992; *Corrieu et al.*, 1994; *King*, 1995; *Thoraval and Richards*, 1997]. The argument for a substantial increase of viscosity with depth was further bolstered by disparate analyses of the stability of both the hot spot reference frame [*Richards*, 1991] and long-term rates of polar wander [e.g., *Sabadini and Yuen*, 1989; *Spada et al.*, 1992; *Richards et al.*, 1997; *Steinberger and O'Connell*, 1997], and the planform of mantle convection [e.g., *Zhang and Yuen*, 1995; *Bunge et al.*, 1996].

The apparent inconsistency between inferences of viscosity based on GIA and convection studies led to suggestions that the mantle viscosity may have transient effects over time scales that characterize these two processes (GIA, $10^3 - 10^5$ yr; mantle convection,

$10^7 - 10^8$ yr). These arguments led to a flurry of articles dealing with the influence of transient viscosity on predictions of GIA observables [e.g., *Sabadini et al.*, 1985; *Peltier*, 1985; *Peltier et al.*, 1986; *Yuen et al.*, 1986]. However, the necessity of invoking such transients in the viscous response was weakened by two independent lines of GIA research. First, *Nakada and Lambeck* [1989] argued that differential late Holocene sea-level high-stands in the Australian region required a viscosity that increased by ∼two orders of magnitude from the upper to lower mantle. Second, *Mitrovica* [1996] demonstrated that many previous arguments for an isoviscous mantle [e.g., *Wu and Peltier*, 1983; *Tushingham and Peltier*, 1991] were biased by a serious misinterpretation of the so-called Haskell constraint on viscosity. In particular, these studies (in contrast, for example, to *Nakada and Lambeck* [1989]) associated the average of 10^{21} Pa s cited above with the upper mantle defined to end at 670 km depth, rather than with the depth range of 1000-1500 km indicated by the associated resolving kernel [*Parsons*, 1972; *Mitrovica*, 1996].

The growing appreciation that GIA data may be compatible with a significant increase of viscosity with depth in the mantle has led to efforts to reconcile GIA and convection data sets with a single profile of mantle viscosity. The first joint inversions of such data sets were performed by *Forte and Mitrovica* [1996] and *Mitrovica and Forte* [1997]. Using non-linear, iterative Occam inversion procedures, these studies analysed decay times associated with the post-glacial uplift of Hudson Bay and Fennoscandia, and long-wavelength free-air gravity anomalies associated with mantle convection. The inversions generated a set of viscosity models, characterized by a ∼two order of magnitude increase with depth, that simultaneously reconciled both data sets. We concluded, in these earlier studies, that a transient viscosity need not be invoked to reconcile this specific subset of the GIA and convection data sets. This conclusion was supported by *Peltier* [1996], who argued that a viscosity profile he determined by inverting postglacial decay times and *McConnell's* [1968] Fennoscandian relaxation spectrum qualitatively matched a profile that had been shown by *Forte et al.* [1993] to fit a variety of convection data sets.

In recent work, we explored the thermochemical structure of the deep mantle on the basis of a viscosity profile determined through an inversion a broad set of convection-related observables, including free-air gravity anomalies, the excess ellipticity of the core-mantle-boundary inferred from geodetic studies of the period of the free-core-nutation, and plate motions [*Forte and Mitrovica*, 2001]. These inversions were characterized by two viscosity maxima within the lower mantle, at 1000 km and 2000 km depth. The deepest of the two was found to suppress all but the longest horizontal wavelengths of the present-day flow in the bottom 1000 km of the lower mantle, thereby providing a simple interpretation for the 'red' spectrum of seismically-inferred heterogeneity in this region. In this paper we report on the results of an effort to extend our joint inversions [*Forte and Mitrovica*, 1996; *Mitrovica and Forte*, 1997] to include this larger set of convection data and an updated set of GIA observables. In our original work we did not include *McConnell's* [1968] relaxation spectrum because of serious concerns regarding the accuracy of the strandline data upon which it was based [*Wolf*, 1997]. In the present analysis we adopt a newly derived relaxation spectrum [*Wieczerkowski et al.*, 1999]. We also incorporate a set of post-glacial decay times from sites in Fennoscandia and Hudson Bay. The latter are based on a recent reappraisal of the sea-level record in Richmond Gulf and James Bay, Canada [*Mitrovica et al.*, 2000].

We have two goals in the present analysis. The first is to explore the extent to which the viscosity inference in *Forte and Mitrovica* [2001] is refined by the addition of GIA-related constraints. The detailed implications of this refinement for analyses of the thermochemical structure of the deep mantle will be treated elsewhere. Second, we investigate whether a radial profile of mantle viscosity can be found that simultaneously reconciles our new, and much more extensive, data base of GIA and convection observables. The existence of such a solution clearly has bearing on the continued debate over the issue of a transient mantle viscosity and hence the question of the microphysical mechanisms which govern steady-state mantle creep.

2. RESULTS

Our non-linear, iterative inversions are based on the Occam algorithm described by *Constable et al.* [1987]. The Occam procedure weights the individual data points by their respective uncertainties, and yields the smoothest possible model that provides an acceptable (χ^2) fit to the data. Following our earlier work [*Forte and Mitrovica*, 1996; *Mitrovica and Forte*, 1997], we parameterize the inversions in terms of the logarithm of viscosity (see below) in 13 constant-viscosity layers stretching from the surface to CMB. The upper and lower mantle contain 5 and 8 such layers, respectively.

In the next two subsections we review, in detail, the GIA and convection data sets and discuss the relative

sensitivity of these data to variations in the mantle viscosity profile.

2.1. GIA Data Sets

Geophysical observables related to GIA are generally sensitive to both mantle viscosity and the space-time history of the Late Pleistocene ice cover. Accordingly, inferences of viscosity derived from GIA data sets may be biased, sometimes significantly so, by errors in the assumed ice model. In this article we avoid this fundamental problem by using a rather specialized subset of GIA data that is demonstrably insensitive to uncertainties in the surface mass loading.

The first subset involves post-glacial decay times that characterize the uplift history near the center of previously glaciated regions. It has long been known that uplift curves obtained from geological survey in such regions are characterized by a simple exponential form [*Andrews*, 1970; *Walcott*, 1972, 1980]. Let us denote a set of I relative sea-level markers at a given site by $(\text{RSL}(t_i), t_i;$ for $i = 1, I)$, where t_i denotes the age of the ith marker. *Walcott* [1980] introduced the following parameterization for this set of data:

$$\text{RSL}(t_i) = A[\exp(-t_i/\tau) - 1] + C \qquad (1)$$

where (A, τ) are the amplitude and decay time that yield a 'best-fit' of the form (1) through the data. *Walcott* [1980] added the term C to reflect the uncertainty in absolute height of RSL curves determined by survey.

Mitrovica resurrected the use of *Walcott's* [1980] form (1) by demonstrating that predictions of such decay times, as long as they were based on data limited to the local post-glacial time window, were relatively insensitive to assumptions regarding the detailed ice history [*Mitrovica and Peltier*, 1993a, 1995; *Mitrovica and Forte*, 1997]. This is in contrast to the amplitude, A, and especially the raw RSL data, $\text{RSL}(t_i)$. Decay times determined from Canadian and Fennoscandian uplift curves are now widely cited in GIA studies of mantle viscosity [*Peltier*, 1994, 1996, 1998; *Forte and Mitrovica*, 1996; *Mitrovica et al.*, 2000; *Fang and Hager*, 2002].

In the present study we will adopt decay times determined from RSL records in Angerman River, Sweden, and two sites in the southern Hudson Bay region: Richmond Gulf and James Bay. In regard to Angerman River, we have adopted the 'corrected' RSL data set presented by *Lambeck et al.* [1990]. Applying the estimation procedure described, in detail, in *Mitrovica et al.* [2000], we have determined a decay time from these data of 4.9 ± 0.9 kyr. This constraint is consistent with estimates from independent analyses [e.g., *Peltier*, 1998; *Fang and Hager*, 2002]. Unfortunately, estimates of the post-glacial decay time from Hudson Bay are far more contentious.

RSL constraints from the Hudson Bay region were tabulated by *Tushingham and Peltier* [1991]. However, *Mitrovica and Peltier* [1993a] cautioned that the curves for a number of sites in this data base were generated by subjective fitting of temporally-sparse survey points, and they argued that these should not be used for decay time estimates. The Richmond Gulf data set appearing in the *Tushingham and Peltier* [1991] data base was an exception to this rule, since it was based on constraints provided by *Hillaire-Marcel* [1980]. This curve served as the basis for *Peltier's* [1994] estimate of 7.6 kyr for the decay time; a value later confirmed, using the same data set, by *Mitrovica and Peltier* [1995].

Peltier [1998] has recently provided a detailed recompilation of RSL constraints from a number of sites in southeastern Hudson Bay, and, on this basis, he derived decay times of 3.426 kyr for the entire region and 3.399 for James Bay. He furthermore argued that the estimate of 7.6 kyr for the Richmond Gulf decay time, which he inexplicably traced to *Mitrovica* [1996], was suspect because it was based on 'very few (uncalibrated) carbon dates'.

Mitrovica et al. [2000] revisited the same issue with their own recompilation of RSL data from Richmond Gulf and James Bay. Their analysis revealed a serious error in the *Tushingham and Peltier* [1991] entries for Richmond Gulf, involving an unnecessary correction from carbon to sidereal time. Furthermore, they raised several criticisms concerning the *Peltier* [1998] estimate of decay times. First, they demonstrated that *Peltier's* [1998] use of a single, composite sea-level curve for the entire southeastern Hudson Bay introduced potentially significant error into the decay time estimate and obscured significant variation in the independent data sets from Richmond Gulf and James Bay. Second, they argued that *Peltier's* [1998] mixing of data types (e.g., specific shells, driftwood, etc.) on a single sea-level curve was not consistent with the assumptions inherent to the form (1). The argument for using a single data type when computing decay times from such curves may be traced to *Walcott* [1980], and it has recently also been reiterated by *Fang and Hager* [2002]. Finally, *Mitrovica et al.* [2000] questioned several examples of *Peltier's* [1998] data selection.

Mitrovica et al. [2000] derived decay time estimates of 2.4 ± 0.4 kyr for James Bay and 5.3 ± 1.3 kyr for Richmond Gulf. The latter is reasonably consistent with a more recent estimate of 6.3 ± 1.0 kyr for Richmond

Gulf derived by *Fang and Hager* [2002]. The origin of the disagreement between the post-glacial decay times for Richmond Gulf and James Bay is unclear, but it is likely due to errors in the observational record. Rather than simply combining these estimates into some hybrid constraint for the region, our analysis will include both decay time estimates from southeast Hudson Bay. In this case, it will be instructive to compare a-posteriori predictions based on our inverted viscosity profiles to these individual constraints.

The second class of GIA data included in the inversion is the revised relaxation spectrum for Fennoscandia derived by *Wieczerkowski et al.* [1999]. The relaxation spectrum represents the decay time of the post-glacial uplift of Fennoscandia as a function of the wavelength (or, alternatively, the spherical harmonic degree) of the deformation. Specifically, the new spectrum constrains the decay times between degrees ~10-50 and 60-75. Under the assumption of post-glacial free decay, the relaxation spectrum provides a constraint on viscosity that is, at least in principle, independent of the loading history [e.g., *McConnell*, 1968]. Furthermore, for the spherical harmonic degrees above 10-15 that are sampled by the relaxation spectrum, the uplift at each degree is generally dominated by a single mode of viscoelastic relaxation; hence, predictions of the relaxation spectrum may be simply approximated by the decay time of the fundamental (M0) mode of relaxation [*Mitrovica*, 1997].

McConnell's [1968] original derivation of the relaxation spectrum was based on strandline data derived by *Sauramo* [1958]. Specifically, under the assumption of axisymmetry about the center of Fennoscandia uplift, *McConnell* [1968] Hankel-transformed a set of strandlines of distinct ages and then found the best-fitting exponential decay times through the amplitudes as a function of wavenumber. The accuracy of the *Sauramo* [1958] data set has been questioned by a number of authors, including *McConnell* [1968], *Mitrovica and Peltier* [1993b] and, in particular, *Wolf* [1997]. This motivated *Wieczerkowski et al.* [1999] to generate a revised relaxation spectrum based on more recent shoreline reconstructions [e.g., *Donner*, 1980].

Our forward predictions of GIA observables are based on spherically symmetric, self-gravitating, Maxwell viscoelastic Earth models. The elastic structure and density profile of these models is given by the seismic model PREM [*Dziewonski and Anderson*, 1981], and the viscoelastic normal modes are computed using the formalism described by *Peltier* [1974]. As discussed above, predictions of the relaxation spectrum are taken to be equivalent to the decay times of the fundamental 'M0' mode of viscoelastic relaxation [*Peltier*, 1976]. Relative sea-level predictions are computed using the pseudo-spectral algorithm described by *Mitrovica and Peltier* [1991]. These calculations adopt a version of the ICE-3G model of late Pleistocene deglaciation [*Tushingham and Peltier*, 1991] that has been modified to include a linear glaciation phase of duration 90 kyr. The decay time for a given site is then computed by fitting the form (1) (without the term C) to the predicted RSL curve. This decay time represents, effectively, a weighted sum of the multi-normal mode decay times associated with the impulse response of the Earth model. This weighting is not sensitive to the detailed geometry of the ice load; however, it is dependent on the broad spatial scale of the ice cover over the site of interest.

The non-linear Occam methodology requires, at each iteration, a set of Frechet kernels relating arbitrary depth-dependent variations in the viscosity profile to perturbations in predictions of each observable. Following our previous work, we will parameterize the analysis in terms of the logarithm of the decay times in order to weaken the non-linearity of the inversions. Accordingly, we can write, for the ith observable:

$$\delta \log \tau_i = \int_{b/a}^{1} \text{FK}_i[\nu(r); r] \delta \log \nu(r) \, dr \qquad (2)$$

where τ represents either a site-specific decay time from Fennoscandia or Hudson Bay, or the decay time for a specific degree in the relaxation spectrum, b and a are the dimensional radii of the core-mantle boundary and the surface, respectively, and r is the radius non-dimensionalized by a. In equation (2), $\nu(r)$ is the radial profile of mantle viscosity, and the dependence of the Frechet kernel, FK_i, on this profile is made explicit.

Frechet kernels for the site-specific decay times are computed numerically by perturbing the viscosity in each of the 13 radial layers that define our 'model' and comparing the decay times computed on the basis of this set of models with the results from the unperturbed solution. In contrast, analytical expressions exist for the Frechet kernels associated with the decay times of individual normal modes of viscoelastic relaxation [*Peltier*, 1976]. As an example, Fig. 1 shows a set of such kernels for the M0 mode (and hence for the relaxation spectrum) over a range of degrees that sample the observational constraints. These specific kernels are generated for an Earth model with an 80 km elastic lithosphere overriding an isoviscous, 10^{21} Pa s mantle. (The kernels involve a set of several hundred radial nodes within the mantle; mean values within any region of the mantle, for example within the 13 model layers treated in our

inversions, can be computed by applying the appropriate volumetric averaging.)

The Frechet kernels reflect the depth-dependent sensitivity of a particular datum to variations in the radial viscosity profile. Not surprisingly, the kernels in Fig. 1 peak at progressively shallower depths as the harmonic degree is increased. These kernels have the property that [*Peltier*, 1976]:

$$\int_{b/a}^{1} \text{FK}_i[\nu(r); r] \, dr = 1 \qquad (3)$$

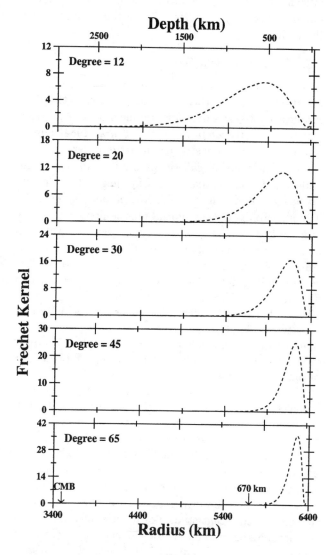

Figure 1. Frechet kernels (defined in equation 2) associated with the decay time of the fundamental mode of relaxation, M0, at a set of spherical harmonic degrees (as labelled on each frame). The kernels are computed using an Earth model with an 80 km elastic lithosphere overlying an isoviscous, 10^{21} Pa s, mantle. The kernels satisfy the normalization given by equation (3).

Thus, a uniform, order of magnitude increase in viscosity produces an order of magnitude increase in the decay times defining the relaxation spectrum.

We have not shown Frechet kernels for decay times associated with the post-glacial uplift of Angerman River, Sweden and southeast Hudson Bay (see, for example, Fig. 5 of *Mitrovica and Forte*, 1997); however, for the viscosity profile used to generate Fig. 1, these have forms similar to the M0 kernels at degree 12 and 20, respectively. We conclude that the set of GIA data incorporated in the present inversions will provide significant resolving power in the top half of the mantle, but little constraint on viscosity in the bottom half of the region.

2.2. Convection Data Sets

A number of studies investigating the dynamical implications of the earliest very-long wavelength ('low-resolution') 3-D seismic tomographic models [e.g., *Dziewonski*, 1984; *Woodhouse and Dziewonski*, 1984] demonstrated that the density perturbations derived from these models yield realistic predictions of the large-scale mantle convective circulation [e.g., *Hager et al.*, 1985; *Forte and Peltier*, 1987; *Ricard and Vigny*, 1989]. The plausibility of these initial mantle-flow calculations was verified by the good agreement between the predicted convection-related observables (e.g., nonhydrostatic geoid anomalies, tectonic plate motions) and the corresponding data. The tomography-based convective flow models require, as necessary inputs, the scaling coefficient between perturbations of seismic velocity and density and knowledge of the effective viscosity of the mantle. Convection-related surface data therefore provide constraints on the long-term rheological behaviour of mantle minerals.

The resolution of 3-D mantle structure provided by global seismic tomography models has significantly improved over the past few years, mainly as a result of the rapidly increasing size, diversity and quality of global seismic data sets [e.g., *Li and Romanowicz*, 1996; *Grand et al.*, 1997; *van der Hilst et al.*, 1997; *Ekström and Dziewonski*, 1998; *Ritsema et al.*, 1999]. The advent of these new 'high-resolution' global tomography models has motivated recent studies of mantle dynamics using tomography-based mantle flow models [e.g., *Steinberger and O'Connell*, 1998; *Forte*, 2000; *Panasyuk and Hager*, 2000; *Forte and Mitrovica*, 2001].

The convection-related data sets we will employ here to constrain the mantle viscosity profile are the satellite-inferred global free-air gravity anomalies [*Marsh et al.*, 1990], the observed tectonic plate motions [*De Mets et al.*, 1990], the dynamic or excess ellipticity of the core-mantle boundary (CMB) inferred from space geodetic determinations of the free-core nutation period [*Gwinn et al.*, 1986; *Mathews et al.*, 1999], and the dynamic surface topography obtained by isostatic reduction of the observed topography [*Mooney et al.*, 1998; *Forte and Perry*, 2000].

Our predictions of the convection-related surface observables are based on the theory of buoyancy-induced flow in a spherical, compressible, self-gravitating, mantle [*Forte and Peltier*, 1991]. This theory has recently been reformulated by *Forte* [2000] such that the calculation of viscous flow in the mantle now depends entirely on the natural logarithm of mantle viscosity rather just viscosity itself, as in previous formulations [e.g., *Richards and Hager*, 1984; *Ricard et al.*, 1984; *Forte and Peltier*, 1987,1991]. From the perspective of the viscosity inverse problem, this new theoretical formulation is particularly advantageous as it allows us to resolve very large radial gradients in the mantle viscosity profile [e.g., *Forte and Mitrovica*, 2001].

An important aspect of our mantle flow model is that the plate motions are predicted rather than imposed. The theory we use to couple plate motions to underlying mantle flow [*Forte and Peltier*, 1994] provides a dynamically consistent way to incorporate the tectonic plates which are assumed to behave as rigid bodies. The buoyancy driven mantle flow is constrained only by the present-day plate geometry.

The relationship between surface convection observables and internal density anomalies is expressed in the spectral domain defined by spherical harmonic basis functions. The theoretical dependence of convection data on density anomalies $\delta\rho_\ell^m(r)$ of harmonic degree and order ℓ-m, is expressed in terms of spectral kernel functions as follows:

$$\delta g_\ell^m = \frac{g_o}{R_o} \frac{3(\ell-1)}{(2\ell+1)\bar{\rho}} \int_b^a G_\ell \left(\ln\left[\frac{\nu(r)}{\nu_o}\right]; r \right) \delta\rho_\ell^m(r)\, dr,$$
$$\text{(4a)}$$

$$\delta a_\ell^m = \frac{1}{\Delta\rho_{mo}} \int_b^a A_\ell \left(\ln\left[\frac{\nu(r)}{\nu_o}\right]; r \right) \delta\rho_\ell^m(r)\, dr, \quad \text{(4b)}$$

$$(\nabla_H \cdot \mathbf{v})_l^m = \frac{g_0}{\nu_0} \int_b^a D_\ell \left(\ln\left[\frac{\nu(r)}{\nu_o}\right]; r \right) \delta\rho_\ell^m(r)\, dr, \quad \text{(4c)}$$

$$\delta b_2^0 = \frac{1}{\Delta\rho_{cm}} \int_b^a B_2 \left(\ln\left[\frac{\nu(r)}{\nu_o}\right]; r \right) \delta\rho_2^0(r)\, dr, \quad \text{(4d)}$$

where δg_ℓ^m, δa_ℓ^m, $(\nabla_H \cdot \mathbf{v})_l^m$, δb_2^0 are the spherical harmonic coefficients of the free-air gravity anomalies, the dynamic surface topography, the horizontal divergence of the tectonic plate motions, and the excess CMB ellipticity, respectively. The spectral kernel functions corresponding to these data are G_ℓ, A_ℓ, D_ℓ, B_2 and, in addition to radius, they depend on the logarithm of the relative (dimensionless) viscosity $\nu(r)/\nu_o$, where ν_o is a reference scaling value. Other geophysical constants appearing (4a-d) are: the mean gravitational acceleration at Earth's surface $g_o = 9.82$ m/s^2 (982,000 mGal), Earth's mean radius $R_o = 6371$ km, Earth's mean density $\bar{\rho} = 5.515$ Mg/m^3, the density jump across the mantle-ocean boundary $\Delta\rho_{mo} = 2.2$ Mg/m^3, and the density jump across the CMB $\Delta\rho_{cm} = -4.43$ Mg/m^3. As is evident in (4a-d), the predicted convection observables, with the exception of the horizontal plate divergence in (4c), do not depend the absolute value of mantle viscosity.

Density anomalies $\delta\rho$ are derived from the relative seismic shear velocity anomalies $\delta V_s/V_s$ provided by the global tomography model obtained by Grand [*Grand et al.*, 1997]. This tomography model is parameterized radially in terms of 22 layers and horizontally in terms of $2° \times 2°$ cells. We computed the equivalent spherical harmonic representation of the seismic anomalies in each layer and truncated the harmonic expansion at degree $\ell = 32$. We hereafter refer to this spherical harmonic representation of Grand's model as 'GRAND32'. To convert the seismic anomalies into equivalent density anomalies we employed a velocity-to-density scaling coefficient $d\ln\rho/d\ln V_s(r)$ derived by *Karato* [1993] and subsequently modified on the basis of geodynamic constraints [*Forte and Woodward*, 1997]. This initial $d\ln\rho/d\ln V_s(r)$ scaling coefficient is updated and refined in the course of the iterative viscosity inversions described below.

Frechet kernels for the convection data are calculated on the basis of the theory presented by *Forte* [2000]. These kernels relate perturbations of the predicted convection observables to perturbations of the radial viscosity profile as follows:

$$\delta p_i = \int_{b/a}^1 FK_i[\nu(r);r]\,\delta\log\nu(r)\,dr \qquad (5)$$

where δp_i is the predicted variation in the ith convection-related surface observable (consisting of any one of the harmonic coefficients of the free-air gravity anomalies, dynamic surface topography, plate divergence, or excess CMB ellipticity in (4a-d) above) and FK_i is the corresponding Frechet or sensitivity kernel. As is evident in (5), these sensitivity kernels depend on the viscosity profile and they will change from one iteration to the next in the viscosity inversions presented below.

The Frechet kernels FK_i in (5) are calculated numerically by solving the perturbed flow equations [*Forte*, 2000] in which viscosity perturbations are introduced

at a series of radii across the mantle. The mean values of FK_i, in each of the 13 layers employed in the viscosity inversions below, are calculated by averaging across each layer interval. In Fig. 2 we present the convection-data Frechet kernels for an isoviscous (10^{21} Pa s) mantle using the density anomalies derived from the GRAND32 tomography model. At the longest wavelengths ($\ell = 2$), the convection data are generally most sensitive to viscosity variations in the top and bottom portions of the mantle. At shorter wavelengths (e.g., $\ell = 8$) the sensitivity to viscosity variations is almost evenly distributed across the mantle, except for the plate divergence where we observe a more clearly focused sensitivity to viscosity changes in the upper mantle. At the shortest wavelengths ($\ell = 32$) the sensitivity is confined to the upper mantle and, in the case of plate divergence, it is strongly concentrated in the shallowest mantle layer. The convection data thus have the potential to provide good resolution of the viscosity profile throughout most of the mantle.

2.3. Viscosity Profiles

We have performed a large series of Occam inversions distinguished on the basis of the starting viscosity model, relative weighting of convection and GIA data, velocity-to-density scalings applied to the seismic models, etc. We focus here on one subset of these inversions that was based on starting models that included a low-viscosity notch within the 70 km layer just above 670 km depth. Physical arguments for the appearance of a thin low-viscosity region above the boundary between the upper and lower mantle are reviewed, for example, in *King and Masters* [1992], *Forte et al.* [1993] and *Pari and Peltier* [1995]. A discussion of our complete set of inversion results will appear in future work.

Including a thin low-viscosity zone (henceforth LVZ) in the inversions was motivated by prior work that suggested that such a feature improves fits to the convection observables [e.g., *Forte et al.*, 1993]. A thin low-viscosity layer at this location has also been shown to influence certain subsets of GIA data [*Milne et al.*, 1998]. Since the Occam methodology penalizes roughness, a thin LVZ would not be expected to emerge from inversions that do not include this feature at the outset; accordingly, we considered a suite of inversions in which an LVZ was imposed a-priori.

In Fig. 3 we present results from two joint inversions of the GIA and convection data sets described above. The profile given by the solid line, which we will denote as I1, has an a-posteriori viscosity within the LVZ of 0.66×10^{20} Pa s. The second profile, I2 (dotted line), has

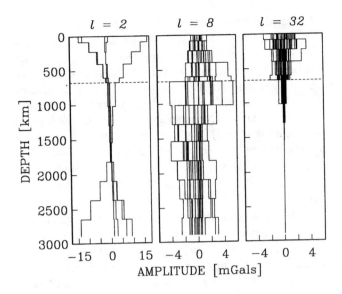

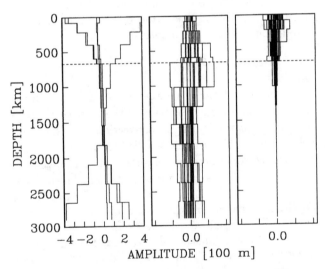

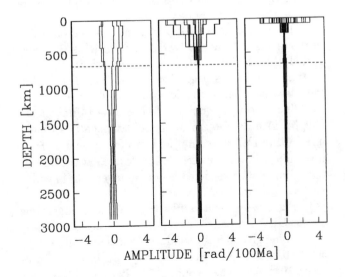

a viscosity of 0.77×10^{19} Pa s within this region. These results are characteristic of the larger set of inversions based on the presence of a LVZ. Both models show a significant, 2-3 order of magnitude, increase of viscosity from the base of the lithosphere to 2000 km depth and a region of lower viscosity from 2000 km depth to the CMB. The mean values of the model I1 within the upper and lower mantle are 4.3×10^{20} Pa s and 6.5×10^{21} Pa s, respectively. Analogous averages for model I2 are 3.9×10^{20} Pa s and 11.0×10^{21} Pa s.

Models I1 and I2 provide fits to the convection data, including long wavelength free-air gravity harmonics, plate motions and the excess CMB ellipticity, that are comparable to those we obtained in an inversion of these data alone [*Forte and Mitrovica*, 2001]. The fits obtained by these models to the GIA data, specifically the Fennoscandian relaxation spectrum and post-glacial decay times at Angerman River, Sweden, and Hudson Bay, are reviewed on Figs. 4 and 5. Both models I1 and I2 provide excellent fits to all degrees of the relaxation spectrum and to the Angerman River decay time. Figs. 4 and 5 illustrate the inconsistency of the observational constraints on the Richmond Gulf and James Bay decay times; in this case, both inverted models yield a decay time close to ~5 kyr, in accord with the Richmond Gulf observation. In general, we have found that the inversions invariably produce viscosity models that fit the Richmond Gulf constraint, rather than the shorter decay time observed at James Bay, and this tendency may reflect on the accuracy of the latter. However, models can be found which better fit the James Bay observations; these are characterized by a moderately lower viscosity between 670-1300 km depth than the values evident in Fig. 3 (see the discussion below).

We have found that between degrees ~10-20 the postglacial response predicted for models I1 and I2 is not strongly dominated by a single normal mode. The solid

Figure 2. Frechet kernels for convection-related observables calculated for an isoviscous (10^{21} Pa s) mantle and using the mantle density anomalies derived from the GRAND32 tomography model. (*top*) Frechet kernels for the predicted free-air gravity anomalies, for various spherical harmonic degrees ℓ (as labeled). (Note: the scale for the horizontal axis for the degree 8 and 32 kernels is 3 × smaller than for the degree 2 kernels.)(*middle*) Frechet kernels for dynamic surface topography. (Note: the scale for the horizontal axis for the degree 8 and 32 kernels is 4 × smaller than for the degree 2 kernels.) (*bottom*) Frechet kernels for the predicted horizontal divergence of the tectonic plate velocities.

Depth (km)

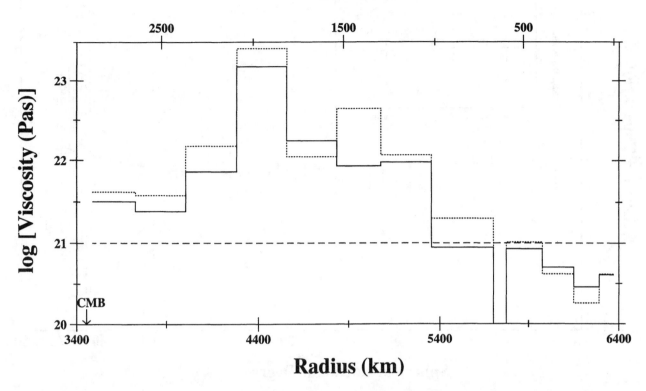

Radius (km)

Figure 3. Results from the joint inversion of GIA and convection data. The solid and dotted lines represent the inverted, 13-layer, radial viscosity profiles I1 and I2 described in the text. These models have a value of 0.66×10^{20} Pa s and 0.77×10^{19} Pa s, respectively, in the 70 km thick layer just above 670 km depth.

lines on the main frame of Figs. 4 and 5 represent the dominant mode at each degree. The other modes of significant strength cluster close to the dominant mode and hence the lumped multi-mode post-glacial response for these models will also satisfy the relaxation spectrum constraint.

As we discussed in the context of the Frechet kernels in Fig. 1, the GIA data provide a significant constraint on the inverted models within the top half of the mantle. As an example, the mean value of the models within the top ~1000 km of the mantle is close to 10^{21} Pa s, and thus they both satisfy the so-called Haskell constraint on viscosity (which is equivalent to fitting the decay time at Angerman River; *Mitrovica* [1996]). The decay time at Hudson Bay has a deeper resolving kernel. Accordingly, as we discussed above, one can improve the fit to the James Bay decay time, at the expense of the fit to the decay time at Richmond Gulf, by lowering the viscosity in the top ~600-800 km of the lower mantle. This process is, however, limited by the constraint

imposed by the Fennoscandian relaxation spectrum. If one lowers the viscosity too much in this shallow lower mantle region, then the predicted decay times at the lowest (below ~ 20) degrees on the relaxation spectrum will become too short to satisfy the observational constraint.

The inverted models, I1 and I2, may be compared to the model we recently derived on the basis of convection data alone [*Forte and Mitrovica*, 2001]. The latter model was characterized by large viscosity peaks at ~ 1000 and 2000 km depth, and a deep viscosity minimum within the upper mantle. The requirement for a viscosity minimum is accommodated, at least in part, by the LVZ at the base of the upper mantle in the inverted models. Furthermore, the GIA constraints appear to have either removed the shallow lower mantle viscosity peak (e.g., model I1) or shifted a lower amplitude version of this peak to greater (~ 1500 km) depths (e.g., model I2). The net result is that the joint inversions are characterized by a relatively gradual increase

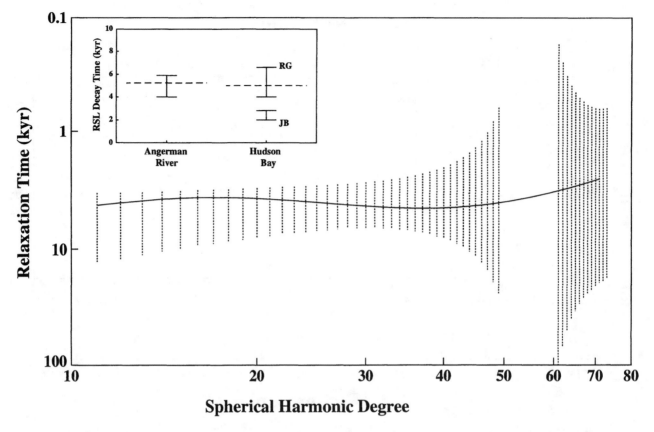

Figure 4. The fit of the inverted model I1 (see Fig. 3) to the set of GIA data. The vertical dotted lines in the main frame represent the constraint on the Fennoscandian relaxation spectrum derived by *Wieczerkowski et al.* [1999] and the solid line is the forward prediction generated using the inverted, 13-layer, viscosity model I1 (with the exception that an 80 km elastic lithosphere has been imposed). The inset provides the observational constraint (see text) on the decay times of post-glacial uplift at Angerman River, Sweden, and two sites in Hudson Bay (Richmond Gulf, RG, and James Bay, JB). In this case, the forward predictions based on model I1 are given by the dashed line.

of viscosity within the lower mantle and a dominant viscosity maximum at 2000 km depth. *Forte and Mitrovica* [2001] argued that the high-viscosity deep mantle peak acted to 'redden' the heterogeneity spectrum in the bottom 1000 km of the mantle, in agreement with observations from seismic tomography. The deep high-viscosity zone inferred by *Forte and Mitrovica* [2001] remains a robust feature in our joint inversions.

3. SUMMARY

We have presented a subset of results from our recent efforts to jointly invert an extensive set of data related to the glacial isostatic adjustment process and mantle convection. These results confirm a conclusion from our earlier work based on a more limited set of GIA and convection data; namely, that both data sets can be reconciled by a single profile of mantle viscosity. This weak-

ens arguments for the necessity of invoking transient mantle rheology across the time scales that characterize the two geophysical processes. We caution, however, that this conclusion requires re-assessment each time an additional data set is incorporated into the inverse analysis.

Our inverted profiles are characterized by a viscosity increase that exceeds two orders of magnitude from the base of the lithosphere to 2000 km depth. The profiles are, in fact, dominated by a high-viscosity peak located ~ 1000 km above the CMB, and they include a relatively weak upper mantle underlain by a thin low-viscosity notch at the base of the transition zone. In future work we present results from a much broader set of inversions. We furthermore provide a detailed examination of the implications of the inverted models for a suite of applications related to both the GIA and convection processes.

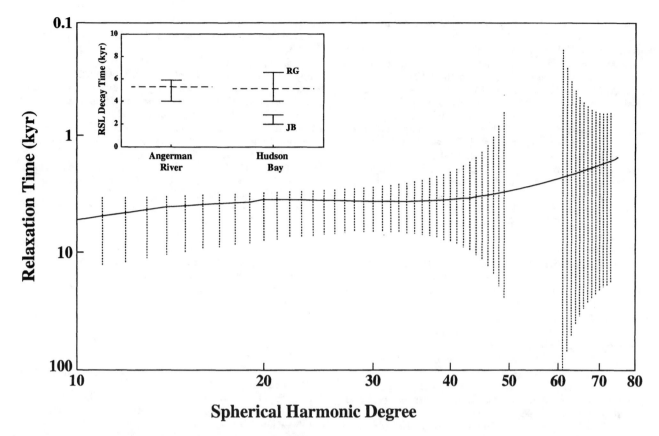

Figure 5. As in Fig. 4, with the exception that the forward predictions are generated using the inverted 13-layer viscosity model I2 (see Fig. 3).

Acknowledgments. The authors gratefully acknowledge funding from the Natural Sciences and Engineering Research Council of Canada and the Canadian Institute for Advanced Research (Earth Systems Evolution Program).

REFERENCES

Andrews, J. T., *A Geomorphological Study of Postglacial Uplift With Particular Reference to Arctic Canada*, Oxford Univ. Press, New York, 1970.

Bunge, H.-P., M. A. Richards, and J. R. Baumgardner, Effect of depth-dependent viscosity on the planform of mantle convection, *Nature, 379*, 436–438, 1996.

Cathles, L. M., The Viscosity of the Earth's Mantle, *Ph.D. Thesis*, Princeton Univ., 1971.

Cathles, L. M., *The Viscosity of the Earth's Mantle*, Princeton Univ. Press, Princeton, N.J., 1975.

Constable, S. C., R. L. Parker, and C. G. Constable, Occam's inversion: A practical algorithm for generating smooth models from electromagnetic sounding data, *Geophys., 52*, 289–300, 1987.

Corrieu, V., Y. Ricard, and C. Froidevaux, Converting mantle tomography into mass anomalies to predict the Earth's radial viscosity, *Phys. Earth Planet. Inter., 84*, 3–13, 1994.

De Mets, C.R., Gordon, R.G., Argus, D.F., and S. Stein, Current plate motions, *Geophys. J. Int., 101*, 425–478, 1990.

Donner, J., The determination and dating of synchronous late Quaternary shorelines in Fennoscandia, in *Earth Rheology, Isostasy and Eustasy*, edited by N. A. Mörner, pp. 285–293, Wiley, Chichester, 1980.

Dziewonski, A. M., Mapping the lower mantle: determination of lateral heterogeneity in *P* velocity up to degree and order 6, *J. Geophys. Res., 89*, 5929–5952, 1984.

Dziewonski, A. M., and D. L. Anderson, Preliminary reference Earth model (PREM), *Phys. Earth Planet. Inter., 25*, 297–356, 1981.

Ekström, G., and A.M. Dziewonski, The unique anisotropy of the Pacific upper mantle, *Nature, 394*, 168–172, 1998.

Fang, M., and B. H. Hager, On the apparent exponential relaxation curves at the central regions of the last Pleistocene ice sheets, *this volume*, 2002.

Forte, A.M. Seismic-geodynamic constraints on mantle flow: Implications for layered convection, mantle viscosity, and seismic anisotropy in the deep mantle, in *Earth's Deep Interior: Mineral Physics and Tomography From the Atomic to the Global Scale, Geophys. Monogr. Ser.* **117**, edited by S.–i. Karato et al., AGU (Washington, DC), pp. 3–36, 2000.

Forte, A. M., and J. X. Mitrovica, A new inference of mantle viscosity based on a joint inversion of post-glacial rebound data and long-wavelength geoid anomalies, *Geophys. Res. Lett.*, *23*, 1147–1150, 1996.

Forte, A. M., and J. X. Mitrovica, Deep-mantle high-viscosity flow and thermochemical structure inferred from seismic and geodynamic data, *Nature*, *410*, 1049–1056, 2001.

Forte, A. M., and W. R. Peltier, Plate tectonics and aspherical Earth structure: The importance of poloidal-toroidal coupling, *J. Geophys. Res.*, *92*, 3645–3679, 1987.

Forte, A. M., and W. R. Peltier, Viscous flow models of global geophysical observables, 1, Forward problems, *J. Geophys. Res.*, *96*, 20,131–20,159, 1991.

Forte, A.M., and W.R. Peltier, The kinematics and dynamics of poloidal-toroidal coupling of mantle flow: The importance of surface plates and lateral viscosity variations, *Adv. Geophys.*, *36*, 1–119, 1994.

Forte, A.M., and H.K.C. Perry, Geodynamic evidence for a chemically depleted continental tectosphere, *Science*, *290*, 1940–1944, 2000.

Forte, A.M., and R.L. Woodward, Global 3D mantle structure and vertical mass and heat transfer across the mantle from joint inversions of seismic and geodynamic data, *J. Geophys. Res.*, *102*, 17981–17994, 1997.

Forte, A. M., W. R. Peltier, and A. M. Dziewonski, Inferences of mantle viscosity from tectonic plate velocities, *Geophys. Res. Lett.*, *18*, 1747–1750, 1991.

Forte, A. M., A. M. Dziewonski, and R. L. Woodward, Aspherical structure of the mantle, tectonic plate motions, nonhydrostatic geoid, and topography of the core–mantle boundary, in *Dynamics of the Earth's Deep Interior and Earth Rotation, Geophys. Monogr. Ser.*, vol. 72, edited by J.-L. Le Mouël, D. E. Smylie, and T. Herring, pp. 135–166, AGU, Washington, D.C., 1993.

Forte, A. M., R. L. Woodward, and A. M. Dziewonski, Joint inversions of seismic and geodynamic data for models of three–dimensional mantle heterogeneity, *J. Geophys. Res.*, *99*, 21,857–21,877, 1994.

Grand, S.P., R.D. van der Hilst, and S. Widiyantoro, Global seismic tomography: A snapshot of convection in the Earth, *GSA ·Today*, *7*, 1–7, 1997.

Gwinn, C.R., T.A. Herring, and I.I. Shapiro, Geodesy by radio interferometry: studies of the forced nutations of the Earth, 2, Interpretation, *J. Geophys. Res.*, *91*, 4755–4765, 1986.

Hager, B. H., Subducted slabs and the geoid: Constraints on mantle rheology and flow, *J. Geophys. Res.*, *89*, 6003–6015, 1984.

Hager, B. H., and R. W. Clayton, Constraints on the structure of mantle convection using seismic observations, flow models, and the geoid, in *Mantle Convection: Plate Tectonics and Global Dynamics*, edited by W.R. Peltier, pp. 657-763, Gordon and Breach, Newark, N.J., 1989.

Hager, B.H., R.W. Clayton, M.A. Richards, R.P. Comer, and A.M. Dziewonski, Lower mantle heterogeneity, dynamic topography and the geoid, *Nature*, *313*, 541–545, 1985.

Haskell, N. A., The motion of a fluid under a surface load, 1, *Physics*, *6*, 265–269, 1935.

Hillaire-Marcel, C., Multiple component postglacial emergence, eastern Hudson Bay, Canada, in *Earth Rheology, Isostasy and Eustasy*, edited by N.-A. Morner, pp. 215–230, John Wiley, New York, 1980.

Karato, S. Importance of anelasticity in the interpretation of seismic tomography, *Geophys. Res. Lett.*, *20*, 1623–1626, 1993.

King, S. D., Radial models of mantle viscosity: Results from a genetic algorithm, *Geophys. J. Int.*, *122*, 725–734, 1995.

King, S. D., and T. G. Masters, An inversion for the radial viscosity structure using seismic tomography, *Geophys. Res. Lett.*, *19*, 1551–1554, 1992.

Lambeck, K., P. Johnston, and M. Nakada, Holocene glacial rebound and sea-level change in NW Europe, *Geophys. J. Int.*, *103*, 451-468, 1990.

Li, X.-D., and B. Romanowicz, Global mantle shear-velocity model developed using nonlinear asymptotic coupling theory, *J. Geophys. Res.*, *101*, 22245–22272, 1996.

Marsh, J.G., et al. The GEM–T2 gravitational model, *J. Geophys. Res.*, *95*, 22043–22071, 1990.

Mathews, P., B.A. Buffet, and T.A. Herring, What do nutations tell us about the Earth's interior?, *Eos Trans. AGU*, *80* (46), Fall Meeting suppl., 19, 1999.

McConnell, R. K., Viscosity of the mantle from relaxation time spectra of isostatic adjustment, *J. Geophys. Res.*, *73*, 7089–7105, 1968.

Milne, G.A., J.X. Mitrovica, and A.M. Forte, The sensitivity of GIA predictions to a low viscosity layer at the base of the upper mantle, *Earth Planet. Sci. Lett.*, *154*, 265–278, 1998.

Mitrovica, J. X., Haskell [1935] revisited, *J. Geophys. Res.*, *101*, 555–569, 1996.

Mitrovica, J. X., Reply to comment on 'The Inference of mantle viscosity from an inversion of the Fennoscandian relaxation spectrum, by L. Cathles and W. Fjeldskaar', *Geophys. J. Int.*, *128*, 493–498, 1997.

Mitrovica, J. X., and A. M. Forte, Radial profile of mantle viscosity: Results from the joint inversion of convection and postglacial rebound observables, *J. Geophys. Res.*, *102*, 2751–2769, 1997.

Mitrovica, J. X., and W. R. Peltier, On post-glacial geoid subsidence over the equatorial oceans, *J. Geophys. Res.*, *96*, 20,053–20,071, 1991.

Mitrovica, J. X., and W. R. Peltier, A new formalism for inferring mantle viscosity based on estimates of post-glacial decay times: Application to RSL variations in N.E. Hudson Bay, *Geophys. Res. Lett.*, *20*, 2183–2186, 1993a.

Mitrovica, J. X., and W. R. Peltier, The inference of mantle viscosity from an inversion of the Fennoscandian relaxation spectrum, *Geophys. J. Int.*, *114*, 43–63, 1993b.

Mitrovica, J. X., and W. R. Peltier, Constraints on mantle viscosity based upon the inversion of post-glacial uplift data from the Hudson Bay region, *Geophys. J. Int.*, *122*, 353–377, 1995.

Mitrovica, J. X., A. M. Forte, and M. Simons, A reappraisal of postglacial decay times from Richmond Gulf and James Bay, Canada, *Geophys. J. Int.*, *142*, 783–800, 2000.

Mooney, W.D., G. Laske, and T.G. Masters, CRUST 5.1: A global crustal model at 5° × 5°, *J. Geophys. Res.*, *103*, 727–747, 1998.

Nakada, M., and K. Lambeck, Late Pleistocene and Holocene sea-level change in the Australian region and mantle rheology, *Geophys. J. Int.*, *96*, 497–517, 1989.

Panasyuk, S.V., and B.H. Hager, Inversion for mantle viscosity profiles constrained by dynamic topography and the geoid, and their estimated errors, *Geophys. J. Int.*, *143*, 821–836, 2000.

Pari, G., and W. R. Peltier, The heat flow constraint on mantle tomography-based convection models: Towards a geodynamically self-consistent inference of mantle viscosity, *J. Geophys. Res.*, *100*, 12,731–12,751, 1995.

Parsons, B. D., Changes in the Earth's shape, Ph.D. thesis, Cambridge Univ., Cambridge, England, 1972.

Peltier, W. R., The impulse response of a Maxwell Earth, *Rev. Geophys.*, *12*, 649–669, 1974.

Peltier, W. R., Glacial isostatic adjustment, II, The inverse problem, *Geophys. J. R. Astron. Soc.*, *46*, 669–706, 1976.

Peltier, W. R., New constraint on transient lower mantle rheology and internal mantle buoyancy from glacial rebound data, *Nature*, *318*, 614–617, 1985.

Peltier, W. R., Ice age paleotopography, *Science*, *265*, 195–201, 1994.

Peltier, W. R., Mantle viscosity and ice-age ice sheet topography, *Science*, *273*, 1359–1364, 1996.

Peltier, W. R., Postglacial variations in the level of the sea: Implications for climate dynamics and solid-earth geophysics, *Rev. Geophys.*, *36*, 603-689, 1998.

Peltier, W. R., and J. T. Andrews, Glacial isostatic adjustment, I, The forward problem, *Geophys. J. R. Astron. Soc.*, *46*, 605–646, 1976.

Peltier, W. R., R. A. Drummond, and A. M. Tushingham, Post-glacial rebound and transient lower mantle rheology, *Geophys. J. R. Astron. Soc.*, *87*, 79–116, 1986.

Ricard, Y., and C. Vigny, Mantle dynamics with induced plate tectonics, *J. Geophys. Res.*, *94*, 17,543–17,559, 1989.

Ricard, Y., L. Fleitout, and C. Froidevaux, Geoid heights and lithospheric stresses for a dynamic Earth, *Ann. Geophys.*, *2*, 267–286, 1984.

Ricard, Y., C. Vigny, and C. Froidevaux, Mantle heterogeneities, geoid and plate motion: A Monte Carlo inversion, *J. Geophys. Res.*, *94*, 13,739–13,754, 1989.

Richards, M. A., Hot spots and the case for a high viscosity lower mantle, in *Glacial Isostasy, Sea-Level and Mantle Rheology*, *NATO ASI Ser. C*, vol. 334, edited by R. Sabadini, K. Lambeck, and E. Boschi, pp. 571–588, Kluwer Acad., Norwell, Mass., 1991.

Richards, M. A., and B. H. Hager, Geoid anomalies in a dynamic Earth, *J. Geophys. Res.*, *89*, 5987–6002, 1984.

Richards, M. A., Y. Ricard, C. Lithgow-Bertelloni, G. Spada and R. Sabadini, An explanation for the long-term stability of Earth's rotation axis, *Science*, *275*, 372–375, 1997.

Ritsema, J., H.J. van Heijst, and J.H. Woodhouse, Complex shear velocity structure imaged beneath Africa and Iceland, *Science*, *286*, 1925–1928, 1999.

Sabadini, R., and D. A. Yuen, Mantle stratification and long-term polar wander, *Nature*, *339*, 373–375, 1989.

Sabadini, R., D. A. Yuen, and P. Gasperini, The effects of transient rheology on the interpretation of lower mantle rheology, *Geophys. Res. Lett.*, *12*, 361–365, 1985.

Sauramo, M., Die geschichte der Ostsee, *Acta. Geogr. Helsinki*, *14*, 334-348, 1958.

Spada, G., Y. Ricard, and R. Sabadini, Excitation of true polar wander by subduction, *Nature*, *360*, 452–454, 1992.

Steinberger, B., and R. J. O'Connell, Change of the Earth's rotation axis inferred from the advection of mantle density heterogeneities, *Nature*, *387*, 169–173, 1997.

Steinberger, B., and R.J. O'Connell, Advection of plumes in mantle flow; implications for hot spot motion, mantle viscosity and plume distribution, *Geophys. J. Int.*, *132*, 412–434. 1998.

Thoraval, C., and M. A. Richards, The geoid constraint in global geodynamics: Viscosity structure, mantle heterogeneity models, and boundary conditions, *Geophys. J. Int.*, *131*, 1–8, 1997.

Tushingham, A. M., and W. R. Peltier, ICE-3G: A new global model of late Pleistocene deglaciation based upon geophysical predictions of postglacial relative sea level change, *J. Geophys. Res.*, *96*, 4497–4523, 1991.

van der Hilst, R.D., S. Widiyantoro, and E.R. Engdahl, Evidence for deep mantle circulation from global tomography, *Nature*, *386*, 578–584, 1997.

Walcott, R. I., Late Quaternary vertical movements in Eastern North America: Quantitative evidence of glacio-isostatic rebound, *Rev. Geophys. Space Phys.*, *10*, 849–884, 1972.

Walcott, R. I., Rheological models and observational data of glacio-isostatic rebound, in *Earth Rheology, Isostasy and Eustasy*, edited by N.-A. Morner, pp. 3–10, John Wiley, New York, 1980.

Wieczerkowski, K., J. X. Mitrovica, and D. Wolf, A revised relaxation-time spectrum for Fennoscandia, *Geophys. J. Int.*, *139*, 69–86, 1999.

Wolf, D., Notes on estimates of the glacial-isostatic decay spectrum for Fennoscandia, *Geophys. J. Int.*, *127*, 801–805, 1997.

Woodhouse, J. H., and A. M. Dziewonski, Mapping the upper mantle: three–dimensional modeling of earth structure by inversion of seismic waveforms, *J. Geophys. Res.*, *89*, 5953–5986, 1984.

Wu, P., and W. R. Peltier, Glacial isostatic adjustment and the free air gravity anomaly as a constraint on deep mantle viscosity, *Geophys. J. R. Astron. Soc.*, *74*, 377–450, 1983.

Wu, P., and W. R. Peltier, Pleistocene deglaciation and the Earth's rotation: A new analysis, *Geophys. J. R. Astron. Soc.*, *76*, 753–791, 1984.

Yuen, D. A., R. Sabadini, P. Gasperini, and E. V. Boschi, On transient rheology and glacial isostasy, *J. Geophys. Res.*, *91*, 11,420–11,438, 1986.

Zhang, S., and D. A. Yuen, The influences of lower mantle viscosity stratification on 3D spherical-shell mantle convection, *Earth Planet. Sci. Lett.*, *132*, 157–166, 1995.

J. X. Mitrovica, Department of Physics, University of Toronto, 60 St. George Street, Toronto, Canada M5S 1A7. (e-mail: jxm@physics.utoronto.ca)

A. M. Forte, Department of Earth Sciences, Biological and Geological Building, University of Western Ontario, London, Canada N6A 5B7. (e-mail: aforte@uwo.ca)

On the Apparent Exponential Relaxation Curves at the Central Regions of the Last Pleistocene Ice Sheets

Ming Fang and Bradford H. Hager

Department of Earth Atmospheric & Planetary Sciences, Massachusetts Institute of Technology

Combined analysis of spherical wavelets and relaxation diagrams reveals strong signatures of single harmonic relaxations in the exponential-like relative sea level curves near the central region of the two major Pleistocene ice sheets. The relative sea level near the center of the Laurentide ice sheet characterizes harmonic degree 9, and the relative sea level near the center of the Fennoscandian ice sheet represents a strong signal of harmonic degree 16. Independent estimates of the apparent exponential decay times are 6.3 ± 1.0 kyr for the center of the Laurentide ice dome and 4.6 ± 0.7 kyr for the center of Fennoscandia ice dome. These new estimates together with the wavelet results indicate a monitonic decrease of relaxation times in the 9-to-16 degree band in the relaxation spectrum, a major contradiction with McConnell [1968] spectrum. Viscosity models preferred respectively by the revised McConnell's [1968] spectrum and the recent Wieczerkowski et al [1999] spectrum are presented.

1. INTRODUCTION

The global pattern of postglacial relative sea level (RSL) as a function of time, t, exhibits a strong dependence on the distribution of ice loads [e.g., Tushingham & Peltier 1992; Fang and Hager, 1996; 1999]. Of particular interest among the sites available are those near the center of two major ice loads: the Laurentide ice sheet and the Fennoscandia ice sheet. The RSL curves at these sites exhibit characteristics of exponential functions. This observation laid the foundation for Haskell's [1935] classic analysis to derive the relaxation time τ of a single exponential mode, $\exp(-t/\tau)$, for the characteristic wavelength of the Fennoscandia ice load. To express the problem more clearly, let us call Haskell's [1935] analysis Problem 1 which is to relate the exponential-like RSL curves in the central region of the ice sheets to a single harmonic relaxation. To this end the real ice dome has to be distorted into a harmonic

undulation extending to infinity. An opposite approach, let us call it Problem 2, is to preserve the physical integrity of the ice dome [Mitrovica, 1996]. The so called decay time, T, in Mitrovica's [1996] analysis is identical with Haskell's relaxation time, τ, in the sense that they are obtained from fitting the same curves by essentially identical expressions (different only in spatial and temporal references). But there is a fundamental difference: the decay time obtained by fitting a RSL curve corresponds to a lumped signal of contributions from all harmonic degrees.

A natural question that arises here is whether the ice-ocean-earth system can achieve the best of both Problem 1 and Problem 2: that is, to relate the exponential like RSL curves to a single harmonic relaxation without sacrificing the physical integrity of the ice dome. This is analogous to the problem Heisenberg encountered, which leads to his famous uncertainty principle. According to the uncertainty principle, the answer is no, because Problem 1 is to seek an exact harmonic localization and Problem 2 is to seek an exact spatial localization. It is impossible to achieve both simultaneously. On the other hand, also according to Heisenbergs' uncertainty principle, a compromise state could be reached between Problem1 and Problem 2 at the expense of ambiguities in both spatial and harmonic local-

Ice Sheets, Sea Level and the Dynamic Earth
Geodynamics Series 29
10.1029/029GD13

ization. The robustness of a spatial or a harmonic localization depends on the physics of the system. In fact, an ice dome can be viewed as a natural spherical window over the surface. If the solid Earth response to the ice load is laterally homogeneous, the rebound signal at the center of an ice dome would represent an average effect of the entire ice load and could be characterized by the size of the ice sheet. In the harmonic domain, this ice window creates a bandpass filter. The relationship between the window size and the bandwidth is governed by the uncertainty principle. Recently, the concept of spatial-harmonic localization has been applied to the study of post glacial rebound, either in an explicit manner [Simons, 1995; Simons & Hager, 1997] or an implicit manner [Mitrovica, 1996].

Recent publications by Peltier [1998, 1999] have raised some controversy over the estimates of the decay time near the center of Laurentide ice sheet. His recent values, 3.4 kyr to 4.7 kyr, are markedly smaller than the previous estimate of 7.6 kyr [Peltier, 1994; Mitrovica and Peltier, 1995]. Even if the mean solar ages of some data points might have been mistaken as ^{14}C ages in those analysis [Mitrovica et al, 2000], the corrected values should not be this low. The difference is primarily due to two steps added in Peltier's [1998, 1999] analysis: combination of the data points stretching about 600km across the Richmond Gulf and James Bay areas; lowering the elevations by about 6 meters of all the data points, obtained from various sources like shells, driftwood, etc., for a storm beach correction. Mitrovica et al [2000] conducted a simulation of the data points by forward modeling the RSL curves at those observation sites within the broad spatial region across Richmond Gulf and James Bay areas. The predicted RSL clearly fall into two distinctive groups, indicating the a composite data set only blurs the information rather than enhancing the accuracy of estimates. Mitrovica et al [2000] also re-emphasize, after Walcott [1980], the importance of using the same type of marine samples in compiling the RSL database. The new estimates of the decay time from their improved analysis is 4.0 to 6.6 kyr.

In this paper, we use a locally supported isotropic spherical wavelet (ISW) of Freeden & Windheuser [1996], together with a relaxation diagram analysis to explore the physical nature of the apparent exponential curves at the center of the two major ice domes. It becomes evident through this combined analysis that the two apparent exponential curves indeed characterize two single harmonics. We present a brief review of the geomorphology of the beach processes to argue that shell samples should not be treated the same way as driftwood in terms of the storm beach corrections. In fact, it is clearly shown by the data used by Peltier [1999] that the shell samples are systematically downwashed while the driftwood samples are systematically upwashed. A new estimate of the doublet of decay amplitude A and decay time T is provided. We demonstrate by calculating the 3D misfit surface that there is

no strong preference for a specific doublet (A,T), instead, there is an elongated narrow region of good-fits in the parameter (A,T) plane. The direct cause of such a lack of robustness is the nonlinearity of the exponential function. As a result, error bars obtained on the basis of linearized variance analysis may seriously underestimate the true uncertainties in the estimated parameters (A,T). All these results are used collectively to set up a constraint on the relaxation time spectrum in the 9-to-16 degree band. We introduce two viscosity models preferred respectively by the revised McConnell [1968] relaxation spectrum and the recent Wieczerkowski et al [1999] spectrum.

Throughout this paper, the PREM model [Dziewonski & Anderson 1981] is taken as the reference model for a Maxwell viscoelastic Earth. The viscosity structure is also assumed to be laterally homogeneous.

2. WAVELET ANALYSIS

We adopt a strategy similar to Simons & Hager [1997] to compute the ISW spectrograms for the RSL along the 60° latitude profile. The 60° latitude profile is not a great circle on the Earth's surface. The reason we take this small circle is that it is approximately the latitude for the centers of both the Laurentide and Fennoscandia ice sheets. The historical site of Angerman land in Sweden is located at about 63° latitude, while the widely used Hudson Bay site of Richmond Gulf, Quebec, is at about 57° latitude. Thus, we can localize both of the major ice sheets simultaneously.

RSL curves are essentially the time history of the geoid relative to the present mean sea level. Therefore, our analysis of RSL can be extended to the entire surface [Dahlen, 1976; Mitrovica & Peltier, 1991; Johnston, 1993; Fang & Hager, 1999]. Since there is no direct observation of the history of the geoid off the shorelines, the only way we can reconstruct the RSL on the surface is through modeling. We adopt ice model ICE-1A, which is a slightly modified version of ICE-1 by Peltier & Andrews [1976]. The major change was to add the Antarctic coverage from Nakada & Lambeck [1987]. The uncertainty in the spatial and time history of Pleistocene ice sheets is an active subject across a wide range of disciplines. Alternative models have been proposed in the attempt to further refine the ICE-1 model [Wu & Peltier, 1983; Tushingham & Peltier, 1991; Peltier, 1994]. None of the existing ice models deviates significantly from the geological constraints adopted by ICE-1. Particularly, the location of central regions are consistent among all the models. The radial viscosity profile is a major uncertainty in our modeling of RSL. Fortunately, the pattern of the spectrogram is almost independent of viscosity. The physics of this observation is simple: the response of the solid Earth is laterally homogeneous if the rheology of the Earth is laterally homogeneous, thus

Table 1. Parameters used in creating the viscosity profile displayed in Fig. 1. Here r denotes radius. Viscosities η are normalized by $10^{21}\,\mathrm{Pa\,s}$.

r_a	6371 km	Earth surface
r_{LVZ}	6171 km	Low viscosity zone
r_{660}	5711 km	660 km discontinuity
$r_{D''}$	3671 km	Top of D'' layer
r_{CMB}	3480 km	Core mantle boundary
η_{660}^{-}	0.6	On lower mantle side
η_{660}^{+}	5.6	On upper mantle side
η_{min}	0.2	$\eta(r_{LVZ})$
η_{max}	50	$\eta(r_{D''})$

the heterogeneity of the RSL signal is mainly controlled by the irregular distribution of the ice sheets over the surface.

To compromise between the limited resolving power of the RSL data and detailed microphysical modeling [e.g. Ranalli, 1991, 1998], we developed a phenomenological parameterization for the radial viscosity profile [Fang and Hager, 1996, 1999]

$$
\eta(r)=\begin{cases}
\eta_{min}+\left(\eta_{660}^{+}-\eta_{min}\right)\left(\dfrac{r_{LVZ}-r}{r_{LVZ}-r_{660}}\right)^{2}\cdot \\[4pt]
\qquad \left(3-2\dfrac{r_{LVZ}-r}{r_{LVZ}-r_{660}}\right) & r_{660}<r \\[10pt]
\eta_{max}\left(\dfrac{r_a-r_{D''}}{r_a-r}\right)^{\alpha}\exp\left[-\beta\dfrac{r^{\varepsilon}-r_{D''}^{\varepsilon}}{r_a^{\varepsilon}-r_{D''}^{\varepsilon}}\right] & r_{D''}\le r<r_{660} \\[10pt]
\eta_{max}\left(1-\dfrac{r_{D''}-r}{r_{D''}-r_{CMB}}\right)^{2} & r_{CMB}\le r\le r_{D''}
\end{cases}
\tag{1}
$$

This seemingly complicated parameterization contains only five commonly used adjustable physical quantities: the thickness of the lithosphere (not included in eq. (1)), the minimum and the maximum, η_{min}, η_{max} viscosity in the profile, and the viscosities flanking the 660 km boundary, η_{660}^{-}, η_{660}^{+} on the lower mantle side and the upper mantle side, respectively. The layer boundaries in (7) coincide with the major seismic discontinuities in the mantle (Table 1). One of the three parameters, α, β and ε in (1) has to be chosen as a free parameter to ensure a good fit in curvature to Ranalli's [1991, 1998] lower mantle viscosity profiles derived from microphysical considerations. Here we choose $\varepsilon=5.5$. The remaining two are determined by the peak viscosity, η_{max}, and the lower mantle side η_{660}^{-}. Fig. 1 displays a typical viscosity profile created using parameterization (1).

The wavelet spectrogram of the predicted RSL is calculated at 9 kyr. B.P. (Fig. 2). The major portion of the ice sheets at this time has already melted into the sea, thus, what we see in Fig. 2 is mainly the viscoelastic relaxation of the solid Earth. The half angle of support (left annotation in Fig. 2) represents the spatial localization of different scales, while the center wave number (right annotation in Fig. 2) is the central harmonic degree of the wavelet bandpass filter. It characterizes the harmonic localization. A striking feature in Fig. 2 is the harmonic localization at about degree 9 in the central region of the Laurentide ice sheet, while in the central area of the Fennoscandia ice sheet, the harmonic is localized at degree 16. This difference in harmonic localization is due to the fact that the Laurentide ice sheet covers a broader area than the Fennoscandian. As far as the central region is concerned, a broader spatial coverage is equivalent to a poorer spatial localization, which will result in a sharper harmonic localization (Fig. 3). The bandwidth of the center of the Laurentide ice sheet in Fig. 3 is significantly narrower than that of the Fennoscandia ice sheet. This is just another confirma-

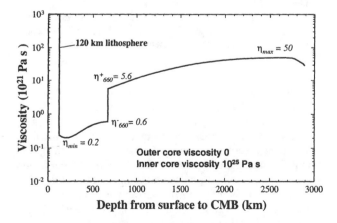

Figure 1. A viscosity profile generated based on our phenomenological parameterization. This parameterization is given in detail in Fang and Hager [1999], and the formula is also presented in the text (equation 7) with the parameters listed in Table 1.

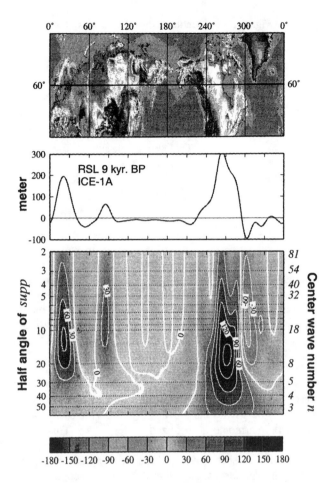

degree 16. We will see later on that to some extent, they can even be characterized by single degrees, 9 and 16, respectively. In his original analysis, Haskell [1935] treated the uplift of the beaches at the mouth of Angerman river in Sweden as a single sinusoidal harmonic in a 2-D half space with a wavelength about 3,000 km. The wavelength of spherical harmonic degree 16 is about 2,500 km, qualitatively consistent with Haskell's [1935] wavelength.

3. RELAXATION DIAGRAM ANALYSIS

The wavelet results discussed above expose the problem of bias: unless the Earth responds to surface mass redistribution uniformly at all harmonic degrees, the mantle viscosity $\eta \sim 1.0 \times 10^{21}$ Pa.s. inferred from the rebound signal at the central region of the Fennoscandia ice sheet by Haskell [1935] has a strong signature of relaxation at harmonic

Figure 2. Predicted relative sea level (RSL) 9 kyr before present time (BP) along the 60° latitude and the wavelet spectrogram of the same RSL. Since the static sea level coincides with the geoid, we extend the meaning of sea level to the land as the geoid variation thus the wavelet transform for sea level can be performed all over the globe. The calculation is based on the modified ice model ICE-1A and the viscosity model displayed in Fig. 1. As discussed in the text, the pattern of the wavelet spectrogram is identical for a wide range of viscosity models, and the central area of the major ice sheets are insensitive to the details of the ice history.

tion of Heisenberg's uncertainty principle [see Breeden & Windheuser, 1997 for a quantitative measure for the isotropic spherical wavelets]. We have calculated the same spectrogram for different epochs up to 2 kyr. B.P. (the RSL is identically zero at the present time of 0 kyr. B.P.), and found that the pattern of the spatial-harmonic localization shown in Fig. 2 and 3 remains unchanged. The implication of this analysis is the following. Near the central region of the Laurentide ice sheet, loading can be characterized by a narrow band of harmonics centered at degree 9 [also see Simon & Hager, 1997], and the relaxation observed near the central region of the Fennoscandia ice sheet can be characterized by a narrow band of harmonics centered at

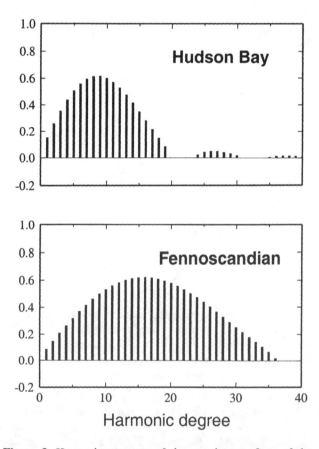

Figure 3. Harmonic spectrum of the wavelet transform of the same RSL as in Fig. 2 over the Hudson Bay and Fennoscandia areas. To calculate the Hudson Bay spectrum, for example, we take the angle of support as the size of the Laurentide ice sheet and center the wavelet at the center of the ice sheet. This bandpass-filter-like spectrum corresponds to the harmonic localization of RSL over the Hudson Bay area. The same interpretation applies to the Fennoscandia spectrum.

degree 16. One issue that has been discussed before is the effective depth of this Haskell value [McKenzie, 1967; O'Connell, 1971; Mitrovica, 1996]. A rule of thumb established from these previous investigations is that the penetration depth is about equal to the radius of the applied load. It is readily shown by using a perturbation technique [Mitrovica, 1996] or a perturbation free "anatomy" technique [Fang & Hager, 1996] that the penetration depth is highly dependent upon the viscosity structure. For a "soft" upper mantle with a considerable increase in viscosity across the 660 km boundary, there is a strong channel flow effect such that the perturbation can hardly penetrate deeper than the 660 km boundary. For a uniform viscosity mantle, on the other hand, the penetration is much deeper, and even the fluid core will be perturbed and contributes as much as 10% of the total observed RSL.

It was once believed that the viscoelastic response of the Earth at each harmonic degree is dominated by few prominent exponential modes in the form of $\exp(-t/\tau_i)$ where the subscript i indicates those prominent modes. Fang and Hager [1994, 1995] have demonstrated by presenting a suite of admissible viscosity models that this is not true in general. As shown in Fig. 4a and 4c, the prominent mode associated with the two-layer viscosity model completely disperses into a continuous distribution of "modes" when the upper mantle viscosity profile becomes continuous. In this situation, the concept of discrete modes is no longer valid. This "continuous mode" becomes a fundamental issue due to the fact an individual relaxation mode corresponds to a pole in the Laplace domain [e.g. Peltier, 1974]. It has been recognized that for a layered Earth model, each viscosity discontinuity will generate two viscosity modes [Fang and Hager, 1995]; each density jump will generate one buoyancy mode [Peltier, 1976; Han and Wahr, 1995], and the compressibility will generate a series of modes [Wu and Peltier, 1982; Vermeersen and Sabadini, 1997]. As a result, the number of poles far exceeds the number of layers. A pole has to be isolated. When a layered model tends to be continuous, the number of layers becomes uncountable. The mathematical structure of the resulting continuous distribution of poles is not clear in the context of classical mathematical physics. This continuous distribution of poles is closely related to the so called "singular bound" problem of integrating the equation of motion in the Laplace domain [Fang and Hager, 1995]. Normally we can treat the "continuous poles" as a branch cut in the complex plane by deforming the integral path [Fang and Hager, 1995], or perform the Laplace inversion in the entire complex plane in order to avoid encountering singular bounds [Fang and Hager, 1994]. A straightforward method of obtaining the solution directly in the time domain has been developed by Hanyk et al [1996, 1998].

In spite of the complexity of the mode structures. there is one property that is quite consistent for all the viscosity models investigated by us. For a time function $f(t)$ we

can define the complex Fourier spectrum in the complex Fourier domain $\omega = \omega_1 + i\omega_2$ [Fang and Hager, 1994]

$$F(\omega_1, \omega_2) = \int_0^\infty \left[f(t)e^{\omega_2 t} \right] e^{-i\omega_1} \, dt \qquad (2)$$

Let us fix ω_2 as a free parameter, then (8) can be interpreted as a conventional Fourier spectrum of a new function $f(t)e^{\omega_2 t}$. We have found consistently that for a vertical love number, h_n, the imaginary part of the Fourier spectrum always has a single peak regardless of the viscosity profile (Fig. 4b; also see Fang and Hager, 1994 Fig. 4). One can easily prove that, if a system is under a single exponential relaxation, $\exp(-t/\tau)$, the imaginary part of its

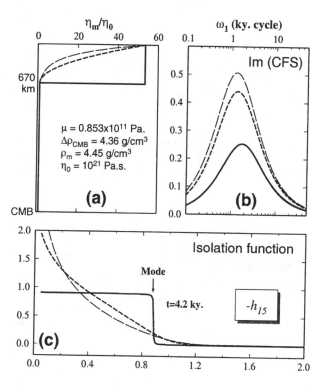

Figure 4. Demonstration of the complete dispersion of a single relaxation mode into a continuous distribution. The isolation function is designed by Fang and Hager [1995] to unfold the contribution from the "singular bound" piece by piece in a cumulative manner. We can see from panels (a) and (c) that, for these particular viscosity models, the isolation functions, i.e., cumulative mode distributions, are correlated with the viscosity profiles: A sharp contrast in viscosity profile corresponds to a sharp contrast in the mode distribution and a continuous viscosity profile corresponds to a continuous mode distribution. Despite the complexity of the mode distribution, the imaginary part of the complex Fourier spectra all have a single peak as shown in panel (b). The same line type in all three panels corresponds to the same viscosity model.

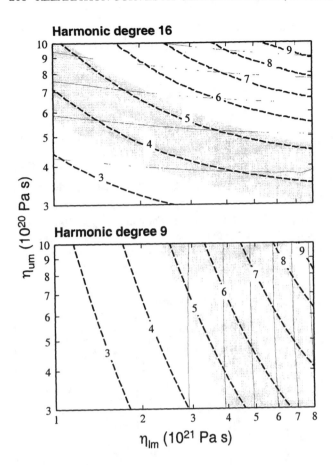

Figure 5. Relaxation diagrams of degree 16 and degree 9 calculated to compare with the decay time diagrams of Mitrovica [1996] for Fennoscandia and Hudson Bay. Two layer viscosity models with a fixed lithosphere at 120 km are set. The lower mantle viscosity η_{lm} and upper mantle viscosity η_{um} form a 2D parameter plane on which the contours of constant relaxation times are plotted. The shaded areas are adopted from Mitrovica's [1996] decay time diagrams for comparison. The unit of the contours is in kyr. Thin dot lines are the relaxation diagrams for degree 25 (top) and degree 4 (bottom).

Fourier spectrum defined by (2) will have a single peak as demonstrated in Fig. 4, and the relaxation time τ can be determined by the peak ω_{1peak} of the imaginary part of the Fourier spectrum

$$\tau = \frac{1}{\omega_{1peak} + \omega_2} \qquad (3)$$

We use relation (3) to define the average relaxation time for the vertical Love number for an arbitrary viscosity model. This definition is also valid for the gravitational Love number, k_n. In fact, one can easily prove that for a homogeneous Earth model, the gravitational Love number, k_n, is strictly proportional to the vertical Love number, h_n. Incidentally, relation (3) does not apply to the tangential Love number, l_n. As demonstrated by Fang and Hager [1994, 1995; see also Mitrovica and Davis, 1995], horizontal motion is more sensitive to the viscosity structure than the vertical motion. For an arbitrary viscosity model, neither the time domain solution nor the Fourier spectrum of a horizontal Love number, l_n, exhibit any resemblance to a single harmonic relaxation. The major problem with tangential motion is the lack of direct geological indicators to trace back the history of the last glacial cycle. Nonetheless, modern space geodetic techniques have provided valuable information on the present day crustal deformation which can be extrapolated backwards to constrain the rheology of the deep interior [James and Morgan, 1990; James and Lambert, 1993; Mitrovica et al, 1994, Peltier, 1995].

The significant bandwidth in the harmonic domain (Fig. 3) raises doubts about whether the signals flanking the central harmonic could cancel each other to single out the central harmonic. We look into this issue by comparing the decay time diagram of Mitrovica [1996] with the relaxation time diagram in Fig. 5. Mitrovica and Peltier [1993; 1995] parameterized the RSL curves in the central region of the major ice sheets by a single exponential form

$$H(t) = A\left(\exp(t/T) - 1\right) \qquad (4)$$

where $H(t)$ denotes the height of the RSL at a specific site, A and T are the parameters determined by fitting the RSL curve at the site. The time, t, here is measured as positive backwards before the present time. Obviously, the RSL $H(t)$ is a lumped signal of all harmonic degrees governed by the so called sea level equation [Farrell and Clark, 1976]. To distinguish between the parameter T and the relaxation time for a single harmonic degree, τ, we follow Mitrovica [1996], calling T a decay time. Assuming a two-layer viscosity model with fixed lithosphere and ice model, one can find that the decay time is a function of the upper mantle viscosity η_{um} and the lower mantle viscosity η_{lm},

$$T = T\left(\eta_{um}, \eta_{lm}\right) \qquad (5)$$

For a fixed value of T, equation (5) determines an iso-contour in the $\log \eta_{um} \sim \log \eta_{lm}$ plane, called the decay time diagram. In creating decay time diagrams, Mitrovica [1996] summed up harmonic degrees up to 256 to calculate the predicted RSL, $H(t)$, at the centers of disk ice loads simulating the Laurentide and Fennoscandia ice sheets. He then fit the observed RSL curves over the last 9 kyr at Angermanland Sweden, the center of the Fennoscandia ice sheet, and over the last 6.5 kyr at Richmond Gulf, Southeastern Hudson Bay, the center of the Laurentide ice sheet.

For comparison, we generate a relaxation time diagram, Fig. 5, following a procedure similar to Mitrovica's [1996] decay time diagram: assuming a two-layer viscosity model with a fixed lithosphere

$$\tau = \tau(\eta_{um}, \eta_{lm}) \qquad (6)$$

For a constant value of τ, equation (6) determines an iso-contour in the $\log\eta_{um} \sim \log\eta_{lm}$ plane, called a relaxation time diagram.

We pick the degree 16 relaxation diagram to compare with the decay time diagram at Angerman land, and the degree 9 relaxation time diagram to compare with the decay time diagram at Richmond Gulf. Simultaneous matches in slopes can be found in both the degree 16 (Angerman land) and degree 9 (Richmond Gulf) cases. Slopes are in fact the most diagnostic factor in identifying the harmonic degree in the relaxation diagram: the transition from lower degrees through intermediate degrees to higher degrees is very regular and can be clearly identified from the slopes of the iso-contours (compare the thin lines with the thick lines in Fig. 5). Thus, this simultaneous match in slopes strongly suggests that the decay times characterizing the exponential-like RSL histories in the central region of a major ice sheets are indeed akin to the relaxation times of single harmonic degrees.

In addition to the good fit in slopes, the degree 16 relaxation diagrams (Fig. 5) match especially well with their Angermanland counterparts [Mitrovica's (1996) Fig. 4a] indicating a strong boost of degree 16 as a result of strong cancellations among the flanking degrees in the bandpass filter. As with the degree 9 relaxation diagrams compared to the Richmond Gulf decay diagrams, there are noticeable shifts in the relaxation diagram with reference to the decay diagram. For instance, the 6 kyr. contour of the relaxation diagram locates approximately at the place of 6.6 kyr. contour of the decay diagram, making about a 10% shift in isovalue. The largest of such shifts is about 20% by the 3 kyr. contour. These discrepancies represent relatively larger ambiguity in relating the RSL to a single harmonic, as a result of poorer cancellation among the flanking degrees in the bandpass filter. Mitrovica and Peltier [1995] have found through numerical tests that if we only consider the time range in which the RSL change is dominated by free rebound, the decay time T is relatively insensitive to the details of the ice history. This observation can be easily explained if we interpret the decay times as strong signatures of the relaxation times of single degrees, since the relaxation times are independent of ice models. Conversely, this observation indirectly supports our conclusion that the decay times, approximately, represent single harmonic degrees.

4. RSL DATA

Estimates of the decay time at Angermanland, Sweden, have been consistent over the years since Haskell [1935]. In contrast, estimates of the decay time at Richmond Gulf, Canada, are controversial [e.g. Peltier, 1998, 1999]. Angermanland is characterized by the rhythmites of the varved sediments in the deltas at the mouth of the Angerman river. This is due to seasonal river regimes, in combination with a high rate of glacio-isostatic uplift, such that non-glacial varves have been formed and deposited in accordance with the continuing regression of the sea. Each rhythmite accords to a single mean-solar year. (Incidentally, the astronomical time synchronous to the seasonal variation of the terrestrial climate system is the mean-solar time or the calendar time rather than sidereal time). Floating varve chronologies have been associated with the retreat of major ice sheets in a number of sites. Only in Sweden, however, has it been possible to relate the varve chronology to the present time [Liden, 1938] and to assess the errors in the data through extensive rechecking [Nilsson, 1968; Wenner, 1968; Fromm, 1970]. The unique location of Angermanland enabled Liden [1938] to associate the varves with the average of shorelines 2-5 meters above the delta surfaces. Thus the RSL curve dated by the varve age, or mean-solar age, can be constructed. The weakness in this derivation of the RSL data is that the varve sequences hardly continue up to the present time, thus extrapolation has to be involved to calibrate the identified varve sequences. As it turned out, this difficulty is minor and can be eased by cross-checking with the ^{14}C dating [Tauber, 1970; Lundqvist, 1975].

The RSL data in the Richmond Gulf area was derived mainly from the ^{14}C dating of marine organic samples deposited in the sediments of the uplifted beach ridges. The controversy over the decay time has every thing to do with the fact that the ancient shorelines where the samples are found and dated do not necessarily represent the contemporaneous sea levels associated with the samples. Even if the samples are found *in situ* as occasionally are some mollusk species, the species usually have a considerable range of depth of occurrence [e.g. Sutherland, 1983]. By far the most complete collection of available ^{14}C data for the Southeast Hudson Bay region has been compiled by Hillaire-Marcel and published in Peltier [1998]. The majority of the samples accumulated in the Richmond Gulf area ($55.3° \sim 56.4°$ N, $76.5° \sim 77.6°$ W) are detrital shells of the popular blue mussels, *Mytilus edulis*, and driftwood [Hillaire-Marcel, 1976; Walcott, 1980; Allard and Tremblay, 1983]. Peltier [1998] used the entire collection of data in the Southeast Hudson Bay region from Hillaire-Marcel's compilation to obtain a decay time of 3.4 kyr. Two steps have to be added in the analysis in order to reach this value: combination of the data points across the Richmond

Gulf and James Bay areas, and lowering the elevations by about 6 meters of all the data points, obtained from various sources like shells, driftwood, etc., for a storm beach correction.

As demonstrated by Mitrovica et al [2000], the near 600 km stretch of coverage by the combined sites gives rise to two distinctive groups of RSL curves that can be attributed to, respectively, the Richmond Gulf and James Bay areas. Furthermore, usable data points in the James Bay area tend to cluster at older ages (>5 kyr.), while the data points in the Richmond Gulf area tends to group at young ages (<5 kyr.). This uneven time distributions makes the estimate of the decay time from the combined data set strongly biased toward the James Bay's value [see Mitrovica et al, 2000].

As for the lowering of the elevation of all the data points uniformly by 6 meters to counterbalance the storm beach off-set, we note that driftwood samples collected by Allard and Trembley [1983] are relatively young, ranging from modern time to less than 800 yr. B.P. They are apparently upwashed by about 5 meters, if the uplift rate of these young beaches is in the vicinity of 10 mm/yr. On the other hand, the shell samples appear to be systematically downwashed. In fact, a good portion of the data was eliminated by Peltier [1998] in his analysis for their abnormally low elevations. This observation raises the question of the validity of such a uniform storm-beach off-set.

The style of sedimentation of driftwood is different from that of shells. Driftwood samples are found buried, half-buried, and exposed in the Richmond Gulf area [Allard and Trembley, 1983]. A piece of wood can sit unburied for hundreds of years only in places isolated from wave actions. Let us assume that the wood was carried downstream by rivers to the sea and washed ashore by longshore currents. Then, the only natural agency to transport the driftwood back to inland places out of the reach of wave action is the "exceptional storm," which can cast and recast the floating wood up significantly away from the reach of the plunging breakers [Allard and Trempley, 1983].

For shell deposits, the situation is far more complicated and far from conclusive in terms of tracing the path of the shells during the beach forming processes. There are two interrelated outstanding issues: beach ridge formation and sedimentation, neither of which seems to have clear answers. A review of the early studies of beach formation was given by Johnson [1919]. A recent remark by Carter [1986] demonstrates how little we have advanced since the early time. Even for shingle beaches, where storms are conceived as a constructive force, the formation of beach ridges are not necessarily synchronous with the blast of the plunging breakers [Orford, 1977].

During the onshore-offshore transport by wave action, sediments are constantly subjected to various degrees of selective sorting. For instance, fine grain sands are seldom found in shingle beaches; they tend to be backwashed to the deep sea by wave actions because the fine grain sands are much easier to be placed in autosuspension in the water when mixed with gravels [Komar, 1976 chapter 11]. The sorting of the broken shell fragments mixed with clasts of similar size or greater is much more complicated and still unclear. It appears that the shell fragments tend to be sorted offshore, because it is easier with shell fragments to maintain autosuspension in the water than with the rock fragments. Shell fragments in the Richmond Gulf area are commonly found in the gravel under sizable boulders and thus the contrast in grain size is large [Hillaire-Marcell, 1976; Allard and Trembley, 1983]. If the detrital shells were deposited during an exceptional storm, it is very likely that they had been sorted and reworked rather than found *in situ*. Another possible way of transporting the shells inland is by the push of sea ice [Allard and Trembley, 1983]. Ice-push ridges are found on the Arctic beaches [Owen, et al, 1970].

Either storm-built or ice pushed, the time scale for the build-up of the contemporary shingle beach ridges is rather short, just in the scale of days or months. This time scale also applies to the contemporary sand beach ridges built-up during storm-recovery phases [Carter, 1988, pp 121]. In contrast, the time period of a series of Holocene beach ridges in the eastern Hudson Bay area is in the range of 45 yr. [Fairbridge and Hillaire-Marcel, 1977]. This time scale of periodicity is consistent with other noticeable Holocene beach ridges, like the Darss forland in Germany, about 20 to 50 yr. [Johnson, 1919], and the coast de Nayarit, Mexico, about 12 to 20 yr. [Curray and Moore, 1964]. As argued by Johnson 80 years ago [1919], even with the existence of cyclicity in the climate system, it is extremely unlikely that there is only one exceptional storm in each cycle, and these successive exceptional storms always blow in the right direction and deposit the right amount of sediments to form the parallel beach ridges that could survive thousands of years. An implication of Johnson's argument along with the observed time lag between the storm related build-up for the contemporary beach ridges and the period of Holocene beach ridges is that erosion is constantly associated with the formation of the Holocene beach ridges.

The current shingle beach ridge piled up by storms over bed rock in the Manitounuk island in Richmond Gulf area is 4.2 meters above sea level [Allard and Trembley, 1983]. Most of the driftwood samples are found on the young beach ridges up to 500 year of age, and virtually none of the shell samples of the same young ages have been found at these beach ridges. In fact most of the shell samples found at the young beaches are unreasonably older than the driftwood samples, thus, have to be eliminated by Peltier [1998] in his estimate of the decay time. This observation has two major implications: (1) the driftwood has been well sorted on top of the sediments on the beach ridges, they should be lowered in order to relate them to their contemporaneous beaches, (2) the shells have been either well sorted off-shore or downwashed by erosion, they have to be raised rather than lowered in order to correct them to their contemporaneous beaches. It is incorrect to lower the elevations uniformly on both driftwood and shell data points. Walcott [1980] and recently Mitrovica et al [2000] advocated the use of same type of samples in compilation of the RSL database. This would make it much easier to control the systematic errors brought about by the morphology. Here, we take a pragmatic procedure in the light of the fact that driftwood samples appear upwashed and shell samples appear downwashed, that is to respect the data as they are, and eliminate those exceptionally displaced from their contemporaneous beach ridges.

5. DECAY TIME ESTIMATES

We interchangeably use the varve age and the mean-solar age for the direct age indicator of the varve data points at the Angermanland site. "Direct ages" of the shell and driftwood samples in the Richmond Gulf area are the ^{14}C ages obtained from radiocarbon dating. The major problem in calibrating the ^{14}C age to the mean-solar age is that the calibration curves established under the control of tree-rings for materials formed in isotopic equilibrium with atmospheric CO_2 can not be applied to the marine samples because of the reservoir-atmosphere offset of the ^{14}C activities. In principle, the $^{234}U/^{230}Th$ dating of corals (Bard et al, 1990) provides a direct calibration for the marine samples, but the measuring errors in the $^{234}U/^{230}Th$ ratio and the ^{14}C accelerator mass spectrometry determinations usually smear its resolution such that the bidecadal chronological detail achieved for tree-rings is impossible [Stuiver and Braziunas, 1993]. The calibration for marine samples in the program CALIB 3.0 described by Stuiver and Reimer [1993] is based on carbon-reservoir model calculations that utilize the bidecadal atmospheric [tree-ring] data as an input [Stuiver and Braziunas, 1993]. The climate-influenced mixing parameters in these calculations, such as the air-sea CO_2 gas exchange rate, the oceanic vertical diffusion coefficient, and the air-land biospheric carbon uptake/release rate, have been assumed constant. The carbon-reservoir model calculations have been tested against the coral $^{234}U/^{230}Th$ calibration of Bard et al [1990]. Nonetheless, we shall not consider the calibration by the CALIB 3.0 program for marine samples as free of assumptions, especially for detailed corrections on the scale of centuries. These short-term variations can not be accounted for by the well controlled long-term trend of the air-sea offset, thus heavily depend upon the model calculations.

Least-square based curve fitting on a fixed form of function is elementary in principle. In our problem here, the data set is nonlinear to the estimated doublet (A,T) in the form of equation (4), and, beside their generic error bars, the data points are generally inadequate in number and

poor in distribution as a time sequence. To cope with this situation, we employ a "deterministic" approach instead of a Monte Carlo scheme. Suppose we have a total of N data points in age-height pairs $(t \pm \varepsilon_t, H \pm \varepsilon_H)$ where ε_t and ε_H are the corresponding standard errors. To simulate the randomness of the distance l_i between the i th point and the curve of a trial doublet (A,T), we pick up 12 additional subpoints within each error-bar-rectangle (Fig. 6). A single l_i is obtained by averaging the 13 vectors of distances. The misfit can be expressed as

$$\sigma^2(A,T) = \sum_{i=1}^{N} l_i^2 \qquad (7)$$

The so-called "deterministic" approach means a grid by grid systematic search in the A-T plane to locate the minimum of σ, which corresponds to the best-fit for (A,T).

There are two types of uncertainties the estimated parameters (A,T) suffer from, namely, non-random misfit error and random error. The non-random misfit results from dynamic processes unaccounted for and an imperfect physical model. The inadequate number and poor distribution of the data points make it impossible to have a quantitative measure for such non-random misfits. The random errors in A and T are referred to as the uncertainties induced by data errors. We assume throughout this paper that the measurement errors in age and height, ε_t and ε_H, are random, uncorrelated, and independent of position. In contrast, the induced random errors in the parameters (A,T) are correlated, position dependent, and worst of all, nonlinear functions of the standard errors of the data ε_t and ε_H. We take three steps to estimate the induced random errors in (A,T): first, consider the linearized random errors, ε_A

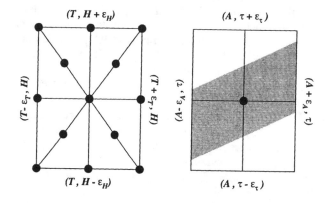

Figure 6. Left, a cartoon of the 13 points chosen from the error bar rectangle in the age-height plane. The data point is at the center of the error bar rectangle. Right, a cartoon of the error bar rectangle in the estimated parameter (A,T) plane. The shaded area demonstrates the distribution of good-fits.

and ε_T, then, analyze the nonlinear effects, and finally, piece the two effects together to set up the range of uncertainties in (A, T).

The linearized random errors in (A, T), at the i th data point (t_i, H_i) can be obtained by a linearized covariance analysis, the random errors ε_A and ε_T are calculated by the following weighted averages

$$
\begin{cases}
\varepsilon_A^2 = \dfrac{N}{\sum_{i=1}^{N}\left(\exp(t_i/T)-1\right)^2}\varepsilon_H^2 + \dfrac{A^2\sum_{i=1}^{N}\exp(2t_i/T)}{T^2\sum_{i=1}^{N}\left(\exp(t_i/T)-1\right)^2}\varepsilon_t^2 \\[4mm]
\varepsilon_T^2 = \dfrac{T^4\sum_{i=1}^{N}\exp(-2t_i/T)}{A^2\sum_{i=1}^{N}t_i^2}\varepsilon_H^2 + \dfrac{T^2 N}{\sum_{i=1}^{N}t_i^2}\varepsilon_t^2
\end{cases} \quad (8)
$$

Generally speaking, there will always be some good-fits in the neighborhood of the best-fit in the parameter A-T plane. By good-fits we mean a set of doublets (A, T) for which the curves in the form (4) match the best-fit curve so well such that we could not discriminate between the best-fit and a good-fit according to the resolving power of the data. For a nonlinear problem, the distribution of good-fits may extend along certain directions in the A-T plane beyond the range of the linearized random error bars. Thus, directional distribution of good-fits is a diagnostic feature of the non-linear effect (if the distribution of good-fits is not oriented, then the error ellipse determined by the linearized random errors ε_A and ε_T would be adequate to set the range of uncertainties in the estimated parameters).

The oldest records in age-height pairs for the uplifted shorelines set up lower bounds of complete deglaciation at those specific sites, but not necessarily the limits for a single exponential decay at the central region of the ice sheets, for the adjacent ice blocks may not yet have completely melted away and the time span for the free land uplift may not be long enough to eliminate the memory of earlier viscous relaxation. Mitrovica [1996] has demonstrated, based on forward modeling, that if the data set is perfect, the time window for a single exponential decay at Angermanland could be 10 to 11 kyr (varve age). He took a time window of 9 kyr as a conservative estimate.

Here we take the time window for the Angermanland site as 8.1 kyr (varve age) for the following reason. As reported by Tauber [1970], from 3 kyr to 8 kyr B.P. the ^{14}C ages show deviations of the same magnitude as is found in comparison between ^{14}C dates and the tree-ring chronology. For older ages recorded, about 8 kyr to 12.5 kyr, there seems to be an agreement between the ^{14}C age and the varve age. This apparent conformity contains two major uncertainties. On the physical ground, it precludes significant oscillation in the atmospheric ^{14}C activity within this period which has yet to be confirmed by more thorough studies. On the data grounds, the quality of older varve ages is in general poorer than the ^{14}C age [Sutherland, 1983], hence the apparent conformity may be merely apparent. The 8.1 kyr window is conveniently terminated at a point of the revised varve data (Fromm, 1970).

For the Southeast Hudson Bay region, a natural time window can be taken as 6 kyr ^{14}C age, about 6.8 kyr mean-solar age or calibrated (Cal) age [Mitrovica and Peltier, 1995; Mitrovica, 1996]. This is the time when the eastern coast of Hudson Bay began to emerge out of the Tyrrell Sea [e.g. Dyke and Prest, 1987]. A large number of marine data, primarily radiocarbon dated shorelines, indicate that the coast of the Hudson Bay region was almost totally ice free at about 7.9 kyr B.P. (^{14}C age). This seems to be the main reason that Peltier [1998, 1999] set up the time window at 7.9 kyr B.P. (^{14}C age). On the other hand, continental information from tracing the moraine complexes in Eastern Canada and New England has shown that during the same period the residual Hudson Bay ice, as a result of the collapse of the ice center, surged the glacial Lake Ojibway and then occupied the James Bay lowland [Hillaire-Marcel and Occhietti 1980]. The causally related deglaciation along the coasts and southward ice surges isolated and stabilized the New Quebec sub ice sheet in northern Quebec next to the Tyrrell Sea [Andrews and Peltier, 1976; Hillaire-Marcel and Occhietti 1980]. The New Quebec ice remained active during the subsequent centuries as evidenced by moraine distributions [Vincent, 1977]. The final melting of the New Quebec ice may have been by 6,800 B.P.(^{14}C age), if one takes the oceanic volume as close to present value by 6,800 B.P. (^{14}C age) [Fairbridge, 1976]. These geological constraints obtained from field surveys seem to have been incorporated into the reconstruction of ice models. For instance, the ICE-4G model advocated by Peltier [1994] contains a significant bulk of ice from 8 kyr BP to 6 kyr B.P. (^{14}C age). Therefore, the 6 kyr time window is more reasonable than the 7.9 time window to ensure free rebound of the entire Hudson Bay region, including both coastal and inland areas.

We list in Table 2 the results of (A, T) at Richmond Gulf with three different data sets under various preprocessing schemes, including with/without age calibration, with/without driftwood data, and with/without time window. Data 1 in Table 2 is collected from the recent compilation by Hillaire-Marcel from various sources, published in Peltier [1998]. Here we follow Peltier [1998] in eliminating the obviously downwashed shell samples marked with stars in the data list. Among the 48 data points, 23 are derived from driftwood samples. We do not include the data points from the James Bay area, because they are derived from mixed shells rather than a single species, and most of the data points from the James Bay area are scattered and older than allowed by our time window. Data 2 came from another compilation of Hillaire-

Table 2. Estimated doublets [A,T] based on three different data sets for both ^{14}C and calibrated ages for Richmond Gulf Quebec in the Southeast Hudson Bay region. The units are in meter and kyr. respectively. Data1 represents the recent compilation from various sources by Hillaire-Marcel cited by Peltier (1998); Data2 denotes another compilation by Hillarie-Marcel (1980); and Data3 is a subset of Data1 containing only the data points from Hillaire-Marcel (1976) and Walcott (1980).

	^{14}C age $T_{max} < \infty$	^{14}C age $T_{max} \leq 6.0$kyr	Cal. age $T_{max} \leq \infty$	Cal. age $T_{max} \leq 6.8$kyr
Data1 (48 points)	[44.80, 4.24]	[49.20, 4.56]	[64.32, 5.84] [55.60, 5.36]*	**[70.56, 6.32]** [59.80, 5.68]*
Data2 (10 points)	[61.44, 5.04]	[67.36, 5.36]	[72.48, 6.16]	[98.24, 7.76] [96.2, 7.6]**
Data3 (7 points)	[46.60, 4.24]	[51.40, 4.56]	[75.20, 6.32]	[97.80, 7.76]

* With driftwood data eliminated   ** Mitrovica and Peltier (1995).

Marcel [1980]. Among the 10 data points, 5 are directly derived from shell samples (these 5 direct indicators are included in Data 1) and the rest are indirectly derived based on the assumption that beach ridges were formed at this location according to a 45 kyr cyclicity [Fairbridge and Hillaire-Marcel, 1977]. Data 2 is the data set from which Mitrovica and Peltier [1995] obtained the decay time of 7.6 kyr using the Monte Carlo technique. As shown in Table1, this estimate of 7.6 kyr is very close to the best-fit obtained with our method. Data 3 is a subset of Data1 containing the 5 direct indicators in Data 2 plus 2 more points compiled by Craig and Walcott and cited by Walcott [1980] based on blue mussel or *mytilus edulis* samples.

A diagnostic comparison can be made in the Cal age between Data1 without driftwood (let us call it Data 1*) and Data3, which is simply the old age portion (> 1.7 kyr Cal age) of Data 1*. The resulting doublets (A, T) from Data 3 are systematically larger than that from Data1* regardless of windowing. This observation suggests that the downwash or upwash by the beach processes and erosion are not uniform for samples of different ages. The young shells (< 1.7 kyr Cal age) were systematically downwashed relative to the old shells, indicating that erosion in the last 1.7 kyr (Cal age) has been more severe than earlier. As mentioned above we take the windowed Data 1 as they are in estimating the doublet (A, T). Inclusion of the apparently upwashed driftwood offsets the inconsistency between young and old shells in terms of a single exponential representation by 0.6 kyr in decay time. The result for Richmond Gulf (RG) is

$$(A,T)_{RG} = (70.6, 6.3) \tag{9}$$

(printed in bold in Table 2). To estimate the linearized random errors in (A, T), we follow Dyke's suggestion (partly adopted in Peltier, 1998 without reference to its origin) to take the height errors as ±0.5m for elevations <10 m and ±5% for elevations > 10 m. The standard errors in age and height are $\varepsilon_t^2 = 0.013$kyr^2 and $\varepsilon_H^2 = 4.1$m^2. We obtain from equation (8)

$$\varepsilon_A = \pm 5\,\text{m}, \qquad \varepsilon_T = \pm 0.5\,\text{kyr} \tag{10}$$

For Angermanland, we consider two data sets: Liden's [1938] orginal varve data and its revised version by adding Fromm's [1970] corrections (Table 3). We can see by comparing Table 1 and Table 2 that the scatter caused by windowing and different choices of data sets in the estimates is noticeably smaller in Table 3 than in Table 2. So, there should be no surprise that our result for Angermanland (AL)

$$(A,T)_{AL} = (32.4, 4.6) \tag{11}$$

is consistent with previous authors [Mitrovica and Peltier, 1995; Peltier, 1999]. The linearized random errors in (A,T) obtained from equation (8) for Angermanland are

$$\varepsilon_A = \pm 3\,\text{m}, \qquad \varepsilon_T = \pm 0.4\,\text{kyr} \tag{12}$$

Results of the curve fittings are displayed in Fig. 7.

Interestingly, the simultaneously estimated parameters A and T in Table 2, 3 are correlated, in the sense that a larger A corresponds to a larger T and vice versa. This feature can be better understood by calculating the 3D σ surfaces on the A-T plane. Fig. 8 and Fig. 9 display the σ surfaces on which our results (9) and (11) are obtained. As it turns out, the σ surface is much like a sheet of paper with two diagonal corners pulled up creating a trough along the two opposing diagonal corners. The lowest point in the trough

Table 3. Estimated doublets (A, T) based on two different data sets for both ^{14}C and varve ages for Angermanland Sweden. The units are in meter and kyr. respectively. Here Liden (1938) represents Liden's orginal data set, Corrected means that corrections by Fromm (1970) have been added to Liden's orginal data.

	^{14}C age $T_{max} < \infty$	^{14}C age $T_{max} \leq 7.4\text{kyr}$	Varve age $T_{max} \leq \infty$	Varve age $T_{max} \leq 8.1\text{kyr}$
Liden (1938)	[26.40, 3.76]	[24.43, 3.62]	[33.20, 4.40]	[39.00, 4.88]
Corrected			[28.80, 4.24]	**[32.40, 4.56]**

corresponds to the best-fit. The shaded areas projected in the A-T planes represent a misfit of $\sigma/\sqrt{N} \leq 0.5$. It means that if the ages are perfect, the misfit at a data point will on average not exceed the smallest error in height, 0.5 m. One can easily distinguish the ^{14}C doublets in Fig. 8 and Fig. 9, as the ^{14}C doublets are systematically off the axis of the trough while the Cal age or varve age doublets distribute along the axis. In a positive sense this alignment of the Cal age doublets along the axis of the trough of the σ surface in Fig. 8 demonstrates an effective control in the estimates of (A, T) by the shell data points (> 1.7 kyr Cal age) shared by all the data sets in Table 2. On the other hand, the narrowly elongated good-fit areas on the σ surfaces in both Fig. 8 and 9 confirm our speculation that total uncertainties in the estimated (A, T) will exceed the linearized error bars (10) and (12) along the axis of the trough of the σ surfaces.

We choose the unwindowed $(T_{max} < \infty)$ estimates of (A, T) along with their mirror points (with respect to the official results (9) and (11) along the axis of the troughs of the σ surfaces) as the natural limits for the good-fits (Fig. 10). By incorporating the linearized error bars (16) and (18), we obtain, separately, the total uncertainties for A and T as shown in Fig. 10. The complete results are

$$(A, T)_{RG} = (70.6 \pm 12\,\text{m}, 6.3 \pm 1.0\,\text{kyr}) \quad (13)$$

$$(A, T)_{AL} = (32.4 \pm 7\,\text{m}, 4.6 \pm 0.7\,\text{kyr}) \quad (14)$$

The estimated errors in our results (13) and (14) are comparable with the recent result of Mitrovica et al [2000], but several times larger than those estimated by Peltier [1999].

6. RELAXATION SPECTRUM

We have come to the realization through wavelet and decay time analysis that the decay times at Richmond Gulf and Angermanland can serve as preliminary estimates for the average relaxation times at single harmonic degrees 9

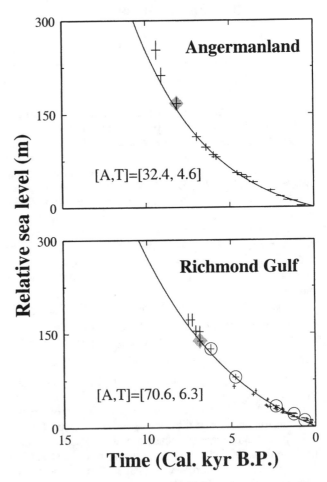

Figure 7. Best fits to the RSL data using the exponential formula $A[\exp(t/T) - 1]$ at the Angermanland site and the Richmond Gulf site. The data sets have been discussed in detail in the text. The shaded diamonds indicate the windows that we set up to ensure free relaxation. The circled points are not used in curve fitting and are plotted only for comparison. These data points derived by Hillaire-Marcel [1980] have not been based upon dating of shell samples but rather upon assigning ages to specific beach ridges under the assumption of 45 year periodicity in the beach formation process. We tried adding these data points in estimating the doublet (A, T); the difference is insignificant.

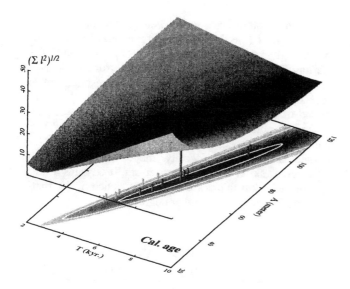

Figure 8. 3D misfit surface as a function of (A, T) for the Richmond Gulf data set, the same data set used to create the best fit in Fig. 7. The long bar represents the best fit for this specific data set (the bold letter in Table 2). The small lighter bars are best fits of other data sets of ^{14}C ages listed in Table 2, and the darker small bars are best fits for other data sets of Cal. ages listed in Table 2. Projected on the 2D surface is the 0.5 misfit zone, and the inner contours in the projection demonstrate that the location of the good-fits is a narrow area along the axis of the trough of the misfit surface.

and 16, respectively. But the wavelet analysis has not provided any information for the relaxation times between degree 9 and 16. A common procedure to fill in the gap is by linear interpolation. There are two evidences in support of such linear interpolation.

The first evidence comes from Simons and Hager's [1997] analysis. For clarity, let us, for the moment, consider a forward model and stay away from the observations. We plot in the background (the grey dots) in Fig. 11 the viscosity profile proposed by Simons and Hager [1997] along with its relaxation spectrum calculated using our relation (3). A comparison between the predicted global transfer function F_n for this viscosity model (the red curve in Simons and Hager's [1997] Fig. 2b) and the relaxation spectrum τ_n (grey inverse triangle in our Fig. 11) shows a remarkable resemblance in terms of the inflection characteristics within the 4-to-50 harmonic band. This similarity is not fortuitous. The predicted global transfer function F_n is weighted spatial average of the local correlation (in both space and harmonics) between the predicted glacially relaxed gravity field and the gravity field due to instantaneous deglaciation as if there had been no viscous relaxation [Simons and Hager, 1997]. Therefore, information of the gravitational relaxation is conveyed in the transfer function F_n. Similar to the wavelet case, the variable n in the func-

tion F_n denotes the central degree of the n th band-pass filter. We may take Fig. 3 for a rough idea of the band-pass filters centered at degree 9 and degree 16, respectively. The similarity in the inflection characteristics between the F_n curve and the τ_n curve in the 4-to-50 harmonic band has another significant implication. As we know, the transfer function F_n when interpreted for a single harmonic n, suffers from ambiguities due to the band width factor governed by Heisenberg's uncertainty principle. The relaxation times τ_n, on the other hand, are exactly localized in the harmonic domain. Thus, the similarity between the F_n curve and the τ_n curve indicates that gravitational relaxation at the central degree dominates the function F_n. This observation also reinforces our decay time analysis, and validates the replacements of the relaxation times with the decay times.

Now, let us replace the predicted gravity field with the observed gravity field in otherwise the same analysis. We revise the McConnell [1968] spectrum by imposing the linearly interpolated 9-to-16 harmonic band (Fig. 11). Immediately, we find that the "observed" transfer function F_n (the black curve in Simons and Hager's [1997] Fig. 2b) fits well with the linearly interpolated 9-to-16 harmonic band. Furthermore, a viscosity model (solid line in Fig 11) that simultaneously fits McConnell's [1968] spectrum at higher degrees (> 16) and our linearly interpolated 9-to-16 band

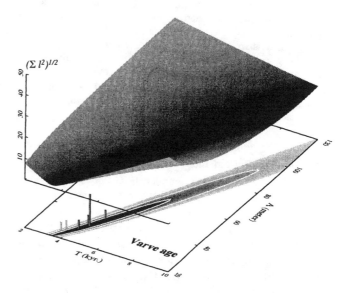

Figure 9. Same as Fig. 8, except for the Fennoscandia data set, the same data set used to create the best fit in Fig. 7. The long bar represents the best fit for this specific data set (the bold letter in Table3). The small lighter bars are best fits of other data sets of ^{14}C ages listed in Table 3, and the darker small bars are best fits for other data sets of Cal ages listed in Table3.

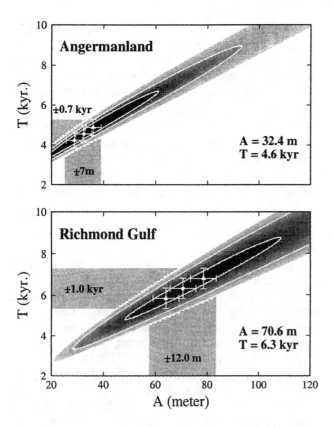

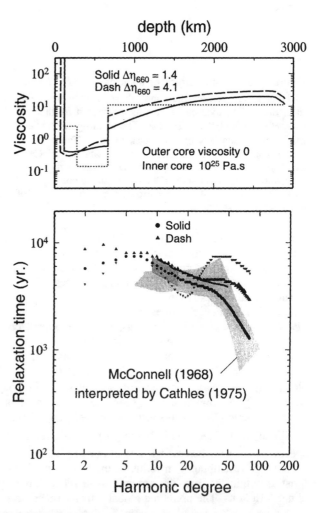

the elevated strandlines in the Fennoscandian region. We plot in Fig. 11 (the thick dark gray line mixed with the upper triangles) the spectrum of Wieczerkowski et al [1999], starting from degree 10. We omit their error bars to avoid confusion. As shown in Fig. 11, this new estimate of the relaxation spectrum systematically follows the upper bound of our linearly interpolated 9-to-16 band. The two individual estimates at degree 10 and 16 are about 7.1 kyr and 5.5 kyr respectively. Also plotted in Fig. 11 (the dashed line) is the viscosity model preferred by the Wieczerkowski et al's [1999] spectrum. It contains a lithosphere of 70 km thick, a direct consequence of the flatter trend of the spectrum at higher degrees.

Figure 10. Visualization of the uncertainties in the estimated parameters and T. Projections onto the A-T plane are the same as in Fig. 8 and 9, respectively. triangles in each panel are the best fit for the unwindowed otherwise the same data set and its mirror point with respect to the best fit point (dot) along the axis of the trough. Error bars on each point (triangle and dot) are estimated according to linearized covariance analysis. The uncertainty area for the doublet (A, T) is a narrow ellipse covered by the error bars. If we consider individual parameter A or T as we often do, the uncertainty has to be the projection of the error ellipse as illustrated by the shades.

also fits remarkably well the "observed" global transfer function F_n within the 4-to-50 harmonic band. Incidentally, the satellite gravity model JMG-2G [Nerem et al, 1994] terminates at degree 70 while the transfer function F_n of Simons and Hager [1997] has to truncate at degree 50 in order to maintain the resolution restricted by the Nyquist limit [see Simons, 1995 for detail]. For the same reason F_n, has to start at degree 4 instead of degree 2. Wavelet analysis is free from the Nyquist problem, thus our analysis starts from the lowest degree 2 all the way to the upper limit of resolution.

The second evidence in support of the linearly interpolated 9-to-16 band comes from recent results by Wieczerkowski et al [1999]. These authors employed a spherical Earth, in contrast to McConnell's [1968] infinite half space model, in their inference of the relaxation spectrum from

Figure 11. Relaxation spectrum and the preferred viscosity models. McConnell's [1968] spectrum with its error bars as interpreted by Cathles [1975] is plotted in the light gray area. Our revision is overlapped as the darker parallelogram. The thick line mixed with the upper triangles represents Wieczerkowski et al's [1999] spectrum without error bars. Plotted in the background as grey dots and inverse triangles are the viscosity model proposed by Simons and Hager [1997] and its relaxation spectrum.

Although each method, including ours, has its limitation, results are remarkably consistent among the three independent studies of Simons & Hager [1997], Wieczerkowski et al [1999], and this paper that the relaxation spectrum in the 9-to-16 band is not constant, a major contradiction to McConnell's [1968] spectrum. We save the complete spectrum analysis for elsewhere, and simply point out that the 9-to-16 band is sensitive to the viscosity structure in the vicinity of the 660 km discontinuity. A monotonic decrease of relaxation time as a function of degree within the 9-to-16 band indicates either a significant viscosity jump at the 660 km boundary followed by a flat viscosity structure in the deeper interior or a moderate jump at the 660 km boundary followed by a continuous increase in viscosity in the deeper interior (see Fig. 11). The significantly longer relaxation time of degree 9 than that of degree 16 has another important implication in connection with the recognition that the degree 9 and 16 relaxation times correspond to the apparent exponential RSL curves at the central region of the major ice sheets. It is a manifestation of the domination of the ice sheet size in the spatial-harmonic localization. Dynamically, this is only possible when the viscoelastic response of the solid Earth to the ice load is laterally uniform on the spatial scale of the major ice sheets. We conclude at this point that the lateral variation in viscosity at the length scale of the major ice sheets is significantly weaker than its radial variation.

7 DISCUSSION

A basic assumption in Simons & Hager's [1997] analysis is that a portion (not necessarily all) of the gravity signal in the Hudson Bay area is due to incomplete and still ongoing glacial isostatic adjustment. A potential argument against this assumption is that the clear correlation between the ice model and the gravity field in the Hudson Bay area revealed by Simons and Hager's [1997] analysis does not exist in the Gulf of Bothnia, although there is a gravity low associated with the Fennoscandia ice sheet. It is low in amplitude compared both to the gravity anomalies in Europe and to the Hudson Bay anomaly.

This result is not unexpected. first, The scale of the Fennoscandia ice sheet is smaller in both surface distribution and total volume than the Laurentide ice sheet. Second, it has a shorter relaxation time. Third, it is flanked by tectonically active regions such as the Alps and the Iceland hotspot. This situation makes it even more difficult for the observed free-air gravity to be coherent with the ice dome. A convincing evidence is shown in Fig. 12 where the predicted present day gravity disturbance has reached more than 50% of that observed over the Hudson Bay area. The topography of the Laurentide ice sheet is clearly identified in the wavelet spectrogram of the predicted gravity as we compare Fig. 12 with Fig. 2. On the other hand, the characteristics of the Fennoscandia ice sheet identified in the

spectrogram of the RSL (Fig. 2) are lost completely in the spectrogram of the gravity (Fig. 12). The predicted gravity in Fig. 12 is supposedly the upper bound of the glacial related gravity signal, and there is no non-glacial related gravity signal to blur it out. Yet the gravity over the Gulf of Bothnia has been relaxed to such an extend that the characteristics of the Fennoscandia ice sheet are unrecognizable in the midst of other signals. We therefore conclude that Simons and Hager's [1997] theory is not useful for the Gulf of Bothnia because of a low signal to noise ratio.

Half of the age-height RSL data points derived by Hillaire-Marcel [1980] have been based not upon dating of the shell samples, but rather upon assigning ages to specific beach ridges under the assumption of 45 year periodicity in

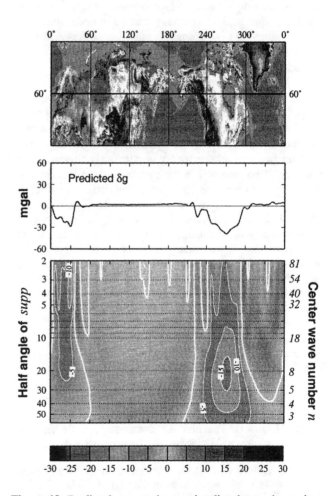

Figure 12. Predicted present day gravity disturbance due to incomplete glacial isostatic adjustment along the 60° latitude, and the wavelet spectrogram of the predicted disturbance. The setting and annotations for the wavelet spectrogram are the same as in Figure 2. The calculation is based on the modified ice model ICE-1A and a viscosity model identical to that displayed in Fig. 1 except that the elastic lithosphere is set to 200 km in order to boost the rebound signal to its upper bound.

the beach formation process. As shown in Fig. 7, these data points (circled) fit our exponential curve quite well. While there is no strong climate record to indicate the existence of such a 45 year cycle of storminess as proposed by Fairbridge and Hillaire-Marcel [1977] for the formation of a series of Holocene beach ridges on the east side of Hudson Bay, this 45 year cyclicity on the phenomenological level is not an unreasonable assumption. Periodicity of beach ridges has been commonly observed all over the world on prograding beaches with various sediments, ranging from fine grain sands to gravel mixtures. The exact process of the formation of the ridges is still unknown because of the volatility of the beach processes. Nonetheless, the periodicity enables these ridges to provide information on the timing of their formation. There have been 121 distinguishable dune beach ridges (beach ridges capped by sand dunes) in the Darss foreland in Germany that have been built in a period estimated to be from 3 kyr to 6 kyr [Johnson, 1919]. It means an average of 25 to 50 years for the construction of each ridge. The 185 distinguishable shingle beach ridges dated back to 8.3 kyr (Cal age) in Richmond Gulf [Fairbridge and Hillaire-Marcel, 1977] indicates an average of 45 years for the construction of each ridge. The major uncertainty associated with the 45 year periodicity assumption lies in the fact that the building of ridges is neither uniform in rate nor necessarily continuously forward. But this type of uncertainty, in the eastern Hudson Bay area at least, appears no greater than the uncertainties in associating the marine samples with their contemporaneous sea levels. This simple procedure of counting the number of ridges to determine the ages is very useful in the areas lacking in marine deposits. It also helps to improve the distribution of the data points by filling in the gaps where other dating methods are not applicable. We believe that it is unwise to reject those indirectly dated but very useful data points.

Acknowledgment: We thank Drs. Mark Simons and Jerry Mitrovica for reviews. This work is supported by NASA Program NAG5-6352 and NSF grant EAR-9627859.

REFERENCES

Allard, M. and G. Tremblay, La dynamique littorale des iles Manitounukdurant l'Holocene, Z. Geomorphol., *47*, 61-95, *1983*.

Andrews, J. T. and W. R. Peltier, Collapse of the Hudson Bay ice center and glacio-isostatic rebound, *Geology, 4*, 73-75, *1976*.

Bard, E., B. Hamelin, R. G. Fairbank, and A. Zindler, Calibration of the 14C timescales over the past 30,000 years using mass spectrometric U-Th ages from Barbados corals, *Nature, 345, 405-410, 1990.*

Currey, J. R. and Moore D. G., A Holocene regressive littoral sand, costa de Nayarit, Mexico, in *Deltaic and Shallow Marine Deposit,* ed. Van Straatern, L. M. J. U, Elsevier, Amsterdam, pp. 76-82, 1964.

Cater, R. W. G., The morphodynamics of beach-ridge formation: Magilligan, Northern Ireland, *Marine Geology*, pp. 191—214, 1986.

Cater, R. W. G., *Coastal Environments an introduction to the physical, ecological and cultural system of coastlines,* Acad. Press, 1988.

Cathels, L. M., *The viscosity of the Earth's Mantle,* Princeton Univ. Press, 1975.

Dahlen, F. A., The passive influence of the oceans upon the rotation of the earth, *Geophys. J. R. astron. Soc., 46, 363-406, 1976.*

Dyke A. S. and V. K. Prest, Late Wisconsin and Holocene retreat of the Laurentide ice sheet, *Geol. Surv. Canada,* Map 1702A, 1987.

Dziewonski, A. M. and D. L Anderson, Preliminary reference Earth model, *Phys. Earth Planet. Inter., 25, 297-356, 1981.*

Fairbridge R. W. and C. Hillaire-Marcel, An 8,000-yr paleoclimatic record of the "Double-Hale' 45-yr solar cycle, *Nature, 268, 413-416, 1977.*

Fairbridge R. W., Shellfish -eating preceramic indians in coastal Brazil, *Science, 191, 353-359, 1976*

Fang, M. and B. H. Hager, A singularity free approach to postglacial rebound calculations, *Geophys. Res. Lett., 21, 2131-2134, 1994.*

Fang, M. and B. H. Hager, The singularity mystery associated with a radially continuous Maxwell viscoelastic structure, *Geophys. J. Int., 123, 849-865, 1995.*

Fang, M. and B. H. Hager, The sensitivity of post-glacial sea level to viscosity structure and ice-load history fro realistically parameterized viscosity profiles, *Geophys. Res. Lett., 23, 3787-3790, 1996.*

Fang, M. and B. H. Hager, Postglacial sea level: energy method, *Global & Planet. Change, 20, 125-156, 1999.*

Fang, M., M. Simons, and B. H. Hager, The Hudson BAy gravity anomaly and postglacial rebound, (Abstract) *EOS, Trans. American Geophy. Union,* 76, No. 46, 158, 1995.

Farrell, W. E. and J. A. Clark, On postglacial sea level, *Geophys. J. R. astr. Soc., 46, 647-667, 1976.*

Freeden W. and U. Windhuser, Spherical wavelet transform and its discretization, *Adv. comput. Math., 5, 51-94, 1996.*

Freeden W. and U. Windhuser, Combined spherical harmonic and wavelet expansion- a future concept in Earth's gravitational determination, *Appl. Comput. Harmonic Anal.,* 4, 1-37, 1997.

Fromm, E., An estimate of errors in the Swedish varve chronology, in *Radiocarbon Variations and Absolute Chronology,* ed. I. U. Olsson, Nobel Symp. 12, Stockholm, pp. 163-172, 1970.

Hager, B. H., Mantle viscosity: a comparison of models from postglacial rebound and from the geoid, plate driving force, and advected heat flux, in *Glacial Isostasy Sea-Level and Mantle Rheology,* ed. R. Sabadini, K. Lambeck, and E. Boschi, Nato ASI Ser. C334, Kluwer Acad. Publi., pp. 493-513, 1989.

Haskell, N. A., 1935. Th motion of a fluid under a surface load,1, *Physics, 6, 265-269, 1935.*

Han, D. and J. Wahr, The viscoelastic relaxation of a realistically stratified earth, and a further analysis of post glacial rebound, *Geophys. J. Int., 120, 287-311, 1995.*

Hanyk, L., D. A. Yuen, and Matyska, C., Initial-value and modal approaches for transient viscoelastic responses with complex viscosity profiles, *Geophys. J. Int., 127, 348-362, 1996.*

Hanyk, L., Matyska, C., and D. A. Yuen, Initial-value approach for viscoelastic responses of the Earth mantle, in *Dynamics of the Ice Age Earth,* ed. P. Wu, Trans. Tech. Publi. Swizerland, pp. 135-154, 1998.

Hillaire-Marcel, C., La deglaciation et le relevement isostatique a l'est de la baie Hudson Bay, *Cah. Geogr. Que. 20*, 185-220, 1976.

Hillaire-Marcel, C., Multiple component postglacial emergence, Eastern Hudson Bay, Canada, in *Earth Rheology, Isostasy and Eustasy*, ed. N. Möner, pp. 215-230. John Wiley & Sons, New York, 1980.

Hillaire-Marcel C. and S. Occhietti, Chronology paleogeography and paleoclimatic significance of the late and post-glacial events in Eastern Canada, *Z. Geomorph.*, 24, 373-392, 1980.

James, T. S., The Hudson Bay free-air gravity anomaly and glacial rebound, *Geophys. Res. Lett.*, 19, 861-864, 1992.

James, T. S and W. J. Morgan, Horizontal motions due to postglacial rebound, *Geophys. Res. Lett.*, 17, 957-960, 1990.

James, T. S. and Lambert, A, A comparison of VLBI data with the ICE-3G glacial rebound model, *Geophys. Res. Lett.*, 20, 871-874, 1993.

Johson, D. W., *Shore processes and Shoreline Development*, reprinted in 1972 by Hafner Publi. Company, New York, 1919.

Johnston, P., The effect of spatially non-uniform water loads on prediction of sea level change, *Geophys. J. Int.*, 114, 615-634, 1993.

Komar, P. D., *Beach processes and Sedimentation*, Prentice-Hall, Inc. New Jersey, 1976.

Lambeck, K., P. Johnston, and M. Nakada, Holocene glacial rebound and sea-level change in NE Europe, *Geophys. J. Int.*, 103, 451-468, 1990.

Liden, R., Den senkvartära strandförskjutningens förlopp och kronolgi i Angermanland, *Geol. Fören. Stockholm Förhendl.*, 60, 397-404, 1938.

Lundqvist, J., Ice recession in central Sweden, and the Swedish time scale, *Boreas*, 4, 47-54, 1975.

McConnell, R. K., Isostatic adjustment in a layered Earth, *J. Geophys. Res.*, 70, 5171-5188, 1965.

McConnell, R. K., Viscosity of the mantle from relaxation time spectra of isostatic adjustment, *J. Geophys. Res.*, 73, 7089-7105, 1968.

McKenzie, D. P., The viscosity of the manrtle, *Geophys. J. R. Astr. Soc.*, 14, 297-305, 1967.

McKenzie, D. P. Some remarks on heat flow and gravity anomaly, *J. Geophys. Res.*, 72, 6261-6273, 1967b.

Mitrovica, J. X, Haskell [1935] revisited, *J. Goephys. Res.*, 101, 555-569, 1996.

Mitrovica, J. X. and R. W. Peltier, A new formalism for inferring mantle viscosity based on estimates of post-glacial decay time: application to RSL variations in N.E. Hudson Bay, *Geophys. Res. Lett.*, 20, 2183-2186, 1993.

Mitrovica, J. X., and W. R. Peltier, On postglacial geoid subsidence over the equatorial oceans, *J. Geophys. Res.*, 96, 20,053-20,071, 1991.

Mitrovica, J. X., J. L. Davis, and I. I. Shapiro, A spectral formulism for computing three dimensional deformations due to surface loads, 2, present day glacial isostatic adjustment, *J. Geophys. Res.*, 99, 7075-7101, 1994.

Mitrovica, J. X., J. L. Davis, Some comments on the 3-D impulse response ot a Maxwell viscoelastic Earth, *Geophys. J. Int.*, 120, 227-234, 1995.

Mitrovica, J. X. and W. R. Peltier, Constraints on mantle viscosity based upon the inversion of post-glacial uplift data from the Hudson Bay region, *Geophys. J. Int. 122*, 353-377, 1995.

Mitrovica, J.X., A. M. Forte, and M. Simons, A re-app[raisal of post glacial decay times from Richmond Gulf and James Bay, Canada, Geophys. J. Int. *142*, 783-800, 2000.

Nakada, M and K. Lambeck, Glacial rebound and relative sea-level variation: a new appraisal, *Geophys. J. R. Astr. Soc.*, 90., 171-224, 1987.

Nerem, R. S., et al, Gravity model development for TOPEX/POSEIDON: joint gravity models 1 and 2, *J. Geophys. Res.*, 99, 24,421-24,447, 1994.

Nilsson, E., The late-Quaternary history of Southern Sweden, *Kungl. Sv. Vetensk, Akad. Handl.*, Ser. 4, 12, No.1, 1968.

O'Connell, R. J., Pleistocene glaciation and the viscosity of the lower mantle, *Geophys. J. R. Astr. Soc.*, 23, 299-327, 1971.

Orford, J. D. A proposed mechanism for storm beach sedimentation, *Earth Surf. Proce.*, 2, 381-400, 1977.

Owens E. H. and S. B. McCann, The role of ice in the Arctic beach environment with special references to Cape Ricketts, Southwest Devon Island, Northwest Territories, Canada, *Amer. Jour. Sci.*, 268, 397-414, 1970.

Pari, G. and W. R. Peltier, The free air gravity constraint on subcontinental mantle dynamics, *J. Geophys. Res.*, 101, 28,105-28,132, 1996.

Peltier, W. R., Glacial isostatic adjustment, II, the inverse problem, *Geophys. J. R. Astr. Soc.*, 46, 669-706, 1976.

Peltier, W. R., Ice age Palaeotopography, *Science*, 265, 195-201, 1994.

Peltier, W. R., VLBI baseline variations for the ICE-4G model of postglacial rebound, *Geophys. Res. lett.*, 22, 465-468, 1995.

Peltier, W. R., Postglacial variations in the level of the sea: implications for climate dynamics and solid-Earth geophysics, *Rev. Geophys.*, 36, 603-689, 1998.

Peltier, W. R., Global sea level rise and glacial isostatic adjustment, *Global & Planet. Change*, 20, 93-123, 1999.

Peltier, W. R. and J. T. Andrews, Glacial isostatic adjustment I, the forward problem, *Geophys. J. R. astro. Soc.*, 46, 605-646, 1976.

Peltier, W. R., A. M. Forte, J. X. Mitrovica, and A. M. Dziewonski, Earth's gravitational field: seismic tomography resolves the enigma of the Laurentian anomaly, *Geophys. Res. Lett.*, 19, 1555-1558, 1992.

Ranalli, G., The microphysical approach to mantle rheology, in Sabidini et al (Editors) *Glacial Isostasy, Sea-level, and Mantle rheology*, Kluwer Acad. Publi. pp. 493-513, 1991.

Ranalli, G., Inference on mantle rheology from creep laws, in *Dynamics of the Ice Age Earth*, ed. P. Wu, Trans. Tech. Publi. Swizerland, pp. 323-340, 1998.

Simons, M., Localization of gravity and topography: constraints on the tectonics and mantle dynamics of Earth and Venus, *Ph.D. Thesis*, MIT, 1995.

Simons, M. and B. H. Hager, Localization of the gravity field and the signature of glacial rebound, *Nature*, 390, 500-504, 1997.

Suiver, M. and T. Braziunas, Modeling atmospheric ^{14}C inferences and ^{14}C ages of marine samples to 10,00 BC, *Radioacarbon*, 35 , 137-189, 1993.

Suiver, M. and P. J. Reimer, Extended ^{14}C data base and revised calib. 3.0 ^{14}C age calibration program, *Radiocarbon*, 35, 215-230, 1993.

Sutherland, D. G., The dating of former shoreline, in *Shorelines and Isostasy*, ed. D. E. Smith and A. G. Dawson, Acad. Press., pp. 129-157, 1983.

Tauber, H., The Scandinavian varve chronology and C14 dating, in *Radiocarbon Variations and Absolute Chronology*, ed. I. U. Olsson, Nobel Symp. 12, Stockholm, pp. 173-196, 1970.

Tushingham, A. M. and W. R. Pelter, ICE-3G: a new global model of Wurm-Wisconsin deglaciation using a global data base of relative sea level history, *J. Geophys. Res. 97, 3285-3304, 1991*.

Tushingham, A. M. and W. R. Pelter, Validation of the ICE-3G model of Wurm-Wisconsin deglaciation using a global data base of relative sea level histories, *J. Geophys. Res., 97, 3285-3304, 1992*.

Vermeersen, L. L. A. and R. Sabadini, A new class of stratified viscoelastic models by analytical techniques, *Geophys. J. Int., 129, 531-570, 1997*.

Vicent J. S., Le Quarternaire recent de la region due cours inferieur de la Grande rivere Quebec, *Com. Geol. Can, Etude 20, 76-19, 1977*.

Walcott, R. E., Rheology models and observational data of gla-cio-isostatic rebound, in *Earth Rheology, Isostasy and Eustasy*, ed. N. Möner, pp. 3-10. John Wiley & Sons, New York, 1980.

Wenner, C.-G., Comparison of varve chronology, pollen analysis, and radio-carbon dating, *Acta Univ. Stockh., Stockholm Contrib. in Geol., 18, 75-97, 1968*.

Wieczerkowski, K, J. X. Mitrovica, and D. Wolf, A revised relaxation-time spectrum for Fennoscandia, *Geophys. J. I., 139, 69-86, 1999*.

Wu, P. and W. R. Peltier, Viscous gravitational relaxation, *Geophys. J. R. Astr. Soc., 70, 435-485, 1982*.

Wu, P. and W. R. Peltier, Glacial isostatic adjustment and the free air gravity anomaly as a constraint on deep mantle viscosity, *Geophys. J. R. astron. Soc., 74, 377-449, 1983*.

Postglacial Induced Surface Motion, Gravity and Fault Instability in Laurentia: Evidence for Power Law Rheology in the Mantle?

Patrick Wu

Department of Geology & Geophysics, University of Calgary, Calgary, Alberta, Canada

A 3D finite element model of postglacial readjustment is used to calculate the earth's response due to realistic ice and eustatic water loads on stratified incompressible viscoelastic Maxwell flat-earths with nonlinear rheology. From the comparison of the predicted and observed sealevel data around Laurentia, earth models with nonlinear rheology that give the best fit to the observations are found to be: a) model with linear 10^{21} Pa-s upper mantle and nonlinear lower mantle with $A^* = 10^{-36}$ Pa^{-3} s^{-1}; b) models with thin (<300 km) nonlinear zones ($A^* = 3 \times 10^{-35}$ Pa^{-3} s^{-1}) above a linear 10^{21} Pa-s mantle. Other geophysical and geodetic observables - namely uplift rate, horizontal velocity, free-air gravity, the rate of change of gravity, fault instability and earthquake onset timing have also been calculated. The direction of horizontal velocity outside the former ice margin is found to be extremely diagnostic of these two earth models. Other observables are also found to be useful in discriminate mantle rheology.

1. INTRODUCTION

Since the dynamics of the earth is determined by mantle rheology, an important question is whether the flow law in the mantle is linear (Newtonian) or nonlinear (power-law). Microphysics is unable to answer this question definitively because the deformation mechanism in the mantle is poorly constrained and the transition condition between linear and nonlinear creep is uncertain by an order of magnitude [*Ranalli* 1998]. However, results from high temperature creep experiments of relevant rock material suggest that nonlinear rheology prevails in the shallow part of the upper mantle [*Goetze & Kohlstedt* 1973] while the lower mantle may be linear [*Karato & Li* 1992; *Li et al.* 1996].

An alternate approach is to infer mantle rheology from geophysical observations e.g. postglacial sealevels and other related geophysical and geodetic data. However, for surface loading problems on a nonlinear medium, no ana-

lytical solution exists. Thus simplifying assumptions were introduced in early investigations (e.g. *Post & Griggs* 1973; *Brennen* 1974; *Crough* 1977; *Yokokura & Saito* 1978; *Nakada* 1983) to make the problem tractable. However, as pointed out by *Wu* [1992], some of these assumptions are invalid whereas others are too restrictive. The method preferred by *Wu* [1998a; 1999] and also used in this paper is the Finite Element (FE) method. This method is advantageous since none of the simplifying assumptions that undermined earlier work has to be made and the solution is much more rigorous. The effects of power-law rheology on vertical motion and their physics are discussed in [*Wu* 1992, 1993]. As reviewed in *Wu* [1998a], results of these finite element studies with simple loading histories show that nonlinear mantle cannot explain the sea-level data immediately outside the Laurentian ice margin (e.g. Boston). This is due to the stress induced "low viscosity channel" in the nonlinear mantle directly underneath the ice (see equation 3), which creates a "viscously stationary zone" (VSZ) that in turn acts as a hinge line for vertical displacement. Thus, nonlinear mantles cannot explain the observed land emergence followed by submergence which is characteristic of the data in the "Relative Sealevel Transition Zone" (e.g. near Boston) and which is normally in-

Ice Sheets, Sea Level and the Dynamic Earth
Geodynamics Series 29
10.1029/029GD14

terpreted to be due to the inward migration and collapse of the peripheral bulge associated with "deep flow".

Combining microphysics, seismic anisotropy observations and postglacial rebound modeling with simple ice histories, *Karato & Wu* [1993] suggested that the top 300 km of the mantle may be nonlinear but the flow law becomes linear below that depth. Such an earth model has recently been shown by *Wu* [1999] to give a better fit to the sealevel data in and around Laurentia than the model with a linear 10^{21} Pa-s uniform mantle when realistic ice histories are used. *Wu* [1999] also studied a large number of earth models with nonlinear rheology in the upper mantle, lower mantle or both. From that study, another earth model was found to give even better fit to the sealevel data in and around Laurentia than the model proposed by *Karato & Wu* [1993]. Contrary to the expectation of *Karato & Li* [1992] and *Li et al.* [1996], this model has a nonlinear lower mantle below a linear 10^{21} Pa-s upper mantle and 150 km thick elastic lithosphere.

However, in *Wu*'s [1999] investigation, the creep parameter in the nonlinear zone or in the nonlinear lower mantle was fixed to A*=3.33x10^{-35} Pa^{-3} s^{-1}; the sensitivity of the sealevel predictions to changes in this value of A* has not been investigated. Furthermore, only sealevel data were considered in that study. Recent investigations of the postglacial readjustment process include observations of uplift rates, horizontal velocities, changes in the geopotential field, the Earth's rotational motion and fault instability. Therefore it is desirable to have predictions for some of these other observables, as they may provide extra constrains on mantle rheology.

The purpose of the present paper is to: 1) review some of the comparisons between predicted and observed sealevels for the earth models in *Wu* [1999]; 2) study the sensitivity of the creep parameter A* for the model with nonlinear rheology in the upper 300 km or in the lower mantle; 3) extend the results of *Wu* [1999] to include predictions of uplift rate, horizontal velocity, free air gravity anomaly, the rate of change of gravity and fault instability.

Before we proceed, it is worth summarizing some of the complexities in the study of the glacial isostatic adjustment (GIA) process with power-law rheology, so that the reader may appreciate the difficulties and limitations of this type of study. First of all, the principle of superposition and the popular spectral technique (e.g. [*Peltier* 1998]) cannot be applied because the problem is nonlinear. To overcome this, the finite element (FE) technique is used [*Wu* 1992]. Since many FE packages are available commercially, one may think that such packages can be easily adopted for the modeling of the GIA process. However, one must realize that most commercial finite element codes are written for engineering purposes and they do not include the important effects of buoyancy (i.e. pre-stress advection term) nor self-gravitation. In this study, only the effect of buoyancy has been included. The inclusion of self-gravitation in a spherical FE model has been a challenge that has only been solved recently [*Wu* 2001 (in preparation)]. The gravitationally self-consistent sealevel calculation (e.g. [*Mitrovica & Peltier* 1991]) which has worked well for linear rheology also do not work in this case and has to be redeveloped for the spherical self-gravitating FE model. A second complication exists for power-law rheology because the effective viscosity is also determined by the history of loading (see equation 3 below) and perhaps the ambient tectonic stress too. This coupling of ice/water load and earth rheology means that one can no longer infer mantle rheology independent from the ice and water load history. In particular, the coupling between effective viscosity and load means that if the ice thickness is doubled, then the induced displacement will not be doubled as in the case of linear rheology. However, the problem is less severe if the change in the local ice thickness is small (<10%). Under such circumstance, the associated change in relative sealevels can be predicted with linear theory (see [*Wu* 1993]). Anyhow, the parameter space for earth rheology no longer just includes viscosity, but must now include creep parameter A*, creep exponent n, the load history and perhaps the ambient tectonic stress too. Further complication arise due to the many uncertainties that plague current ice models - many of these are based on incomplete terminal moraine data and ice flow observations, inadequate theoretical ice profiles (because of uncertain parameters such as basal stress conditions) and relative sealevel data which is also dependent on mantle rheology! Thus the parameter space to be investigated is very large and complex indeed. Since it is impossible to explore the parameter space thoroughly in a single study, we should be contented to restrict our investigation with an ice model that fits a lot of geological constraints. The one chosen is the ICE3G model [*Tushingham & Peltier* 1991] which has produced acceptable fit to many sealevel data around Laurentia for linear rheology and which is better received by field geologists than the more recent ICE4G model. We have constructed another ice model based on geological constraints and power-law rheology, but that will be reported in another paper. Clearly, our results and conclusions may be affected by the load model employed and the reader should keep this important point in mind. Finally, one should be aware that study of GIA with power-law is still at its infancy. Unlike the case for linear rheology, where small changes in the timing or amplitude of the relative sealevels matters, the main concern for power-law studies until recently is still on first order effects – e.g. whether one can have land emergence rather than gence rather than submergence around Boston during the last 8 ka. With the simple earth and ice models employed, it is not worthwhile to include the finer effects, since they will be the subjects in future investigations.

2. THE MODEL

In this paper, all earth models are isotropic, incompressible viscoelastic flat-earths with power-law rheology and no self-gravitation. This flat-earth approximation without self-gravitation gives better sealevel predictions than non-self-gravitating spherical earths because the effect of self-gravitation is partially compensated by the effect of sphericity [Amelung & Wolf 1994]. Wu & Johnston [1998] further demonstrated that the flat-earth approximation without self-gravitation is adequate in describing the relative sealevel histories during the last 8 ka provided that the sealevel site lies within 800 km from the edge of the Laurentian ice margin at glacial maximum (where the effect of sphericity remains small). Compressibility affects the amplitude and the timing of the sealevel curves, but these effects are second order because they are not large enough to change land submergence into land emergence or vice versa to affect the existence of the RSL Transition Zone just outside the former ice margin. The effects of compressibility are more important in horizontal motion. However, the effects of internal compressibility are not completely understood even in the case of linear rheology (e.g. [Klemann et al. 2000]), thus incompressibility is assumed in this study.

It is further assumed that the rebound process sees the steady state creep and that there is little or no interaction between rebound stress and the ambient tectonic stress [Karato 1998]. The steady-state creep assumption has been questioned by Karato [1998] but the problem of transient creep and tectonic-rebound stress interaction will be addressed in future papers.

For a viscoelastic material, the deformation can be decomposed into the elastic response and the creep response. The elastic strain is related to the stress by Hooke's law while the steady state creep law is given by:

$$\dot{\varepsilon}_{ij}^{C} = A^{*} \, \sigma'^{n-1}_{E} \, \sigma'_{ij} \qquad (1)$$

Here A^{*} is the creep parameter, σ'_{ij} are the deviatoric stress components and σ'_{E} is the equivalent deviatoric stress with

$$\sigma'_{E} = \sqrt{\tfrac{1}{2} \, \sigma'_{ij} \, \sigma'_{ij}} \qquad (2)$$

For linear rheology, n equals 1. For nonlinear rheology, n normally lies between 2 and 6 for mantle rocks, but in this

paper n is taken to be 3. For later discussion and comparison of results with other papers, it is useful to define the effective viscosity as:

$$\eta_{eff} = \frac{1}{3 \, A^{*} \, \sigma'^{(n-1)}_{E}} \qquad (3)$$

Note that this differs from the usual definition by a factor of 3/2 (see [Ranalli 1987], p.75-80 for discussion). Equation (3) says that large stress level (e.g. from the load) will result in low effective viscosity.

The earth models considered here all contain a 150 km thick elastic lithosphere overlying a stratified viscoelastic mantle and an inviscid fluid core. Lithospheric thickness under North America is poorly constrained, thus an intermediate value between the traditional value of 100 km [Walcott 1970] and the 200 km proposed by [Peltier 1984] is adopted. This assumption is adequate for our purpose here because the effects of lithospheric thickness are not significant enough to affect the existence of the RSL Transition Zone. The elastic parameters of the earth models are listed in Table 1 and the rheology parameters are given in Tables 2 to 5.

The 3D Finite Element Model used to calculate the deformation of the earth due to glacial loading and unloading is composed of 10 layers with each layer consisting of 1088 three-dimensional 8-node elements covering an area of 240,000 km by 240,000 km. However, only results in the central region near the surface are intended to be useful. The grid there is the densest and includes 14x12 elements on the surface, each with horizontal dimensions of 340 km x 323 km, which is close to the resolution of the ICE3G model. The details of the Finite Element model can be found in [Wu & Johnston 1998] who also showed that the relative displacement curves computed with the FE method on a 3D flat-earth are in excellent agreement with the relative sealevel curves (during the last 8 ka) computed with the gravitationally self-consistent sealevel equation [Mitrovica & Peltier 1991] for a spherical self-gravitating earth.

The ice model is adapted from the ICE3G model of [Tushingham and Peltier 1991]. It consists of the Laurentian, Cordillera, Innuition and Greenland ice-sheets. Several saw-tooth glacial cycles that have slow buildup time of 90 ka but rapid deglacial time of 10 ka are assumed to precede the deglaciation history given by ICE3G. Increasing the number of glacial cycles does not significantly affect our results. Eustatic ocean loading due to melting of all the ice sheets in ICE3G is also included.

Table 1. Elastic Parameters of stratified earth models

	Depth	Density (kg/m^3)	Young's Modulus (Pa)	Poisson Ratio
Lithosphere	0 - 150 km	3475	1.92×10^{11}	0.5
Layer 1	150 - 420 km	3475	2.16×10^{11}	0.5
Layer 2	420 - 670 km	3616	3.25×10^{11}	0.5
Lower Mantle	below 670 km	3888	6.61×10^{11}	0.5

Table 2a. Rheologic Structure of the uniform models

Name of Model:	U22	U34	U35	U36
Lithospheric thickness (km)	150	150	150	150
Stress exponent n in mantle	1	3	3	3
A* in the mantle (Pa^{-3} s^{-1})	3.33×10^{-22}	3.33×10^{-34}	3.33×10^{-35}	3.33×10^{-36}
Ice thickness Scale factor h	1	1	1	1
χ^2 within ice margin	3.9	13.2	7.7	7.8
χ^2 outside ice margin	6.7	13.1	15.4	26.7
Total χ^2	5.4	13.2	11.8	17.8

Table 2b. Rheologic Structure of the uniform models with different ice scaling factor h

Name of Model:	U34b	U35b	U36b
Lithospheric thickness (km)	150	150	150
Stress exponent n in mantle	3	3	3
A* in the mantle (Pa^{-3} s^{-1})	3.33×10^{-34}	3.33×10^{-35}	3.33×10^{-36}
Ice thickness Scale factor h	2.5	1.5	1.5
χ^2 within ice margin	9.5	5.5	3.8
χ^2 outside ice margin	20.1	14.2	26.4
Total χ^2	15.2	10.1	15.8

3. RESULTS

3.1 Relative Sea Levels

In this subsection, the observed sea-level data for 27 sites with have lengthy records (13 of them lie inside the former ice margin while 14 of them lie outside, see Fig. 1a) are compared to the relative displacement curves predicted by different earth models of Table 2-5. These sealevel data have been collected by the Toronto group (e.g. Appendix B in [*Tushingham & Peltier* 1992]) and the ^{14}C age have been converted to sidereal age by the Calib 3.0 program [*Bard*, 1988]. To quantify the comparison, χ^2 (Chi-square) statistics are computed for each model. Here,

$$\chi^2 = \frac{1}{M} \sum_{n=1}^{M} \left(\frac{\zeta_{observed} - \zeta_{predicted}}{s} \right)^2 \qquad (4)$$

where $\zeta_{observed}$, $\zeta_{predicted}$, s and M are the observed sealevels, the predicted sealevels, the standard deviation of the error in height and the number of observations respectively. Since we are particularly interested in the misfit for sites outside the former ice margin, χ^2 statistics are also computed separately for the 13 RSL sites inside the ice margin (with a total of 68 data) and for the 14 RSL sites outside (with a total of 76 data) - see the last three rows of Tables 2 to 4.

In the following, we shall give a brief review of the χ^2 statistics of the different rheological models and we shall see that the best fitting models are: 1) a thin nonlinear zone sandwiched between the lithosphere and a linear mantle; and 2) a linear upper mantle and nonlinear lower mantle. The details of these comparisons or explanations can be found in [*Wu* 1999]. The new result in this subsection appears at the end where a search for the interval of acceptance of the creep parameter A* for the two proposed models with nonlinear rheology is described.

Table 3a. Rheologic Structure of models with nonlinear zones

Name of Model:	NLZ300	NLZ300v	NLZ420	NLZ670
Lithospheric thickness (km)	150	150	150	150
Stress exponent n in NLZ	3	3	3	3
A* in the NLZ (Pa^{-3} s^{-1})	3.33x10^{-35}	variable	3.33x10^{-35}	3.33x10^{-35}
Thickness of NLZ (km)	150	150	270	520
Bottom of NLZ at depth (km)	300	300	420	670
Stress exponent n in mantle	1	1	1	1
A* in the mantle (Pa^{-3} s^{-1})	3.33x10^{-22}	3.33x10^{-22}	3.33x10^{-22}	3.33x10^{-22}
Ice thickness Scale factor h	1	1	1	1
χ^2 within ice margin	4.0	See Fig. 3a	4.4	5.2
χ^2 outside ice margin	4.8	See Fig. 3a	5.7	14.5
Total χ^2	4.4	See Fig. 3a	5.1	10.2

Table 3b. High viscosity Lower Mantle with Nonlinear Zones

Name of Model:	NLZ420A	NLZ420B	NLZ670A
Lithospheric thickness (km)	150	150	150
Stress exponent n in NLZ	3	3	3
A* in the NLZ (Pa^{-3} s^{-1})	3.33x10^{-35}	3.33x10^{-35}	3.33x10^{-35}
Thickness of NLZ (km)	270	270	520
Bottom of NLZ at depth (km)	420	420	670
n in Upper Mantle (below NLZ)	1	1	-
A* in Upper Mantle (Pa^{-3} s^{-1})	3.33x10^{-23}	3.33x10^{-22}	-
n in Lower Mantle (below 670 km)	1	1	1
A* in Lower Mantle (Pa^{-3} s^{-1})	3.33x10^{-23}	3.33x10^{-23}	3.33x10^{-23}
Ice thickness Scale factor h	1	1	1
χ^2 within ice margin	6.2	5.6	5.3
χ^2 outside ice margin	92.7	31.2	64.2
Total χ^2	52.2	19.2	36.6

Let us begin with models whose rheology is uniform in the mantle. Among the earth models in Table 2a, model U22, which has a linear 10^{21} Pa-s uniform mantle, gives the smallest χ^2 (total χ^2 and χ^2 inside or outside the ice margin). The value of χ^2 increases dramatically as the mantle rheology becomes nonlinear - and this is mainly due to the misfit outside the ice margin. This can be seen in Fig. 1b where a visual comparison between the data and the predictions of some earth models at six representative sites are shown. For sites immediately outside the ice margin (e.g. Boston & Brigantine), nonlinear mantles (e.g. U35) grossly underestimate the amplitude of submergence - in fact, models U36 even predicts land emergence instead of submergence around Boston (see[Wu 1999]). As discussed in [Wu 1999], the reason of this is that nonlinear rheology results in a VSZ that exists immediately outside the ice margin near Boston. For sites near the center of rebound

(e.g. Ottawa Island & Churchill), the nonlinear models U34, U35 & U36 also underpredict the height of the 8 ka BP beach because the rebound rate was much higher near the end of deglaciation when the remaining load resulted in high stress and thus small effective viscosity underneath (see equation 3). Another reason for the poor misfit in nonlinear models for data near the center of rebound is because ICE3G was constructed with a linear rheological model with a viscosity profile close to that in U22. To see if adjustments in ice thickness for the nonlinear earth models can give better predictions for relative sealevels (migration of the ice margin is fixed by the isochrone data of terminal moraines and cannot be changed), [Wu 1999] scaled the ice thickness in ICE3G by a factor h to minimize the misfit to sealevel data within the ice margin. The factor h and the resulting χ^2 are given in Table 2b. The thicker ice from scaling do give better fit to the data within the ice

Table 4. Linear Upper Mantle with Nonlinear Lower Mantle

Name of Model:	NLLM21	NLLM22	NLLM22v	NLLM23
Lithospheric thickness (km)	150	150	150	150
n in Upper Mantle (below NLZ)	1	1	1	1
A* in Upper Mantle ($Pa^{-3}\,s^{-1}$)	3.33×10^{-21}	3.33×10^{-22}	3.33×10^{-22}	3.33×10^{-23}
n in Lower Mantle (below 670 km)	3	3	3	3
A* in Lower Mantle ($Pa^{-3}\,s^{-1}$)	3.33×10^{-35}	3.33×10^{-35}	variable	3.33×10^{-35}
Ice thickness Scale factor h	1	1	1	1
χ^2 within ice margin	6.8	5.5	See Fig. 3b	5.2
χ^2 outside ice margin	46.4	3.1	See Fig. 3b	31.5
Total χ^2	27.8	4.2	See Fig. 3b	19.2

margin, but misfit outside the ice margin didn't improve much for models U35b & U36b and actually deteriorated for model U34b (Table 2b). The sealevel curves predicted by model U35b at the six sites are also shown in Fig.1b from which it is clear that no further modification of ice thickness for this earth model can explain all these sealevel data simultaneously.

Since earth models with nonlinear mantles cannot explain all the sealevel data inside and outside the ice margin simultaneously, the next step is to follow *Karato & Wu's* [1993] suggestion by confining nonlinear rheology to a zone below the lithosphere. Models NLZ300, NLZ420 and NLZ670 (Table 3a) have nonlinear zones extending from the base of the lithosphere down to 300, 420 and 670 km depth respectively. Further below, the mantle is linear and has an effective viscosity of 10^{21} Pa-s. The models in Table 3b have in addition a high viscosity lower mantle. In models NLZ420B & NLZ670A, the viscosity in the lower mantle below 670 km is 10^{22} Pa-s, and in NLZ420A, the viscosity below 420 km depth is 10^{22} Pa-s. The sealevel curves predicted by model NLZ300 at the six representative sites are also shown in Fig.1b. Inspection of Table 3a shows that models with a thin nonlinear zone (NLZ300, NLZ420) give χ^2 values slightly better than that for model U22. This confirms the conclusion of *Karato & Wu* [1993] that a nonlinear zone with thickness less than 300 km is

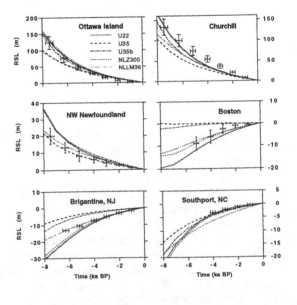

Figure 1. a) Map showing location of the relative sealevel sites where Chi-Square statistics are computed. b) Visual comparison between the predicted and observed data at six representative sites in Fig.1a.

permitted by the sealevel data. Table 3b also shows that for the given ice model, thick nonlinear zones or high viscosity lower mantle give much poorer fit to the sealevel data.

In Table 4, we consider earth models with linear upper mantle but nonlinear lower mantle below 670 km depth. In models NLLM21, NLLM22 and NLLM23, the value of the creep parameters in the lower mantle are fixed at 3.33×10^{-35} Pa^{-3} s^{-1}, but the viscosity of the linear upper mantle are 10^{20}, 10^{21} and 10^{22} Pa-s respectively. The χ^2 statistics for these models are much larger than that for model U22 except for model NLLM22 that is actually smaller. In fact, model NLLM22 has the lowest χ^2 statistic for all the models considered so far.

In the NLZ and NLLM models discussed so far, creep parameter A* in the nonlinear zone or in the nonlinear lower mantle is fixed at 3.33×10^{-35} Pa^{-3} s^{-1}. If the mantle does have a 300 km thick nonlinear zone beneath the lithosphere as in model NLZ300v (Table 3a) or a nonlinear lower mantle as in model NLLM22v (Table 4), then one may ask: "What is the interval of acceptance of the creep parameter A* in the nonlinear zone or in the nonlinear lower mantle?" In order to investigate this, the series of models NLZ300v and NLLM22v are used to compute postglacial sealevels. The value of the creep parameter in the nonlinear lower mantle studied ranges from $3. \times 10^{-34}$ Pa^{-3} s^{-1} to 3×10^{-37} Pa^{-3} s^{-1}. The sealevel curves predicted by model NLLM36 which has A*$=3 \times 10^{-36}$ Pa^{-3} s^{-1} in the lower mantle is plotted in Fig.1b and the χ^2 statistics for all of them are plotted in Fig.2. For models with a 300 km thick nonlinear zone (Fig. 2a), it can be seen that the best χ^2 statistic is for A*$=3.33 \times 10^{-35}$ Pa^{-3} s^{-1}. For the series NLLM22v that has a nonlinear lower mantle, sealevel data outside the ice margin are best fit by model NLLM22v with A*$=5 \times 10^{-37}$ Pa^{-3} s^{-1}, but sealevel data within the ice margin are best fitted with A*$=3.33 \times 10^{-36}$ Pa^{-3} s^{-1}. Overall, the model with the best total-χ^2 statistics is one with the value of A* around 10^{-36} Pa^{-3} s^{-1}, but in reality, models with A* between 10^{-35} to 10^{-36} Pa^{-3} s^{-1} all give comparable fits to the sealevel data.

3.2 Present-day Velocities

From the preceding subsection, we saw that both models NLLM22v and NLZ300 are preferred by the relative sealevel data, although the series NLLM22v achieve slightly better χ^2 statistics. However, due to recent advances in space-geodetic techniques such as GPS, VLBI, SLR and satellite altimetry, there are other geophysical observable related to the postglacial readjustment process, which may provide further discrimination among these rheological models. The ones that will be investigated in this subsection are the present-day uplift rate and horizontal velocity. In view of the large number of earth models

considered, we shall present here and in the next subsections only four representative models, which show the best χ^2 statistics from each group. They are model U22 which has a linear 10^{21} Pa-s mantle, model U35b which has a uniform nonlinear mantle with thick ice, model NLZ300 which has a thin nonlinear channel under the lithosphere and model NLLM36 which is model NLLM22v with A* = 10^{-36} Pa^{-3} s^{-1} and n=3 in the lower mantle.

Fig.3 shows the present-day uplift rate predicted by these four models. The pattern of uplift for all of them looks similar since that is mainly determined by the ice model. However their amplitudes are different - model U35b has the smallest present-day rate of uplift since rapid uplift occurred near the end of deglaciation. Model U22 and NLZ300 have similar uplift rates. Model NLLM36 has the largest maximum and minimum uplift rates (10 mm/a to -2 mm/a) among these models.

Fig. 4 shows the present-day horizontal velocities of these models. Note that these are computed under the assumptions of incompressibility and flat-earth geometry -

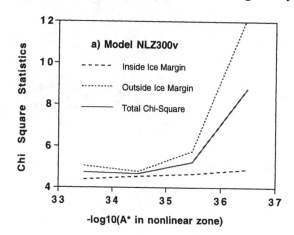

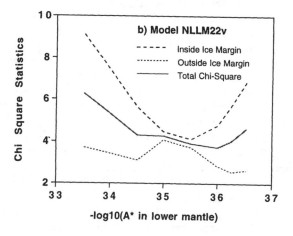

Figure 2. Chi-Square statistics as a function of creep parameter A* within a) nonlinear zone in model NLZ300v (Table 3a) and b) nonlinear lower mantle in model NLLM22v (Table 4).

Present day Uplift Rate (mm/a)

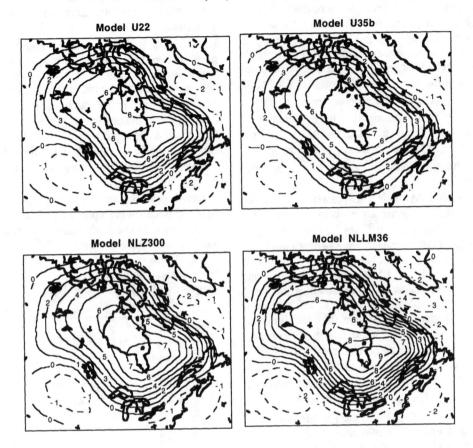

Figure 3. Present day uplift rate predicted by the four representative earth models. Contours in mm/a. Solid contours for positive values, dashed contours for negative values.

both of which may have non-negligible effects on horizontal motion [*Klemann et al.* 2000; *O'Keefe & Wu* 2001]. Thus, these results are not directly comparable to those obtained by *Mitrovica et al.* [1994] nor *Peltier*[1998] (which uses ICE4G) although their results do show some resemblance with ours here. However, the main point here is to investigate the effects of power-law rheology, which presumably will carry over even when compressibility and sphericity are included. In Fig. 4, the arrows give the direction of the velocity vector and their lengths are proportional to the magnitude, which is also contour-plotted on the same diagram. Both models U22 and NLZ300 predict that the motion is directed outwards from Hudson Bay, however the peak velocity in model U22 is about 50% larger than that in model NLZ300. On the other hand, the pattern of horizontal motion in models U35b and NLLM36 are quite complex - just south of Hudson Bay, the motion is directed outwards from it just as in model U22, but further south, the motion reversed and is directed towards Hudson Bay. However, the magnitude of the horizontal motion of model NLLM36 is about twice that for U35b despite of its

thicker ice. Thus, although both model NLZ300 and NLLM36 provide good fits to the sealevel data, these two models can be easily discriminated from each other on the basis of the direction of horizontal motion just outside the former ice margin. (Recent observed directions of horizontal motion in eastern Canada and U.S. [MacMillan & Ma, 1999] tend to support the NLLM36 model over the NLZ300/U22 models.) In addition, models NLZ300 and U22 can be discriminated from each other if the magnitude of horizontal motion can be determined with accuracies less than 0.5 mm/a.

3.3 Gravity Anomalies and the Rate of Change of Gravity

In this subsection, we discuss present-day free air gravity anomalies and the rate of change of gravity for the four earth models in the last subsection. The gravity anomaly is derived from the current amount of uplift remaining by assuming a gravity-displacement ratio of 0.20 mGal/m. Fig. 5 shows that the shape of the free air gravity anomalies are similar but their magnitudes are different. Of these models,

Present day Horizontal Velocity (mm/a)

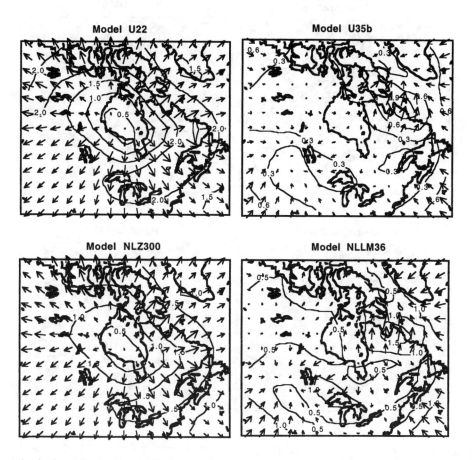

Figure 4. Present day horizontal velocity predicted by the four representative earth models. Contours in mm/a. Solid contours for positive values, dashed contours for negative values.

model NLLM36 is able to predict -30 to -40 mgals over Hudson Bay, which is close to the observed value of -30 to -47 mgal. Models U35b predicts about -25 to -30 mgals, which is slightly smaller than the observed value. Model U22 only predicts -8 to -12 mgal and model NLZ300 predicts -8 to -10 mgal over Hudson Bay, thus the magnitudes for U22 and NLZ300 are considerable less than the observed value. However, the observed magnitude (around -30 to -47 mgal) may have contributions from both postglacial rebound and dynamic topography induced by large-scale mantle convection [*Peltier et al.* 1992], which can account for about -25 mgal of the observed anomaly. If *Peltier et al* [1992] are correct, then the observed free air gravity favors models NLZ300 and U22 rather than NLLM36 or U35b.

Finally, the current rate of change of gravity can be obtained by multiplying the uplift rate by -0.20 mGal/m. From Fig. 3, one can infer that the peak rate of change in gravity is around the south tip of Hudson Bay, which has value of about -1.4 µgal/a for model U35b, -1.5 µgal/a for

models U22 and NLZ300 and -2.0 µgal/a for model NLLM36. Since repeated absolute gravity measurements have been made in Churchill by the Geological Survey of Canada with the JILA-2 apparatus since about 1987 [*Tushingham et al.* 1991], it is useful to give the predicted values there. The values for models U22, NLZ300, U35b and NLLM36 at Churchill are: -1.22, -1.27, -1.28 and -1.5 µgal/a respectively.

3.4 Changes in Fault Stability Margin (dFSM)

Recent studies [*Wu & Hasegawa* 1996, 1998b] have shown that postglacial faults in Eastern Canada are likely caused by glacial unloading which ended around 9 ka BP and the current intraplate earthquake activities in Eastern Canada are likely triggered by the stresses due to postglacial rebound although tectonic stress is also needed to keep the pre-existing faults near the condition of failure [Wu & Hasegawa 1996]. From the Mohr-Coulomb Failure criteria, earthquake or fault potential can be defined. The

Gravity Anomalies (mgal)

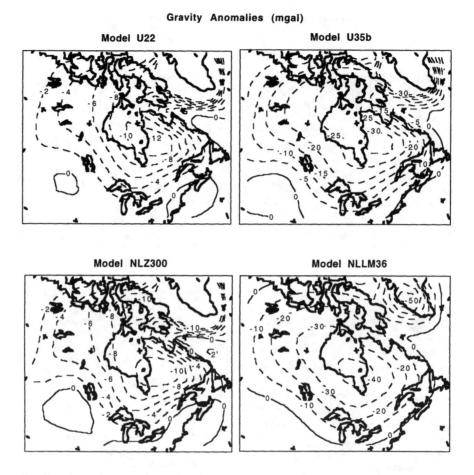

Figure 5. Present day free-air gravity anomaly predicted by the four representative earth models. Contours in mgal. Solid contours for positive values, dashed contours for negative values.

quantity calculated is dFSM or change in Fault Stability Margin at time t compared to the initial value before the onset of glaciation (see *Wu & Hasegawa* [1996] for details of its computation). This quantity varies in space and time as rebound stress changes. For sites underneath the ice, glacial loading usually shifts the Mohr circle away from failure, thus dFSM becomes positive and pre-existing faults that are initially near failure become stabilized. However, following deglaciation, dFSM underneath the ice becomes negative, meaning that the Mohr circle moves towards failure, thus pre-existing faults near failure can become reactivated. The time when dFSM at a location first becomes negative is the onset time. When dFSM is negative, its magnitude is an indicator of the amount of rebound stress available to trigger earthquakes. The mode of failure is determined by whether the maximum, intermediate or minimum principal stress is near vertical. For all models, thrusting is predicted and this agrees with the observed mode of failure (see *Wu* [1998b]).

The spatial distribution of dFSM in Eastern Canada today for the four representative earth models is shown in

Fig. 6. It can be seen that model NLLM36 predicts the largest magnitude of dFSM while model NLZ300 predicts the smallest magnitude. Since the onset timing of earthquakes is directly observable, we shall consider that next. The predicted variation of dFSM at Charlevoix, Quebec (47.5°N, 70.1°W) and Wabash Valley, Indiana (38.5°N, 87 °W) for the four earth models are shown in Fig. 7. Inspection of Fig. 7a shows that models U22, NLZ300 and NLLM36 all predict the onset time at Charlevoix to be around 9.2 ka BP, while model U35b predicts it to be 8.5 ka BP. Not that within the 1 ka uncertainty in age determination, all these predictions agree with the observed timing of 9 ka BP in Lac Temiscouata near Charlevoix. While the onset time within the former ice margin is not very sensitive to mantle rheology (as in Charlevoix), this is not true for sites outside the ice margin. An example is shown in Wabash Valley, where the onset time for models U22 and NLZ300 is predicted to be 8.3 ka BP, but that for models U35b and NLLM36, the onset times are 7 and 4.9 ka BP respectively. Unfortunately, the paleoearthquake there, dated as 8 to 1 ka BP by paleo-liquefaction research, has

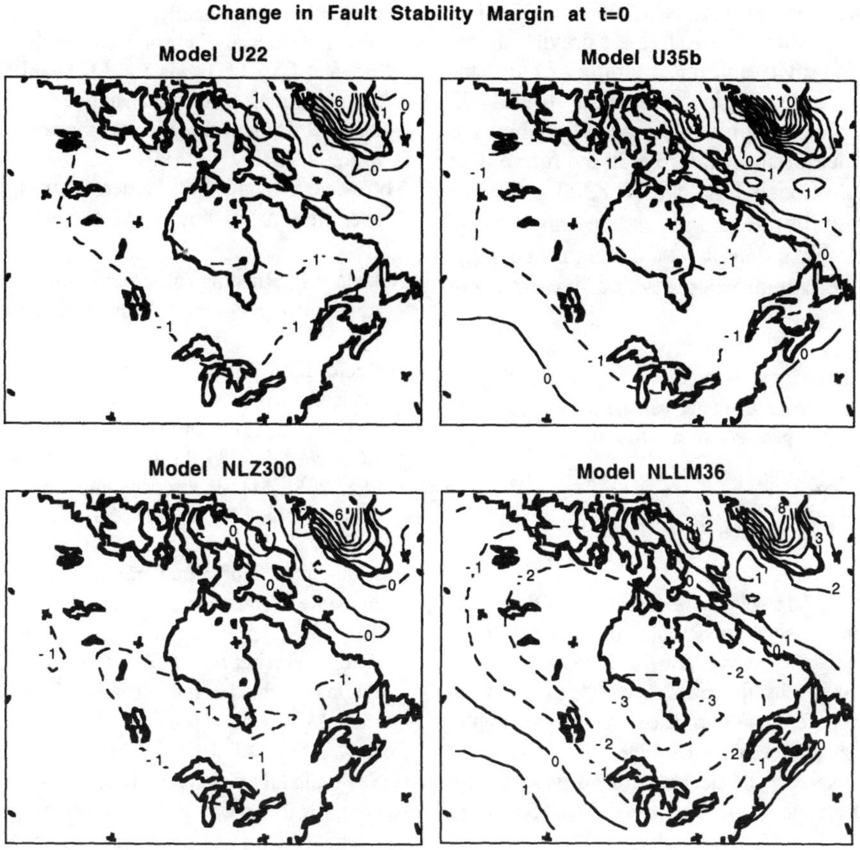

Change in Fault Stability Margin at t=0

Model U22

Model U35b

Model NLZ300

Model NLLM36

Figure 6. Change in Fault Stability Margin (dFSM) today as predicted by the four representative earth models. Contours in MPa. Solid contours for positive values, dashed contours for negative values.

rather large uncertainties. However, if more precise paleoearthquake onset time becomes available for sites outside the ice margin, then they can also be used to discriminate mantle rheology.

4. CONCLUSIONS

In this paper we compared the predicted and observed sealevel data around Laurentia for a number of earth models with nonlinear rheology when the ICE3G model is assumed. It is found that the ones that give the best fit are: a) models with linear 10^{21} Pa-s upper mantle and nonlinear lower mantle with $A^* = 10^{-36}$ Pa^{-3} s^{-1}. b) models with thin (<300 km) nonlinear zones ($A^* = 3 \times 10^{-35}$ Pa^{-3} s^{-1}) above a linear mantle. For this second model, the viscosity of the mantle is taken to be 10^{21} Pa-s, but this does not exclude the possibility of an increase in the deeper part of the lower mantle viscosity as proposed in *Peltier* [1996] or *Forte & Mitrovica* [1996].

Uplift rate, horizontal velocity, free-air gravity, the rate of change of gravity, fault instability and earthquake onset timing have also been calculated for four earth models - the

two best fitting models NLZ300 and NLLM36 plus two reference models, namely the uniform linear mantle model U22 and the uniform nonlinear mantle model U35b with thick ice. It is found that the direction of horizontal velocity outside the former ice margin is particularly useful in distinguishing models with nonlinear rheology in the lower mantle, e.g. NLLM36 and U35b from models with linear

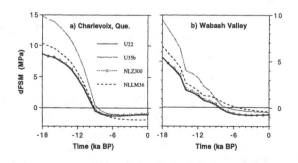

Figure 7. Change in Fault Stability Margin (dFSM) as a function of time as predicted by the four representative earth models. Charlevoix is within the ice margin at glacial maximum but Wabash Valley is outside the ice margin.

rheology in the lower mantle, e.g. NLZ300 or U22. The other observable, such as uplift rate, free-air gravity anomaly, rate of change of gravity and onset timing of fault stability (outside the ice margin) are also useful in discriminating these rheologic models especially when the accuracies of these measurement improve in the future. For example, even though models U22 and NLZ300 give very similar signatures in uplift rate and in relative sealevels for most sites (except Boston), they can be distinguished from their magnitudes in horizontal velocities and free-air gravity.

Acknowledgments. This work is supported by a grant from NSERC (Canada). The finite element calculations were performed with the ABAQUS package from Hibbitt, Karlsson and Sorensen Inc.

REFERENCES

Amelung, F. and D. Wolf, Viscoelastic perturbations of the earth: significance of the incremental gravitational force in models of glacial isostasy, *Geophys. J. Int.*, 117, 864-879, 1994.

Bard, E., Correction of accelerator mass spectrometer ^{14}C ages measured in planktonic foraminifera: Paleoceanographic implications, *Paleoceanography*, 3, 635-645, 1988.

Brennen, C., Isostatic recovery and the strain rate dependent viscosity of the earth's mantle. *J. Geophys. Res.*, 79, 3993-4001, 1974.

Crough, S.T., Isostatic rebound and power-law flow in the asthenosphere, *Geophys.J.R.astr.Soc.*, 50, 723-738, 1977.

Forte, A.M. and J.X. Mitrovica, New inferences of mantle viscosity from joint inversion of long-wavelength mantle convection and postglacial rebound data, *Geophys.Res.Lett.*, 23, 1147-1150, 1996.

Gasperini, P., D.A. Yuen & R. Sabadini, Postglacial rebound with a non-Newtonian Upper Mantle and a Newtonian Lower Mantle Rheology. *Geophys. Res. Lett.* 19, 1711-1714, 1992.

Goetze C. and D.L. Kohlstedt, Laboratory study of dislocation climb and diffusion in olivine, *J. Geophys. Res.*, 78, 5961-5971, 1973.

Karato, S., Micro-Physics of Post Glacial Rebound, *in Dynamics of the Ice Age Earth: A Modern Perspective*, edited by P.Wu, pp.351-364, Trans Tech Publ., Switzerland, 1998.

Karato, S. and P. Li, Diffusion Creep in Perovskite: implications for the rheology of the Lower Mantle. *Science*, 255, 1238-1240, 1992.

Karato, S. and P. Wu, Rheology of the Upper Mantle: a synthesis, *Science*, 260, 771-778, 1993.

Klemann, V., P. Wu & D. Wolf, Compressible Viscoelasticity: a comparison of plane-earth solutions, *EOS*, 81, F326, 2000.

Li, P., Karato, S., and Z. Wang, High-temperature creep of fine-grained polycrystalline CaTiO3. *Phys. Earth Planet. Inter.*, 95, 19-36, 1996.

MacMillan, D. and C. Ma, NASA Goddard Space Flight Center's VLBI Earth orientation series number er1122, 1999. (Data available electronically at lupus.gsfc.nasa.gov/plots/map/jpg/North_American_Eastern_Coast.jpg)

Mitrovica, J.X., J.L Davis and I.I. Shapiro, A spectral formulism for computing three-dimensional deformations due to surface loads, 2, Present-day glacial isostatic adjustment, *J. Geophys. Res.*, 99, 7075-7101, 1994.

Mitrovica, J.X. and W.R. Peltier, On postglacial geoid subsidence over the equatorial oceans, *J. Geophys. Res.*, 96, 20053-20071, 1991.

Nakada, M., Rheological structure of the earth's mantle derived from glacial rebound in Laurentide. *J.Phys.Earth*, 31, 349-386, 1983.

O'Keefe, K. & P. Wu, Effect of mantle structure on postglacial induced horizontal displacement, this volume, 2001.

Peltier, W.R., The thickness of the continental lithosphere, *J.Geophys.Res.*, 89, 11303-11316, 1984.

Peltier, W.R., Mantle viscosity and Ice-Age ice sheet topography, *Science*, 273, 1359-1364, 1996.

Peltier, W.R., A space geodetic target for mantle viscosity discrimination: Horizontal motions induced by glacial isostatic adjustment, *Geophys. Res. Lett.*, 25, 543-546, 1998.

Peltier, W.R., A.M. Forte, J.X. Mitrovica, A.M. Dziewonski, Earth's gravitational field: seismic tomography resolves the enigma of the Laurentian anomaly, *Geophys. Res. Lett.*, 19, 1555-1558, 1992.

Post, R.L. and D.T. Griggs, The Earth's mantle: Evidence of non-Newtonian flow, *Science*, 181, 1242-1244, 1973.

Ranalli, G., Rheology of the Earth: deformation and flow processes in geophysics and geodynamics, Allen & Unwin, Boston, 1987.

Ranalli, G., Inferences on Mantle Rheology from Creep Laws *in Dynamics of the Ice Age Earth: A Modern Perspective*, edited by P.Wu, p.323-340, Trans Tech Publ., Switzerland, 1998.

Tushingham, A.M., and W.R. Peltier, Ice-3G: a new global model of late Pleistocene deglaciation based upon geophysical predictions of post-glacial relative sea-level change, *J.Geophys.Res.*, 96, 4497-4523, 1991.

Tushingham, A.M., and W.R. Peltier, Validation of the ICE-3G model of Wurm-Wisconsin deglaciation using a global data base of relative sea-level histories, *J.Geophys.Res.*, 97, 3285-3304, 1992.

Tushingham, A.M., A.Lambert, J.O. Liard and W.R. Peltier, Secular gravity changes: measurements and predictions for selected Canadian sites, *Can.J. Earth Sci.*, 28, 557-560, 1991.

Walcott, R.I., Isostatic response to loading of the crust in Canada, *Can.J. Earth Sci.*, 7, 716-727, 1970.

Wu, P., Deformation of an incompressible viscoelastic flat earth with power law creep: a finite element approach. *Geophys.J.Int.*, 108, 136-142, 1992.

Wu, P., Post-glacial Rebound in a Power-law medium with Axial symmetry and the existence of the Transition zone in Relative Sea Level data. *Geophys.J.Int.*, 114, 417-432, 1993.

Wu, P., Postglacial rebound Modeling with Power Law Rheology. *in Dynamics of the Ice Age Earth: A Modern Perspective*, edited by P.Wu, pp.365-382, Trans Tech Publ., Switzerland, 1998a.

Wu, P., Intraplate earthquakes and postglacial rebound in Eastern Canada and Northern Europe. *in Dynamics of the Ice Age Earth: A Modern Perspective*, edited by P.Wu, pp.603-628, Trans Tech Publ., Switzerland, 1998b.

Wu, P., Modelling postglacial sea levels with power-law rheology and a realistic ice model in the absence of ambient tectonic stress, *Geophys. J. Int.,* 139, 691-702, 1999.

Wu, P. & H. Hasegawa, Induced stresses and fault potential in Eastern Canada due to a realistic load: a preliminary analysis, *Geophys. J. Int.*, 127, 215-229, 1996.

Wu, P. and P.. Johnston, Validity of using Flat-Earth Finite Element Models in the study of Postglacial Rebound, *in Dynamics of the Ice Age Earth: A Modern Perspective*, edited by P.Wu, pp.191-202, Trans Tech Publ., Switzerland, 1998.

Yokokura, T., and M. Saito, Viscosity of the upper mantle as non-Newtonian fluid. *J.Phys.Earth*, 26, 147-166, 1978.

Patrick Wu, Department of Geology & Geophysics, University of Calgary, 2500 University Dr. N.W., Calgary, Alberta T2N 1N4, Canada. (email: ppwu@ucalgary.ca)

The Convective Mantle Flow Signal in Rates of True Polar Wander

Bernhard Steinberger

Institut für Meteorologie und Geophysik, Johann Wolfgang Goethe-Universität, Frankfurt am Main, Germany

Richard J. O'Connell

Department of Earth and Planetary Sciences, Harvard University, Cambridge, Massachusetts

We investigate changes of the rotation axis caused by mantle convection. In the first part, the coupled problem of viscoelastic deformation and rotational dynamics is solved for simple Earth models in order to compute how the rotation axis changes following emplacement of non-hydrostatic excess masses. It is shown that both direct integration and a computationally more effective "quasi-static integration" give virtually identical results. With the latter method, results of eigenmode and time domain approach were compared, with little difference found. Although the number of viscoelastic relaxation eigenmodes is infinite, for an adiabatic Earth mantle only two eigenmodes need be considered for an approximately correct description, and an even simpler steady-state solution can be used. The results indicate that for our best estimates of present-day mantle properties, the maximum speed of polar motion is about 1 degree per million years, and during Cenozoic times the rotation axis has always followed closely the axis of maximum non-hydrostatic moment of inertia imposed by advection of mantle density heterogeneities. The latter was calculated for a number of tomographic models and inferred flow fields. Results indicate on average a slow motion of about 5 degrees in 60 Ma roughly towards Greenwich, which is not in conflict with paleomagnetic results. Only one of the models additionally predicted a faster motion prior to about 80 Ma in a direction similar to what is inferred from paleomagnetism.

1. INTRODUCTION

1.1. Overview

While the geodetically observed rates of polar motion [*Dickman*, 1977] are believed to be to a large part due to glacial events, the longer-term polar motion may largely be due to redistributions of masses caused by convection in the Earth's mantle [e.g., *Ricard and Sabadini*, 1990], and convection may still cause a non-negligible contribution to present-day polar wander. Here we will elaborate on this issue, and for this purpose we present a numerical model for changes of the rotation axis caused by a convecting mantle and compare results with observations of true polar wander. Some important results of this model have already been published [*Steinberger and O'Connell*, 1997]. Here we expand this work essentially in two ways: Firstly, the model of viscoelastic

Ice Sheets, Sea Level and the Dynamic Earth
Geodynamics Series 29
10.1029/029GD15

relaxation that is used to compute changes of the rotation axis following a change in the moment of inertia is introduced in more detail. In this section some of the printing errors that occurred in our previous publication will also be pointed out. We will reiterate the conclusion that normally the rotation axis follows imposed changes of the moment of inertia tensor. Secondly, although only changes of the degree two non-hydrostatic geoid cause changes in long-term Earth rotation, we will compute time change of the geoid up to higher degree. Through this additional information, we are better able to pin down which features, when advected, are most responsible for the change in nonhydrostatic moment of inertia. Furthermore, we compute time changes of the moment of inertia tensor for a greater number of tomographic models, thus corroborating the robustness of our results.

Because the Earth is not rigid, it makes sense to define the axis of rotation of a reference frame relative to an inertial frame of reference. In this work, the calculations of the "rotation axis" will without further mention always refer to "Tisserand's mean axes of the body". This reference frame is characterized by zero net rotation when integrated over the entire mantle, and will be also referred to as "mean mantle" reference frame.

When comparing our computations with observations of true polar wander, we need to bear in mind that, besides the obvious axial dipole hypothesis, the comparison also assumes that the mean mantle reference frame in which our calculations are performed and which cannot be directly constrained by observations, and the "hotspot reference frame" which is based on observations and to which the observed "true polar wander" refers, are essentially the same. If hotspots move in a convecting mantle, this is not necessarily the case, and we have previously computed how the observed "true polar wander" curve changes when it is converted to a mean mantle reference frame, taking computed hotspot motions into account [Steinberger and O'Connell, 2000]. Since the difference turned out to be small, it will be disregarded for the purpose of this paper.

1.2. Changes of the Earth's Rotation Axis - Dynamic Modelling

The subject of long-term changes in the Earth's rotation axis and its relation to mass redistributions inside the Earth has a rather long history. The idea that the rotation axis might significantly change relative to some reference frame tied to the solid Earth, and that, for example, poles might move to where the equator used to be, and vice versa, was proposed much earlier than plate tectonics.

Darwin [1876] made the first quantitative attempt to deal with changes of the Earth's axis of rotation due to geological changes. Darwin tried to solve the problem of how the rotation axis changes if there is an uplift of material at some area at the Earth's surface. He arrived at the result that a change in the rotation axis is possible. However his work was marred by several errors. The first one is an algebraic error discovered by *Lambert* [1931]. The correction is given by *Jeffreys* [1952] on page 343. With this correction, the conclusion is reversed, and the rotation axis should not move a significant amount. In addition, Darwin made another error in assuming that the axis of the geoid (as defined by *Gold* [1955]) moves towards the rotation axis at a rate proportional to the separation of the axis of figure (following *Gold* [1955] defined as the principal axis of the inertia tensor with the largest moment of inertia) and the rotation axis. This error is pointed out by *Munk* [1956]. His paper reviews Darwin's paper, its errors, and the discussions it sparked during the 55 years until the first error was detected.

The first qualitatively correct treatment is given by *Gold* [1955]. He describes how an excess mass that is added to the Earth leads to a slow deformation of the Earth and a change of the rotation axis, causing the excess mass to move slowly toward the equator (without being displaced relative to the solid Earth). This process is described in Figure 1, where four excess masses are added, such that the center of mass does not change, and only one non-diagonal inertia tensor element changes due to the excess masses.

Gold's paper inspired *Burgers* [1955] to treat the problem quantitatively. Burgers considers a homogeneous viscoelastic sphere, the moments of inertia of which are changed by a small amount. He arrives at a solution, which consists of small oscillations ("wobble") of the rotation axis and a long-term mean motion. The subject is reviewed by *Inglis* [1957].

Munk and MacDonald [1960] discuss how an approximate solution for a layered sphere may be obtained using the concept of Love numbers assuming a Maxwell viscoelastic rheology. More recently, calculations for a layered viscosity structure have been done. *Sabadini et al.* [1982] treat the changes of the Earth's rotation in response to growing and melting of large ice sheets. They look at an Earth model composed of an elastic lithosphere, a viscoelastic mantle and an inviscid core, i.e. a three-layer-model. They treat the equations by doing a Laplace transform and calculating eigenmodes of the relaxation. *Wu and Peltier* [1984] perform a similar kind of analysis, but they arrive at a different result. They claim that *Sabadini et al.* [1982] did an invalid

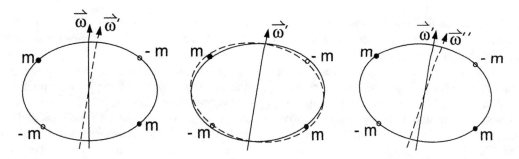

Figure 1. Interplay between deformation and change of the rotation axis of the Earth. First picture: Non-hydrostatic excess masses cause the rotation axis of the Earth to change (from the solid to the dashed line). Second picture: The change in rotation axis causes the Earth to deform (from solid to dashed outline). Third picture: This change in shape causes the rotation axis to move again. The process would cease once it has shifted the excess masses to the equator, but this is only approached asymptotically.

approximation. *Sabadini et al.* [1984] provide a more general treatment and show that the two formulations are equivalent to some extent. They also point out that a solution requires in addition to the eigenmodes of viscoelastic relaxation another set of eigenmodes arising from the coupling of viscoelastic relaxation and rotational dynamics. *Mitrovica and Milne* [1998] give a detailed comparison of the two approaches by *Sabadini et al.* [1982] and *Wu and Peltier* [1984] and show, using a generalized theory, that the two formalisms yield essentially equivalent predictions.

Sabadini and Yuen [1989] state that the rate of change in the rotation axis depends on the viscosity structure and on the chemical structure (i.e. non-adiabatic density variations). They emphasize that the change of the rotation axis over long times can put constraints on both chemical stratification and viscosity distribution in the mantle and that there are trade-offs between the style of mantle chemical stratification and the magnitude of lower mantle viscosity. *Ricard et al.* [1992] investigate how the nature of the 670-km-discontinuity affects the rate of change of the rotation axis. In their calculations, a phase change yields a large rate of change, whereas the rate is drastically reduced for chemically stratified models. They use the linearized Liouville equations valid for small displacements and express viscoelastic deformation in terms of eigenmodes. *Spada et al.* [1992] compute changes of rotation due to subduction using an asymptotic expansion for large time of the nonlinear Liouville equations for an Earth model consisting of an elastic lithosphere, a Maxwell viscoelastic mantle and an isostatic core. They show that a viscosity increase with depth is required to achieve realistic rates of polar wander. Using a viscous quasi-fluid approximation, *Ricard et al.* [1993] further discuss the change of the rotation axis induced by a downgoing cold slab, depending on the viscosity structure.

More recently, *Fang and Hager* [1995] have shown that for a realistic Earth model with continuously varying density, viscosity and elastic parameters, viscoelastic relaxation cannot be solely expressed in terms of discrete eigenmodes, and the error made by only considering discrete modes depends on the exact viscosity structure. This problem arises because of the so-called "Maxwell singularities". For this reason, *Hanyk et al.* [1995] proposed a "time-domain approach for the transient responses in stratified viscoelastic Earth models". On the other hand, Boschi et al. [1999] showed how the problem of Maxwell singularities can be avoided, and that hence the eigenmode approach is still valid.

Here we develop an independent and entirely self-contained algorithm which allows us to use both time-domain and eigenmode approach to calculate changes of the Earth's rotation axis for a rather general viscoelastic Earth model in an efficient manner. The algorithm is developed and results are presented in section 2. We will first show that a quasi-static approximation, i.e. disregarding differences between the axis of figure and the rotation axis, which give rise to the "Chandler wobble" [*Ricard et al.* 1993] and direct integration give virtually identical results. We thus verify that the derivation and numerical implementation of the equations for quasi-static integration has been done correctly. We will compare the results of the time domain and eigenmode approach and show that for the cases considered here differences are much smaller than other uncertainties in the model. We will also discuss in which cases a "steady-state approach", which allows an even more effective computation, is suitable. In the end, we will calculate changes in the rotation axis caused by advection of realistic density anomalies inferred from seismic tomography in a realistic flow field [*Hager and O'Connell,* 1979, 1981]. Our approach is similar to the one pursued by *Richards et al.* [1997], except that they use models of subduction history instead of mantle flow models based on seismic tomography in order to compute changes of the moment of inertia tensor. Thus their approach only includes the effect of downgoing slabs, whereas our approach implicitly captures all sources of the changing density distribution. The difference is probably significant, because mantle flow is likely not only driven by downgoing slabs but also includes active upwellings [*Gurnis et al.*, 2000].

Along a different line of thought, *Goldreich and Toomre* [1969] discuss some statistical aspects of the change of the rotation axis in time. They make the assumption that the equatorial bulge does not hinder a change in the rotation axis, and thus the rotation axis always follows the axis of maximum non-hydrostatic moment of inertia. They then simulate a randomly evolving, almost spherical body. They find, that the rotation axis changes rapidly when two of the principal non-hydrostatic moments of inertia become almost equal, because then the principal axes of the non-hydrostatic inertia tensor move most rapidly. The procedures outlined in this work allow us to calculate how rapid this change may be, and obtain the maximum speed of change in the case when the two larger principal moments of inertia become equal or almost equal ("inertial interchange"). These results were already shown by *Steinberger and O'Connell* [1997] where it was also shown that under normal circumstances the rotation axis will indeed follow the axis of maximum non-hydrostatic moment of inertia very closely, thus justifying the assumption made by *Goldreich and Toomre* [1969].

1.3. Changes of the Earth's Rotation Axis - Observational Aspects

Early work suggested that in the mean lithospheric frame there occurs no significant change of the rotation axis [*McElhinny*, 1973; *Jurdy and van der Voo*, 1974], whereas there is a change of the hotspot frame with respect to the rotation axis [*Duncan et al.*, 1972]. This led to the "mantle roll" hypothesis [*Hargraves and Duncan*, 1973]: The rotation axis is fixed relative to the mean lithosphere, whereas the lower mantle "rolls" relative to both of them. *Jurdy* [1981] points out that the previous works cannot be compared directly, because they use different datasets. By using the same dataset in both cases she rules out that the different results are due to different data. She finds no significant motion of the rotation pole relative to the mean lithosphere, but a large motion of 10 - 12 degrees relative to the hotspots since early Tertiary, thus confirming the mantle roll hypothesis.

More recently however this hypothesis has been questioned. *Gordon and Jurdy* [1986] showed that both reference frames are more similar than previously thought and suggest that both reference frames and the paleomagnetic axis might be in relative motion, however with the paleomagnetic-hotspot motion probably larger than the relative motion of mean lithospheric and hotspot reference frame. *Gordon and Livermore* [1987] extended the analysis into the late Cretaceous and reported that in both the mean lithospheric and in the hotspot reference frame the rotation axis has shifted by $10° - 20°$ in the same direction during the last 100 m.y.

Published true polar wander paths usually show the motion of the pole in the African hotspot reference frame [*Andrews*, 1985, *Besse and Courtillot*, 1991, *Prevot et al.*, 2000]. Despite some differences, recent results agree on the following general features:

- Fast polar motion of a few degrees during the past few Myr, roughly towards Greenland. This is most likely due to glacial effects.

- Slow motion (if any) of a few degrees at most (i.e., less than the few degrees uncertainty of a paleomagnetic determination) in a similar direction, during the Tertiary.

In this work, we will compare the calculated motion of the rotation axis with the most recent results of *Besse and Courtillot* [2000, pers. comm.] and *Prevot et al.* [2000]. Both these results also agree on faster motion prior to about 60 to 90 Ma in a roughly opposite direction, but *Tarduno and Smirnov* [2001] argue that the rotation axis has moved by no more than $\sim 5°$ over the last 130 million yr and that the apparent polar shift in reality represents hotspot motion. The required relatively fast hotspot motion does not agree with modelling results [e.g. *Steinberger*, 2000], however models of mantle flow, and hence hotspot motion, become rather unreliable prior to the Tertiary: This is also evidenced by the predictions of polar motion due to mantle flow presented in section 3.

2. CHANGE OF THE EARTH'S ROTATION AXIS DUE IMPOSED CHANGES OF THE INERTIA TENSOR

2.1. Overview

In this section we will develop a quantitative treatment of the process that was first qualitatively explained by *Gold* [1955]). This requires a coupled solution of two problems: How a change of the rotation axis causes a deformation of the Earth, and how a deformation of the Earth causes a change of the rotation axis. The first part is a continuum mechanics problem, the second part is a problem of rotational dynamics.

We will first develop a formalism to solve the first problem separately: For a given change of the rotation vector, what is the deformation that results for an Earth model consisting of a mantle with Maxwell viscoelastic rheology overlying a hydrostatic core? The formulation will first be given in the Laplace transform domain, where the governing equations can be written in the form of a linear system of ordinary differential equations. This allows us to compute eigenfunctions of viscoelastic decay. However we do not restrict ourselves to the eigenfunction approach. We will therefore transform the equations back into time domain and show how any viscoelastic deformation consists of an immediate elastic deformation corresponding to the change in the rotation axis and a slow viscoelastic deformation corresponding to the deviation of the actual shape from equilibrium shape. We will show how to compute the equilibrium shape, the immediate elastic deformation following a change in rotation and the slow viscoelastic deformation in the time domain. Our formulation is similar to the equations for viscous flow given by *Hager and O'Connell* [1979, 1981]. A far more detailed derivation of this formalism is given by *Steinberger* [1996], whereas here only the principal steps are recapitulated.

This formalism will then be combined with the equations of motion of a rotating body in a rotating frame of reference (Euler equations, often also referred to as Liouville equations in the case of a deformable body). A method to integrate the Euler equations efficiently will be presented. This method will be used to calculate changes in the Earth's rotation axis for various cases. For simplicity a distribution of excess masses as in Figure 1 will be used.

2.2. Deformation of the Earth due to a Change in Rotation – Formulation for Viscoelastic Relaxation in the Mantle in the Laplace Transform Domain

Time derivatives in the constitutive relationship of a Maxwell-body can be eliminated by performing a Laplace transform. By doing so, each time derivative operator $\frac{\partial}{\partial t}$ is replaced by the Laplace transform variable s.

Changes of the rotation vector only excite displacements $\delta \mathbf{u}$, stress anomalies $\delta \mathbf{T}$, potential anomalies $\delta \varphi$ and density anomalies $\delta \rho$ of spherical harmonic degree two and zero. Since we only deal with changes of the rotation axis without changing the rotation rate, we can neglect degree zero terms. Therefore the spherical harmonic expansions simplify to

$$\delta u_r = \sum_{m=-2}^{2} u_{1,2m} Y_{2m}$$

$$\delta u_\theta = \sum_{m=-2}^{2} u_{2,2m} \frac{\partial Y_{2m}}{\partial \theta}$$

$$\delta u_\phi = \sum_{m=-2}^{2} u_{2,2m} \frac{1}{\sin \theta} \frac{\partial Y_{2m}}{\partial \phi}$$

$$\delta T_{rr} = \frac{\alpha_0}{r} \sum_{m=-2}^{2} u_{3,2m} Y_{2m}$$

$$\delta T_{r\theta} = \frac{\alpha_0}{r} \sum_{m=-2}^{2} u_{4,2m} \frac{\partial Y_{2m}}{\partial \theta}$$

$$\delta T_{r\phi} = \frac{\alpha_0}{r} \sum_{m=-2}^{2} u_{4,2m} \frac{1}{\sin \theta} \frac{\partial Y_{2m}}{\partial \phi}$$

$$\delta \varphi = \frac{\alpha_0}{\rho_{00} r} \sum_{m=-2}^{2} u_{5,2m} Y_{2m} \qquad (1)$$

$$\frac{\partial \delta \varphi}{\partial r} = \frac{\alpha_0}{\rho_{00} r^2} \sum_{m=-2}^{2} u_{6,2m} Y_{2m}$$

where ρ_{00} and α_0 are numerical constants, and Y_{2m} are (unnormalized) spherical harmonics.

After omitting indices $2m$ and combining $u_1 \ldots u_6$ to a vector $\mathbf{u}$ (which is not identical to $\delta \mathbf{u}$), the governing equations are transformed into a matrix equation for spherical harmonic degree two:

$$\frac{d\mathbf{u}}{dr} = \frac{1}{r} \mathbf{M} \cdot \mathbf{u} + \mathbf{b}_1. \qquad (2)$$

Without loss of generality it is assumed that the rotation vector $\boldsymbol{\omega}$ is aligned with the z-axis. Then the vector $\mathbf{b}_1$, which depends on ω^2, only appears for order $m = 0$. Explicit expressions for $\mathbf{b}_1$ and the matrix $\mathbf{M}$ are given by *Steinberger* [1996].

2.3. Equations for Viscoelastic Relaxation in the Time Domain

The equation (2) can be symbolically restored to the time domain by replacing s by $\frac{\partial}{\partial t}$ after algebraic rearrangement. We can thus obtain

$$\frac{\partial}{\partial t} \left(\frac{d\mathbf{u}}{dr} - \frac{1}{r} \mathbf{M}_2 \mathbf{u} \right) =$$
$$\frac{\mu}{\eta} \left(-\mathbf{M}_3 \frac{d\mathbf{u}}{dr} + \frac{1}{r} \mathbf{M}_4 \mathbf{u} + \mathbf{b}_3 \right) + \frac{\partial}{\partial t} \mathbf{b}_1 \qquad (3)$$

$\mathbf{M}_2$ is analogous to $\mathbf{M}$ in eqn. (2) in the purely elastic case. The vector $\mathbf{b}_3$ depends on ω^2; and only appears for order $m = 0$; $\frac{\partial}{\partial t} \mathbf{b}_1$ depends on $\omega \cdot \frac{d\boldsymbol{\omega}}{dt}$, and since we neglect changes in the magnitude of $\boldsymbol{\omega}$, it only appears for orders $m = 1$ and $m = -1$. Explicit expressions for $\mathbf{b}_3$ and $\frac{\partial}{\partial t}\mathbf{b}_1$ and the matrices $\mathbf{M}_2$, $\mathbf{M}_3$ and $\mathbf{M}_4$ are given by *Steinberger* [1996].

We solve equations (3) by splitting $\mathbf{u}$ into an equilibrium value $\mathbf{u}_{eq}$ and a departure from equilibrium $\mathbf{u}_{ne}$. Equations for the hydrostatic equilibrium shape [*Clairaut*, 1743] can be given in the form of a 2×2 matrix equation compatible with eqn. (2):

$$\frac{d}{dr}\begin{pmatrix} u_5 \\ u_6 \end{pmatrix} = \frac{1}{r}\mathbf{M}_{eq}\begin{pmatrix} u_5 \\ u_6 \end{pmatrix} + \mathbf{b}_2 \qquad (4)$$

$$u_3 = \rho_0/\rho_{00} \cdot u_5 + b_{14}$$

$$u_4 = 0,$$

where ρ_0 is the density of the reference Earth model. $\mathbf{b}_2$ and b_{14}, which depend on ω^2, only appear for order $m = 0$, however when computing the time derivative of eqn. (4), $\frac{\partial}{\partial t}\mathbf{b}_2$ and $\frac{\partial}{\partial t}b_{14}$ only appear for orders $m = 1$ and $m = -1$. Explicit expressions for $\mathbf{b}_2$, $\frac{\partial}{\partial t}\mathbf{b}_2$, b_{14}, $\frac{\partial b_{14}}{\partial t}$, and $\mathbf{M}_{eq}$, are given by *Steinberger* [1996]. These equations only depend on density structure, not on compressibility, rigidity or viscosity. In the case of a nonadiabatic density structure, expressions for u_1 and u_2 in terms of u_5, u_6 and ω^2, can be found for the mantle [*Steinberger*, 1996]. For an adiabatic density structure, equilibrium values of u_1 and u_2 are not uniquely defined, but the same expressions may still be used. The expression for u_1 then means the elevation of the equipotential surface from its spherical reference shape.

Hence (3) can be brought into the form

$$\frac{\partial}{\partial t}\left(\frac{d\mathbf{u}}{dr} - \frac{1}{r}\mathbf{M}_2\mathbf{u}\right) = \frac{\partial}{\partial t}\mathbf{b}_1 + \frac{\mu}{\eta}\left(-\mathbf{M}_3\frac{d\mathbf{u}_{ne}}{dr} + \frac{1}{r}\mathbf{M}_4\mathbf{u}_{ne}\right) \qquad (5)$$

(Note that $\mathbf{u}$ still appears on the left-hand-side, whereas $\mathbf{u}_{ne}$ appears on the right-hand-side). The change in shape therefore consists of two parts: An immediate elastic change and a slow viscoelastic change (corresponding to the second and third term on the r.h.s.)

In the mantle and crust we therefore solve eqn. (5) by expressing the change in $\mathbf{u}$ as sums of two terms: $\frac{\partial\mathbf{u}}{\partial t} = \dot{\mathbf{u}}_a + \dot{\mathbf{u}}_b$, with

$$\frac{d\dot{\mathbf{u}}_a}{dr} - \frac{\mathbf{M}_2\dot{\mathbf{u}}_a}{r} = \frac{\partial\mathbf{b}_1}{\partial t} \quad \text{and} \qquad (6)$$

$$\frac{d\dot{\mathbf{u}}_b}{dr} - \frac{\mathbf{M}_2\dot{\mathbf{u}}_b}{r} = \frac{\mu}{\eta}\left(-\mathbf{M}_3\frac{d\mathbf{u}_{ne}}{dr} + \frac{\mathbf{M}_4\mathbf{u}_{ne}}{r}\right)(7)$$

Similarly we solve eqn. (4) in the core by splitting the change in $\mathbf{u}_2$ and u_3: $\frac{\partial\mathbf{u}_2}{\partial t} = \dot{\mathbf{u}}_{2,a} + \dot{\mathbf{u}}_{2,b}$ and $\frac{\partial u_3}{\partial t} = \dot{u}_{3,a} + \dot{u}_{3,b}$, with

$$\frac{d\dot{\mathbf{u}}_{2,a}}{dr} - \frac{1}{r}\mathbf{M}_{eq}\dot{\mathbf{u}}_{2,a} = \frac{\partial\mathbf{b}_2}{\partial t}, \quad \dot{u}_{3,a} = \frac{\rho_0}{\rho_{00}}\dot{u}_{5,a} + \frac{\partial b_{14}}{\partial t}(8)$$

$$\frac{d\dot{\mathbf{u}}_{2,b}}{dr} - \frac{1}{r}\mathbf{M}_{eq}\dot{\mathbf{u}}_{2,b} = 0, \quad \dot{u}_{3,b} = \frac{\rho_0}{\rho_{00}}\dot{u}_{5,b}. \qquad (9)$$

Eqns. (6) and (8) describe the immediate elastic deformation, (7) and (9) the slow viscoelastic deformation.

The viscosity appears in none of the matrices. Therefore, the speed of viscoelastic relaxation is inversely proportional to viscosity: If viscosity is increased by a factor n everywhere, the speed of viscoelastic relaxation decreases by a factor $1/n$ everywhere. With boundary conditions prescribed at the surface and internal interfaces, as well as a regularity condition at the center of the Earth, we are now able to separately solve for equilibrium shape, immediate elastic, and slow viscoelastic deformation. A detailed treatment of these boundary conditions was given by [*Steinberger*, 1996] and is not repeated here.

For the equilibrium shape, we may find two independent solutions which satisfy the regularity condition at the center, and can construct a general solution with one free parameter by linear superposition. The free parameter is determined by matching the gravity boundary condition (corresponding to "all sources of gravity within") at the surface.

For the elastic deformation we can similarly find a general solution with one free parameter in the core. In the mantle, we can find a general solution with four free parameters, which satisfies the boundary conditions $\dot{u}_{3,a} = \rho_0/\rho_{00}\dot{u}_{5,a} + \frac{\partial}{\partial t}b_{14}$ and $\dot{u}_{4,a} = 0$ below the core-mantle boundary. The free parameters (five in total) are determined by matching the three surface boundary conditions (no normal or tangential stress; all sources of gravity within) and enforcing continuity of potential and gravity across the core-mantle boundary. There is no continuity requirement for displacement, since it is undetermined in the core. The theory of elastic deformation has been established long ago, in the context of calculating the free oscillations of the Earth. *Alterman et al.* [1959] use a formalism similar to ours.

For a known right-hand-side vector the solution for $\dot{u}_b$ can be obtained in an analogous way. However in this case, the right-hand side does not only depend on the current change of rotation, but on the whole previous history of rotation. Starting from given initial values of $\mathbf{u}$ in the mantle and $\mathbf{u}_2$ in the core, values at any time can be calculated with a numerical time integration. At each time integration step, $\dot{u}_b$ and $\dot{u}_{2,b}$ are calculated with a radial integration analogous to the elastic case. These time derivatives are then used to calculate changes in $\mathbf{u}$ and $\mathbf{u}_2$. $\mathbf{u}$ and $\mathbf{u}_2$ and their time derivatives are expressed in terms of radial basis functions, and the time integration is performed on the expansion coefficients. For basis functions, we either use eigenfunctions of viscoelastic decay, which are described in the next section, or Chebychev polynomials. If eigenfunctions are used, the equations decouple for each coefficient. Results on how the inertia tensor would change over time due to slow viscoelastic deformation following an instantaneous change in rotation after initial hydrostatic equilibrium will be presented in section 2.5 together with results on changes in the rotation axis obtained by integrating the Euler equations.

2.4. Eigenfunctions of Viscoelastic Decay

If a solution can be expressed in terms of exponentially decaying eigenfunctions $\mathbf{u}(t) = \mathbf{u}_0 e^{-st}$, we can calculate these by solving eqn. (2) without the inhomogeneous term:

$$\frac{d\mathbf{u}}{dr} = \frac{1}{r}\mathbf{M} \cdot \mathbf{u} \qquad (10)$$

Solutions to this equation only exist for certain decay rates. For example, in the case of a viscous, incompressible mantle with layers of constant density we can, at any time, expand $\mathbf{u}_{ne}$ in terms of eigenfunctions of viscoelastic (in this case: viscous) decay. In this case, the displacements of the boundaries from the equilibrium value drive the flow towards equilibrium. In the case of n layers, there are n eigenfunctions, and we can express the boundary displacements in terms of eigenfunctions exactly. For example, in the case of a mantle of constant density underlain by a core of higher constant density, there are two eigenfunctions. These modes have been termed M0 and C0 in the literature [Peltier, 1976]. The corresponding decay times (inverse of decay constant s) depend on viscosity structure. Figure 2 shows the radial displacement u_1 as a function of radius for the eigenmodes of several viscous Earth models. Whereas a decrease in upper mantle viscosity leads to only a slight decrease in decay times, a viscosity increase in a substantial part of the lower mantle leads to a substantial increase in decay times. Figure 3 shows sketches of the displacement field for the two modes M0 and C0.

Each chemical boundary with a density jump within the mantle introduces an additional mode. These modes are characterized by long decay times and called M1, M2, ... [Peltier, 1976]. An example is shown in the bottom right panel of Figure 2, where at depth 670 km a jump in density equal to the density jump in PREM is introduced. One question that we will address later is whether the presence of the M1 mode etc. with its long decay time will significantly affect the maximum speed at which the Earth's rotation axis may change.

In the case of a compressible mantle with Maxwell rheology, further modes and series of modes arise. These modes are apparently first reported by *Han and Wahr*

[1995], where they are termed "compressible modes". They were also found by *Vermeersen et al.* [1996], and we have previously given examples of viscoelastic mode spectra as well [*Steinberger*, 1996]. These are not included here, because we show that despite all the complication, in many cases for an approximately adiabatic Earth model (such as PREM) consideration of only the two fundamental modes M0 and C0 leads to a good approximation in computing true polar wander.

2.5. Integrating the Euler Equations for a Rotating, Self-Gravitating Viscoelastic Body

In the absence of external torques, the equation governing changes of the rotation vector can be written in the form

$$\frac{dJ_{il}}{dt} \cdot \omega_l + J_{il} \cdot \frac{d\omega_l}{dt} = -\varepsilon_{ijk}\omega_j \cdot J_{km}(t) \cdot \omega_m(t). \qquad (11)$$

ω_i are the components of the angular velocity vector, which describes the rotation of the frame of reference in "absolute space" (note that for rotation the concept of "absolute space" makes sense, whereas it makes no sense for translatory motion).

The inertia tensor can be expressed in terms of the five degree two components of $u_5(r_E)$ as defined in equation (1) [*Steinberger*, 1996]:

$$\mathbf{J} = J_{ss}\mathbf{I} + \frac{r_E^2 \alpha_0}{\rho_{00}G} \cdot \begin{pmatrix} T_{11} & T_{12} & T_{13} \\ T_{21} & T_{22} & T_{23} \\ T_{31} & T_{32} & T_{33} \end{pmatrix} \qquad (12)$$

J_{ss} is the moment of inertia for the spherically symmetric reference shape, G is the gravity constant,

$$\begin{aligned}
T_{11} &= -\tfrac{1}{3}u_{20,5}(r_E) + 2u_{22,5}(r_E) \\
T_{12} &= 2u_{2-2,5}(r_E) \\
T_{13} &= u_{21,5}(r_E) \\
T_{21} &= 2u_{2-2,5}(r_E) \\
T_{22} &= -\tfrac{1}{3}u_{20,5}(r_E) - 2u_{22,5}(r_E) \\
T_{23} &= u_{2-1,5}(r_E) \\
T_{31} &= u_{21,5}(r_E) \\
T_{32} &= u_{2-1,5}(r_E) \\
T_{33} &= \tfrac{2}{3}u_{20,5}(r_E)
\end{aligned}$$

and the first index group of u is referring to the spherical harmonic (20, 21, 2-1, 22 or 2-2) and the second index to the component 5.

$\mathbf{J}$ can be split up into a hydrostatic equilibrium part and a non-hydrostatic part. We also split the *total* non-hydrostatic part up into (1) a contribution *imposed* by mantle convection – this would be the only part if the Earth wasn't rotating, and (2) a contribution caused by the change of the rotation axis and the immedi-

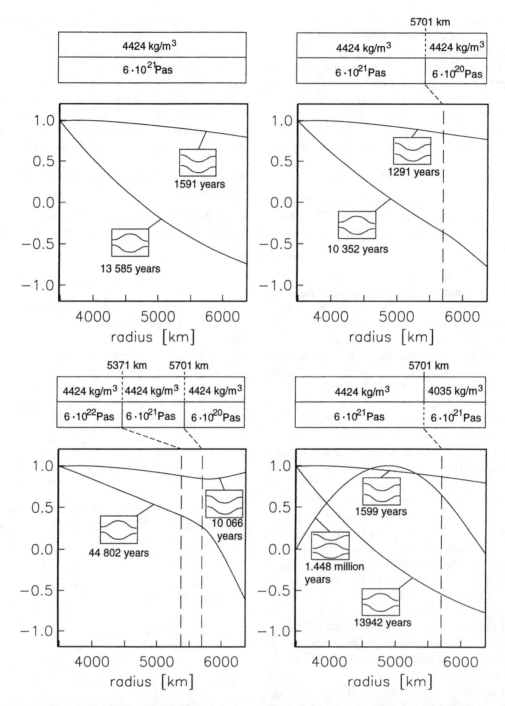

Figure 2. Radial displacement of eigenmodes as function of radius for several Earth models. All models consist of a core (radius 3480 km) of density 10 989 kg m^{-3} in hydrostatic equilibrium and a viscous incompressible mantle. Mantle density and viscosity structure is indicated for each model. All functions are normalized to a maximum value of 1. Corresponding displacement fields are indicated by the sketches in the small rectangles; corresponding decay times are shown below the rectangles.

ate elastic and delayed viscoelastic adjustment to the new equilibrium shape. In this section we will calculate changes of the rotation axis caused by initially *imposed* non-hydrostatic inertia tensor elements, which may be

due to excess masses such as in Figure 1 or sinking "slablets" such as discussed by *Ricard et al.* [1993]. The axis of maximum hydrostatic moment of inertia is by definition parallel to the rotation axis, and one

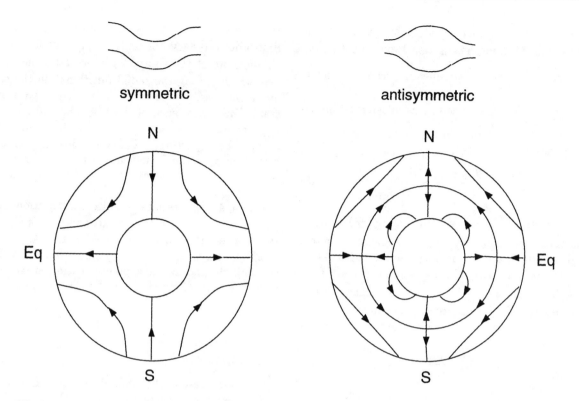

symmetric   antisymmetric

Figure 3. Sketch of displacement field (degree two, order zero) for the viscous modes M0 (left) and C0 (right) (polar cross section with core radius not to scale).

of the principal axes of the *total* non-hydrostatic inertia tensor always follows the rotation axis closely. This latter statement is actually not self-evident. *Spada et al.* [1996a, b] show that for Venus, which is rotating much more slowly than Earth, significant separations of more than 1° may occur, but for the Earth, a separation of only about 0.001″ can be expected to be caused by mass redistributions due to mantle flow, whereas the observed separation related to the Chandler wobble is about 0.3″. Usually, it will be the principal axis with the maximum total non-hydrostatic moment of inertia (axis of figure), which is approximately aligned with the rotation axis, but not always, as was pointed out by *Ricard et al.* [1993]: Because of delays in the adjustment of the equatorial bulge, it may be one of the other principal axes for some period of time. However this approximate alignment does not imply an alignment of the rotation axis with the axis of maximum *imposed* non-hydrostatic moment of inertia; due to the same reason of delayed adjustment of the equatorial bulge, a misalignment may result.

We will now present several methods of integrating the Euler equations:

- Direct numerical integration, using no further approximations

- "Quasi-static" numerical integration, assuming the rotation axis is always exactly parallel to the axis of figure.

- Analytical "steady-state" solution, additionally assuming that the "build-up" of non-equilibrium shape is always exactly compensated by viscoelastic decay.

For direct integration the adopted time steps have to be quite small, since the rotation vector changes with the period of the Chandler wobble ($\approx$ 1 year). We will therefore compare direct and "quasi-static" integration for a simple viscous Earth model and show that there is very good agreement. This justifies using the quasi-static integration. With this method, which is $O(10^3)$ times faster, integration also becomes feasible for more complicated viscoelastic Earth models, which require many more variables to be time-integrated. We will compare several viscous and viscoelastic cases. As it turns out, the speed with which the rotation axis may change mostly depends on the viscosity structure. We will also use an example to show that except for an initial period of adjustment the "steady-state" assumption is valid. In the last part we will therefore use the analytical method to compare several viscous Earth models.

2.5.1. Direct integration. The change of the inertia tensor in eqn. (11) consists of a part due to elastic deformation $\frac{d\mathbf{J}}{dt}\Big|_{el}$ and a part due to viscoelastic deformation $\frac{d\mathbf{J}}{dt}\Big|_{ve}$. The part due to elastic deformation depends only on the change in the rotation vector. We can therefore write

$$\frac{dJ_{il}}{dt} = \frac{\partial J_{il}}{\partial \omega_n}\frac{d\omega_n}{dt} + \frac{dJ_{il}}{dt}\Big|_{ve} \qquad (13)$$

$\frac{\partial J_{il}}{\partial \omega_n}$ is a third-order tensor. If we call $u_5(r_E) = u_{el}$ in the case corresponding to a change from no rotation to a rotation around the z-axis at the present rate (to obtain a solution for this case, in eqns. (6) and (8) $\dot{\mathbf{u}}_a$, $\dot{\mathbf{u}}_{2,a}$, $\dot{u}_{3,a}$, $\frac{\partial}{\partial t}\mathbf{b}_1$, $\frac{\partial}{\partial t}\mathbf{b}_2$ and $\frac{\partial}{\partial t}b_{14}$ have been replaced by $\mathbf{u}$, $\mathbf{u}_2$, u_3, $\mathbf{b}_1$, $\mathbf{b}_2$ and b_{14} respectively) then in a coordinate system with the z-axis aligned with the rotation axis, it is

$$\frac{\partial J_{il}}{\partial \omega_n} = u_{el} \cdot \frac{r_E^2 \alpha_0}{\rho_{00} G} \cdot K_{iln}, \qquad (14)$$

with

$$K_{iln} = \begin{cases} 1 & \text{for } iln = 131, 311, 232, 322 \\ -2/3 & \text{for } iln = 113, 223 \\ 4/3 & \text{for } iln = 333 \\ 0 & \text{otherwise} \end{cases}$$

However we will solve the Euler equations in a coordinate system that is fixed to the body and therefore have to transform them accordingly.

If we insert (13) into (11) we obtain

$$\left(\frac{\partial J_{il}}{\partial \omega_n}\omega_l + J_{in}\right)\frac{d\omega_n}{dt} = -\varepsilon_{ijk}\omega_j \cdot J_{km} \cdot \omega_m - \frac{dJ_{il}}{dt}\Big|_{ve}\omega_l \qquad (15)$$

If we now define M_{pi}^{-1} such that $M_{pi}^{-1}\cdot\left(\frac{\partial J_{il}}{\partial \omega_n}\omega_l + J_{in}\right) = \delta_{pn}$ (i.e. M_{pi} can be calculated by matrix inversion) we have

$$\frac{d\omega_n}{dt} = M_{ni}^{-1} \cdot \left(-\varepsilon_{ijk}\omega_j \cdot J_{km} \cdot \omega_m - \frac{dJ_{il}}{dt}\Big|_{ve}\omega_l\right) \qquad (16)$$

Eqn. (16) enables us to find the change of the rotation vector, given the rotation vector itself, the inertia tensor and the viscoelastic part of its time change. But eqns. (5) and (12) allow us to calculate the inertia tensor and the viscoelastic part of its time change, given the rotation vector and the present values of the vector $\mathbf{u}$ for all five degree-two spherical harmonics. With

this, we can state the Euler equations as a system of first-order ordinary differential equations for the three components of the rotation vector and some parameters which specify the radial functions. In the case of an incompressible viscous medium with n layers of constant density, for each of the five spherical harmonics, n parameters are necessary for specification. For example, the departure of boundary displacements from equilibrium, or the coefficients of the expansion of $\mathbf{u}_{ne}$ in terms of the n eigenfunctions of viscoelastic decay may be chosen.

Figure 4 shows results of a direct integration. As can be seen, even if the rotation axis and axis of figure initially agree, the calculated curve of the pole is not a straight line. It is rather a cycloid curve (a curve describing the motion of a point on the perimeter of a rolling wheel). Comparison of the elastic and viscoelastic cases shows

- an increase of the Chandler wobble period from about 290 days to about 450 days (a well-known effect)

- an increase of the Chandler wobble amplitude by about the same factor

- a slight decrease of the secular drift rate of the pole.

The first two effects are mainly due to immediate elastic deformation, apart from the different density structure. The third effect is only initially present and is reversed for longer time integration, as we will see. Direct comparison for the same density structure is not possible, since an incompressible mantle of constant density corresponds to an adiabatic viscoelastic mantle. The PREM lower mantle, which is used here in the viscoelastic case is approximately adiabatic.

2.5.2. Quasi-static integration. Following *Ricard et al.* [1993] we now assume that the rotation axis is aligned with one of the principal axes of the inertia tensor, which shall be the z-axis (therefore $\omega_1 = \omega_2 = 0$; $\omega_3 =: \omega_0$). Eqn. (16) can then be simplified to

$$\frac{d\omega}{dt} = \frac{\frac{\omega_0}{J_3^p - J_1^p}}{1 - \frac{\omega_0}{J_3^p - J_1^p}\cdot\frac{u_{el}\cdot r_E^2\alpha_0}{\rho_{00}G}}\begin{pmatrix} \frac{dJ_{13}}{dt}\Big|_{ve} \\ \frac{dJ_{23}}{dt}\Big|_{ve} \\ \frac{4}{3}\frac{dJ_{33}}{dt}\Big|_{ve} \end{pmatrix} \qquad (17)$$

[*Steinberger*, 1996]. J_3^p and J_1^p are the principal moments of inertia. The third component is approximately zero, since mostly the direction and not the magnitude of ω changes. Eqn. (17) may be trans-

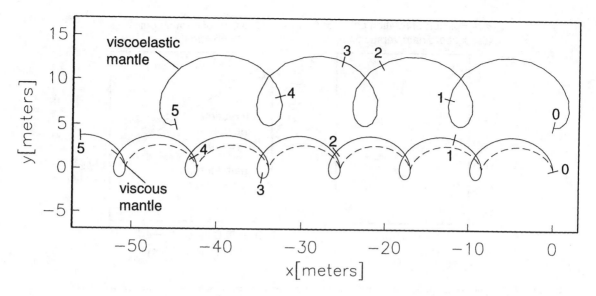

Figure 4. Direct integration of Euler equations over a period of 5 years, showing the Chandler wobble and a secular drift of the pole, due to non-hydrostatic inertia tensor elements $J_{13} = J_{31} = 10^{33}$ kg m^2. Results for a viscous incompressible mantle ($\rho_0 = 4424$ kg m^{-3}) and a core in hydrostatic equilibrium ($\rho_0 = 10\,989$ kg m^{-3}) are contrasted with a result for a viscoelastic mantle (parameters for ρ_0, μ and k as in PREM layer 4 = bulk lower mantle) and a core in hydrostatic equilibrium (ρ_0 as in PREM). $\eta = 6 \cdot 10^{21}$ Pa s in the mantle, $r_{CMB} = 3486$ km, $r_E = 6371$ km in both cases. For better visibility, curves are drawn offset by 5 meters in y direction. Numbers on curves indicate time [in years]. Solid lines: 2 meters initial distance (on Earth surface) between rotation axis and axis of figure; dashed line: zero initial distance.

formed into any other coordinate system. It shows that the immediate elastic deformation has the effect of simply magnifying the rate of change of ω by a factor $\left(1 - \dfrac{\omega_0}{J_3^p - J_1^p} \cdot \dfrac{u_{el} \cdot r_E^2 \alpha_0}{\rho_{00} G}\right)^{-1}$, which is a constant number somewhat bigger than 1 for each Earth model and related to the Love numbers. For example, for the PREM Earth model, we calculate a value of 1.485376. Also, it shows that the speed at which the rotation axis changes is directly proportional to the speed of viscoelastic relaxation, which is itself inversely proportional to viscosity: If the viscosity is everywhere increased by a factor n, the speed at which the rotation axis moves will decrease by a factor $1/n$. Unlike the case of direct integration, the rotation vector is now uniquely determined by the shape, therefore the components of the rotation vector are not integrated as independent variables, but were rather calculated from the inertia tensor at each time step.

Figure 5 compares results from direct and quasi-static integrations. The left panel shows that there are virtually no differences in the long-term behavior. This result, which can be reproduced for any viscous Earth model, serves as justification to use henceforth the quasi-static integration. In the viscoelastic case,

direct integration is extremely slow with the present algorithm, and was thus not performed for any extended periods of time. The right panel shows that both curves disagree by a few meters after the pole moved by about 150 km. The initial Chandler wobble is damped down to a steady-state magnitude of less than 10 cm with a damping time of about 5000 years. This is a much longer time than the actual damping time of the wobble, indicating that the anelastic effects acting on short timescales are more important causes of damping than viscous effects in the mantle.

Some results of quasi-static integration were presented by *Steinberger and O'Connell* [1997], Figure 3. This figure essentially showed that the speed at which the rotation axis moves is mainly determined by lower mantle viscosity. In this figure, the axis labels of the right panels in the second and third row were misprinted. They should correctly be 6×10^5 and 1.2×10^6. In the caption, the mantle viscosity assumed in the first, second and fourth row, was given incorrectly; it is $6 \cdot 10^{21}$ Pas. The fourth row was calculated for a viscous rheology (not viscoelastic, as wrongly written in the caption).

In Figure 6 we compare results of the time domain and the eigenmode approach, for a calculation of vis-

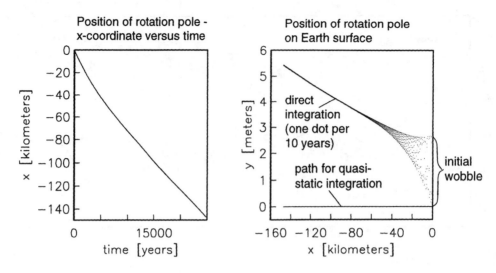

Figure 5. Direct and quasi-static integration of Euler equations over a period of 25 000 years, for a viscous Earth model and excess masses as in Figure 4. In the left panel, the two curves are virtually identical. For the direct integration, the rotation axis is initially coinciding with the axis of figure.

coelastic relaxation towards a new equilibrium shape. For an instantaneous change of the rotation axis of 1 degree, we plot the corresponding non-diagonal inertia tensor element as a function of time. In the eigenmode approach, we fit only radial displacement on chemical boundaries with a density jump (CMB, surface, internal boundaries) using an appropriate number of eigenmodes of viscoelastic decay. For two boundaries (surface and CMB), we use modes M0 and C0, for three boundaries (surface, 670 km, CMB) we use M0, C0, M1. For four boundaries (surface, 400 km, 670 km, CMB) we use M0, C0, M1, M2, neglecting all the other modes. This means, we treat a viscoelastic compressible mantle consisting of adiabatic layers entirely analogous to a viscous mantle, where no other modes occur, with the only difference that we also take the immediate elastic deformation into account. The left panel is for constant elastic parameters throughout the mantle, such that the viscoelastic decay spectrum only consists of discrete lines, whereas in the right panel the elastic parameters vary continuously. In both cases, the curves for time domain and eigenmode approach are in very good agreement. Also, the left and right panel are in most places similar, indicating again that the exact shape of the functions $\mu(r)$ and $k(r)$ has only very minor influence on the viscoelastic decay. In the top panels, the relaxation during the first 10 000 years is shown magnified and on a linear scale, demonstrating that on the left side the two curves are almost indistinguishable (the same would be the case for the dashed or dotted lines), whereas on the right side there is a difference of a few per cent (which would be similar for the dotted lines).

2.5.3. Steady-state solution. As we have seen in the previous section, it is mainly the viscosity structure which determines the speed at which the rotation axis may change. Also, starting from an initial state, after some time a steady state is approached in which the buildup of non-hydrostatic shape is compensated by viscoelastic decay. This is illustrated in Figure 7, which shows the rate of change of the rotation vector as a function of time, normalized by a factor which takes into account that the non-hydrostatic mass anomalies, which are originally at 45°, move to a different latitude in the process. This normalizing factor is approximately 1 for small values ω_2/ω_3. A horizontal line corresponds to steady-state. The figure shows that both for a viscous incompressible mantle with constant density and an adiabatic compressible mantle steady state is approached after approximately the decay time of mode C0 (in this case, a few times 10^4 years). However if there is a chemical boundary with a density jump within the mantle the steady state approximation is less valid on this time scale - the line maintains a finite slope. This is due to the mode M1 with its very long decay time. Also, the figure shows that the steady state is approached faster for a viscous rheology, due to the shorter relaxation times of eigenmodes. The transition from an initial state during which the pole moves faster for a viscous rheology (depicted in Figure 4) to a steady state in which the pole moves faster for the viscoelastic rheology is shown at very short times on the figure.

We outline here an analytical steady-state solution for the speed at which the rotation axis changes for a viscous rheology with layers of constant density. This

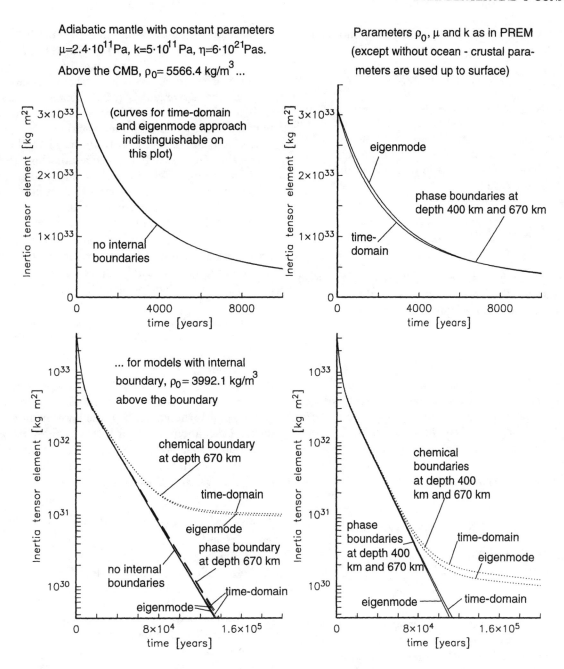

Figure 6. Change of the nondiagonal inertia tensor element due to deformation following an imposed instantaneous change of the rotation axis by 1 degree — comparison of time domain and eigenmode approach for various cases of a viscoelastic mantle and a core (with PREM density structure) in hydrostatic equilibrium. In the top panels, only the continuous lines (for the same models as below) are plotted for better visibility.

solution can also be applied to the eigenmode approximation for a viscoelastic body with adiabatic layers, which was described above, by simply replacing equilibrium radial displacement with equilibrium radial displacement minus radial elastic deformation.

If the rotation axis is parallel to the z-axis, the equilibrium shape is degree two order zero only, and we

can express the equilibrium radial displacement $u_{eq}(r_j)$ (w.r.t. spherical symmetry) at radii $r_j, j = 1, \ldots n$ in terms of the radial displacement of the n eigenfunctions of viscous decay $u_i(r_j), i = 1 \ldots n$. r_j are the radii of boundaries with density jumps, e. g. r_1 is the core-mantle-boundary, r_n the surface, and the number of eigenfunctions always matches exactly the number of boundaries in this case. We can therefore find equilibrium coefficients $c_{i0}, i = 1 \ldots n$ such that

$$u_{eq}(r_j) = c_{i0} u_i(r_j) \omega_0 \qquad (18)$$

If the rotation axis changes, a degree two order one component in the spherical harmonic expansion of the equilibrium shape is created. Hence an equal and opposite deviation u_{ne} of the actual shape from equilibrium shape is created. The change in non-equilibrium shape due to this process is

$$\left. \frac{du_{ne}(r_j)}{dt} \right|_1 = -u_i(r_j) c_{i0} \left\{ \begin{array}{l} \dfrac{d\omega_x}{dt} \text{ for } Y_{21} \\[2mm] \dfrac{d\omega_y}{dt} \text{ for } Y_{2-1} \end{array} \right\} \qquad (19)$$

[*Steinberger*, 1996]. This is compensated by viscoelastic relaxation: The corresponding contribution to the change of the non-equilibrium shape is

$$\left. \frac{du_{ne}(r_j)}{dt} \right|_2 = s_i c_i u_i(r_j) \omega_0 \qquad (20)$$

where s_i is the decay constant of mode i, and c_i is the expansion coefficient of the degree two order one part of the actual shape minus equilibrium shape in terms of eigenfunctions, and where, against the usual convention, the summation index i appears three times. Without loss of generality we now assume polar motion in the x-direction. Assuming steady-state we set

$$\left. \frac{du_{ne}(r_j)}{dt} \right|_1 + \left. \frac{du_{ne}(r_j)}{dt} \right|_2 = 0$$

in eqns. (19) and (20), and thus obtain

$$c_{i0} \cdot \frac{d\omega_x}{dt} \cdot \frac{1}{\omega_0} = s_i c_i \qquad (21)$$

Furthermore, the rotation axis is always very nearly parallel to the axis of figure. If $J_{13,j}$ is the inertia tensor element corresponding to the j-th eigenfunction and $\mathbf{J}^0$ is the part of the inertia tensor due to initially imposed non-hydrostatic excess masses, this means

$$c_j J_{13,j} + J_{13}^0 = 0$$

Using eqn. (21) we can find an expression for the change in $\boldsymbol{\omega}$:

$$\frac{d\omega}{dt} = -\frac{\omega_0 J_{13}^0}{\sum\limits_j c_{j0} J_{13,j}/s_j}. \qquad (22)$$

For the cases depicted in Figure 7 we obtain the following values:

viscous rheology $\qquad\qquad\qquad$ $1.7386 \cdot 10^{-18}$ s^{-2}

viscoelastic rheology $\quad$ solid line $\quad$ $1.9427 \cdot 10^{-18}$ s^{-2}
$\qquad\qquad$ ” $\qquad\qquad$ dashed line $\quad$ $2.0307 \cdot 10^{-18}$ s^{-2}
$\qquad\qquad$ ” $\qquad\qquad$ dotted line $\quad$ $1.0486 \cdot 10^{-18}$ s^{-2}

There is good agreement in the first three cases, whereas the agreement is rather poor in the last case. This indicates again that in this case a steady state is not approached over the time scale shown here, because in this case the mode M1 with its very long decay time is present. Once steady state is approached, the speed of polar wander is significantly reduced in the case of a chemical boundary, but before that, results with and without chemical boundary are similar.

The steady-state solution can also be used for a general viscoelastic rheology without the eigenmode approximation, except that the proportionality constant c_{tpw} relating $\frac{d\omega}{dt}$ and $\omega_0 J_{13}^0$ in an equation equivalent to eqn. (22) has to be determined by a numerical integration.

Some results obtained with the steady-state approach for a viscous rheology were already shown by *Steinberger and O'Connell* [1997]. The main conclusion there was that, except for the unusual case of inertial interchange, the rotation axis always follows the axis of the maximum *imposed* non-hydrostatic moment of inertia very closely, unless the viscosity of the lower mantle is much higher than about 10^{23} Pa s (which most researchers would not consider very likely). The results can also be used to estimate the speed of inertial interchange true polar wander: We had shown that, if mass anomalies are emplaced slowly, the maximum rate of polar wander $\frac{d\omega}{dt}\big|_{max} \propto \sqrt{\dot{J}_{13} c_{tpw}}$, whereas for instantaneous emplacement, we obtain $\frac{d\omega}{dt}\big|_{max} \propto J_{13} c_{tpw}$; it is always the slower one of these two rates which is appropriate. For large-scale mantle flow, the first formula is appropriate: for $\dot{J}_{13} = 10^{33}$kg m^2/20Myr and a mantle viscosity $6 \cdot 10^{21}$ Pas, the fourth panel of Figure 3 of that paper had shown a maximum rate of about 30 degrees in 4 Ma. From Figure 2 of that paper, it can be inferred that c_{tpw} is approximately inversely proportional to lower mantle viscosity. If we therefore use the lower mantle viscosity $4 \cdot 10^{22}$ Pas of the viscosity model employed in the next section, and a growth rate $\dot{J}_{13} = 3 \cdot 10^{32}$kg m^2/50Myr which turns out to be a typical value for the models tested in the next section, we obtain a rough estimate for the maximum speed of polar wander driven by mantle convection of about 1 degree per Myr, much slower than inferred in some recent papers [*Kirschvink et al.*, 1997, *Prevot et al.*, 2000, *Sager and Koppers*, 2000].

Numerical models indicate that plume heads may rise significantly faster than typical mantle flow speeds, [*Larsen and Yuen*, 1997, *van Keken*, 1997]. Using the second formula, which is appropriate in this case, and

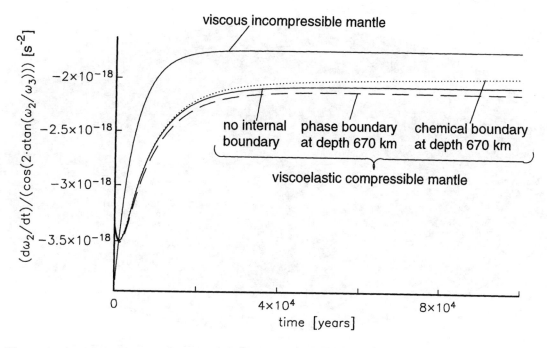

Figure 7. Approach toward steady state for the same models as in the first and second row of *Steinberger and O'Connell* [1997], Figure 3. A horizontal line corresponds to steady state.

reasonable numbers, we find that a fast rising plume-head cannot cause more than a few degrees of true polar wander during the few Ma of its ascent.

3. CHANGE OF AXIS OF MAXIMUM NON-HYDROSTATIC MOMENT OF INERTIA CAUSED BY ADVECTION OF MANTLE DENSITY HETEROGENEITIES

A more realistic distribution of non-hydrostatic mass anomalies than the ones used so far can be inferred from results of seismic tomography. The method of *Hager and O'Connell* [1979, 1981] for calculating flow in a viscous spherical shell, and its extension to calculate the advection of mantle density heterogeneities has been previously explained in detail [*Steinberger and O'Connell*, 1998], and this is not reiterated here. How this method is used to compute true polar wander was outlined by *Steinberger and O'Connell* [1997]. Here we just reiterate a few important points:

- Because the observed geoid is related to the *total* inertia tensor, according to eqn. (12), the degree two order one coefficients of the observed geoid very nearly vanish. This is, however, not the case for the degree two order one coefficients of the geoid calculated from tomographic Earth models, unless we specifically choose scaling factors to convert seismic velocity to density anomalies

such as to satisfy this condition. When using a tomographic model to calculate advection of density heterogeneities and corresponding changes of the degree two geoid, we therefore usually add the changes to the *observed* present degree two geoid, rather than the calculated one, in order to calculate the past degree two geoid.

- Although unrealistic, a free upper boundary has been shown to yield the best geoid predictions [*Thoraval and Richards*, 1997]. On the other hand, imposing plate motions as boundary conditions does not yield a good geoid prediction unless layering of the flow is imposed artificially [*Čadek and Fleitout*, 1999] but we prefer not to do that. Thus, although somewhat inconsistent, we regard it most appropriate to compute changes of the geoid, and hence changes of the rotation axis, by combining geoid kernels that represent a free upper boundary [*Panasyuk et al.*, 1996] with a flow field computed with imposed time-dependent plate motions [*Gordon and Jurdy*, 1986; *Lithgow-Bertelloni et al.*, 1993] as boundary condition.

- We directly compare the paleomagnetic axis with the computed axis of maximum non-hydrostatic moment of inertia, and neglect the difference between *imposed* and *total* non-hydrostatic moment of inertia. This approach is justified as long as

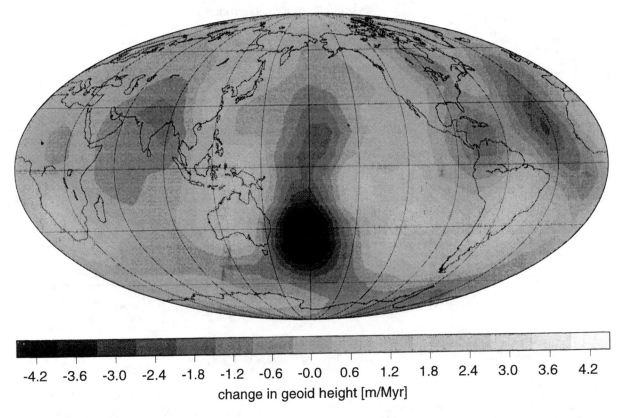

Figure 8. Predicted change of the geoid – degrees 2-12 for the same model as in *Steinberger and O'Connell* [1997], Figure 1. Geoid kernels were computed with a code written by S. V. Panasyuk.

the axis of maximum *imposed* non-hydrostatic moment of inertia moves sufficiently slow (i.e., much slower than the maximum speed of polar wander, estimated in the previous section to be about 1 degree per Myr). The paleomagnetic results indicate that this was the case at least during the Tertiary.

All calculations are done for a compressible mantle with no phase boundaries. All flow fields are expanded up to spherical harmonic degree and order 31. Figure 8 shows the predicted present-day geoid change for the same model of true polar wander shown by *Steinberger and O'Connell* [1997]. It appears that sinking of the geoid centered at the northern tip of New Zealand that is predicted here is the feature most responsible for predicted Cenozoic polar wander towards Greenland: The pole moves such as to maximize the moment of inertia tensor. If the time scale over which mass redistributions occur are sufficiently slow, this means polar motion tends to move regions of sinking geoid to the poles and regions of rising geoid towards the equator. For the

comparatively fast mass redistributions associated with recent glaciations however, the delay of the response cannot be neglected: At present the pole is moving towards Hudson Bay, where a large ice mass melted, and hence the geoid was sinking about 20,000 years ago, but is currently rising due to post-glacial rebound. When caused by mantle flow, a sinking of the geoid can result from either the sinking of dense material or the rising of material of low density in a region where the geoid kernel decreases with depth (here, the mid-mantle). In the case of this model, it is a combination of both effects: downgoing cold material apparently related to subduction of the Tonga slab, and an upwelling in the neighboring Pacific. However not all the tomographic models tested lead to this prominent feature of the predicted geoid change.

Figure 9 compares the computed true polar wander for another model with recent paleomagnetic true polar wander curves. These do not start at the present pole, because even for very recent times, the computed paleomagnetic pole is offset from the present pole. The implied recent fast polar motion is most likely due to

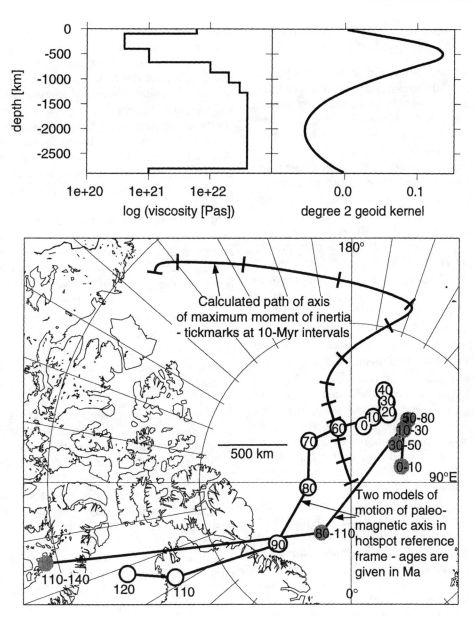

Figure 9. True polar wander – geodynamic calculations and paleomagnetic results. The calculation is done for the viscosity model shown; the degree 2 geoid kernel was computed with a code written by S. V. Panasyuk. Mantle density anomalies below 220 km are inferred from the latest model of *Grand* [2001; pers. comm], (similar to *Grand et al.* [1997]; available via anonymous ftp amazon.geo.utexas.edu), using a conversion factor from seismic velocity to density anomalies $(\delta\rho/\rho)/(\delta v_s/v_s) = 0.3$. Density variations above 220 km depth are disregarded. Paleomagnetic results are from *Besse and Courtillot* [2000; pers. comm.] (black circles) and *Prevot et al.* [2000] (grey dots).

glacial effects, which are not modelled here. The two paleomagnetic results shown here and our modelling agree in two aspects:

- Slow polar motion of only a few degrees during the Tertiary roughly towards Greenwich.

- Faster polar motion prior to that. This faster motion ends at around 60 Ma and changes direction at around 80 Ma in our geodynamic model, and it ends at 50 or 80 Ma according to the paleomagnetic results. The direction of faster polar motion prior to about 80 Ma predicted from the geodynamic model differs by about 30 degrees from the average direction inferred from paleomagnetism.

Such a result was however only obtained for this particular model. Figure 10 shows results for a larger number of tomographic models: For a constant conversion factor below 220 km (solid circles) the models all predict similar direction and magnitude of polar motion. One problem with such predictions is, of course, that we are not able to adequately predict the present-day geoid, unless we modify viscosity structure and/or conversion factors from seismic velocity to density anomalies $(\delta\rho/\rho)/(\delta v_s/v_s)$ separately for each model. Particularly, matching the non-hydrostatic excess flattening of the Earth is a problem, which has been previously discussed by *Forte et al.* [1993, 1995]. A better match can be obtained by choosing a particular viscosity profile, but at the expense of deteriorating the fit to other geoid coefficients. Here we include some results where we modify the conversion factors only within reasonable bounds (Table 1) such that $J_{13} = 0$ and $J_{23} = 0$ and the polar axis is actually the predicted rotation axis. This is only done for those three models, for which the predicted axis of figure is already within 15 degrees of the rotation axis for a constant $(\delta\rho/\rho)/(\delta v_s/v_s) = 0.2$ below 220 km depth. For the light shaded dots, we then still adjust the remaining non-zero inertia tensor elements, for the dark shaded dots no further adjustments are done. Obviously this is a very preliminary approach, but it shows that directions of polar motion stay broadly similar, regardless of the procedure taken. For the dark shaded dots, faster polar wander is predicted, because here the predicted differences between the polar and equatorial non-hydrostatic moments of inertia are much smaller than observed, in agreement with the results of *Forte et al.*, [1993]. These results indicate that the procedure of adding the difference between predicted and observed moments of inertia does not fundamentally undermine our results.

4. DISCUSSION

We have developed a formalism for viscoelastic relaxation combined with the Liouville equations to calculate changes in the rotation axis. Calculations were done for a very simple non-hydrostatic mass distribution, consisting of only two positive and two negative excess masses. This assumed mass distribution (no dynamic compensation, no sinking) is far simpler and less realistic than previously assumed by others [e.g., *Spada et al.*, 1992, *Ricard et al.*, 1993], but it still contains the essential physics of true polar wander and makes results

more easily comprehensible, we believe. A direct integration with no further approximations was compared with a "quasi-static" approximation, where the rotation axis is assumed to be always exactly parallel to the axis of maximum moment of inertia, leading to almost indistinguishable results. An even simpler steady-state solution was derived. This solution may be used over timescales longer than the decay time of the slowest-decaying mode. For an adiabatic mantle, this will be around 10,000 – 100,000 years (depending on viscosity), for an internal chemical boundary, it may be more than a million years.

Since our approach is independent of previous work, benchmark comparisons would be important, but would require a more dedicated effort: Our results obtained so far cannot be directly compared with results of true polar wander due to glacial effects, such as compiled by *Mitrovica and Milne* [1998], because of the different loading history: Whereas these results were obtained for a "saw-tooth" loading history to mimic the effect of repeated glaciations, we use an instantaneously imposed constant mass load, or a linear increase of loading. Also, these models usually contain an elastic lithosphere, which is not included in our models. The most successful predictions of the geoid due to mantle flow use a viscous lithosphere, with viscosity intermediate between the low-viscosity upper mantle and the high-viscosity lower mantle, and a free upper boundary [e.g., *Thoraval and Richards*, 1997]. We therefore regard that, with more realistic models of the lithosphere with lateral variations only beginning to emerge, this is also most appropriate for modelling true polar wander on the timescales of mantle convection. The lithosphere will then not play an important role in determining rates of polar wander and is hence completely neglected here. We will qualitatively compare our results with those compiled by *Mitrovica and Milne* [1998] for a high deep mantle viscosity $\gtrsim 10^{22}$ Pas, because in this case, the elastic lithosphere will play a less important role: E.g. comparison of models a, b, c and d of *Ricard et al.*, [1992], Table 2 shows that in the case of an isoviscous mantle with 10^{21} Pas, the presence of a lithosphere increases decay times of modes M0 and C0 by factors 3.8 and 2.4, whereas for a lower mantle viscosity of $3 \cdot 10^{22}$ Pas, the M0 relaxation time remains the same and the C0 relaxation time increases by a factor of 1.6. Hence in the second case, the presence of the lithosphere will reduce polar wander speed by a much smaller amount than in the first case. A direct comparison is also difficult with published results that account for dynamic compensation and partly use a more

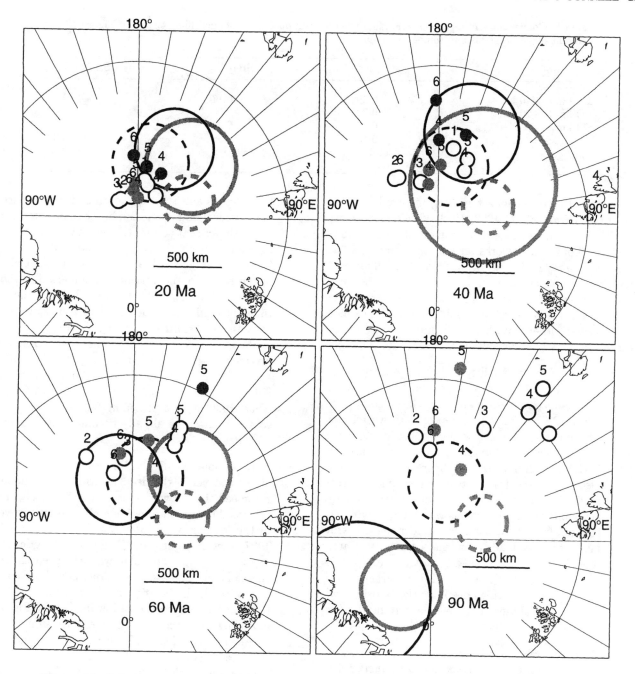

Figure 10. Pole positions at 20, 40, 60 and 90 Ma – geodynamic calculations and paleomagnetic results. Calculations are done for the viscosity model shown in Figure 9 using various tomographic models to infer mantle density. The models are indicated by numbers: 1 = S12WM13 *Su et al.* [1994]; 2 = S20A *Ekström and Dziewonski* [1998], isotropic part; 3 = S20RTS *Ritsema and Van Heijst* [2000], 4 = SAW24B16 *Mégnin and Romanowicz* [2000], 5 = SB4L18 *Masters et al.* [2000], 6 = latest model *Grand* [2001; pers. comm]; Positions indicated by solid circles are calculated for a constant $(\delta\rho/\rho)/(\delta v_s/v_s) = 0.2$ below 220 km depth; Positions indicated by light and dark shaded filled circles are calculated for a conversion factor varying with depth, as listed in Table 1, such that predicted geoid coefficients C_{21} and S_{21} vanish. The difference between actual and calculated present-day degree-two geoid coefficients is added for solid circles and light shaded filled circles; no adjustments are done for dark shaded filled circles. Density variations above 220 km depth are disregarded. Paleomagnetic results are from *Besse and Courtillot* [2000; pers. comm.] (large solid circles) and *Prevot et al.* [2000] (large shaded circles). Continuous circles show the pole position at the respective time, with circle radius equal to the 95% confidence interval A_{95}, dashed circles show the pole position for the most recent time interval.

Table 1. Conversion factors $(\delta\rho/\rho)/(\delta v_s/v_s)$ used for computing the shaded filled circles in Figure 10, for different depth intervals and tomographic models.

Model	Grand	SB4L18	SAW24B16
220 km to 410 km	0.29	0.17	0.10
410 km to 670 km	0.28	0.00	0.00
670 km to d_{lm}	0.25	0.20	0.20
d_{lm} to 2900 km	0.15	0.30	0.32
d_{lm}	2600 km	1800 km	2600 km

complicated time-dependent loading history [*Spada et al.*, 1992, *Ricard et al.*, 1992, 1993, *Vermeersen et al.*, 1996] e.g. to model true polar wander driven by subduction. Some other works on long-term polar [*Sabadini and Yuen*, 1989, *Ricard and Sabadini*, 1990] permit a more quantitative comparison, however because of differences in the density and elastic structure of the models, exact agreement with published results cannot be expected either. With that in mind, we compare various aspects of our results with other publications:

- Effects of compressibility on rates of polar wander: From eqn. (17) we see that any difference between viscous incompressible and viscoelastic compressible Earth models is a composite of two effects: The increase of polar wander speed due to immediate elastic deformation (about 50 % for the PREM Earth model), and differences in the slow viscous or viscoelastic rate of change of the inertia tensor in both cases. The two effects can counteract, and for the models shown in Fig. 7 with no internal boundary, the polar wander speed is still faster by about 20 % for the viscoelastic Earth model in the case of steady state, whereas it is about 10 % slower in Fig. 2 of *Mitrovica and Milne* [1998], for a high deep-mantle viscosity, and can be several tens of % slower according to *Vermeersen et al.* [1996].

- Rates of polar wander and deep mantle viscosity: Our results [*Steinberger and O'Connell*, 1997, Fig. 2, left panel, solid line] can be re-scaled to an upper mantle viscosity 10^{21} Pas (equals 10^{22} P) and then converted to the rotational excitation factor A_1 defined by *Sabadini and Yuen* [1989]. Our model then yields $A_1 \approx 0.15\text{kyr}^{-1}$ for a lower mantle viscosity $\eta_{lm} = 10^{22}$ Pas and $A_1 \approx 0.029\text{kyr}^{-1}$ for $\eta_{lm} = 10^{23}$ Pas, whereas we read from *Ricard and Sabadini*, 1990, Fig. 4 corresponding values $A_1 \approx 0.13\text{kyr}^{-1}$ and $\approx 0.017\text{kyr}^{-1}$. Given their description, we conclude that *Sabadini and Yuen* [1989] wrongly labelled

their curves and that the continuous one is without chemical boundary, and hence to be compared with. Under that assumption we read corresponding values of $A_1 \approx 0.15\text{kyr}^{-1}$ and $\approx 0.019\text{kyr}^{-1}$ from their Fig. 1. We attribute our somewhat faster rates to the absence of an elastic lithosphere, which significantly increases the decay times of modes M0 and C0, and thus slows polar wander [*Ricard et al.*, 1992]. Also the shape of the curves is somewhat similar: A_1 is roughly inversely proportional to η_{lm} for η_{lm} between 10^{21} and 10^{22} Pas, and the curves become somewhat flatter for larger η_{lm}. However, *Sabadini and Yuen* [1989] show a flattening for $\eta_{lm} > 10^{23}$ Pas, whereas our curve already becomes flatter above 10^{22} Pas.

- Rates of polar wander and the thickness of a high-viscosity layer in the deep mantle: In our model, the speed of polar wander is roughly inversely proportional to the thickness of a high-viscosity layer in the lower mantle [*Steinberger and O'Connell*, 1997, Fig. 2, right panel]. For a high viscosity below 1400 km, the polar wander speed should therefore be increased by about one third compared to a high viscosity in the entire mantle below 670 km. *Mitrovica and Milne* [1998] (Fig. 3) also show that their results are quite similar regardless of the depth of the viscosity boundary, but that *Peltier and Jiang* [1996] obtain polar wander speeds that differ by orders of magnitude for the two cases.

- Effect of a chemical boundary on rates of polar wander: Our results indicate that before steady-state is reached, the rates of polar motion with and without chemical boundary are similar, but in steady-state a chemical boundary can significantly reduce speed of polar motion. For the models shown here, it is reduced by about a factor of two, whereas in the corresponding models (with constant mantle viscosity) shown by *Saba-*

dini and Yuen [1989], Fig. 1 and *Ricard and Sabadini* [1990] Fig. 4 it is reduced by a factor $\approx$ 3.5 and $\approx$ 5.5 respectively. This reduction is due to the presence of the mode M1 with its long relaxation time. If we re-scale our model shown in Fig. 3 to a viscosity of 10^{21} Pas and hence divide relaxation times by 6, we find a M1 relaxation time of 241 kyr, almost the same as *Ricard et al.* [1992], for the corresponding model e of their Table 2. This table also shows that the M1 relaxation time gets several times longer if an elastic lithosphere is added to the model, thus the effect of the chemical boundary on rates of polar wander is stronger in the models with elastic lithosphere. The effect of a chemical boundary on rates of polar wander is further discussed by *Mitrovica and Milne* [1998].

In agreement with results obtained by others, our results show that the rotational equations can be accurately integrated in an effective manner, and computational limitations do not pose a problem. While the mechanism of true polar wander can hence be considered well understood, and numerical methods for its computation well developed, the actual true polar wander of the Earth and its relation to the Earth's geodynamic evolution remain poorly known and controversial. A better understanding of actual true polar wander is not merely of academic interest, as it may be linked to other issues, such as the evolution of life on Earth, as e.g. proposed by *Kirschvink et al.* [1997].

Here we have shown geodynamic models of true polar wander for a number of tomographic models. They generally agree on a polar motion of the order of 5 degrees during the past 60 Ma roughly towards Greenwich, which is not in conflict with paleomagnetic results. In order to compute such a slow polar motion in roughly such a direction, a degree two geoid kernel that reverses sign with roughly equal magnitude on either side is required. Such a geoid kernel is also needed to successfully explain the present-day geoid. Calculations for viscosity structures which yield qualitatively different geoid kernels yield polar motions too large to be consistent with the observations. If a larger conversion factor from seismic velocity to density is assumed, an increased magnitude of polar motion results, but the direction stays more or less the same.

For the viscosity structure adopted here, the computed recent polar motion due to mantle convection typically amounts to about 10 % of that observed geodetically, but if the viscosity of the lower mantle is $\leq 10^{22}$ Pa s, it may be up to about 40 % [*Steinberger and O'Connell*, 1997]. Based on the figures shown by *Mitro-*

vica and Milne [1998], and references cited therein, a 10-20 % decrease of polar wander rate due to glacial effects may quite drastically change inferences on deep mantle viscosity. On the other hand, mantle flow contribution to rotational variations $\dot{J}_2$ computed from the same models typically turn out to be well below 1 % of the observed rate, i.e. there is no significant effect.

Prior to about 60 Ma, the computations apparently become less reliable and the differences between our various geodynamic model results increase. Only for the tomographic model of *Grand*, [2001, pers. comm.], a faster polar motion in a direction similar to what is inferred from paleomagnetism was computed. This geodynamic model predicts true polar wander speeds of not more than about 0.5 degrees per Myr. The paleomagnetic results disagree on the speed of polar motion: Whereas *Besse and Courtillot* [2000, pers. comm.] report speeds of no more than about 0.5 degrees per Myr, i.e. well below our estimated "speed limit" of about 1 degree per Myr, *Prevot et al.* [2000] obtain higher speeds exceeding 5 degrees per Myr around 80 Ma. *Sager and Koppers* [2000] use Pacific data and obtain 3–10 degrees per Myr around 84 Ma, however these results have been questioned by *Tarduno and Smirnov* [2001] and *Cottrell and Tarduno* [2000]. Our estimated speed limit is of course not a stringent bound, since we cannot exclude that mantle convection has been more vigorous and mantle viscosity lower than in our model at some time in the past, however this is unlikely during the last 100 Myrs.

A possibly even faster true polar wander event earlier in Earth history has been proposed by *Kirschvink et al.*, [1997], but questioned by *Torsvik et al.*, [1998]. It is presently not possible to set up an actualistic model of an event so far back in time – over timescales longer than $\approx$ 100 Ma models of subduction history as well as the advection of mantle density heterogeneities become unreliable, and a different approach has to be used: *Richards et al.* [1999] therefore combine mantle convection models with the rotational dynamics described by *Ricard et al.* [1993]. While this approach does not allow to model the actual true polar wander further back into the past, it can yield insight into questions such as which properties an Earth model must have such that the frequencies of "inertial interchanges" (i.e. that the maximum and intermediate non-hydrostatic principal moments of inertia become equal) are close to what is observed (i.e. not more than once every few hundred Myr).

We anticipate that better constraints on both mantle density anomalies and mantle viscosity will give more reliable predictions of true polar wander in the near future. Tomographic models are now beginning to agree on their large-scale features, and we have shown here that a number of models generally yield similar predictions of true polar wander. The "observed" true polar wander that dynamic model predictions are compared to is also becoming more reliably constrained. Besides paleomagnetic data, sea level variations have recently been proposed as an observable to detect true polar wander [*Mound and Mitrovica*, 1998], particularly for events of inertial interchange true polar wander [*Mound et al.*, 1999].

Furthermore we expect that it will be possible to explain parts of tomographic anomalies in terms of subduction history in a dynamically consistent model. This should give better constraints on mantle viscosity and also facilitate a comparison between true polar wander predictions from tomography (as done here) and from subduction history [*Richards et al.*, 1997] and hence also to better assess the effect of dynamic upwellings on true polar wander. In this way, input parameters to convection models and their (in a statistical sense) predicted true polar wander [*Richards et al.*, 1999] will also become better constrained.

Acknowledgments. We like to thank Ming Fang and Bert Vermeersen for discussions about viscoelastic relaxation, Svetlana Panasyuk for a routine to compute geoid kernels, all the authors of the models used for supplying their models, especially Steve Grand for his latest yet unpublished tomographic model, and Jean Besse for his latest yet unpublished true polar wander model. Reviews by Giorgio Spada and Jerry X. Mitrovica helped to clarify a number of points, and especially put this work in the context of previous publications. This work was partially supported by NSF grants EAR-9205930 and EAR-9814666.

REFERENCES

Alterman, Z. H., H. Jarosch, and C. L. Pekeris, Oscillations of the earth, *Proc R. Soc. London, Ser. A, 252*, 80-95, 1959.

Andrews, J. A., True polar wander: An analysis of Cenozoic and Mesozoic paleomagnetic poles, *J. Geophys. Res., 90*, 7737-7750, 1985.

Besse, J., and V. Courtillot, Revised and synthetic apparent polar wander paths of the African, Eurasian, North American and Indian plates, and true polar wander since 200 Ma, *J. Geophys. Res., 96*, 4029-4050, 1991.

Boschi, L., J. Tromp and R. J. O'Connell, On Maxwell singularities in postglacial rebound, *Geophys. J. Int., 136*, 492-498, 1999.

Burgers, J., Rotational motion of a sphere subject to viscoelastic deformation, 1, 2, 3, *Konikl. Ned. Akad. Wetenschap. Proc., 58*, 219-237, 1955.

Čadek, O., and L. Fleitout, A global geoid model with imposed plate velocities and partial layering. *J. Geophys. Res., 104*, 29055-29075, 1999.

Clairaut, M., Théorie de la figure de la terre, tirée des principes de l'hydrostatique, Paris, *Ches David fils*, 1743.

Cottrell, R. D., and J. A. Tarduno, Late Cretaceous true polar wander: not so fast, *Science, 288*, www.sciencemag.org/cgi/content/full/288/5475/2283a, 2000.

Darwin, G., On the influence of geological changes on the earth's axis of rotation, *Phil. Trans. Roy. Soc. London, A, 167*, 271-312, 1877.

Dickman, S. R., Secular trends in the Earth's rotation pole: Consideration of the motion of latitude observatories, *Geophys. J. R. Astron. Soc., 51*, 229-244, 1977.

Duncan, R. A., N. Petersen, and R. B. Hargraves, Mantle plumes, movement of the European plate and polar wandering, *Nature, 239*, 82, 1972.

Dziewonski, A. M., and D. L. Anderson: Preliminary Reference Earth Model, *Phys. Earth Planet. Inter., 25*, 297-356, 1981.

Ekström, G., and A. M. Dziewonski, The unique anisotropy of the Pacific upper mantle, *Nature, 394* 168-172, 1998.

Fang, M., and B. H. Hager, The singularity mystery associated with a radially inhomogeneous Maxwell viscoelastic structure, *Geophys. J. Int., 123*, 849-865, 1995.

Forte, A. M., A. M. Dziewonski, and R. L. Woodward, Aspherical structure of the mantle, tectonic plate motions, nonhydrostatic geoid, and topography of the core-mantle boundary, in *Dynamics of the Earth's Deep Interior and Earth Rotation, Geophys. Monogr. Ser., 72*, edited by J.-L. Le Mouël, D. E. Smylie, and T. Herring, pp. 135-166, AGU, Washington, D. C., 1993.

Forte, A. M., J. X. Mitrovica, and R. L. Woodward, Seismic-geodynamic determination of the origin of excess ellipticity of the core-mantle boundary, *Geophys. Res. Lett., 22*, 1013-1016, 1995.

Gold, T., Instability of the earth's axis of rotation, *Nature, 175*, 526-529, 1955.

Goldreich, P., and A. Toomre, Some remarks on polar wandering. *J. Geophys. Res., 74*, 2555-2569, 1969.

Gordon, R. G, and D. M. Jurdy, Cenozoic global plate motions, *J. Geophys. Res., 91*, 12389-12406, 1986.

Gordon, R.G. and R. A. Livermore, Apparent polar wander of the mean lithosphere reference frame, *Geophys. J. R. astr. Soc., 91*, 1049-1057, 1987.

Grand, S. P., R. D. Van der Hilst, and S. Widiyantoro, Global seismic tomography: A snapshot of convection in the Earth, *GSA Today, 7*, 1-7, 1997.

Gurnis, M., J. X. Mitrovica, J. Ritsema, and H.-J. van Heijst, Constraining mantle density structure using geological evidence of surface uplift rates: The case of the African superplume, *Geochem., Geophys., Geosys., 1*, 1999GC000035, 2000.

Hager, B. H., and R. J. O'Connell, Kinematic models of large-scale mantle flow, *J. Geophys. Res., 84*, 1031-1048, 1979.

Hager, B. H., and R. J. O'Connell: A simple global model of plate dynamics and mantle convection, *J. Geophys. Res., 86*, 4843-4867, 1981.

Han, D., and J. Wahr, The viscoelastic relaxation of a realistically stratified Earth, and a further analysis of postglacial rebound, *Geophys. J. Int., 120*, 287-311, 1995.

Hanyk, L., J. Moser, D. A. Yuen, and C. Matyska, Time-domain approach for the transient responses in stratified viscoelastic Earth models, *Geophys. Res. Lett.*, *22*, 1285-1288, 1995.

Hargraves, R. B., and R. A. Duncan, Does the mantle roll? *Nature*, *245*, 361-363, 1973.

Inglis, D., Shifting of the earth's axis of rotation, *Rev. Mod. Phys*, *29*, 9-19, 1957.

Jeffreys, H., *The Earth*, Cambridge University Press, New York, 1952.

Jurdy, D., and R. Van der Voo, A method for the separation of true polar wander and continental drift including results for the last 55 m.y., *J. Geophys. Res.*, *79*, 2945-2952, 1974.

Jurdy, D., True polar wander, *Tectonophysics*, *74*, 1-16, 1981.

Kirschvink, J. L., R. L. Ripperdan, and D. A. Evans, Evidence for a large-scale reorganization of Early Cambrian continental masses by inertial interchange true polar wander, *Science*, *277*, 541-545, 1997.

Lambert, W., F. Schlesinger, and E. Brown, The variation of latitude, *Bull. 78 U.S. Nat. Res. Coun.*, *16*, 245, 1931.

Larsen, T. B., and D. A. Yuen, Fast plumeheads: Temperature-dependent versus non-Newtonian rheology, *Geophys. Res. Lett.*, *24*, 1995-1998, 1997.

Lithgow-Bertelloni, C., M. A. Richards, Y. Ricard, R. J. O'Connell, and D. C. Engebretson, Toroidal-poloidal partitioning of plate motions since 120 ma, *Geophys. Res. Lett.*, *20*, 375-378, 1993.

Livermore, R. A., F. J. Vine, and A. G. Smith, Plate motions and the geomagnetic field, II, Jurassic to Tertiary, *Geophys. J. R. Astron. Soc.*, *79*, 939-961, 1984.

Masters, G., G. Laske, H. Bolton, and A. Dziewonski, The relative behavior of shear velocity, bulk sound speed, and compressional velocity in the mantle: implications for chemical and thermal structure, in *Seismology and Mineral Physics, Geophys. Monogr. Ser.*, *117*, edited by S. Karato, pp. 63-87, AGU, Washington, D. C., 2000.

McElhinny, M. W., Mantle plumes, paleomagnetism and polar wandering, *Nature*, *241*, 523, 1973.

Mégnin, C., and B. Romanowicz, The shear velocity structure of the mantle from the inversion of of body, surface and higher modes waveforms, *Geophys. J. Int*, *143*, 709-728, 2000.

Mitrovica, J. X., and G. A. Milne, Glacial-induced perturbations in the Earth's rotation: A new appraisal, *J. Geophys. Res.*, *103*, 985-1005, 1998.

Mound, J. E., and J. X. Mitrovica, True polar wander as a mechanism for second-order sea-level variations, *Science*, *279*, 534-537, 1998.

Mound, J. E. , J. X. Mitrovica, D. A. D. Evans, and J. L. Kirschvink, A sea-level test for inertial interchange true polar wander events, *Geophys. J. Int*, *136*, F5-F10, 1999.

Munk, W., Polar wandering: a marathon of errors, *Nature*, *177*, 551-554, 1956.

Munk, W., and G. J. F. MacDonald, *The Rotation of the Earth*, Cambridge University Press, New York, 1960.

Panasyuk, S. V., B. H. Hager, and A. M. Forte, Understanding the effects of mantle compressibility on geoid kernels, *Geophys. J. Int.*, *124*, 121-133, 1996.

Peltier, W. R., Glacial-isostatic adjustment – II. The inverse problem, *Geophys. J. R. astr. Soc.*, *46*, 669-705, 1976.

Peltier, W. R., and X. Jiang, Glacial isostatic adjustment and Earth rotation: Refined constraints on the viscosity of the deepest mantle, *J. Geophys. Res.*, *101*, 3269-3290, 1996.

Prevot, M., P. Camps, and M. Daignières, Evidence for a 20° tilting of the Earth's rotation axis 110 million years ago, *Earth Planet. Sci. Lett*, *179*, 517-528, 2000.

Ricard, Y., and R. Sabadini, Rotational instabilities of the Earth induced by mantle density anomalies, *Geophy. Res. Lett.*, *17*, 627-630, 1990.

Ricard, Y., R. Sabadini, and G. Spada, Isostatic deformations and polar wander induced by redistribution of mass within the Earth, *J. Geophys. Res.*, *97*, 14223-14236, 1992.

Ricard, Y., G. Spada, and R. Sabadini, Polar Wandering on a dynamic Earth, *Geophys. J. Int.*, *113*, 284-298, 1993.

Richards, M. A., Y. Ricard, C. Lithgow-Bertelloni, G. Spada, and R. Sabadini, An explanation for Earth's long-term rotational stability *Science*, *275*, 372-375, 1997.

Richards, M. A., H.-P. Bunge, Y. Ricard, and J. R. Baumgardner, Polar Wandering in mantle convection models, *Geophys. Res. Lett.*, *26*, 1777-1780, 1999.

Ritsema, J., and H. J. Van Heijst, Seismic imaging of structural heterogeneity in Earth's mantle: Evidence for large-scale mantle flow, *Science Progress*, *83 (3)* 243-259, 2000.

Sabadini, R. D., D. A. Yuen, and E. Boschi, Polar wander and the forced responses of a rotating, multilayered, viscoelastic planet, *J. Geophys. Res.*, *87*, 2885-2903, 1982.

Sabadini, R., D. A. Yuen, and E. Boschi, A comparison of the complete and truncated versions of the polar wander equations *J. Geophys. Res.*, *89* 7609-7620, 1984.

Sabadini, R., and D. A. Yuen, Mantle stratification and long-term polar wander, *Nature*, *339*, 373-375, 1989.

Sager, W. W., and A. A. P. Koppers, Late Cretaceous polar wander of the Pacific plate: Evidence of a rapid true polar wander event, *Science*, *287*, 455-459, 2000.

Spada, G., Y. Ricard, and R. Sabadini, Excitation of true polar wander by subduction, *Nature*, *360*, 452-454, 1992.

Spada, G., R. Sabadini, and E. Boschi, The spin and inertia of Venus, *Geophys. Res. Lett.*, *23*, 1997-2000, 1996a.

Spada, G., R. Sabadini, and E. Boschi, Long-term rotation and mantle dynamics of the Earth, Mars and Venus, *J. Geophys. Res.*, *101*, 2253-2266, 1996b.

Steinberger, B. M., Motion of hotspots and changes of the Earth's rotation axis caused by a convecting mantle, Ph.D. thesis, 203 pp., Harvard Univ., Cambridge, Mass., Jan. 1996.

Steinberger, B., and R. J. O'Connell, Changes of the Earth's rotation axis owing to advection of mantle density heterogeneities, *Nature*, *387*, 169-173, 1997.

Steinberger, B., and R. J. O'Connell, Advection of plumes in mantle flow: Implications for hotspot motion, mantle viscosity and plume distribution, *Geophys. J. Int.*, *132*, 412-434, 1998.

Steinberger, B., and R. J. O'Connell, Effects of mantle flow on hotspot motion, in *The History and Dynamics of Global Plate Motions, Geophys. Monogr. Ser.*, *121*, edited by M. A. Richards, R. G. Gordon, and R. D. van der Hilst, pp. 377-398, AGU, Washington, D. C., 2000.

Steinberger, B., Plumes in a convecting mantle: Models and observations for individual hotspots, *J. Geophys. Res.*, *105*, 11,127–11,152, 2000.

Tarduno, J. A., and A. V. Smirnov, Stability of the Earth with respect to the spin axis for the last 130 million years, *Earth Planet. Sci. Lett., 184,* 549-553, 2001.

Thoraval, C., and M. A. Richards, The geoid constraint in global geodynamics: viscosity structure, mantle heterogeneity models and boundary conditions, *Geophys. J. Int., 131,* 1-8, 1997.

Torsvik, T. H., J. G. Meert, and M. A. Smethurst, Polar wander and the Cambrian, *Science, 279,* www.sciencemag.org/cgi/content/full/279/5347/9a, 1998.

Su, W., R.L. Woodward, and A. M. Dziewonski, Degree 12 model of shear velocity heterogeneity in the mantle, *J. Geophys. Res., 99,* 6945-6980, 1994.

van Keken, P., Evolution of starting mantle plumes: A comparison between numerical and laboratory models, *Earth Planet. Sci. Lett., 148,* 1-11, 1997.

Vermeersen, L. L. A., R. Sabadini, and G. Spada, Compressible rotational deformation, *Geophys. J. Int., 126,* 735-761, 1996.

Wu, P., and W. R. Peltier, Pleistocene glaciation and the Earth's rotation: A new analysis, *Geophys. J. R. astr. Soc., 76,* 753-791, 1984.

B. Steinberger, Institut für Meteorologie und Geophysik, Johann Wolfgang Goethe-Universität, Feldbergstr. 47 60323 Frankfurt am Main, Germany. (e-mail: steinber@geophysik.uni-frankfurt.de

R. J. O'Connell, Department of Earth and Planetary Sciences, Harvard University 20 Oxford Street, Cambridge, MA 02138.
(e-mail: oconnell@geophysics.harvard.edu)

Determination of Viscoelastic Spectra by Matrix Eigenvalue Analysis

Ladislav Hanyk and Ctirad Matyska

Department of Geophysics, Faculty of Mathematics and Physics, Charles University, Prague, Czech Republic

David A. Yuen

University of Minnesota Supercomputing Institute and Department of Geology and Geophysics, Minneapolis

This study is devoted to the eigenvalue method for computing the normal modes of spherically symmetric self-gravitating viscoelastic Earth models. We employ the approach of the method of lines to the governing partial differential equations, i.e., we discretize the equations in space. This results in a system of time-dependent ordinary differential equations of the form $d\mathbf{Y}/dt = \mathbf{BY}$, which is fundamentally different from the modal approach where the time-dependence is dealt first. Using the finite differences in the grid space, we have conducted the eigenvalue analysis of the matrix $\mathbf{B}$, which yields simultaneously a full spectrum including the classical relaxation modes (M0, L0, C0), the stable dilatation modes and the unstable Rayleigh-Taylor modes. The Maxwell eigenvalues appear as a special class of degenerated modes with zero strength. However, there exists also a class of modes which are the by-products of the spatial discretization of the partial differential equations being investigated. Recognition of these discretization modes is easy due to their sensitivity to the particular kind of discretization employed, as well as the grid point density. The advantage of the eigenvalue procedure is that normal modes of realistic elastically compressible models with radially dependent viscosity profiles, which are characterized by complicated "continuous" relaxation spectra, can be found straightforwardly by a standard matrix eigenanalysis without transforming the problem to the tedious task of finding roots of a secular determinant in the Laplacian plane, which is the classical approach. Moreover, the computational speed of the eigenvalue method is extremely fast (of the order of a few seconds on current GHz processors for 100 radial grid points) and can be used for future work in nonlinear inversion. A series of numerical results starting from simple incompressible layered models and finishing with realistic PREM-based models with complicated viscosity profiles is presented.

Ice Sheets, Sea Level and the Dynamic Earth
Geodynamics Series 29
Copyright 2002 by the American Geophysical Union
10.1029/029GD16

1. INTRODUCTION

Geodynamic processes with short to intermediate timescales ranging from postseismic deformation from large earthquakes, on the period of months, to postglacial rebound, on the order of thousands of years, exhibit substantial deviation from elasticity due to both transient creep and viscoelastic processes [*Karato*, 1998]. For both anelastic and linear viscoelastic rheology, the normal mode approach can be applied in the same manner as for elastic free oscillations of the Earth by the correspondence principle [*Yuen and Peltier*, 1982]. The correspondence principle allows one to use the solution from the elastic problem with the Lamé parameters becoming dependent on the Laplacian variable s, depending on the particular type of rheology [*Christensen*, 1982]. This classical modal theory has remained the mainstay of the geophysical community, since it was introduced by *Peltier* [1974], because of its efficiency for simple models with layers of constant density, shear modulus and viscosity.

In the last several years, increased numerical difficulties have been encountered with more complex models where steep viscosity profiles have been introduced. E.g., *Hanyk et al.* [1995; 1996; 1998] showed numerically that finding even hundreds of modes need not lead to finding satisfactory accurate viscoelastic response and proposed a technique based on direct time-integration of ordinary differential equations. *Fang and Hager* [1995] dealt with viscosity profiles which vary continuously with depth due to the pressure dependence of mantle rheology. In some depth range the Laplace-transformed parameters become singular; to invert a response in the time domain without the effect of this singularity, they proposed a new integration contour in the complex Laplacian domain. *Spada et al.* [1992], *Wu and Ni* [1996] and *Boschi et al.* [1999] showed that the singular factors can be factorized from the analytically expressed secular determinant to prevent the occurrence of the Maxwell singularities. However, this regularization has been achieved only for layered incompressible models. *Han and Wahr* [1995] discussed the possibility of misinterpretation in finding modes by root-finding procedures of the secular determinant in the Laplacian domain due to false-zero crossings, referring to jumps between $\pm\infty$. Moreover, they also pointed to infinitely dense sets of modes in the Laplacian spectra of compressible models. In the case of the simplest layered models, these infinite set of compressible modes can be found analytically; there are not only the dilatation (D) modes, either stable [*Vermeersen et al.*, 1996] or unstable [*Vermeersen and Mitrovica*, 2000], but also the unstable Rayleigh-Taylor (RT) modes [*Hanyk et al.*, 1999]. The unstable RT modes are also present in realistic continuously varying models with a subadiabatic gradient of density [*Plag and Jüttner*, 1995], such as the PREM [*Dziewonski and Anderson*, 1981]. Effects of a non-adiabatic density gradient on the gravitational stability of the viscoelastic responses have been studied by *Nakada* [1999].

Owing to the complicated nature of the normal mode spectrum for realistic Earth models with sharp variations of the elastic and viscous properties, the traditional approach of finding the roots to the secular determinant within the framework of the Laplacian domain becomes too cumbersome for using this in nonlinear inversion. Thus we need new numerical methods for this purpose. *Tromp and Mitrovica* [1999, 2000] presented a new normal-mode formalism, which is based upon eigenfunction expansions. However, relaxation times are still found by means of the clasical root-finding procedure, which can fail if the distance between neighbouring roots is infinitesimal, and is computationally expensive if a large number of modes should be found. *Hanyk et al.* [2000] have recently developed a new theoretical approach, which does not make use of the Laplacian transform. Instead, the problem is now cast as a matrix eigenvalue problem by discretizing the radial spatial variable. Such an eigenvalue approach in grid space allows one to employ the power of standard eigenvalue routines, which is a well-studied problem in numerical analysis. With this eigenvalue approach, we can easily obtain the whole spectrum in one fell swoop.

The aim of this study is to write all steps of this new theory in detail and to demonstrate its applicability to a variety of Earth models. In the next section we will formulate fundamental governing equations in the spatial-time domain as well as their discretization by means of spherical harmonics in the case of spherically symmetric models. The resultant two-dimensional partial differential equations (PDEs), with radius and time being the variables, can then be discretized in the radius, leaving a set of ordinary differential equations (ODEs) in time. Then we show that this system of ODEs can be solved by means of eigenvalue analysis from which we can retrieve both the spectra and the associated eigenvectors in the radial grid. The third section deals with the details concerning the relationships between the order of the finite-difference technique and the structure of the matrices, whose eigenvalues yield the complete set of relaxation times. We display the numerical results in the fourth section. We will first show that our technique provides precisely well-known spectral groups for both incompressible and compressible layered models. We will concentrate on the problem of distinguishing relevant modes from those which are a result of radial discretization of rheology and density profiles. Lastly we show results from more realistic models with elastic parameters taken from the PREM and continuously varying viscosity profiles with variations of several orders of magnitude along the radius.

2. FORMULATION OF THE EIGENVALUE PROBLEM

2.1. PDEs in the 4-D Space-Time Domain

We consider a self-gravitating, compressible, non-rotating continuum initially in hydrostatic equilibrium. It is conventional to decompose total fields, such as the position vector, the stress tensor and the gravitational potential, into initial and incremental parts. The incremental fields are employed for description of infinitesimal, quasi-static, gravitational-viscoelastic perturbations of the initial fields. Physical quantities and PDEs given below conform to the standard form of gravitational viscoelastodynamics [*Peltier*, 1974], also referred to as the material-local form of the linearized field theory [*Wolf*, 1997].

The initial state of the continuum is described in terms of the initial Cauchy stress tensor τ_0, the initial gravitational potential φ_0, the initial density distribution ϱ_0 and the forcing term f_0 by the momentum equation and the Poisson equation, respectively,

$$\nabla \cdot \tau_0 + f_0 = 0, \tag{1}$$
$$\nabla^2 \varphi_0 - 4\pi G \varrho_0 = 0, \tag{2}$$

where G is the Newton gravitational constant. We assume the initial stress to be hydrostatic, $\tau_0 = -p_0 I$, where p_0 denotes the mechanical pressure and I is the unit diagonal tensor, and identify the force f_0 with the gravity force per unit volume, $f_0 = -\rho_0 \nabla \varphi_0$. For the spherically symmetric density, $\varrho_0 = \varrho_0(r)$, eqn. (2) is reduced to

$$g_0' + 2g_0/r - 4\pi G \varrho_0 = 0, \tag{3}$$

where $g_0 e_r = \nabla \varphi_0$ stands for the gravitational acceleration and the prime $'$ denotes differentiation with respect to r. The boundary conditions of the initial fields required at the surface and all internal boundaries are the continuity of the normal initial stress, the gravitational potential and the normal component of its gradient; moreover, the tangential stress vanishes at the surface and at liquid boundaries.

The incremental fields include the displacement $\mathbf{u}$, the incremental Cauchy stress tensor τ, the incremental gravitational potential φ_1 and the incremental density ϱ_1. For the incremental fields the adoption of the concept of Lagrangian or Eulerian formulations of a field

becomes necessary, the former relating the current value of a field at the material point to its initial position, the latter relating the field to the current, local position. Leaving the derivation of the following equations for specialized monographs [e.g., *Wolf*, 1997], we assume that if τ is in Lagrangian description and φ_1 and ϱ_1 are in Eulerian description, then within this rather conventional casting the momentum equation and the Poisson equation for infinitesimal, quasi-static perturbations will take the form

$$\nabla \cdot \tau + f = 0, \tag{4}$$
$$\nabla^2 \varphi_1 - 4\pi G \varrho_1 = 0, \tag{5}$$

where the forcing term f and the Eulerian incremental density ϱ_1 are

$$f = -\varrho_0 \nabla \varphi_1 - \varrho_1 \nabla \varphi_0 - \nabla(\varrho_0 \mathbf{u} \cdot \nabla \varphi_0), \tag{6}$$
$$\varrho_1 = -\nabla \cdot (\varrho_0 \mathbf{u}). \tag{7}$$

Introducing the dynamic viscosity η and the elastic Lamé parameters λ and μ, related to the bulk modulus K by $K = \lambda + \frac{2}{3}\mu$, the constitutive relation of the Maxwell rheology reads

$$\dot{\tau} = \dot{\tau}^E - \xi(\tau - K\nabla \cdot \mathbf{u} I), \tag{8}$$
$$\tau^E = \lambda \nabla \cdot \mathbf{u} I + \mu\left[\nabla \mathbf{u} + (\nabla \mathbf{u})^T\right], \tag{9}$$
$$\xi = \mu/\eta, \tag{10}$$

where the dots above letters denote differentiation with respect to time and the superscript T indicates the transpose operation. τ^E is the auxiliary stress quantity formally identical to the elastic stress tensor; we introduce this symbol because of availability of spherical expansions of expressions with τ^E which become necessary later. We note that other types of anelastic rheology, such as the standard linear solid or Burgers' body rheology [e.g., *Yuen and Peltier*, 1982] can be implemented in the same way as the Maxwell rheology, with suitable time-differentiation of the stress and strain tensors.

The internal boundary conditions for the incremental fields require continuity of the displacement, the incremental stress, the incremental gravitational potential and its gradient, $[\mathbf{u}]_-^+ = 0$, $[\tau]_-^+ = 0$, $[\varphi_1]_-^+ = 0$ and $[\nabla\varphi_1]_-^+ = 0$, respectively. At liquid boundaries only zero tangential stress is required, $n \cdot \tau = (n \cdot \tau \cdot n)n$, where n is the unit vector normal to the boundaries; both radial and tangential displacements can be discontinuous. If the surface is loaded with the surface density γ_L, the surface boundary conditions for the incremental stress and the gradient of the incremental gravitational potential require the equilibrium with the load, $[n \cdot \tau]_-^+ = -g_0 \gamma_L n$ and $[n \cdot (\nabla \varphi_1 + 4\pi G \rho_0 \mathbf{u})]_-^+ = -4\pi G \gamma_L$.

The momentum equation (4) differentiated with respect to time, the Poisson equation (5) and the constitutive relation (8) can be combined into the system

$$\nabla \cdot \dot{\tau}^E + \dot{f} = \nabla \cdot [\xi (\tau - K\nabla \cdot \mathbf{u}I)], \quad (11)$$

$$\nabla \cdot (\nabla \varphi_1 + 4\pi G \varrho_0 \mathbf{u}) = 0. \quad (12)$$

This is where we drop the Laplacian approach based on the correspondence principle and start developing a new approach.

2.2. Spherical Harmonic Decomposition: PDEs in the 2-D Space-Time Domain

Now we proceed with converting PDEs (11)–(12), governing the response of the viscoelastic Earth mantle to a surface load, into spherically decomposed PDEs with respect to time and radius. We also summarize the relevant boundary and initial conditions. Hereafter we consider the spatial distribution of the parameters and the field variables as follows:

$$\varrho_0 = \varrho_0(r), \ \lambda = \lambda(r), \ \mu = \mu(r), \ K = K(r),$$
$$g_0 = g_0(r), \quad \eta = \eta(r), \quad \xi = \xi(r), \quad (13)$$
$$\mathbf{u} = \mathbf{u}(r, \vartheta, \varphi), \ \varphi_1 = \varphi_1(r, \vartheta, \varphi), \ \tau = \tau(r, \vartheta, \varphi).$$

Let e_r, e_ϑ and e_φ be the unit basis vectors of the spherical coordinates r, ϑ and φ, denoting radius, colatitude and longitude, respectively. We now introduce the definitions for the Legendre polynomials $P_n(x)$, the associated Legendre functions $P_n^m(x)$ and the scalar spherical harmonics $Y_{nm}(\vartheta, \varphi)$,

$$P_n(x) = \frac{1}{2^n n!} \frac{d^n}{dx^n} (x^2 - 1)^n,$$
$$P_n^m(x) = (1 - x^2)^{\frac{m}{2}} \frac{d^m}{dx^m} P^n(x), \quad (14)$$
$$Y_{nm}(\vartheta, \varphi) = (-1)^m \sqrt{\frac{2n+1}{4\pi} \frac{(n-m)!}{(n+m)!}} P_n^m(\cos\vartheta) e^{im\varphi}$$

We define orthogonal vector spherical harmonics $S_{nm}^{(-1)}$, $S_{nm}^{(1)} \equiv \nabla_\Omega Y_{nm}$ and $S_{nm}^{(0)} \equiv e_r \times \nabla_\Omega Y_{nm}$, where ∇_Ω is the tangential gradient operator, which is consistent with the Helmholtz representation of vector fields by the relations [e.g., *Dahlen and Tromp*, 1998]

$$S_{nm}^{(-1)}(\vartheta, \varphi) = Y_{nm} e_r,$$
$$S_{nm}^{(1)}(\vartheta, \varphi) = \partial_\vartheta Y_{nm} e_\vartheta + (\sin\vartheta)^{-1} \partial_\varphi Y_{nm} e_\varphi, \quad (15)$$
$$S_{nm}^{(0)}(\vartheta, \varphi) = -(\sin\vartheta)^{-1} \partial_\varphi Y_{nm} e_\vartheta + \partial_\vartheta Y_{nm} e_\varphi.$$

The notation and normalization of the vector spherical harmonics has been chosen to conform with *Martinec*

[1999]. In the space of square-integrable vector functions defined on the unit sphere, $S_{nm}^{(-1)}$ and $S_{nm}^{(1)}$ form the spheroidal basis, whereas $S_{nm}^{(0)}$ create the toroidal basis.

We begin with spherical harmonic expansions of the following scalar and vector functions [cf. *Backus*, 1967]:

$$\mathbf{u}(t, r) = \sum_{nm} [U_{nm}(t, r) S_{nm}^{(-1)} \quad (16)$$
$$+ V_{nm}(t, r) S_{nm}^{(1)} + W_{nm}(t, r) S_{nm}^{(0)}],$$

$$\varphi_1(t, r) = \sum_{nm} F_{nm}(t, r) Y_{nm}, \quad (17)$$

$$\nabla \cdot \mathbf{u}(t, r) = \sum_{nm} X_{nm}(t, r) Y_{nm}, \quad (18)$$

$$\mathbf{T}_r(t, r) = \sum_{nm} [T_{rr,nm}(t, r) S_{nm}^{(-1)} \quad (19)$$
$$+ T_{r\vartheta,nm}(t, r) S_{nm}^{(1)} + T_{r\varphi,nm}(t, r) S_{nm}^{(0)}],$$

$$\mathbf{T}_r^E(t, r) = \sum_{nm} [T_{rr,nm}^E(t, r) S_{nm}^{(-1)} \quad (20)$$
$$+ T_{r\vartheta,nm}^E(t, r) S_{nm}^{(1)} + T_{r\varphi,nm}^E(t, r) S_{nm}^{(0)}],$$

where t denotes time and $\mathbf{T}_r \equiv e_r \cdot \tau$ and $\mathbf{T}_r^E \equiv e_r \cdot \tau^E$ are the radial stress vectors connected by the relation

$$\dot{\mathbf{T}}_r = \dot{\mathbf{T}}_r^E - \xi (\mathbf{T}_r - K\nabla \cdot \mathbf{u} e_r), \quad (21)$$

that follows from (8). The coefficients satisfy the relations

$$X_{nm} = U_{nm}' + (2U_{nm} - NV_{nm})/r, \quad (22)$$

$$T_{rr,nm}^E = \lambda X_{nm} + 2\mu U_{nm}' \quad (23)$$
$$= \beta U_{nm}' + \lambda (2U_{nm} - NV_{nm})/r,$$

$$T_{r\vartheta,nm}^E = \mu V_{nm}' + \mu (U_{nm} - V_{nm})/r, \quad (24)$$

$$T_{r\varphi,nm}^E = \mu (W_{nm}' - W_{nm}/r), \quad (25)$$

with $\beta = \lambda + 2\mu$, $\gamma = \mu(3\lambda + 2\mu)/\beta$ and $N = n(n+1)$, and an auxiliary coefficient Q_{nm} is defined by

$$Q_{nm} = F_{nm}' + (n+1)F_{nm}/r + 4\pi G \varrho_0 U_{nm}. \quad (26)$$

The prime ′ denotes differentiation with respect to r. Making use of these coefficients, we can write the other necessary spherical harmonic expansions [cf. *Martinec*, 1999],

$$\nabla \cdot \tau^E = \sum_{nm} \left[\left(T_{rr,nm}^E{}' - \frac{4\gamma}{r^2} U_{nm} + \frac{2N\gamma}{r^2} V_{nm} \right. \right.$$
$$+ \frac{4\mu}{r\beta} T_{rr,nm}^E - \frac{N}{r} T_{r\vartheta,nm}^E \right) S_{nm}^{(-1)} + \left(T_{r\vartheta,nm}^E{}' \right.$$
$$+ \frac{2\gamma}{r^2} U_{nm} - \frac{N\gamma + (N-2)\mu}{r^2} V_{nm} + \frac{\lambda}{r\beta} T_{rr,nm}^E$$
$$+ \frac{3}{r} T_{r\vartheta,nm}^E \right) S_{nm}^{(1)} + \left(T_{r\varphi,nm}^E{}' - \frac{(N-2)\mu}{r^2} W_{nm} \right.$$
$$\left. + \frac{3}{r} T_{r\varphi,nm}^E \right) S_{nm}^{(0)} \right], \quad (27)$$

$$f = \sum_{nm} \left[\left(\frac{4\varrho_0 g_0}{r} U_{nm} - \frac{N\varrho_0 g_0}{r} V_{nm} + \frac{(n+1)\varrho_0}{r} F_{nm} \right. \right.$$
$$\left. - \varrho_0 Q_{nm} \right) S_{nm}^{(-1)} - \left(\frac{\varrho_0 g_0}{r} U_{nm} + \frac{\varrho_0}{r} F_{nm} \right) S_{nm}^{(1)} \right], \quad (28)$$

$$\nabla \cdot (\nabla \varphi_1 + 4\pi G \varrho_0 \mathbf{u}) = \sum_{nm} \left[Q'_{nm} - \frac{n-1}{r} Q_{nm} \right.$$
$$+ \left. 4\pi G \frac{(n+1)\varrho_0}{r} U_{nm} - 4\pi G \frac{N\varrho_0}{r} V_{nm} \right] Y_{nm} . \quad (29)$$

From the coefficients of the spherical harmonic expansions of $\mathbf{u}$, φ_1, $\boldsymbol{\tau}$ and $\boldsymbol{\tau}^E$ we construct the 8-element vectors $\mathbf{y}_{nm}(t,r)$ and $\mathbf{y}^E_{nm}(t,r)$,

$$\mathbf{y}_{nm}(t,r) = (U_{nm}, V_{nm}, T_{rr,nm}, T_{r\vartheta,nm},$$
$$F_{nm}, Q_{nm}, W_{nm}, T_{r\varphi,nm}) , \quad (30)$$
$$\mathbf{y}^E_{nm}(t,r) = (U_{nm}, V_{nm}, T^E_{rr,nm}, T^E_{r\vartheta,nm},$$
$$F_{nm}, Q_{nm}, W_{nm}, T^E_{r\varphi,nm}) . \quad (31)$$

As all the introduced spherical harmonic expansions are decoupled with respect to both degree n and order m, we suppress these subscripts hereafter. Elements of $\mathbf{y}$ and $\mathbf{y}^E$ can be trivially decomposed into the spheroidal (elements 1..6) and the toroidal (elements 7..8) parts; the spheroidal elements are ordered in accord with *Peltier* [1974]. We see that $\mathbf{y}$ differs from $\mathbf{y}^E$ only in the coefficients of the stress vectors $\boldsymbol{T}_r$ and $\boldsymbol{T}^E_r$ related by (21),

$$\dot{y}^E_1 = \dot{y}_1 , \qquad \dot{y}^E_2 = \dot{y}_2 ,$$
$$\dot{y}^E_3 = \dot{y}_3 + \xi(y_3 - KX) , \quad \dot{y}^E_4 = \dot{y}_4 + \xi y_4 ,$$
$$\dot{y}^E_5 = \dot{y}_5 , \qquad \dot{y}^E_6 = \dot{y}_6 , \quad (32)$$
$$\dot{y}^E_7 = \dot{y}_7 , \qquad \dot{y}^E_8 = \dot{y}_8 + \xi y_8 ,$$
$$X = y'_1 + (2y_1 - Ny_2)/r .$$

First, we construct the PDEs for the coefficients (23)–(26). By substituting $\mathbf{y}^E$ into the above relations, we assemble four first-order PDEs in t and r,

$$y^{E\prime}_1 = \sum_k a_{1k} y^E_k , \quad (33)$$
$$a_{1,1..8} = \left(-\frac{2\lambda}{r\beta}, \frac{N\lambda}{r\beta}, \frac{1}{\beta}, 0, 0, 0, 0, 0 \right) ,$$
$$y^{E\prime}_2 = \sum_k a_{2k} y^E_k , \quad (34)$$
$$a_{2,1..8} = \left(-\frac{1}{r}, \frac{1}{r}, 0, \frac{1}{\mu}, 0, 0, 0, 0 \right) ,$$
$$y^{E\prime}_5 = \sum_k a_{5k} y^E_k , \quad (35)$$
$$a_{5,1..8} = \left(-4\pi G \varrho_0, 0, 0, 0, -\frac{n+1}{r}, 1, 0, 0 \right) ,$$
$$y^{E\prime}_7 = \sum_k a_{7k} y^E_k , \quad (36)$$
$$a_{7,1..8} = \left(0, 0, 0, 0, 0, 0, \frac{1}{r}, \frac{1}{\mu} \right) ,$$

with $\sum_k$ representing $\sum_{k=1}^8$. After differentiation in time and with the change of variables from $\mathbf{y}^E$ to $\mathbf{y}$ in accord with (32), we obtain four PDEs for elements of $\mathbf{y}$,

$$\dot{y}'_1 - \sum_k a_{1k}\dot{y}_k = \xi[a_{13}(y_3 - KX) \quad (37)$$
$$+ a_{14}y_4 + a_{18}y_8] = \xi a_{13}(y_3 - KX) ,$$

$$\dot{y}'_2 - \sum_k a_{2k}\dot{y}_k = \quad (38)$$
$$= \xi[a_{23}(y_3 - KX) + a_{24}y_4 + a_{28}y_8] = \xi a_{24}y_4 ,$$
$$\dot{y}'_5 - \sum_k a_{5k}\dot{y}_k = \quad (39)$$
$$= \xi[a_{53}(y_3 - KX) + a_{54}y_4 + a_{58}y_8] = 0 ,$$
$$\dot{y}'_7 - \sum_k a_{7k}\dot{y}_k = \quad (40)$$
$$= \xi[a_{73}(y_3 - KX) + a_{74}y_4 + a_{78}y_8] = \xi a_{78}y_8 ,$$

where the zero terms have been discarded in the rightmost expressions. Second, we deal with the momentum equation (11). We start with rewriting (27)–(28) in terms of $\mathbf{y}^E$,

$$\nabla \cdot \boldsymbol{\tau}^E = \sum_{nm} \left[\left(y^{E\prime}_3 - \sum_k b_{3k} y^E_k \right) \boldsymbol{S}^{(-1)}_{nm} \right. \quad (41)$$
$$+ \left(y^{E\prime}_4 - \sum_k b_{4k} y^E_k \right) \boldsymbol{S}^{(1)}_{nm} + \left. \left(y^{E\prime}_8 - \sum_k b_{8k} y^E_k \right) \boldsymbol{S}^{(0)}_{nm} \right] ,$$
$$\boldsymbol{f} = \sum_{nm} \left[-\sum_k c_{3k} y^E_k \boldsymbol{S}^{(-1)}_{nm} - \sum_k c_{4k} y^E_k \boldsymbol{S}^{(1)}_{nm} \right] , \quad (42)$$

where the auxiliary coefficients b_{ik} and c_{ik} are given by

$$b_{3,1..8} = \left(\frac{4\gamma}{r^2}, -\frac{2N\gamma}{r^2}, -\frac{4\mu}{r\beta}, \frac{N}{r}, 0, 0, 0, 0 \right) \quad (43)$$
$$b_{4,1..8} = \left(-\frac{2\gamma}{r^2}, \frac{N\gamma + (N-2)\mu}{r^2}, -\frac{\lambda}{r\beta}, -\frac{3}{r}, 0, 0, 0, 0 \right) \quad (44)$$
$$b_{8,1..8} = \left(0, 0, 0, 0, 0, 0, \frac{(N-2)\mu}{r^2}, -\frac{3}{r} \right) \quad (45)$$
$$c_{3,1..8} = \left(-\frac{4\varrho_0 g_0}{r}, \frac{N\varrho_0 g_0}{r}, 0, 0, -\frac{(n+1)\varrho_0}{r}, \varrho_0, 0, 0 \right) \quad (46)$$
$$c_{4,1..8} = \left(\frac{\varrho_0 g_0}{r}, 0, 0, 0, \frac{\varrho_0}{r}, 0, 0, 0 \right) . \quad (47)$$

Then we differentiate (41)–(42) with respect to t, substitute $\mathbf{y}$ from (32) and discard the zero terms,

$$\nabla \cdot \dot{\boldsymbol{\tau}}^E = \sum_{nm} [\dot{y}'_3 + (\xi(y_3 - KX))'$$
$$- \sum_k b_{3k}\dot{y}_k - b_{33}\xi(y_3 - KX) - b_{34}\xi y_4] \boldsymbol{S}^{(-1)}_{nm}$$
$$+ \sum_{nm} [\dot{y}'_4 + (\xi y_4)'$$
$$- \sum_k b_{4k}\dot{y}_k - b_{43}\xi(y_3 - KX) - b_{44}\xi y_4] \boldsymbol{S}^{(1)}_{nm}$$
$$+ \sum_{nm} [\dot{y}'_8 + (\xi y_8)' - \sum_k b_{8k}\dot{y}_k - b_{88}\xi y_8] \boldsymbol{S}^{(0)}_{nm} \quad (48)$$
$$\dot{\boldsymbol{f}} = \sum_{nm} [-\sum_k c_{3k}\dot{y}_k \boldsymbol{S}^{(-1)}_{nm} - \sum_k c_{4k}\dot{y}_k \boldsymbol{S}^{(1)}_{nm}] . \quad (49)$$

For the right-hand side of (11) we can write

$$\nabla \cdot [\xi (\boldsymbol{\tau} - K\nabla \cdot \mathbf{u}\boldsymbol{I})] =$$
$$= \xi [\nabla \cdot \boldsymbol{\tau} - \nabla(K\nabla \cdot \mathbf{u})] + \nabla\xi \cdot [\boldsymbol{\tau} - K\nabla \cdot \mathbf{u}\boldsymbol{I}]$$
$$= \xi [-\boldsymbol{f} - \nabla(K\nabla \cdot \mathbf{u})] + \xi' [\boldsymbol{T}_r - K\nabla \cdot \mathbf{u}\mathbf{e}_r]$$
$$= \sum_{nm} [\xi\sum_k c_{3k} y_k - \xi(KX)' + \xi'(y_3 - KX)] \boldsymbol{S}^{(-1)}_{nm}$$
$$+ \sum_{nm} [\xi\sum_k c_{4k} y_k - \xi KX/r + \xi' y_4] \boldsymbol{S}^{(1)}_{nm}$$
$$+ \sum_{nm} [\xi' y_8] \boldsymbol{S}^{(0)}_{nm} . \quad (50)$$

Using (48)–(50), we obtain three first-order PDEs with respect to t and r from three scalar components of (11),

$$\dot{y}_3' - \sum_k a_{3k}\dot{y}_k = \xi[-y_3' + b_{33}(y_3 - KX)$$
$$+ b_{34}y_4 + \sum_k c_{3k}y_k] = \xi[-y_3' \qquad (51)$$
$$+ \sum_k a_{3k}y_k - b_{31}y_1 - b_{32}y_2 - b_{33}KX],$$

$$\dot{y}_4' - \sum_k a_{4k}\dot{y}_k = \xi[-y_4' + b_{43}(y_3 - KX)$$
$$+ b_{44}y_4 + \sum_k c_{4k}y_k - KX/r] = \xi[-y_4' \qquad (52)$$
$$+ \sum_k a_{4k}y_k - b_{41}y_1 - b_{42}y_2 - (b_{43}+1/r)KX],$$

$$\dot{y}_8' - \sum_k a_{8k}\dot{y}_k = \xi[-y_8' + b_{88}y_8] = \xi[-y_8'$$
$$+ \sum_k a_{8k}y_k - b_{87}y_7], \qquad (53)$$

where

$$a_{3,1..8} = b_{3,1..8} + c_{3,1..8}, \qquad (54)$$
$$a_{4,1..8} = b_{4,1..8} + c_{4,1..8}, \qquad (55)$$
$$a_{8,1..8} = b_{8,1..8}. \qquad (56)$$

Third, we use the expansion in (29) and differentiate the Poisson equation (12) in time to obtain

$$\dot{y}_6' - \sum_k a_{6k}\dot{y}_k = 0, \qquad (57)$$
$$a_{6,1..8} = \left(-4\pi G\frac{(n+1)\varrho_0}{r}, 4\pi G\frac{N\varrho_0}{r}, 0, 0, 0, \frac{n-1}{r}, 0, 0\right).$$

We note that the analytical differentiation of the Poisson equation in time is not necessary because of its elliptic character. It turns out that it would correspond to index reduction, one of techniques of solving differential-algebraic equations [*Ascher and Petzold*, 1998].

With (37)–(40), (51)–(53), (57) and with the expression (22) for coefficients X, we arrive at the linear system of eight first-order PDEs with respect to t and r for the solution vector $\mathbf{y}$,

$$\boxed{\begin{aligned} \dot{\mathbf{y}}'(t,r) - \boldsymbol{A}_n(r)\,\dot{\mathbf{y}}(t,r) &= \qquad (58) \\ &= \xi(r)\left[\boldsymbol{D}_n(r)\,\mathbf{y}'(t,r) + \boldsymbol{E}_n(r)\,\mathbf{y}(t,r)\right]. \end{aligned}}$$

System (58) is decoupled with respect to degree n, independent of order m, and for each n and m consists of two independent systems, one with 6×6 matrices (e.g., $a_{1..6,1..6}$), connecting the spheroidal coefficients of $\mathbf{u}$, φ_1 and $\boldsymbol{\tau}$, and the other one with 2×2 matrices (e.g.,

$a_{7..8,7..8}$), containing the toroidal coefficients. We can express matrices $\boldsymbol{A}_n(r)$, $\boldsymbol{D}_n(r)$ and $\boldsymbol{E}_n(r)$ explicitly:

$$\boldsymbol{A}_n = \begin{pmatrix} -\dfrac{2\lambda}{r\beta} & \dfrac{N\lambda}{r\beta} & \dfrac{1}{\beta} \\[2mm] -\dfrac{1}{r} & \dfrac{1}{r} & 0 \\[2mm] \dfrac{4\gamma}{r^2} - \dfrac{4\varrho_0 g_0}{r} & -\dfrac{2N\gamma}{r^2} + \dfrac{N\varrho_0 g_0}{r} & -\dfrac{4\mu}{r\beta} \\[2mm] -\dfrac{2\gamma}{r^2} + \dfrac{\varrho_0 g_0}{r} & \dfrac{N\gamma + (N-2)\mu}{r^2} & -\dfrac{\lambda}{r\beta} \\[3mm] -4\pi G\varrho_0 & 0 & 0 \\[2mm] -4\pi G\dfrac{(n+1)\varrho_0}{r} & 4\pi G\dfrac{N\varrho_0}{r} & 0 \\[2mm] 0 & 0 & 0 \\[2mm] 0 & 0 & 0 \end{pmatrix}$$

$$\begin{pmatrix} 0 & 0 & 0 & 0 & 0 \\[2mm] \dfrac{1}{\mu} & 0 & 0 & 0 & 0 \\[2mm] \dfrac{N}{r} & -\dfrac{(n+1)\varrho_0}{r} & \varrho_0 & 0 & 0 \\[2mm] -\dfrac{3}{r} & \dfrac{\varrho_0}{r} & 0 & 0 & 0 \\[2mm] 0 & -\dfrac{n+1}{r} & 1 & 0 & 0 \\[2mm] 0 & 0 & \dfrac{n-1}{r} & 0 & 0 \\[2mm] 0 & 0 & 0 & \dfrac{1}{r} & \dfrac{1}{\mu} \\[2mm] 0 & 0 & 0 & \dfrac{(N-2)\mu}{r^2} & -\dfrac{3}{r} \end{pmatrix} \qquad (59)$$

$$\boldsymbol{D}_n = \begin{pmatrix} -\dfrac{K}{\beta} & 0 & 0 & 0 & 0 & 0 & 0 & 0 \\[2mm] 0 & 0 & 0 & 0 & 0 & 0 & 0 & 0 \\[2mm] \dfrac{4\gamma}{3r} & 0 & -1 & 0 & 0 & 0 & 0 & 0 \\[2mm] -\dfrac{2\gamma}{3r} & 0 & 0 & -1 & 0 & 0 & 0 & 0 \\[2mm] 0 & 0 & 0 & 0 & 0 & 0 & 0 & 0 \\[2mm] 0 & 0 & 0 & 0 & 0 & 0 & 0 & 0 \\[2mm] 0 & 0 & 0 & 0 & 0 & 0 & 0 & 0 \\[2mm] 0 & 0 & 0 & 0 & 0 & 0 & 0 & -1 \end{pmatrix} \qquad (60)$$

$$
\boldsymbol{E}_n =
\begin{pmatrix}
-\dfrac{2K}{r\beta} & \dfrac{NK}{r\beta} & \dfrac{1}{\beta} \\[2mm]
0 & 0 & 0 \\[2mm]
\dfrac{8\gamma}{3r^2} - \dfrac{4\varrho_0 g_0}{r} & -\dfrac{4N\gamma}{3r^2} + \dfrac{N\varrho_0 g_0}{r} & -\dfrac{4\mu}{r\beta} \\[2mm]
-\dfrac{4\gamma}{3r^2} + \dfrac{\varrho_0 g_0}{r} & \dfrac{2N\gamma}{3r^2} & -\dfrac{\lambda}{r\beta} \\[2mm]
0 & 0 & 0 \\[2mm]
0 & 0 & 0 \\[2mm]
0 & 0 & 0 \\[2mm]
0 & 0 & 0
\end{pmatrix}
$$

$$
\begin{pmatrix}
0 & 0 & 0 & 0 & 0 \\[2mm]
\dfrac{1}{\mu} & 0 & 0 & 0 & 0 \\[2mm]
\dfrac{N}{r} & -\dfrac{(n+1)\varrho_0}{r} & \varrho_0 & 0 & 0 \\[2mm]
-\dfrac{3}{r} & \dfrac{\varrho_0}{r} & 0 & 0 & 0 \\[2mm]
0 & 0 & 0 & 0 & 0 \\[2mm]
0 & 0 & 0 & 0 & 0 \\[2mm]
0 & 0 & 0 & 0 & \dfrac{1}{\mu} \\[2mm]
0 & 0 & 0 & 0 & -\dfrac{3}{r}
\end{pmatrix}
\tag{61}
$$

We recall that for compressible Earth models, i.e., for finite values of K and λ, we have introduced the abbreviations $\beta = \lambda + 2\mu$, $\gamma = \mu(3\lambda + 2\mu)/\beta = 3\mu K/\beta$ and $N = n(n+1)$. For the incompressible models, $K \to \infty$, the expressions

$$1/\beta \to 0, \quad \lambda/\beta \to 1, \quad \text{and} \quad \gamma \to 3\mu \tag{62}$$

should be substituted in (59)–(61).

In the limit of the elastic mantle, $\eta(r) \to \infty$ and $\xi(r) \to 0$, PDEs (58) should be consistent with the corresponding equations governing elastic free oscillations in the zero-frequency limit [e.g., *Dahlen and Tromp,* 1998, cf. (8.114)–(8.115) and (8.135)–(8.140)]. It can be found easily that these are satisfied. In the opposite limit of the inviscid mantle, $\eta(r) \to 0$ and $\xi(r) \to \infty$, we obtain the static PDEs

$$\boldsymbol{D}_n \mathbf{y}'(r) + \boldsymbol{E}_n \mathbf{y}(r) = 0, \tag{63}$$

which can be shown to be equivalent with

$$\mathbf{y}'(r) - \boldsymbol{A}_n \mathbf{y}(r) = 0, \tag{64}$$

where $\mu(r) \to 0$ is assumed in $\boldsymbol{A}_n(r)$. PDEs (64), governing the static deformation of the outer core, were discussed recently in, e.g., *Fang* [1998].

Finally, we summarize the valid boundary conditions at the surface, $r = a$, and at the core-mantle boundary, $r = b$, or at $r = 0$, if no liquid core is present in the model. Responses to an arbitrary surface load can be deduced from the response to a surface point mass load with the spherical harmonic representation expressed in the form $\gamma_L(t, \vartheta) = \sum_n \Gamma_n(t) P_{n0}(\cos \vartheta)$, where $\Gamma_n(t) = (2n + 1)/(4\pi a^2) f(t)$ and $f(t)$ describes the time evolution of the load [*Farrell,* 1972]. The surface boundary conditions then can be written as

$$
\begin{aligned}
T_{rr}(t, a) &= -[4\pi/(2n + 1)]^{\frac{1}{2}} \, g_0 \Gamma_n(t), \\
T_{r\vartheta}(t, a) &= 0, \\
Q(t, a) &= -[4\pi/(2n + 1)]^{\frac{1}{2}} \, 4\pi G \Gamma_n(t), \\
T_{r\varphi}(t, a) &= 0,
\end{aligned}
\tag{65}
$$

where the square-rooted terms are the normalization factors due to definition (14). Conditions at the centre of models without the liquid core, $r = b = 0$, require zero values for both the displacement and the incremental gravitational potential,

$$
\begin{aligned}
U(t, 0) &= 0, \\
V(t, 0) &= 0, \\
F(t, 0) &= 0, \\
W(t, 0) &= 0.
\end{aligned}
\tag{66}
$$

For models with the liquid core the boundary conditions at the core-mantle boundary, $r = b > 0$, are according to, e.g., *Wu and Peltier* [1982],

$$
\begin{aligned}
-\varrho_0^- g_0 U(t, b) + T_{rr}(t, b) - \varrho_0^- F(t, b) &= 0, \\
T_{r\vartheta}(t, b) &= 0, \\
-4\pi G \varrho_0^- U(t, b) - (4\pi G \varrho_0^-/g_0 + C) F(t, b) \; + \\
+ \; Q(t, b) &= 0, \\
T_{r\varphi}(t, b) &= 0,
\end{aligned}
\tag{67}
$$

where ϱ_0^- is the density of the core at $r \to b$ and C is a constant defined as $C = Q^-/F^-$. F^- and Q^- are solutions at $r = b$ of ODEs governing the response of the incompressible inviscid core [*Wu and Peltier,* 1982],

$$
\begin{pmatrix} F'(r) \\ Q'(r) \end{pmatrix} =
\begin{pmatrix} \dfrac{4\pi G \varrho_0}{g_0} - \dfrac{n+1}{r} & 1 \\[2mm] \dfrac{8\pi G \varrho_0 (n-1)}{r g_0} & \dfrac{n-1}{r} - \dfrac{4\pi G \varrho_0}{g_0} \end{pmatrix}
\begin{pmatrix} F(r) \\ Q(r) \end{pmatrix}.
\tag{68}
$$

This system is to be solved for the constant C only once by the integration from a point r_0 near the centre with the initial values $F(r_0) = r_0^n$ and $Q(r_0) = 2(n-1)r_0^{n-1}$.

For the load applied at $t = 0$ and kept in effect continuously for $t > 0$, i.e., for the Heaviside time dependence $f(t) = H(t)$, the Maxwell Earth model responds elastically at $t = 0$ and the appropriate initial condition for $\mathbf{y}(t,r)$ thus requires

$$\mathbf{y}(0,r) = \mathbf{y}^E(r). \qquad (69)$$

The elastic solution $\mathbf{y}^E(r)$ can be computed from ODEs

$$\mathbf{y}^{E\,\prime}(r) = \mathbf{A}_n(r)\mathbf{y}^E(r) \qquad (70)$$

again by the integration from the centre. However, explicit evaluation of $\mathbf{y}^E(r)$ is not necessary for the purpose of the present study.

In PDEs (58) and conditions (65)–(69) we arrived at the crucial point of our approach. From these equations we can formulate a purely initial-value problem as well as a corresponding matrix eigenvalue problem. Applying the finite-difference technique, we undertake this in the following paragraphs.

2.3. Discretization in Space: ODEs in Time

It is well known that the viscoelastic responses of compressible Earth models can be characterized by the exponential-like development in time and by the spatial distribution which could be expressed in terms of the spherical Bessel functions. In other words, the behaviour of the solution $\mathbf{y}(t,r)$ of (58) is considerably different in the directions of each independent variable. For such PDEs a method based on discretization in the spatial dimension, referred to as the method of lines (MOL), represents a powerful solution tool [e.g., *Schiesser, 1994*]. Hereafter we consider only the spheroidal part of PDEs (58), so the solution vector $\mathbf{y}(t,r)$ contains 6 spheroidal elements from now on. Let us consider the staggered grids $\{r_i, i = 1, \ldots, J\}$ and $\{x_j, j = 0, \ldots, J\}$,

$$b = x_0 < r_1 < x_1 < r_2 < x_2 < \ldots < r_J < x_J = a, \quad (71)$$

spreading over the Earth mantle, $b \le r \le a$. In order to express PDEs (58) on the grid $\{r_i\}$ by means of $\mathbf{y}(t,r)$ evaluated on the grid $\{x_j\}$, we employ expansions of $\mathbf{y}(t,r)$ and its first derivative evaluated at r_i by means of weighted sums of $\mathbf{y}_j(t) \equiv \mathbf{y}(t,x_j)$,

$$\mathbf{y}(t,r_i) = \sum_{j=0}^{J} \alpha_{ij}^{(0)} \mathbf{y}_j(t), \qquad (72)$$

$$\mathbf{y}'(t,r_i) = \sum_{j=0}^{J} \alpha_{ij}^{(1)} \mathbf{y}_j(t), \qquad (73)$$

where $\alpha_{ij}^{(0)}$ and $\alpha_{ij}^{(1)}$ are the weights given by a choice of r_i and x_js. Using (72)–(73) we obtain $6J$ scalar ODEs in time for $6J+6$ unknown elements of $\mathbf{y}_j$s by expressing PDEs (58) on the grid $\{r_i\}$,

$$\sum_{j=0}^{J} \left[\alpha_{ij}^{(1)} - \mathbf{A}_i \alpha_{ij}^{(0)} \right] \dot{\mathbf{y}}_j(t) = \qquad (74)$$

$$= \sum_{j=0}^{J} \left[\xi_i \left(\mathbf{D}_i \alpha_{ij}^{(1)} + \mathbf{E}_i \alpha_{ij}^{(0)} \right) \right] \mathbf{y}_j(t), \quad i = 1, \ldots, J,$$

where $\mathbf{A}_i = \mathbf{A}_n(r_i)$, $\mathbf{D}_i = \mathbf{D}_n(r_i)$, $\mathbf{E}_i = \mathbf{E}_n(r_i)$ and $\xi_i = \xi(r_i)$. The last 6 necessary equations come from spheroidal boundary conditions (65) at $r = a$,

$$\mathbf{M}_J \mathbf{y}_J(t) = \sqrt{\frac{4\pi}{2n+1}} \begin{pmatrix} -g_0 \Gamma_n \\ 0 \\ -4\pi G \Gamma_n \end{pmatrix}, \quad (75)$$

$$\mathbf{M}_J = \begin{pmatrix} 0 & 0 & 1 & 0 & 0 & 0 \\ 0 & 0 & 0 & 1 & 0 & 0 \\ 0 & 0 & 0 & 0 & 0 & 1 \end{pmatrix}, \quad (76)$$

and the conditions (66) at the centre, $r = b = 0$,

$$\mathbf{M}_0 \mathbf{y}_0(t) = 0, \qquad (77)$$

$$\mathbf{M}_0 = \begin{pmatrix} 1 & 0 & 0 & 0 & 0 & 0 \\ 0 & 1 & 0 & 0 & 0 & 0 \\ 0 & 0 & 0 & 0 & 1 & 0 \end{pmatrix}, \quad (78)$$

or the conditions (67) at the core-mantle boundary, $r = b > 0$,

$$\mathbf{M}_0 \mathbf{y}_0(t) = 0, \qquad (79)$$

$$\mathbf{M}_0 = \begin{pmatrix} -\varrho_0^- g_0 & 0 & 1 & 0 & -\varrho_0^- & 0 \\ 0 & 0 & 0 & 1 & 0 & 0 \\ -4\pi G \varrho_0^- & 0 & 0 & 0 & -4\pi G \frac{\varrho_0^-}{g_0} - C & 1 \end{pmatrix}. \quad (80)$$

For the radially discretized solution vector $\mathbf{Y}(t)$ with $6J + 6$ elements,

$$\mathbf{Y}(t) = \left(\mathbf{y}_0(t), \mathbf{y}_1(t), \ldots, \mathbf{y}_J(t) \right)^T, \qquad (81)$$

ODEs (74) and the boundary conditions (75) and (77) or (79), differentiated in time, form the resulting system of $6J + 6$ ODEs in time,

$$\boxed{\quad \mathbf{P}\dot{\mathbf{Y}}(t) = \mathbf{Q}\,\mathbf{Y}(t), \qquad (82)\quad}$$

where

$$
P = \begin{pmatrix} \boxed{M_0}_{3\times 6} & \boxed{0}_{3\times 6J} \\ \boxed{\alpha_{1j}^{(1)} - A_1\alpha_{1j}^{(0)}}_{6\times(6J+6)} & \\ \vdots & \\ \boxed{\alpha_{Jj}^{(1)} - A_J\alpha_{Jj}^{(0)}}_{6\times(6J+6)} & \\ \boxed{0}_{3\times 6J} & \boxed{M_J}_{3\times 6} \end{pmatrix}, \qquad (83)
$$

$$
Q = \begin{pmatrix} \boxed{0}_{3\times(6J+6)} \\ \boxed{\xi_1\left(D_1\alpha_{1j}^{(1)} + E_1\alpha_{1j}^{(0)}\right)}_{6\times(6J+6)} \\ \vdots \\ \boxed{\xi_J\left(D_J\alpha_{Jj}^{(1)} + E_J\alpha_{Jj}^{(0)}\right)}_{6\times(6J+6)} \\ \boxed{0}_{3\times(6J+6)} \end{pmatrix} \qquad (84)
$$

and $\Gamma_n(t)$ is considered constant in time, i.e., $\dot{\Gamma}_n = 0$. System (82) with the initial condition (69) represents a purely initial-value formulation of the problem of viscoelastic responses of the Earth. For given grids $\{r_i\}$ and $\{x_j\}$, both P and Q are constant matrices. The matrix P is regular for all of the models studied here, so we can write (82) in an equivalent form,

$$
\boxed{\dot{Y}(t) = B\,Y(t), \qquad B = P^{-1}Q,} \qquad (85)
$$

i.e., in the standard form of a linear homogeneous system of ODEs with constant coefficients.

2.4. Modal Decomposition: The Eigenvalue Problem

A solution of ODEs (85), referred to as the fundamental system, can be written [e.g., *Rektorys*, 1994] as a linear combination of the constituents

$$
e^{s_p t} \quad \text{or} \quad R_q(t)e^{s_q t}, \qquad (86)
$$

where s_p is any nondegenerate eigenvalue of B, s_q any Q-degenerate eigenvalue of B and R_q a polynomial of the maximal degree $Q-1$. Thus, eigenanalysis of matrix B can reveal a substantial information about the behavior of the solution of ODEs (85). It is straightforward to see that a generalized eigenvalue problem,

$$
QY = sPY, \qquad (87)
$$

corresponds to ODEs (82), while a standard eigenvalue problem,

$$
BY = sY, \qquad (88)
$$

matches ODEs (85). In the case of regular matrix P both eigenproblems are formally equivalent. However, there are differences in performance and accuraccy of applicable numerical routines, as will be discussed later.

For a given nondegenerate eigenvalue, $s = s_p$, the corresponding eigenvector $\mathbf{Y}_p$ with $6J+6$ elements gathers the discretized $(J+1)$-eigenvectors U_p, V_p, etc. The response of viscoelastic models to the time impulse, i.e., for $f(t) = \delta(t)$, is traditionally expressed by the surface load Love numbers [*Peltier*, 1974],

$$
h(t) = h^E\delta(t) + \sum_p r_p^{(h)}\exp(s_p t), \qquad (89)
$$

$$
l(t) = l^E\delta(t) + \sum_p r_p^{(l)}\exp(s_p t), \qquad (90)
$$

$$
k(t) = k^E\delta(t) + \sum_p r_p^{(k)}\exp(s_p t), \qquad (91)
$$

where h^E, l^E and k^E are the elastic load Love numbers and the sums on the right-hand sides describe the non-elastic response. In order to evaluate the non-elastic part of the Love numbers, formulas (8.1)–(8.3) by *Tromp and Mitrovica* [1999] for the partial modal amplitudes can be applied,

$$
r_p^{(h)} = \frac{\tau}{2s_p}U_p(g_0U_p + F_p), \qquad (92)
$$

$$
r_p^{(l)} = \frac{\tau}{2s_p}V_p(g_0U_p + F_p), \quad \tau = \frac{M}{a}\frac{2n+1}{4\pi}, \qquad (93)
$$

$$
r_p^{(k)} = -\frac{\tau}{2g_0s_p}F_p(g_0U_p + F_p), \qquad (94)
$$

where M is the mass of the Earth and the eigenvectors are normalized in accord with the condition (3.6) by *Tromp and Mitrovica* [1999],

$$
\int_b^a \frac{\mu\xi}{(s_p + \xi)^2}\left[\frac{1}{3}(2U_p' - F_p)^2 + NX_p^2 + \frac{N(n-1)(n+2)}{r^2}V_p^2\right]r^2\,dr = -2s_p. \qquad (95)
$$

In Fig. 1 we summarize both the Laplacian and eigenvalue approaches and compare the differences in the methodologies between both the secular-determinant and the eigenvalue analysis.

LAPLACIAN METHOD

(1.) Laplace transform
$\mathbf{y}(t,r) \Rightarrow \mathbf{y}(r,s)$

(2.) ODEs in space, s – parameter
$d\mathbf{y}(r,s)/dr = \boldsymbol{A}(r,s)\,\mathbf{y}(r,s)$

(3.) Solution of nonlinear secular equation in s
$\det \boldsymbol{M}(s) = 0$
for matrix $\boldsymbol{M}(s)$ built from $\mathbf{y}(a,s)$
at the surface by integration of ODEs

(4.) Evaluation of load Love numbers
from $\mathbf{y}(a,s)$ and $\det \boldsymbol{M}(s)$

EIGENVALUE METHOD

(1.) Spatial discretization (method of lines)
$\mathbf{y}(t,r) \Rightarrow \mathbf{Y}(t) \equiv \{\mathbf{y}(t,r_i), i = 0, \ldots, N\}$

(2.) ODEs in time
$\boldsymbol{P}\,d\mathbf{Y}(t)/dt = \boldsymbol{Q}\,\mathbf{Y}(t)$

(3.) Evaluation of matrices $\boldsymbol{P}, \boldsymbol{Q}$
Solution of linear system $\boldsymbol{B} = \boldsymbol{P}^{-1}\boldsymbol{Q}$
Linear eigenvalue/eigenvector analysis
$(\boldsymbol{B} - s)\,\mathbf{Y} = 0$

(4.) Evaluation of load Love numbers
from eigenvectors

Figure 1. A comparison of the respective sequential steps to be followed in the Laplacian transform method (frames on the left) and the eigenvalue method based on the method of lines (frames on the right) for evaluation of relaxation times and load Love numbers describing the viscoelastic responses to a surface load.

3. NUMERICAL TECHNIQUES

3.1. Pseudospectral vs. Low-order Finite Differences

A choice of a discretization pattern obviously plays a crucial role for the accuracy of the eigenvalue method. In the derivation of ODEs (82) we have made a rather general choice of the staggered grids $\{r_i\}$ and $\{x_j\}$ in (71). Here we will discuss two particular discretization patterns of this form.

The first discretization pattern is based on high-order pseudospectral (PS) schemes for approximation of zero and first derivatives of a function at an arbitrary point by a weighted sum of known values of the function, cf. (72)–(73),

$$\mathbf{y}(t,r_i) = \sum_{j=0}^{J} \alpha_{ij}^{(0)} \mathbf{y}_j(t), \qquad (96)$$

$$\mathbf{y}'(t,r_i) = \sum_{j=0}^{J} \alpha_{ij}^{(1)} \mathbf{y}_j(t), \qquad (97)$$

where $\mathbf{y}_j(t) \equiv \mathbf{y}(t,x_j)$. The weights $\alpha_{ij}^{(0)}$ and $\alpha_{ij}^{(1)}$, in general non-zero for all i and j, can be computed by the algorithm based on a polynomial approximation [*Fornberg*, 1996]. In order to apply the PS schemes, it is necessary to establish a correspondence between the grids $\{r_i\}$ and $\{x_j\}$ and zeroes and extremas of orthogonal polynomials, respectively; we have made the choice of r_is being the roots of Chebyshev polynomials transformed to $\langle b,a \rangle$ and x_js being the interlaced extremas. For models composed of few layers, we have obtained accurate eigenvalues even with very coarse grids. How-

ever, higher grid densities are necessary for realistically stratified models. Then the matrices $\boldsymbol{P}$ and $\boldsymbol{Q}$ become larger and the computational demands of the eigenanalysis of (87), resp. (88) increase quite rapidly.

The other tested discretization pattern is based on low-order finite-difference (FD) schemes. We applied the FD schemes on the equi-spaced staggered grids $\{r_i\}$ and $\{x_j\}$ with $r_i = (x_{i-1} + x_i)/2$ and the constant step-size $h = r_i - r_{i-1} = x_i - x_{i-1}$. For these grids the following second-order (FD2) and fourth-order (FD4) approximations of $\mathbf{y}(t,r_i)$ and $\mathbf{y}'(t,r_i)$ can be found (e.g., by Fornberg's algorithm),

FD2   for $i = 1, \ldots, J$
$$\mathbf{y}(t,r_i) = \tfrac{1}{2}\mathbf{y}_{i-1}(t) + \tfrac{1}{2}\mathbf{y}_{i+1}(t),$$
$$\mathbf{y}'(t,r_i) = [-\mathbf{y}_{i-1}(t) + \mathbf{y}_{i+1}(t)]/h,$$

FD4   for $i = 1$
$$\mathbf{y}(t,r_1) = \tfrac{5}{16}\mathbf{y}_0 + \tfrac{15}{16}\mathbf{y}_1 - \tfrac{5}{16}\mathbf{y}_2 + \tfrac{1}{16}\mathbf{y}_3,$$
$$\mathbf{y}'(t,r_1) = \left[-\tfrac{23}{24}\mathbf{y}_0(t) + \tfrac{7}{8}\mathbf{y}_1(t) + \tfrac{1}{8}\mathbf{y}_2(t) - \tfrac{1}{24}\mathbf{y}_3(t) \right]/h,$$

FD4   for $i = 2, \ldots, J-1$
$$\mathbf{y}(t,r_i) = -\tfrac{1}{16}\mathbf{y}_{i-2} + \tfrac{9}{16}\mathbf{y}_{i-1} + \tfrac{9}{16}\mathbf{y}_{i+1} - \tfrac{1}{16}\mathbf{y}_{i+2},$$
$$\mathbf{y}'(t,r_i) = \left[\tfrac{1}{24}\mathbf{y}_{i-2}(t) - \tfrac{9}{8}\mathbf{y}_{i-1}(t) + \tfrac{9}{8}\mathbf{y}_{i+1}(t) - \tfrac{1}{24}\mathbf{y}_{i+2}(t) \right]/h,$$

FD4   for $i = J$
$$\mathbf{y}(t,r_J) = \tfrac{1}{16}\mathbf{y}_{J-3} - \tfrac{5}{16}\mathbf{y}_{J-2} + \tfrac{15}{16}\mathbf{y}_{J-1} + \tfrac{5}{16}\mathbf{y}_J,$$

$$y'(t, r_J) = \left[\tfrac{1}{24}\mathbf{y}_{J-3}(t) - \tfrac{1}{8}\mathbf{y}_{J-2}(t)\right.$$
$$\left. - \tfrac{7}{8}\mathbf{y}_{J-1}(t) + \tfrac{23}{24}\mathbf{y}_J(t)\right]/h.$$

Although it is easier to implement, the eigenanalysis based on the low-order FD schemes is inferior to the high-order PS schemes in terms of accuracy.

We have also tested the case of overlapping grids $\{r_i\}$ and $\{x_j\}$. The results have been less satisfactory than those obtained on the staggered grids since more discretization modes contaminated the physical part of eigenspectra. Other patterns of discretization can surely be designed, for instance that motivated by the fact that three elements of $\mathbf{y}(t,r)$ correspond to the first derivatives of the other three. In the future we will surely find the most optimal discretization stencil for the purpose of determining the eigenvalues.

3.2. Eigenpackages

In (87) and (88) we have obtained both the generalized and standard eigenvalue problems, respectively. The $(6J+6) \times (6J+6)$ matrices P and Q of the generalized problem are both non-symmetric. While for the PS schemes the matrices are full, they become band diagonal for the FD schemes. There are 8 sub- and 8 superdiagonals for the FD2 scheme and 20 sub- and 20 superdiagonals for the FD4 scheme, as can be deduced from (83)–(84) and (59)–(61). The $(6J+6) \times (6J+6)$ matrix B of the standard problem is non-symmetric and full for all tested schemes. However, a number of CPU operations needed to evaluate B grows with J^3 for PS schemes and only with J for FD schemes. We note that the real-space FD method with up to 12th-order accuracy has been employed to solve eigenvalue problems in solid-state physics for complex systems with thousands of atoms [Stathopoulos et al., 2000].

A synopsis of the available numerical methods, including both direct and iterative techniques for large-scale eigenvalues, can be found in the book by Trefethen and Bau [1997]. Standard numerical libraries (e.g., LAPACK, NAG, IMSL) contain several routines for matrix eigenanalysis. The routines usually perform the following tasks for both standard and generalized problems: (EV) evaluation of all eigenvalues and no eigenvectors, (EVV) evaluation of all eigenvalues and all eigenvectors, (EsV) evaluation of all eigenvalues in a selected interval, (EI) evaluation of an eigenvector to a known eigenvalue. The task (EV) is generally solved by a two-step algorithm based on the reduction to Hessenberg form and the QR algorithm. A similar algorithm is used for the task (EVV). For tasks performing selections, (EsV) and (EI), an algorithm based on inverse iteration is usual-

ly invoked. An extensive package of eigenroutines can be found in both LAPACK and NAG, which contain routines for all tasks mentioned above. In both these libraries there are also routines available for the task (EI) and the generalized problem with band diagonal matrices. IMSL contains routines for the tasks (EV) and (EVV).

We have tested routines for both the standard and generalized eigenproblems (87)–(88). In both cases the total CPU-times were comparable but the accuracy of the solvers of standard eigenproblems was substantially better. The maximal grid density we employed was for $J \approx 300$ with the approximate matrix dimensions of $1,800 \times 1,800$. It is a number of operations which seems to be the main obstacle for higher resolution, as it grows with J^3 (see Table 1), while the memory requirement grows with J^2. Large-scale eigenvalue problems involving pseudospectra are now becoming more common in all fields of science and engineering [e.g., Trefethen, 1999]. For solving the eigenproblems with large matrices (say, up to $10^6 \times 10^6$), other family of methods, based on the Krylov subspace methods, has to be invoked [Saad, 1992]. A package, reffered to as the implicitly-restarted Arnoldi method, is implemented in ARPACK [Lehoucq et al., 1998], a publicly available library for solving large eigenvalue problems.

4. APPLICATIONS

We first demonstrate the numerical ability of the eigenvalue method to catch the eigenspectra of the homogeneous spheres ($a = 6,371$ km, $\varrho_0 = 5,517$ kg m^{-3}, $\mu = 1.4519 \times 10^{11}$ Pa, $\eta = 10^{21}$ Pa s), both incompressible (λ infinite) and compressible ($\lambda = 3.5288 \times 10^{11}$ Pa). The calculations were carried out for selected angular orders up to the degree 120 and using the PS discretization [Fornberg, 1996] on the Chebyshev grids of 30, 60 and 120 layers. The eigenspectrum of the homogeneous incompressible sphere is shown in Fig. 2a. There is on-

Table 1. Computational Times for the Eigenproblem (88) per Spherical Harmonic Degree n

Number of grid layers	Matrix dimension	CPU time [s] (Pentium 1 GHz)
30	186^2	0.05
60	366^2	0.4
120	726^2	3
240	1446^2	25
300	1806^2	50

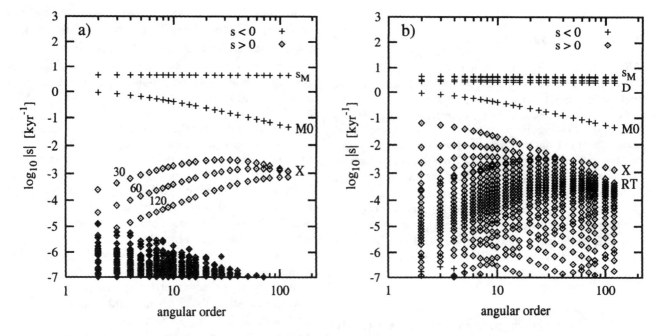

Figure 2. Eigenspectra of the incompressible (panel a) and the compressible (panel b) homogeneous spheres. Both stable and unstable modal branches are plotted (symbols + and ◇ used for negative and positive eigenvalues, respectively). Branches from top to bottom, panel a: Maxwell eigenvalues (s_M), M0 modes, unstable discretization modes (X), panel b: Maxwell eigenvalues, dilatation modes (D), M0 modes and unstable Rayleigh-Taylor modes (RT) crossed by a branch of discretization modes (X). The pseudospectral discretization schemes on the Chebyshev grids have been invoked. The shift of the X-modes towards to longer growth times is demonstrated by presenting the X-branches for the three grid densities (30, 60 and 120 layers) in panel a; the only grid of 30 layers has been used for panel b. The region of $s < 10^{-5}$ kyr^{-1} is contaminated by spurious eigenvalues which result from the iterative character of the applied eigenroutine.

ly one branch of the physical relaxation modes, denoted by M0, the well-known mantle modes, which can be obtained analytically [*Wu and Peltier*, 1982]. We reached the accuracy with more than 10 digits for all tested grid densities. As the analyzed matrices are non-symmetric, their eigenvalues can, in general, be complex. However, all the M0 eigenvalues are real and nondegenerate and thus they yield the classical exponential relaxation.

The branch of eigenvalues denoted by s_M lies in the position of $s = -\mu/\eta$, i.e., of the inverse Maxwell time. These eigenvalues are $3J$-degenerate, so they could yield degenerate normal modes (cf. s_q in (86)). In the Laplacian spectral analysis, they are considered to represent false singularities, which should be removed [*Wu and Ni*, 1996; *Boschi et al.*, 1999]. *Hanyk et al.* [1999] showed numerically by direct integration in the time domain that no significant deformation energy is associated with other than the M0 mode.

The branch denoted by X consists of real positive nondegenerate eigenvalues. Unlike the branches M0 and s_M, the branch X is strongly grid-sensitive. The tenden-

cy of shifting the eigenvalues towards zero is exhibited with increasing number of grid points, which is demonstrated for the grid densities of 30, 60 and 120 layers and which can serve for identification of these modes. As the X-eigenvalues are obviously caused by the radial discretization of the model, we refer to the corresponding modes as the discretization modes. There is also a band of eigenvalues in the region of small s, several orders of magnitude apart from the physical eigenvalues. This is the "numerical noise" produced by the iterative eigenroutines, different for different eigenroutines, grid densities and discretization schemes.

The eigenspectrum of the homogeneous compressible sphere is presented in Fig. 2b for the PS discretization on the Chebyshev grid of 30 layers. In comparison with the spectrum of the incompressible sphere, there are two more groups of modal branches, associated with compressibility. The first group consists of the dilatation or D-modes [*Vermeersen et al.*, 1996], whereas the second group is formed by the Rayleigh-Taylor (RT) modes, generated by the subadiabatic gradient of den-

Mode M0   Mode D3   Mode RT3

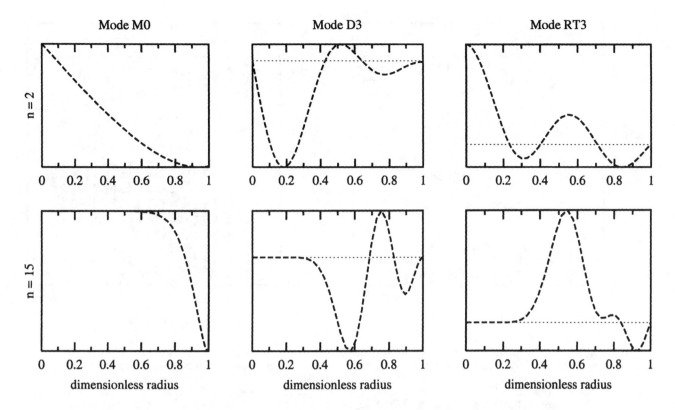

Figure 3. Vertical-displacement eigenvectors U_2 and U_{15} of the homogeneous compressible sphere corresponding to the M0, D3 and RT3 modes computed on the grid of 120 layers using the PS discretization scheme. The vertical bounds correspond with the extremas of the presented eigenvectors. The horizontal zerolines are marked with the dotted lines.

sity [*Hanyk et al.*, 1999]. There are about J D-modes and J RT-modes. The (negative) D-eigenvalues $s^{(i)}$ can be indexed such that $\mathrm{Re}\, s^{(i)} < \mathrm{Re}\, s^{(i+1)}$. About a half of the computed D-eigenvalues is purely real and corresponds to the analytically derived values [*Hanyk et al.*, 1999] with a high accuracy. Since the corresponding eigenvectors are of oscillatory character with the number of oscillations increasing with increasing i, see Fig. 3, the resolution of the grid is not sufficient for higher i and those D-eigenvalues become corrupted (they are complex and the real parts do not fit the analytical values). Similarly, this is also true for the RT modes. We emphasize that all the RT-eigenvalues are positive (or have positive real parts) and describe thus the onset of exponential collapse of the studied system. We also point out that the Maxwell eigenvalues are $2J$-degenerate in this case, that the X-eigenvalues obtained on the equivalent grids are almost identical for both incompressible and compressible spheres and that the low-s spectral region for the compressible sphere is again contaminated by the numerical noise.

The sensitivity of the discretization modes to applied discretization schemes is shown in Fig. 4, where we deal with the core-mantle-lithosphere incompressible model with the following choice of the parameters: $a = 6,371$ km, $b = 3,480$ km, $\varrho_0 = 4,314$ kg m^{-3}, $\mu = 1.4519 \times 10^{11}$ Pa and $\eta = 10^{21}$ Pa s in the mantle, $\varrho_0 = 10,926$ kg m^{-3} in the core and the depth of the lithosphere equal to 120 km. We have studied the eigenspectra of this model for the two different discretizations: a) 20 grid layers in the mantle and 10 grid layers in the lithosphere with the PS scheme employed over the Chebyshev nodal points spread separately in each of the layers, b) 20 layers of equal thickness in the mantle and 10 layers of equal thickness in the lithosphere with the FD2 scheme employed. One can clearly recognize the branches of the modes M0, L0 and C0 as well as the branch of the Maxwell eigenvalues, all insensitive to discretization. On the other hand, the spectral branches of the discretization modes are completely different for the PS and the FD2 scheme. While the discretization spectrum consists from one branch of stable modes and

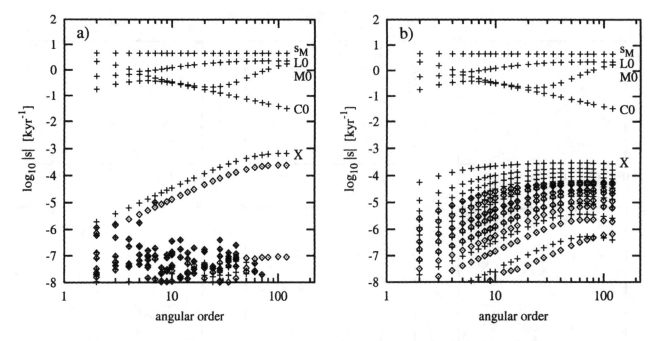

Figure 4. The eigenspectrum of the core-mantle-lithosphere incompressible model evaluated by the PS (panel a) and FD2 (panel b) discretization schemes. The grids of 30 layers (20 in the mantle, 10 in the lithosphere) have been employed in both cases. Besides the classical relaxation modes M0, L0 and C0, there is a branch of the Maxwell eigenvalues (s_M) and the branches of the discretization modes; those with the greatest $|s|$ have been labelled by X. Symbols + and ◊ have been used in the same manner as in Fig. 2.

one branch of unstable modes in the former case, there exist several branches of modes, both stable and unstable, in the latter case. However, the growth time of the first unstable discretization mode produced by the FD2 scheme is by an order of magnitude higher than that of the first unstable X-mode by the PS scheme.

In Fig. 5 we present the eigenspectra for other six simple incompressible models, which were already studied by *Wu and Peltier* [1982]. Model A is the homogeneous sphere discussed above, B has the 120-km thick elastic lithosphere, C has the mantle density 4,314 $kg\,m^{-3}$ and an inviscid core of radius 3,485.5 km and the density 10,926 $kg\,m^{-3}$, D has an outer shell of thickness 195.6 km in which the density is reduced from 5,600 $kg\,m^{-3}$ to 3,200 $kg\,m^{-3}$ and the shear modulus from 1.4519×10^{11} Pa s to 0.7260×10^{11} Pa s, E (F) differ from A by a two-order increase (decrease) of viscosity below (above) the depth of 671 (120) km. The eigenspectra are evaluated by the low-order FD2 discretization with 60 layers employed in each case. We obtained the modal branches visually equivalent with those published by *Wu and Peltier* [1982] in their Fig. 2. There are also groups of the degenerate Maxwell eigenvalues in our eigenspectra. The discretization modes are out of the extent of the panels for the chosen grid densities.

In the eigenvalue method we have a fast, easy and complete tool for finding modes of realistic compressible models with arbitrary radial variations of the density and rheology parameters. The eigenvalues of such a model are shown in Fig. 6. The density and the elastic parameters are taken from the PREM, the elastic lithosphere of 120-km thickness is considered and the viscosity of the upper (lower) mantle is set to 10^{21} (2×10^{21}) Pa s. *Tromp and Mitrovica* [1999] presented a subset of eigenvalues for this model in their Fig. 1. However, there are lots of other modes (D-modes, RT-modes). Here we point out to the RT modes with all the growth times higher than 100 Myr. The physical reason, why these characteristic times of the model instability are much longer in comparison with the homogeneous sphere, is that the PREM is effectively subadiabatic only in the upper mantle above the transition zone. The question whether the mantle can be subadiabatic also in other regions is not satisfactorily resolved but thermal convection models point to the possibility that the lower mantle above the D" layer could also be subadiabatic [*Matyska and Yuen*, 2000], which may result in global instabilities over shorter timescales. There is also a band of the dilatation modes and degenerate Maxwell eigenvalues corresponding to relaxation times between 100 and 1000 years. This band is substantially broad-

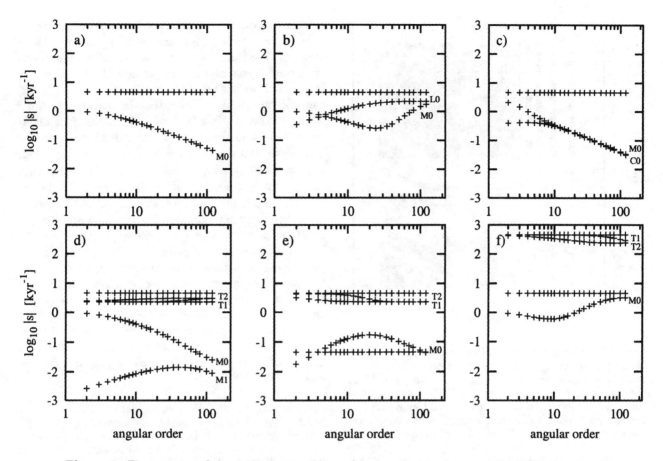

Figure 5. Eigenspectra of the six incompressible models described in the text. The FD2 scheme and equi-spaced grids of 60 layers (for models with two shells: 40 in the lower shell, 20 in the upper shell) have been employed. The extent of the axes and the labelling of the modes has been chosen similarly as in Fig. 2 by *Wu and Peltier* [1982], symbols + and ◇ have been used similarly as in Fig. 2 above.

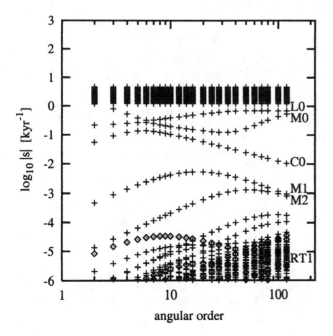

ened if a model with variable viscosity is used. In Fig. 7 we present the eigenspectrum of another PREM-based model with a high-viscosity hill in the lower mantle [*Ricard and Wuming*, 1991; *Forte and Mitrovica*, 2001] and a low-viscosity zone in the upper mantle (viscous profile C2 by *Hanyk et al.* [1996]). The bandwidth of the rectangular region formed by the dilatation modes and the Maxwell eigenvalues spans over four orders of magnitude. One can discern that the M0 and L0 branches

Figure 6. The eigenspectrum of the realistic Earth model (PREM for the density and elastic parameters) with the viscosity of the upper (lower) mantle set to 10^{21} (2×10^{21}) Pa s, discretized on the grid of 240 layers. The classical relaxation modes (M0, L0, C0, M1 and M2) have been labelled as well as the first branch of the unstable RT modes (symbols ◇) on the time scale of 10^5 kyr. There is a lot of discretization X-modes, as the FD2 scheme has been employed (cf. Fig. 4b).

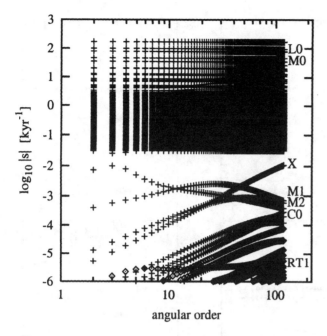

Figure 7. The eigenspectrum of the realistic Earth model (PREM for the density and elastic parameters) with the continuously changing viscous profile in the range from 10^{19} Pa s in the asthenosphere up to more than 10^{23} Pa s in the mid lower mantle, discretized on the grid of 300 Chebyshev layers. The PS scheme has been employed. The classical relaxation modes (M0, L0, C0, M1 and M2) are present but occasionally hidden in the rectangular region formed by the dilatation D-modes and the Maxwell eigenvalues. Within the plotted bounds, there is only one branch of discretization X-modes, which contrasts with Fig. 6. On the other hand, the X-branch interferes with the M1, M2 and C0 branches for higher angular orders.

of the physical modes are buried inside this region. It is like to find needles in a haystack, if one tries to locate these "true" modes by means of the classical root-finding methods.

5. CONCLUDING REMARKS

In this paper we have demonstrated the feasibility of using eigenvalue analysis for determining the relaxation times for both incompressible and compressible self-gravitating stratified viscoelastic Earth models. Such an approach yields the full spectrum of both the relaxation and growth times of a model after discretization, its implementation is easy and computational cost is low. The crucial idea was to recast the governing equations into a mixed set of PDEs (58), and then to apply the method of lines for obtaining the set of ODEs (82) in time. Our previous approach [*Hanyk et al.*, 1995; 1996; 1998] consists in first integration over time and then

solving a two-point boundary value problem in the radius, i.e., we applied the Rothe method [*Rektorys*, 1982]. However, the Rothe method does not allow one to carry out an eigenanalysis of the temporal relaxations. In this regard, we note that the resultant set of ODEs (82) can also be solved by direct time-integration. In other words, the set of ODEs (82) can be employed for a different initial-value approach, where stiff time-integrators can substantially reduce the number of timesteps. Our approach can be adopted for anelastic rheologies, such as the standard linear solid which may be important for lithospheric bending [*Schmalholz and Podladchikov*, 2001]. The understanding of the tectonic impact of the potential energy change and geoid anomalies caused by earthquakes along plate margins can be investigated using the eigenfunctions generated with short time-scale rheology. Computationally this approach can be very fast for summing up cumulatively the gravitational excitation from many large earthquakes [*Tanimoto and Okamoto*, 2001]. This eigenvalue approach can also be useful in understanding crustal instabilities involving folding of lithospheric rocks [*Schmalholz and Podladchikov*, 1999; 2001].

Acknowledgments. We wish to thank Yu. Podladchikov and S. Schmalholz for inspiring discussions, L. Boschi for a careful review and L. L. A. Vermeersen for useful editorial comments. This research has been supported by the Research Project MŠMT J13/98: 113200004, the Grant Agency of the Czech Republic under Nos. 205/00/0906 and 205/00/D113, the Charles University grants 170/1998/B-GEO/MFF and 175/2000/B-GEO/MFF, the geoscience program of the D.O.E. and the geophysics program of the National Science Foundation.

REFERENCES

Ascher, U. M., and L. R. Petzold, *Computer Methods for Ordinary Differential Equations and Differential-Algebraic Equations*, SIAM, Philadephia, 1998.

Backus, G. E., Converting vector and tensor equations to scalar equations in spherical coordinates, *Geophys. J. R. astr. Soc.*, *13*, 71–101, 1967.

Boschi, L., J. Tromp, and R. J. O'Connell, On Maxwell singularities in postglacial rebound, *Geophys. J. Int.*, *136*, 492–498, 1999.

Christensen, R. M., *Theory of Viscoelasticity: An Introduction*, Academic, New York, 1982.

Dahlen, F. A., and J. Tromp, *Theoretical Global Seismology*, Princeton University Press, Princeton, New Jersey, 1998.

Dziewonski, A. M., and D. L. Anderson, Preliminary reference Earth model, *Phys. Earth Planet. Int.*, *25*, 297–356, 1981.

Fang, M., Static deformation of the outer core, in *Dynamics of the Ice Age Earth: A Modern Perspective*, edited by P. Wu, pp. 155–190, Trans. Tech Publ., Zürich, Switzerland, 1998.

Fang, M., and B. H. Hager, The singularity mystery associated with a radially continuous Maxwell viscoelastic structure, *Geophys. J. Int.*, *123*, 849–865, 1995.

Farrell, W. E., Deformation of the earth by surface loads, *Rev. Geophys. Space Phys.*, 10, 761–797, 1972.

Fornberg, B., *A Practical Guide to Pseudospectral Methods*, Cambridge, New York, 1996.

Forte, A. M., and J. X. Mitrovica, Deep-mantle high-viscosity flow and thermochemical stucture inferred from seismic and geodynamic data, *Nature*, *410*, 1049–1056, 2001.

Han, D., and J. Wahr, The viscoelastic relaxation of a realistically stratified earth, and a further analysis of postglacial rebound, *Geophys. J. Int.*, *120*, 287–311, 1995.

Hanyk, L., C. Matyska, and D. A. Yuen, Initial-value approach for viscoelastic responses of the Earth's mantle, in *Dynamics of the Ice Age Earth: A Modern Perspective*, edited by P. Wu, pp. 135–154, Trans. Tech Publ., Zürich, Switzerland, 1998.

Hanyk, L., C. Matyska, and D. A. Yuen, Secular gravitational instability of a compressible viscoelastic sphere, *Geophys. Res. Lett.*, *26*, 557–560, 1999.

Hanyk, L., Matyska, C. and Yuen, D.A., The problem of viscoelastic relaxation of the Earth solved by a matrix eigenvalue approach based on discretization in grid space, *Electronic Geosciences*, 5, http://link.springer.de/link/service/journals/10069/discussion/evmol/evmol.htm, 2000.

Hanyk, L., J. Moser, D. A. Yuen, and C. Matyska, Time-domain approach for the transient responses in stratified viscoelastic Earth models, *Geophys. Res. Lett.*, *22*, 1285–1288, 1995.

Hanyk, L., D. A. Yuen, and C. Matyska, Initial-value and modal approaches for transient viscoelastic responses with complex viscosity profiles, *Geophys. J. Int.*, *127*, 348–362, 1996.

Karato, S.-I., Micro-physics of post glacial rebound, in *Dynamics of the Ice Age Earth: A Modern Perspective*, edited by P. Wu, pp. 351–364, Trans. Tech Publ., Zürich, Switzerland, 1998.

Lehoucq, R. B., D. C. Sorensen, and C. Yang, *ARPACK User's Guide: Solution of Large-Scale Eigenvalue Problems with Implicitly Restarted Arnoldi Methods*, SIAM, Philadelphia, 1998.

Martinec, Z., Spectral, initial value approach for viscoelastic relaxation of a spherical earth with a three-dimensional viscosity – I. Theory, *Geophys. J. Int.*, *137*, 469–488, 1999.

Matyska, C., and D. A. Yuen, Profiles of the Bullen parameter from mantle convection modelling, *Earth Planet. Sci. Lett.*, *178*, 39–46, 2000.

Nakada, M., Implications of a non-adiabatic density gradient for the Earth's viscoelastic response to surface loading, *Geophys. J. Int.*, *137*, 663–674, 1999.

Peltier, W. R., The impulse response of a Maxwell earth, *Rev. Geophys. Space Phys.*, *12*, 649–669, 1974.

Plag, H.-P., and H.-U. Jüttner, Rayleigh-Taylor instabilities of a self-gravitating Earth, *J. Geodynamics*, *20*, 267–288, 1995.

Rektorys, K., *The Method of Discretization in Time and Partial Differential Equations*, Reidel, Dordrecht-Boston-London, 1982.

Rektorys, K., *Survey of Applicable Mathematics*, Kluwer, Dordrecht, 1994.

Ricard, Y., and B. Wuming, Inferring the viscosity and the

3-D density structure of the mantle from geoid, topography and plate tectonics, *Geophys. J. Int.*, *105*, 561–571, 1991.

Saad, Y., *Numerical Methods for Large Eigenvalue Problems*, Manchester University Press, Manchester, England, 1992.

Schiesser, W. E., *Computational Mathematics in Engineering and Applied Science: ODEs, DAEs, and PDEs*, CRC, Florida, 1994.

Schmalholz, S. M., and Yu. Podladchikov, Buckling versus folding: Importance of viscoelasticity, *Geophys. Res. Lett.*, *26*, 2641–2644, 1999.

Schmalholz, S. M., and Y. Y. Podladchikov, Viscoelastic folding: Maxwell versus Kelvin rheology, *Geophys. Res. Lett.*, *28*, 1835–1838, 2001.

Spada, G., R. Sabadini, D. A. Yuen, and Y. Ricard, Effects on post-glacial rebound from the hard rheology in the transition zone, *Geophys. J. Int.*, *109*, 683–700, 1992.

Stathopoulos, A., S. Ogut, Y. Saad, J. Chelikowsky, and H. Kim, Parallel methods and tools for predicting material properties, *Computing in Science and Engineering*, *2*, 19–32, 2000.

Tanimoto, T., and T. Okamoto, Tectonic significance of potential energy change by earthquakes, *J. Geophys. Res.*, in press, 2001.

Trefethen, L. N., *Computation of pseudospectra*, Acta Numerica, Vol. 8, 247–295, Cambridge University Press, 1999.

Trefethen, L. N., and D. Bau, *III. Numerical Linear Algebra*, SIAM, Philadelphia, 1997.

Tromp, J., and J. X. Mitrovica, Surface loading of a viscoelastic earth-II. Spherical models, *Geophys. J. Int.*, *137*, 856–872, 1999.

Tromp, J., and J. X. Mitrovica, Surface loading of a viscoelastic planet-III. Aspherical models, *Geophys. J. Int.*, *140*, 425–441, 2000.

Vermeersen, L. L. A., and J. X. Mitrovica, Gravitational stability of spherical self-gravitating relaxation models, *Geophys. J. Int.*, *142*, 351–360, 2000.

Vermeersen, L. L. A., R. Sabadini, and G. Spada, Compressible rotational deformation, *Geophys. J. Int.*, *126*, 735–761, 1996.

Wolf, D., Gravitational viscoelastodynamics for a hydrostatic planet, Habilitation Thesis, Bayerische Akademie der Wissenschaften, München, Germany, 1997.

Wu, P., and Z. Ni, Some analytical solutions for the viscoelastic gravitational relaxation of a two-layer non-self-gravitating incompressible spherical earth, *Geophys. J. Int.*, *126*, 413–436, 1996.

Wu, P., and W. R. Peltier, Viscous gravitational relaxation, *Geophys. J. R. astr. Soc.*, *70*, 435–485, 1982.

Yuen, D. A., and W. R. Peltier, Normal modes of the viscoelastic earth, *Geophys. J. R. astr. Soc.*, *69*, 495–526, 1982.

L. Hanyk and C. Matyska, Department of Geophysics, Faculty of Mathematics and Physics, Charles University, V Holešovičkách 2, 18000 Praha, Czech Republic. (e-mail: ladislav.hanyk@mff.cuni.cz; ctirad.matyska@mff.cuni.cz)

D. A. Yuen, University of Minnesota Supercomputing Institute, 1200 Washington Avenue South, Minneapolis, MN 55415. (e-mail: davey@krissy.msi.umn.edu)

Compressible Viscoelastic Earth Models Based on Darwin's Law

Detlef Wolf and Guoying Li

GeoForschungsZentrum Potsdam, Kinematics and Dynamics of the Earth, D-14473 Potsdam, Germany

We derive the analytical solution for load-induced perturbations of a spherical, compressible earth model consisting of a viscoelastic mantle and an inviscid core. In contrast to conventional earth models, the density stratification in the mantle is assumed to conform to Darwin's law, which can be shown to be consistent with the assumption of compressibility adopted. We give computational examples in the spherical-harmonic domain and calculate the elastic and viscous amplitudes and the associated relaxation times for Heaviside loading. In the angular domain, we consider an ice load that is circular in plan view and parabolic in cross section with dimensions comparable to the former Laurentide ice sheet of North America. Assuming Heaviside unloading, we calculate the radial displacement, the geoid height and the free-air gravity anomaly. The computational results are compared with those for a spherical, incompressible earth model with homogeneous density in the mantle and the characteristic differences are discussed.

1. INTRODUCTION

The theoretical modelling of glacial-isostatic adjustment (GIA) can be classified into two categories: numerical simulations and analytical solutions. Whereas analytical solutions are useful to gain physical insight, they usually lack the complexity that is necessary for applications. On the other hand, numerical simulations allow the construction of fairly realistic earth models, but cannot be used for a deeper understanding of the physical processes involved.

The first analytical solution for a spherical, incompressible, gravitational–viscoelastic earth that accommodates GIA is due to *Wu and Peltier* [1982], who considered a homogeneous earth. Independently, *Sabadini et al.* [1982] derived a semi-analytical solution

Ice Sheets, Sea Level and the Dynamic Earth
Geodynamics Series 29
Copyright 2002 by the American Geophysical Union
10.1029/029GD17

for the relaxation of multi-layer earth models. Later, *Wolf* [1984] considered the effects due to the lithosphere and presented closed-form solutions for this model. In order to achieve some simplification, the incremental gravitational force (IGF) was neglected in the equilibrium equation. This simplification was later removed by *Amelung and Wolf* [1994], who derived closed-form solutions for an earth model consisting of a viscoelastic mantle and an inviscid core or a viscoelastic mantle and an elastic lithosphere. The problem of multi-layer earth models was further studied by *Spada et al.* [1992], *Vermeersen et al.* [1996a, 1997], *Wu and Ni* [1996] and *Martinec and Wolf* [1998], all of whom derived explicit solutions.

Analytical solutions for compressible earth models are more difficult to derive, with the first work again the investigation by *Wu and Peltier* [1982] for a homogeneous model. The problem was reconsidered by *Vermeersen and Sabadini* [1996b], who derived more explicit formulae for the same earth model. Recently, *Plag and Jüttner* [1995], *Hanyk et al.* [1999], *Wieczerkowski*

Table 1. Parameter values of the Maxwell–viscoelastic earth models H, C and S. The initial density, $\rho^{(0)}$, is in g cm^{-3}, the initial gravity, $g^{(0)}$, is in m s^{-2}, the elastic rigidity, μ_e, is in 10^{11} Pa, and the viscosity, η, is in 10^{21} Pa s. The earth radius is a and the core radius is b

Earth model	H	C	S
$\rho^{(0)}(b^-)$	10.98690	10.98690	10.98690
$\rho^{(0)}(b^+)$	4.44940	5.56645	9.74691
$\rho^{(0)}(a^-)$	4.44940	3.90226	2.60000
$\rho^{(0)}(r), b < r < a$	$Ar^{-\beta}$	$Ar^{-\beta}$	$Ar^{-\beta}$
A	$\rho^{(0)}(a^-)$	$\rho^{(0)}(b^+)b^\beta$	$\rho^{(0)}(a^-)a^\beta$
β	0.00000	0.58738	2.18519
$g^{(0)}(b)$	10.68650	10.68650	10.68650
$g^{(0)}(a)$	9.82025	9.82025	9.82025
μ_e	1.45190	1.45190	1.45190
η	1.00000	1.00000	1.00000

[1999] and *Vermeersen and Mitrovica* [2000] have discussed the analytical solution for homogeneous, compressible earth models and have found that the solution may contain instabilities. Physically, such problems are related to the inconsistent assumptions of compressibility and homogeneous density.

A different approach to compressibility was taken by *Li and Yuen* [1987] and *Wu and Yuen* [1991]. Restricting their solution to viscous earth models and neglecting the IGF, they studied the relaxation of a spherical earth model whose initial density stratification is consistent with the assumption of compressibility. The theory was extended to gravitational–viscoelastic earth models by *Wolf* [1997]. A simplified solution valid for plane earth models without IGF was derived by *Wolf and Kaufmann* [2000].

A characteristic of these solutions is that they are contingent on the assumption of locally incompressible perturbations. This assumption is satisfied closely near the fluid limit of the viscoelastic response and applies less well near the elastic limit, as discussed more thoroughly by *Wolf* [1997] and *Wolf and Kaufmann* [2000]. To understand the essentials of the concept of local incompressibility, we consider an arbitrary Cartesian tensor field, whose Lagrangian representation is $f_{ij...}(\mathbf{X}, t)$. Here, X_i is a particle's position at the initial time, $t = 0$, r_i the particle's position at the current time, t, and $f_{ij...}$ the field value at r_i [*e.g., Wolf*, 1997, 1998]. The initial and current positions can be regarded as first-order tensor fields themselves and are related by

$$r_i(\mathbf{X}, t) = X_i + u_i(\mathbf{X}, t), \qquad (1)$$

where u_i is the particle displacement. The local incre-

mental field, $f_{ij...}^{(\Delta)}(\mathbf{X}, t)$, is defined as the increment of $f_{ij...}(\mathbf{X}, t)$ at a fixed position X_i, whereas the material incremental field, $f_{ij...}^{(\delta)}(\mathbf{X}, t)$, is defined as the increment of $f_{ij...}(\mathbf{X}, t)$ at a particle during its displacement from X_i to r_i. From the definition of local and material increments, it can be shown that, to the first order, they are related by [*e.g., Wolf*, 1997, 1998]

$$f_{ij...}^{(\delta)} = f_{ij...}^{(\Delta)} + f_{ij...,k}^{(0)} u_k. \qquad (2)$$

Here and in the following, the definition $f_{ij...}^{(0)}(\mathbf{X}) := f_{ij...}(\mathbf{X}, 0)$ is used for the initial field and the arguments X_i and t are dropped. Furthermore, $f_{ij...,k} := \partial f_{ij...}/\partial X_k$ is employed for brevity and the usual summation convention applies. The second term on the right-hand side of (2) is called the advective incremental field and describes the contribution to the material increment that is due to the displacement of the particle during the perturbation.

Considering the density, ρ, as a special case, we then have

$$\rho^{(\delta)} = \rho^{(\Delta)} + \rho^{(0)}_{,i} u_i. \qquad (3)$$

Assuming material incompressibility, $\rho^{(\delta)} = 0$ applies such that $\rho^{(\Delta)} = -\rho^{(0)}_{,i} u_i$ follows. Hence, the density at a fixed position changes according to the initial density gradient, with internal buoyancy generated at this position. In the case of local incompressibility, we have $\rho^{(\Delta)} = 0$ and thus $\rho^{(\delta)} = \rho^{(0)}_{,i} u_i$ follows. The density at a displaced particle therefore changes as specified by the initial density gradient, and no internal buoyancy is generated at a fixed position. For the special case of a homogeneous model, $\rho^{(0)}_{,i} = 0$. Then, the advective increment vanishes and there is no distinction between material and local incompressibility.

In this study, we use

$$\rho^{(0)}(r) = Ar^{-\beta} \qquad (4)$$

for the initial density in the mantle as a function of the radial distance, r. This relation is also known as Darwin's law [*Darwin*, 1884; *Bullen*, 1975] and can be shown to be a solution to the Williamson-Adams equation [*Williamson and Adams*, 1923; *Bullen*, 1975] for a gravitating earth model. The parameters A and β can be determined from two constraints. In the following, we require that the total mass of the mantle agrees with that for PREM [*Dziewonski and Anderson*, 1981]. With the correct total mass of the core, this ensures that the correct gravity values apply in the earth model at the surface and at the core–mantle boundary. The other constraint is either that the density at the surface (earth model S) or at the core–mantle boundary (earth model C) is identical with that for PREM. For earth model S, the density jump at the core–mantle boundary and, therefore, the associated buoyancy forces are too small. On the other hand, for earth model C, the density at the earth surface is too large. As a consequence, the associated buoyancy forces are also too large, resulting in significantly reduced radial displacements in the fluid limit. For purposes of comparison, we also consider the case of a homogeneous, incompressible mantle, whose total mass agrees with that for PREM (earth model H). The solution for earth model H has been derived directly by several investigators (see above), but can also be obtained as a special case of our more general solution based on Darwin's law. Hence, earth model H also serves to test the new solution. In all earth models examined, the core is assumed to be homogeneous and inviscid with a density corresponding to the mean core density for PREM. The parameters for the three earth models are listed in Table 1 and the radial density stratifications shown in Fig. 1.

The paper will proceed as follows: We begin by presenting the incremental field equations valid for locally incompressible earth models and derive the general solution for gravitational–viscoelastic perturbations and density stratifications based on Darwin's law (Sec. 2). This is followed by the compilation of the incremental interface conditions and the computation of normal modes (Sec. 3). After this, closed-form expressions required in numerical computations for field quantities are given (Sec. 4). Finally, numerical examples are presented and discussed (Sec. 5).

2. INCREMENTAL FIELD EQUATIONS AND GENERAL SOLUTION

We assume that the earth is initially at hydrostatic equilibrium and consider gravitational–viscoelastic perturbations of its initial state. Supposing locally incompressible perturbations, the incremental equilibrium equation in the Laplace-transform domain reads [*Wolf*, 1997, 1998]

$$t_{ij,j}^{(\delta)} + (p_{,j}^{(0)} u_j)_{,i} + \rho^{(0)} \phi_{,i}^{(\Delta)} = 0, \qquad (5)$$

where $t_{ij}^{(\delta)}$ is the material incremental stress, $p^{(0)}$ the initial pressure and $\phi^{(\Delta)}$ the local incremental potential. We note that no buoyancy term appears in (5) because of the assumption of local incompressibility. Considering only shear relaxation, the constitutive equation in the s domain can be expressed as [*Wolf*, 1997, 1998]

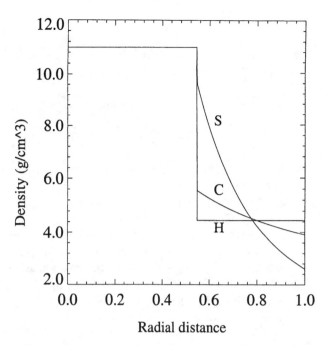

Figure 1. Density profiles in the mantle according to Darwin's law. For earth model C, the density at the core–mantle boundary agrees with the PREM density; for earth model S, the density at the earth surface agrees with the PREM density; for earth model H, the homogeneous mantle density agrees with the mean PREM mantle density. For all earth models, the homogeneous core density agrees with the mean PREM core density. The radial distance is normalized with respect to the earth radius.

$$t_{ij}^{(\delta)} = \left[k - \frac{2}{3}\mu(s) \right] \delta_{ij} u_{k,k} + \mu(s)(u_{i,j} + u_{j,i}), \qquad (6)$$

in which $\mu(s)$ is the s-dependent shear modulus and k the elastic bulk modulus. For Maxwell viscoelasticity, we have

$$\mu(s) = \frac{\mu_e s}{s + \mu_e/\eta}, \qquad (7)$$

where μ_e is the elastic shear modulus and η the viscosity. Since $t_{ij}^{(0)} = -\delta_{ij} p^{(0)}$ for a hydrostatic initial state, we have from (2) the relationship

$$t_{ij}^{(\delta)} = t_{ij}^{(\Delta)} - \delta_{ij} p_{,k}^{(0)} u_k. \qquad (8)$$

This allows us to rewrite (5) and (6) in the forms

$$t_{ij,j}^{(\Delta)} + \rho^{(0)} \phi_{,i}^{(\Delta)} = 0, \qquad (9)$$

$$t_{ij}^{(\Delta)} = -\delta_{ij} p^{(\Delta)} - \frac{2}{3}\mu(s)\delta_{ij} u_{k,k} + \mu(s)(u_{i,j} + u_{j,i}). \quad (10)$$

The incremental potential and continuity equations are, respectively,

$$\phi_{,ii}^{(\Delta)} = 0, \tag{11}$$

$$(\rho^{(0)} u_i)_{,i} = 0, \tag{12}$$

in which

$$
\begin{aligned}
p^{(\Delta)} &= p^{(\delta)} - p_{,i}^{(0)} u_i = -k u_{i,i} - p_{,i}^{(0)} u_i \\
&= -k u_{i,i} - \frac{k}{\rho^{(0)}} \rho_{,i}^{(0)} u_i = \frac{k}{\rho^{(0)}} \rho^{(\Delta)}.
\end{aligned} \tag{13}
$$

We note that (11) is reduced to the Laplace equation as a consequence of the assumption of local incompressibility. Equation (13) shows that $k \to \infty$ is required in order that $p^{(\Delta)} =$ finite for $\rho^{(\Delta)} \to 0$.

To derive the general solution to the incremental field equations (9)–(12), we use their scalar forms in spherical coordinates (r, θ, ϕ) and expand the components of u_i, $t_{ij}^{(\Delta)}$, $\phi^{(\Delta)}$, $p^{(\Delta)}$ and $u_{i,i}$ into spherical harmonics, $Y_l^m(\theta, \phi)$. Since, for the following, only the spheroidal components are of interest, we obtain

$$
u_r = y_1(r,s) Y_l^m, \quad u_\theta = y_3(r,s) \frac{\partial Y_l^m}{\partial \theta},
$$
$$
u_\phi = y_3(r,s) \frac{\partial Y_l^m}{\sin\theta \partial \phi}, \tag{14}
$$

$$
t_{rr}^{(\Delta)} = y_2(r,s) Y_l^m, \quad t_{r\theta}^{(\Delta)} = y_4(r,s) \frac{\partial Y_l^m}{\partial \theta},
$$
$$
t_{r\phi}^{(\Delta)} = y_4(r,s) \frac{\partial Y_l^m}{\sin\theta \partial \phi}, \tag{15}
$$

$$
\phi^{(\Delta)} = y_5(r,s) Y_l^m, \quad p^{(\Delta)} = p(r,s) Y_l^m,
$$
$$
u_{i,i} = d(r,s) Y_l^m, \tag{16}
$$

where, for brevity, the summations over the spherical-harmonic degree, l, and order, m, have been suppressed. We also have the relationships

$$
y_2(r,s) = -p(r,s) - \frac{2}{3}\mu(s) d(r,s) + 2\mu(s) \frac{dy_1(r,s)}{dr}, \tag{17}
$$

$$
d(r,s) = \frac{dy_1(r,s)}{dr} + \frac{2y_1(r,s)}{r} - \frac{l(l+1)y_3(r,s)}{r}, \tag{18}
$$

$$
y_4(r,s) = \mu(s) \left[\frac{dy_3(r,s)}{dr} - \frac{y_3(r,s)}{r} + \frac{y_1(r,s)}{r} \right]. \tag{19}
$$

Upon substitution into the scalar forms of the incremental field equations (9)–(12), we obtain, after some algebraic manipulation, the following set of simultaneous first-order ordinary differential equations:

$$
\frac{dy_1}{dr} = \frac{\beta - 2}{r} y_1 + \frac{l(l+1)}{r} y_3, \tag{20}
$$

$$
\begin{aligned}
\frac{dy_2}{dr} =\ & \frac{12\mu(s)}{r^2}\left(1 - \frac{1}{3}\beta\right) y_1 - \frac{6\mu(s)l(l+1)}{r^2} y_3 \\
& + \frac{l(l+1)}{r} y_4 - \rho^{(0)} y_6,
\end{aligned} \tag{21}
$$

$$
\frac{dy_3}{dr} = -\frac{1}{r} y_1 + \frac{1}{r} y_3 + \frac{1}{\mu(s)} y_4, \tag{22}
$$

$$
\begin{aligned}
\frac{dy_4}{dr} =\ & -\frac{6\mu(s)}{r^2}\left(1 - \frac{1}{3}\beta\right) y_1 - \frac{1}{r} y_2 \\
& + \frac{2\mu(s)}{r^2}\left[2l(l+1) - 1\right] y_3 \\
& - \frac{3}{r} y_4 - \rho^{(0)} \frac{1}{r} y_5,
\end{aligned} \tag{23}
$$

$$
\frac{dy_5}{dr} = y_6, \tag{24}
$$

$$
\frac{dy_6}{dr} = \frac{l(l+1)}{r^2} y_5 - \frac{2}{r} y_6. \tag{25}
$$

In deriving these equations, Darwin's law has been assumed for the radial density stratification and (17) has been used to eliminate $p(r,s)$. We solve (20)–(25) by formulating an eigenvalue problem. For this purpose, we introduce a new set of unkowns, $z_1, ..., z_6$, by

$$
\begin{aligned}
Y &= [y_1, y_2, y_3, y_4, y_5, y_6]^\top \\
&= [r^{1-\beta} z_1, \mu(s) r^{-\beta} z_2, r^{1-\beta} z_3, \\
&\quad \mu(s) r^{-\beta} z_4, \mu(s) z_5, \mu(s) r^{-1} z_6]^\top.
\end{aligned} \tag{26}
$$

In terms of $z_1, ..., z_6$, (20)–(25) become

$$
r\frac{dz_1}{dr} = (2\beta - 3) z_1 + l(l+1) z_3, \tag{27}
$$

$$
\begin{aligned}
r\frac{dz_2}{dr} =\ & 4(3 - \beta) z_1 + \beta z_2 - 6l(l+1) z_3 \\
& + l(l+1) z_4 - A z_6,
\end{aligned} \tag{28}
$$

$$
r\frac{dz_3}{dr} = -z_1 + \beta z_3 + z_4, \tag{29}
$$

$$
\begin{aligned}
r\frac{dz_4}{dr} =\ & -2(3 - \beta) z_1 - z_2 + 2\left[2l(l+1) - 1\right] z_3 \\
& - (3 - \beta) z_4 - A z_5,
\end{aligned} \tag{30}
$$

$$
r\frac{dz_5}{dr} = z_6, \tag{31}
$$

$$
r\frac{dz_6}{dr} = l(l+1) z_5 - z_6. \tag{32}
$$

These equations have solutions of the form

$$
Z = [z_1, z_2, z_3, z_4, z_5, z_6]^\top = r^\lambda Z^{(j)}. \tag{33}
$$

Considering this in (27)–(32) yields the following eigenvalue problem:

$$(E - \lambda I)Z^{(j)} = 0, \qquad (34)$$

in which λ is the eigenvalue, $Z^{(j)}$ the eigenvector and E the coefficient matrix of (27)–(32). Equation (34) can be solved using standard methods. We find

$$Z(r,s) = DF(r)C(s), \qquad (35)$$

where D is a (6×6) matrix with the eigenvectors of (34) as columns, $F(r)$ is a diagonal (6×6) matrix, and $C(s)$ is a (1×6) matrix with elements to be determined. Explicit expressions of D and $F(r)$ are given in Appendix A. If we take $\beta = 0$ in D, (35) reduces to the general solution for a homogeneous, incompressible material.

3. INCREMENTAL INTERFACE CONDITIONS AND SPECIAL SOLUTION

To proceed with the solution, the elements of $C = [C_1, C_2, C_3, C_4, C_5, C_6]$ must be determined using the incremental interface conditions. In general form, they read as

$$[u_i]_-^+ = 0, \qquad (36)$$

$$[\phi^{(\Delta)}]_-^+ = 0, \qquad (37)$$

$$[n_i^{(0)}(\phi_{,i}^{(\Delta)} - 4\pi G\rho^{(0)} u_i)]_-^+ = -4\pi G\sigma, \qquad (38)$$

$$[n_j^{(0)}(t_{ij}^{(\Delta)} - \delta_{ij}\rho^{(0)} g_k^{(0)} u_k)]_-^+ = -g_i^{(0)}\sigma, \qquad (39)$$

where σ is the interface-mass density, n_i the unit normal with respect to the interface, G Newton's gravitational constant and $[f_{ij...}]_-^+$ denotes the jump of $f_{ij...}$ across the interface. Since we consider surface loading, $\sigma \neq 0$ at the earth surface and $\sigma = 0$ at the core-mantle boundary. After transforming (36)–(39) into spherical coordinates and expanding the field quantities into spherical harmonics, the conditions in terms of $z_1, ..., z_6$ become

$$-g^{(0)}(a)a^\beta \sigma_l = \mu(s)z_2(a^-) \\ + \rho^{(0)}(a^-)g^{(0)}(a)az_1(a), \qquad (40)$$

$$0 = \mu(s)[z_2(b^+) + \rho^{(0)}(b^-)b^\beta z_5(b)] \\ - \delta\rho^{(0)}(b)g^{(0)}(b)bz_1(b), \qquad (41)$$

$$0 = z_4(a^-), \qquad (42)$$

$$0 = z_4(b^+), \qquad (43)$$

$$4\pi Ga\sigma_l = \mu(s)[z_6(a^-) + (l+1)z_5(a)] \\ - 4\pi G\rho^{(0)}(a^-)a^{2-\beta}z_1(a), \qquad (44)$$

$$0 = \mu(s)[z_6(b^+) - lz_5(b)] \\ + 4\pi G\delta\rho^{(0)}(b)b^{2-\beta}z_1(b), \qquad (45)$$

where σ_l is the spherical-harmonic coefficient of degree l of the interface-mass density, $\delta\rho^{(0)} := \rho^{(0)}(b^-) - \rho^{(0)}(b^+)$ and the relationship $y_2(b^-) = -\rho^{(0)}(b^-)y_5(a)$ has been used. In the core, $y_5(r) = C_5 r^l$ if the density is constant. Outside the earth, we have $y_5(r) = C_6 r^{-(l+1)}$.

To obtain a solution, we eliminate $z_1(a)$ from (44) by means of (40) and $z_1(b)$ from (45) by means of (41). Using matrix $B(\mu(s), \chi_1, \chi_2, \chi_3, \chi_4)$ specified in Appendix B with arbitrary parameters $\chi_1, ..., \chi_4$, we can write (40)–(45) in matrix form:

$$B(\mu(s), 1, 1, 1, 1)C(s) = V(s), \qquad (46)$$

where

$$V(s) = [-g^{(0)}(a)a^\beta \sigma_l, 0, 0, 0, 0, 0]^\top. \qquad (47)$$

The secular determinant $\det B(\mu(s), 1, 1, 1, 1)$ is a quadratic function of $\mu(s)$:

$$\det B(\mu(s), 1, 1, 1, 1) = a_1 \mu^2(s) + b_1 \mu(s) + c_1, \qquad (48)$$

with

$$a_1 = \det B(1, 1, 0, 1, 0), \qquad (49)$$

$$b_1 = \det B(1, 1, 0, 0, 1) + \\ \det B(1, 0, 1, 1, 0), \qquad (50)$$

$$c_1 = \det B(1, 0, 1, 0, 1), \qquad (51)$$

Eigenvalues are values of s for which

$$a_1 \mu^2(s) + b_1 \mu(s) + c_1 = 0 \qquad (52)$$

applies. From (7), it is obvious that, for Maxwell viscoelasticity, two eigenvalues require that a_1 does not vanish. As will be seen, $a_1 = 0$ only applies to spherical-harmonic degree 1.

Most of the field quantities of geophysical interest can be expressed in terms of y_1, y_3 and y_5. We evaluate these quantities for the earth surface. Using Cramer's rule, we note from (46) that $C_j(s)$ is of the form

$$C_j(s) = \frac{q_j\mu(s) + x_j}{a_1\mu(s)^2 + b_1\mu(s) + c_1}\sigma_l, \qquad (53)$$

where q_j and x_j can be calculated from the determinant obtained by replacing the jth column of $\det[B(s)]$ by $V(s)$. Thus we have

$$
\begin{aligned}
z_i(a,s) &= \sum_{j=1}^{6} C_j D_{ij} F_j(a) \\
&= \frac{1}{a_1 \mu(s)^2 + b_1 \mu(s) + c_1} \\
&\quad \times [\mu(s) \sum_{j=1}^{6} D_{ij} F_j(a) q_j \qquad (54) \\
&\quad + \sum_{j=1}^{6} D_{ij} F_j(a) x_j] \sigma_l \\
&= \frac{N_i \mu(s) + Q_i}{a_1 \mu(s)^2 + b_1 \mu(s) + c_1} \sigma_l.
\end{aligned}
$$

Reconsidering y_i, we have, for Maxwell viscoelasticity,

$$
y_i(a,s) = \frac{N_2^i s^2 + M_2^i s + Q_2^i}{a_2 s^2 + b_2 s + c_2} \sigma_l, \qquad (55)
$$

where

$$
\begin{aligned}
a_2 &= a_1 \mu_e^2 + b_1 \mu_e + c_1, &\qquad (56) \\
b_2 &= b_1 \mu_e s_m + 2 c_1 s_m, &\qquad (57) \\
c_2 &= c_1 s_m^2, &\qquad (58)
\end{aligned}
$$

$$
\left.
\begin{aligned}
N_2^i &= a^{1-\beta}(N_i \mu_e + Q_i) \\
M_2^i &= a^{1-\beta}(N_i \mu_e s_m + 2 Q_i s_m) \\
Q_2^i &= a^{1-\beta} Q_i s_m^2
\end{aligned}
\right\} i = 1, 3, \quad (59)
$$

$$
\left.
\begin{aligned}
N_2^i &= N_i \mu_e^2 + Q_i \mu_e \\
M_2^i &= Q_i \mu_e s_m \\
Q_2^i &= 0
\end{aligned}
\right\} i = 5. \qquad (60)
$$

In these equations, the inverse Maxwell time, $s_m := \mu_e/\eta$ has been used. Denoting the two eigenvalues by $-s_l^1$ and $-s_l^2$, the denominator in (55) can be written as $a_2(s + s_l^1)(s + s_l^2)$. Furthermore, using

$$
h_i(s) := N_2^i s^2 + M_2^i s + Q_2^i \qquad (61)
$$

for the numerator of (55), this equation can be recast as

$$
\begin{aligned}
y_i(a,s) &= \frac{h_i(s)}{a_2(s + s_l^1)(s + s_l^2)} \sigma_l \\
&= \left(y_i^e + \frac{y_i^{v1}}{s + s_l^1} + \frac{y_i^{v2}}{s + s_l^2} \right) \sigma_l,
\end{aligned} \qquad (62)
$$

with

$$
\begin{aligned}
y_i^e &= \frac{N_2^i}{a_2}, \\
y_i^{v1} &= \frac{h_i(-s_l^1)}{a_2(s_l^2 - s_l^1)}, \quad y_i^{v2} = \frac{h_i(-s_l^2)}{a_2(s_l^1 - s_l^2)}.
\end{aligned} \qquad (63)
$$

The final solutions are usually given in terms of the load Love numbers, $h_l(s), l_l(s), k_l(s)$, defined by [e.g., *Munk and MacDonald*, 1960]

$$
\begin{bmatrix} y_1(r,s) \\ y_3(r,s) \\ y_5(r,s) \end{bmatrix} =: \phi_l(r,s) \begin{bmatrix} h_l(r,s)/g^{(0)}(a) \\ l_l(r,s)/g^{(0)}(a) \\ 1 + k_5(r,s) \end{bmatrix}, \qquad (64)
$$

where ϕ_l is the spherical-harmonic coefficient of degree l of the local incremental potential due to the surface-mass density:

$$
\phi_l(r,s) = \frac{4\pi G a}{2l + 1} \left(\frac{r}{a} \right)^l \sigma_l. \qquad (65)
$$

Using (64), the load Love numbers can then be expressed as

$$
h_l(a,s) = h_l^e + \frac{h_l^{v1}}{s + s_l^1} + \frac{h_l^{v2}}{s + s_l^2}, \qquad (66)
$$

$$
l_l(a,s) = l_l^e + \frac{l_l^{v1}}{s + s_l^1} + \frac{l_l^{v2}}{s + s_l^2}, \qquad (67)
$$

$$
k_l(a,s) = k_l^e + \frac{k_l^{v1}}{s + s_l^1} + \frac{k_l^{v2}}{s + s_l^2}. \qquad (68)
$$

A special case is spherical-harmonic degree 1, for which the interface conditions (40) and (44) become linearly dependent [e.g., *Greff-Lefftz and Legros*, 1997]. A consequence is that a new condition must replace one of the interface conditions. In problems where body forces are involved, the origin of the coordinate system is usually placed at the center of mass of the earth. However, when modelling GIA, the common center of mass of the earth and the load is relevant. We therefore place the origin of the coordinate system at this point. This means that the degree-one spherical-harmonic coefficient of the local incremental potential vanishes outside the earth and (40) or (44) can be replaced by

$$
y_5(a,s) = 0. \qquad (69)
$$

As is easily seen, the coefficient a_1 in (48) also vanishes, resulting in only one eigenvalue for spherical-harmonic degree 1. Considering the relationship beween y_5 and z_5, we may regard $s = 0$ as a root of the secular determinant. It is shown however that the viscous amplitude vanishes for this root and, hence, there is no mode associated with this eigenvalue.

Table 2. Inverse relaxation times and load Love numbers for earth model H. l is the spherical-harmonic degree, s_l^1 and s_l^2 (in units of s^{-1}) are inverse relaxation times; h_l^e and h_l^{v1}, h_l^{v2} are elastic and viscous load Love numbers, respectively, for the radial surface displacement; k_l^e and k_l^{v1}, k_l^{v2} are elastic and viscous load Love numbers, respectively, for the local incremental potential; $v1$ and $v2$ refer to mode M0 and mode C0, respectively

l	$s_l^1 10^{11}$	$s_l^2 10^{11}$	$-h_l^e$	$-h_l^{v1} 10^{10}$	$-h_l^{v2} 10^{10}$	$-k_l^e 10^2$	$-k_l^{v1} 10^{11}$	$-k_l^{v2} 10^{11}$
1		1.3507	1.0223		0.0293	100.000		0.0000
2	6.5666	1.2792	0.5758	0.3889	0.1148	30.9796	2.1895	0.4564
3	4.6280	1.3194	0.6106	0.4811	0.1639	22.2178	1.8166	0.5083
4	3.1876	1.3381	0.5988	0.5001	0.2075	16.4459	1.4119	0.5254
5	2.3475	1.3242	0.5955	0.5055	0.2378	13.2082	1.1414	0.5054
6	1.8555	1.2813	0.6031	0.5284	0.2460	11.2607	0.9967	0.4488
7	1.5500	1.2179	0.6157	0.5802	0.2239	9.9446	0.9420	0.3566
8	1.3479	1.1428	0.6293	0.6623	0.1692	8.9633	0.9456	0.2387
9	1.2061	1.0636	0.6422	0.7568	0.0992	8.1823	0.9651	0.1255
10	1.1005	0.9865	0.6538	0.8325	0.0451	7.5353	0.9599	0.0517
20	0.6131	0.5467	0.7153	0.9947	0.0000	4.2228	0.5872	0.0000
50	0.2642	0.2327	0.7593	1.0824	0.0000	1.8197	0.2594	0.0000
100	0.1356	0.1189	0.7754	1.1152	0.0000	0.9337	0.1343	0.0000
500	0.0277	0.0242	0.7887	1.1430	0.0000	0.1907	0.0276	0.0000
800	0.0173	0.0152	0.7900	1.1456	0.0000	0.1194	0.0173	0.0000
1000	0.0139	0.0121	0.7904	1.1465	0.0000	0.0956	0.0139	0.0000

In order to implement the inverse Laplace transform, we consider impulsive loading, for which the Laplace-transformed spherical-harmonic coefficient σ_l is independent of s. Then, (62) can be readily transformed back into the t domain, giving

$$y_i(a,t) = [y_i^e \delta(t) + y_i^{v1} e^{-s_l^1 t} + y_i^{v2} e^{-s_l^2 t}]\sigma_l. \qquad (70)$$

The t domain relations corresponding to (66)–(68) are

$$h_l(a,t) = h_l^e \delta(t) + h_l^{v1} e^{-s_l^1 t} + h_l^{v2} e^{-s_l^2 t}, \qquad (71)$$

$$l_l(a,t) = l_l^e \delta(t) + l_l^{v1} e^{-s_l^1 t} + l_l^{v2} e^{-s_l^2 t}, \qquad (72)$$

$$k_l(a,t) = k_l^e \delta(t) + h_l^{v1} e^{-s_l^1 t} + k_l^{v2} e^{-s_l^2 t}. \qquad (73)$$

We note that the inverse relaxation times, s_l^1 and s_l^2, correspond to the negatives of the eigenvalues. Numerical results for the inverse relaxation times and the load Love numbers for earth models H, C and S are listed in Tables 2–4.

Table 3. Same as Table 2, but for earth model C

l	$s_l^1 10^{11}$	$s_l^2 10^{11}$	$-h_l^e$	$-h_l^{v1} 10^{10}$	$-h_l^{v2} 10^{10}$	$-k_l^e 10^2$	$-k_l^{v1} 10^{11}$	$-k_l^{v2} 10^{11}$
1		1.1844	1.1963		0.0257	100.000		0.0000
2	6.2880	1.1425	0.6801	0.4976	0.1010	31.5306	2.3800	0.3498
3	4.2534	1.1622	0.6962	0.6105	0.1355	22.0090	1.9781	0.3659
4	2.8852	1.1653	0.6687	0.6409	0.1573	16.0426	1.5646	0.3464
5	2.1215	1.1398	0.6560	0.6676	0.1572	12.7403	1.3100	0.2908
6	1.6853	1.0889	0.6572	0.7178	0.1315	10.7566	1.1810	0.2087
7	1.4191	1.0211	0.6651	0.7852	0.0886	9.4194	1.1147	0.1228
8	1.2435	0.9466	0.6749	0.8492	0.0478	8.4291	1.0616	0.0587
9	1.1174	0.8739	0.6845	0.8961	0.0216	7.6478	1.0016	0.0238
10	1.0197	0.8077	0.6931	0.9271	0.0087	7.0068	0.9373	0.0087
20	0.5552	0.4498	0.7386	1.0314	0.0000	3.8243	0.5340	0.0000
50	0.2349	0.1924	0.7697	1.0995	0.0000	1.6178	0.2311	0.0000
100	0.1197	0.0984	0.7808	1.1242	0.0000	0.8246	0.1187	0.0000
500	0.0243	0.0201	0.7898	1.1448	0.0000	0.1675	0.0243	0.0000
800	0.0152	0.0126	0.7907	1.1468	0.0000	0.1048	0.0152	0.0000
1000	0.0122	0.0101	0.7910	1.1475	0.0000	0.0839	0.0122	0.0000

Table 4. Same as Table 2, but for earth model S

l	$s_l^1 10^{11}$	$s_l^2 10^{11}$	$-h_l^e$	$-h_l^{v1} 10^{10}$	$-h_l^{v2} 10^{10}$	$-k_l^e 10^2$	$-k_l^{v1} 10^{11}$	$-k_l^{v2} 10^{11}$
1		0.3350	2.0244		0.0032	100.000		0.0000
2	5.6292	0.3567	1.2364	1.0911	0.0129	35.7623	3.1611	0.0288
3	3.3439	0.3320	1.0671	1.1835	0.0114	21.7725	2.4176	0.0197
4	2.2221	0.3075	0.9392	1.1481	0.0079	14.8178	1.8125	0.0111
5	1.6612	0.2813	0.8744	1.1202	0.0045	11.2593	1.4429	0.0053
6	1.3435	0.2560	0.8440	1.1099	0.0022	9.1869	1.2084	0.0022
7	1.1395	0.2331	0.8297	1.1092	0.0010	7.8248	1.0461	0.0009
8	0.9950	0.2131	0.8226	1.1122	0.0004	6.8447	0.9254	0.0003
9	0.8853	0.1959	0.8187	1.1161	0.0002	6.0947	0.8308	0.0001
10	0.7983	0.1810	0.8162	1.1198	0.0001	5.4971	0.7542	0.0000
20	0.4034	0.1020	0.8055	1.1370	0.0000	2.7786	0.3922	0.0000
50	0.1622	0.0439	0.7978	1.1454	0.0000	1.1172	0.1604	0.0000
100	0.0812	0.0225	0.7950	1.1479	0.0000	0.5594	0.0808	0.0000
500	0.0163	0.0046	0.7927	1.1497	0.0000	0.1120	0.0162	0.0000
800	0.0102	0.0029	0.7925	1.1498	0.0000	0.0700	0.0102	0.0000
1000	0.0081	0.0023	0.7924	1.1499	0.0000	0.0560	0.0081	0.0000

4. GEOPHYSICAL FIELD QUANTITIES

To calculate field quantities related to GIA, we must specify the geometry and history of the load model. For our purposes, it is sufficient to assume an ice sheet that is circular in plan view and parabolic in cross section. The space-dependent and time-dependent parts of the load are supposed to be separable. This means that, for the axially symmetric ice sheet assumed, $\sigma(\theta, t) = \sigma_s(\theta)\sigma_t(t)$ applies for the surface-mass density. Furthermore, in order that the total mass of the load is conserved, we suppose there is no degree-zero spherical-harmonic coefficient in its expansion, so that the total mass of the load vanishes at any time. Based on these assumptions, the space-dependent part of the load can be written as

$$\sigma_s(\theta) = \begin{cases} \sigma_I(\theta) - \sigma_W, & \theta < \alpha \\ -\sigma_W, & \theta > \alpha \end{cases}, \quad (74)$$

where σ_I and σ_W refer to the parabolic (ice) and uniform (water) parts of the load, respectively, and α denotes the angular radius of the parabolic part. It can be expressed as

$$\sigma_I(\theta) = \rho_I h(\theta) = \rho_I h_I \left(1 - \frac{\sin^2 \theta}{\sin^2 \alpha}\right), \quad (75)$$

in which ρ_I is the ice density and $h_I := h(0)$ the maximum ice thickness. The mass of the parabolic part, m, is given by

$$m_I := \rho_I \int_S h(\theta) dS, \quad (76)$$

yielding

$$h_I = \frac{3 \sin^2 \alpha m_I}{2\pi a^2 \rho_I (1 - 3\cos^2 \alpha + 2\cos^3 \alpha)}. \quad (77)$$

The uniform part of the load, σ_W, is determined by the condition that the total mass of the load vanishes:

$$\sigma_W = \frac{m_I}{4\pi a^2}. \quad (78)$$

In our numerical computation, we take $\alpha = 15.6525°$ and $m_I = 1.55 \times 10^{19}$ kg giving $h_I = 3298.54$ m, which are values typical of the former Laurentide ice sheet of North America. Considering that

$$\sigma_s(\theta) = \sum_{l=1}^{\infty} \sigma_l P_l(\theta), \quad (79)$$

the spherical-harmonic coefficients of the expansion are found to be

$$\sigma_l = -\frac{\rho_I h_I}{2\sin^2\alpha} \times$$

$$\begin{cases} -\frac{3}{4}\cos^4\alpha + \frac{3}{2}\cos^2\alpha - \frac{3}{4}, & l = 1, \\ -\cos^5\alpha + \frac{5}{3}\cos^3\alpha - \frac{2}{3}, & l = 2, \\ \frac{(l+1)(l+2)}{(2n+3)(2n+5)}[P_{l+3}(\cos\alpha) - P_{l+1}(\cos\alpha)] \\ +\frac{1}{2l+1}\left[\frac{(l+1)^2}{2l+3} + \frac{l^2}{2l-1}\right] \\ \times[P_{l+1}(\cos\alpha) - P_{l-1}(\cos\alpha)] \\ +\frac{l(l-1)}{(2l-1)(2l-3)}[P_{l-1}(\cos\alpha) - P_{l-3}(\cos\alpha)] \\ -\cos^2\alpha[P_{l+1}(\cos\alpha) - P_{l-1}(\cos\alpha)], & l > 2. \end{cases} \quad (80)$$

With a cut-off degree of 1000, the accuracy is of the order of 10^{-4}.

For the radial displacement, u_r, and the geoid height,

ϵ, the spherical-harmonic cofficients are expressible in terms of y_1 and y_5 or in terms of the load Love numbers $h_l(a,t)$ and $k_l(a,t)$. Considering $\epsilon_l(a,t) := y_5(a,t)/g^{(0)}(a)$ and the relevant formulae, we obtain

$$y_1(a,t) = h_l(a,t)\frac{4\pi aG}{2l+1}\frac{\sigma_l}{g^{(0)}(a)}, \qquad (81)$$

$$\epsilon_l(a,t) = [\delta(t) + k_l(a,t)]\frac{4\pi aG}{2l+1}\frac{\sigma_l}{g^{(0)}(a)}. \qquad (82)$$

The free-air gravity anomaly, $g^{(FA)}$, is defined as the incremental gravity with respect to the geoid. This means that the local incremental gravity, $g^{(\Delta)}$, must be corrected for the geoid height, ϵ. Thus, in terms of spherical-harmonic coefficients,

$$g_l^{(FA)}(r,t) = \frac{dy_5(r,t)}{dr} + \frac{dg^{(0)}(r)}{dr}\epsilon_l(r,t), \qquad (83)$$

which is to be evaluated below the surface-mass load or above the surface-mass load. We obtain, respectively,

$$g_l^{(FA)}(a^-,t) = [(l+2)\delta(t) - (l-1)k_l(t)]\frac{4\pi G}{2l+1}\sigma_l, \qquad (84)$$

$$g_l^{(FA)}(a^+,t) = -(l-1)[\delta(t) + k_l(t)]\frac{4\pi G}{2l+1}\sigma_l. \qquad (85)$$

For the loading history, $\sigma_t(t)$, we assume instantaneous deglaciation. With the time origin, $t = 0$, taken as the instant of deglaciation and $H(t)$ the Heaviside step function, we thus have

$$\sigma_t(t) = 1 - H(t), \qquad (86)$$

which must be convolved with the response to impulsive loading.

5. COMPUTATIONAL RESULTS

We begin by showing computational results in the spectral domain for Heaviside loading. For this, $y_i(a,t)$ given in (70) must be convolved with $H(t)$. Limiting our computations to the radial surface displacement, we obtain

$$\begin{aligned}y_{1N}(a,t) &= -[E_N + V_N^1(1 - e^{-s_l^1 t}) \\ &\quad + V_N^2(1 - e^{-s_l^2 t})]\frac{\sigma_l}{\rho^{(0)}(a)},\end{aligned} \qquad (87)$$

where E_N is the elastic radial amplitude and V_N^1 and V_N^2 are the viscous radial amplitudes associated with mode M0 and mode C0, respectively. We note that the results have been normalized so that, for $t \to \infty$, the relationship

$$T_N := E_N + V_N^1 + V_N^2 = 1 \qquad (88)$$

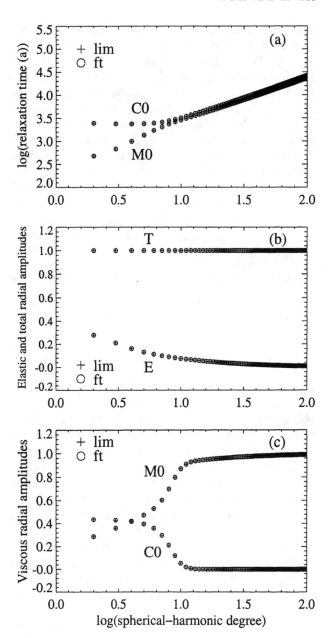

Figure 2. (a) Relaxation times, (b) elastic and total radial amplitudes and (c) viscous radial amplitudes as functions of spherical-harmonic degree. Results apply to the earth surface and to earth model H based on incompressible limit or incompressible field theory. Symbols M0 and C0 denote mantle mode and core mode respectively. All amplitudes are normalized with respect to total the radial amplitude.

applies, where T_N is the total radial amplitude.

Fig. 2 presents tests using the new solution obtained for compressibility. For this, we take the limit $\beta \to 0$ in the solution. This is equivalent to $k \to \infty$ and therefore represents the incompressible limit, for which the mantle density becomes homogeneous. Alternatively, the incompressible solution is obtained more directly by starting from the incompressible field equations [e.g., *Wu and Peltier*, 1982; *Amelung and Wolf*, 1994; *Wolf*, 1994]. We show numerical results for the relaxation times (Fig. 2a), the total and elastic radial amplitudes (Fig. 2b), and the viscous radial amplitudes (Fig. 2c). The results demonstrate that the new solution obtained matches very closely that based on the incompressible field equations. Furthermore, both solutions yield the correct long-time limit, $T_N = 1$.

In Fig. 3, we compare the solutions for earth models H, C and S in the spectral domain. Since, in each earth model, the surface density is different, we abandon the normalization introduced in (87) and henceforth use the relationship

$$y_1(a,t) = -[E + V^1(1 - e^{-s_l^1 t}) + V^2(1 - e^{-s_l^2 t})]\sigma_l. \quad (89)$$

Similarly, relation (88) for the long-time limit is replaced by

$$T := E + V^1 + V^2 = \frac{1}{\rho^{(0)}(a)}. \quad (90)$$

Fig. 3a shows that the difference between the relaxation times for earth models H and C is small. For example, for spherical-harmonic degree 2 and mode M0, the relaxation times are 482.896 a for earth model H and 504.292 a for earth model C. For the same spherical-harmonic degree and mode C0, the relaxation times are 2478.84 a for earth model H and 2775.42 a for earth model C. The reason for this is that the density jumps at the surface and the core–mantle boundary are similar for the two earth models. This is not the case for earth models H and S where, as shown in Fig. 3b, rather different relaxation times result. Thus, for earth model S and spherical-harmonic degree 2, the relaxation times for modes M0 and C0 are 563.312 a and 8890.96 a, respectively.

In Fig. 3c and d, we consider the total and elastic radial amplitudes. In view of its definition, the total radial amplitude is inversely proportional to the surface density, which is confirmed by our results. For spherical-harmonic degree 2, the elastic radial amplitudes are 0.06264 for earth model H, 0.07399 for earth model C and 0.13452 for earth model S (in cgs units). The relatively large difference between earth models H and S is again related to the difference in surface densities. For higher spherical-harmonic degrees, the absolute values and the relative differences become smaller. For example, for spherical-harmonic degree 100, the values are 2.09843×10^{-3} for earth model H and 2.15169×10^{-3} for earth model S.

The sum of the viscous radial amplitudes must equal the difference between the total and elastic radial amplitudes. Of interest is the distribution of the energy between modes M0 and C0. We first notice that the viscous radial amplitude of mode C0 rapidly decreases with increasing spherical-harmonic degree. This is a consequence of the enhanced attenuation with depth for perturbations of shorter wavelength. Thus, for spherical-harmonic degree 2, the viscous radial amplitudes are 0.09768 for earth model H and 0.03921 for earth model S, whereas, for spherical-harmonic degree 15, the amplitudes have diminished to 5.76354×10^{-5} and 4.75096×10^{-7}, respectively. The behaviour of mode M0 follows directly from this and, therefore, its relative importance increases with increasing spherical-harmonic degree. Since the elastic radial amplitude and the viscous radial amplitude for mode C0 approach zero for high spherical-harmonic degrees, the viscous radial amplitude for mode M0 simultaneously approaches the total radial amplitude. The limiting values are 0.2247 for earth model H, 0.2562 for earth model C and 0.3845 for earth model S. For spherical-harmonic degree 2, the viscous radial amplitudes for mode M0 are 0.06443 for earth model H, 0.08610 for earth model C and 0.21088 for earth model S.

We now proceed to the angular domain, where we consider the load model specified in Sec. 4. We begin by testing the accuracy of our solution. For earth model H and before deglaciation, $t < 0$, we obtain from (90) for $\theta > \alpha$

$$\begin{aligned} u_r(a, \theta > \alpha, t < 0) &= \frac{\sigma_W}{\rho^{(0)}(a)} = \frac{m_I}{4\pi a^2 \rho^{(0)}(a)} \\ &= 682.976 \text{ cm}, \end{aligned} \quad (91)$$

whereas our solution yields $u_r(a, \theta, t) = 682.82$ cm for $\theta = 48°$. From (90), it similarly follows for $\theta = 0$

$$\begin{aligned} u_r(a, 0, t < 0) &= -\frac{\sigma(0)}{\rho^{(0)}(a)} = -\frac{\rho_I h_I - \sigma_W}{\rho^{(0)}(a)} \\ &= -734.51 \text{ m}, \end{aligned} \quad (92)$$

which is again close to the value of -734.36 m given by our solution. For earth models C and S, our numerical results also meet this test very well.

Since the earth is assumed to be at hydrostatic equilibrium before deglaciation, the mass of the load, $\sigma(\theta, t < 0)$, is completely compensated by the mass deficit produced by the downward displacement, $\rho^{(0)}(a)u_r(a, \theta, t < 0)$. Therefore, the geoid height is zero and the free-air gravity anomaly refers to the undeformed sur-

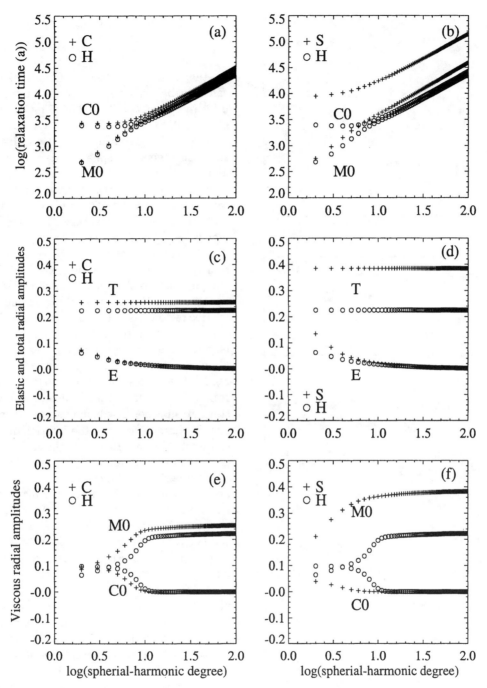

Figure 3. (a, b) Relaxation times, (c, d) elastic and total radial amplitudes and (e, f) viscous radial amplitudes as functions of spherical-harmonic degree. Results apply to the earth surface and to earth models H, C and S. Symbols M0 and C0 denote the mantle mode and core mode, respectively. All amplitudes are in cgs units.

face of the earth. For a point immediately above the surface load, the free-air gravity anomaly is also zero, whereas, for a point immediately below the surface load, it is $-4\pi G\sigma(\theta, t < 0)$. We therefore get

$$
\begin{aligned}
g^{(FA)}(a, \theta > \alpha, t < 0) &= 4\pi G\sigma_W \\
&= 2.54807 \text{ mGal}
\end{aligned}
\tag{93}
$$

in the ice-free region and

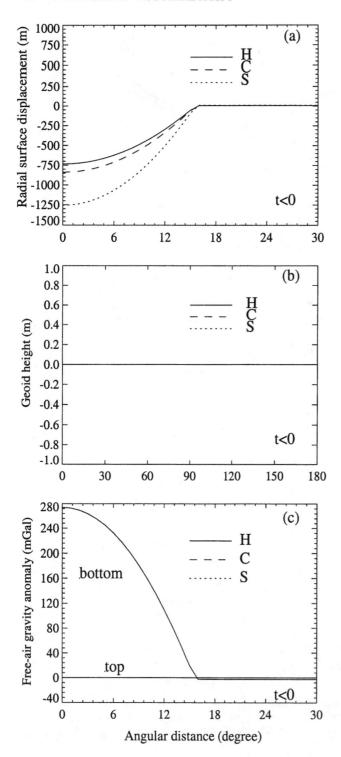

$$g^{(FA)}(a, 0, t < 0) = 4\pi G(\sigma_W - \rho_I h_I)$$
$$= -274.035 \text{ mGal} \qquad (94)$$

on the axis of ice sheet. Equations (93) and (94) can also be used to test our solution, which returns $g^{(FA)}(a, \theta > \alpha, t < 0) = 2.5475$ mGal for $\theta = 48°$ and $g^{(FA)}(a, 0, t < 0) = -273.98$ mGal. On the axis of the ice sheet, the agreement is slightly worse because of the slow convergence of spherical-harmonic series at $\theta = 0$ and $180°$. The test results are shown in Fig. 4.

Next, we consider the relaxation following the instantaneous deglaciation at $t = 0$. We begin with a comparison of the radial surface displacements for earth models H, C and S (Fig. 5). Fig. 5a shows the radial displacement at $t = 0^+$. Since the elastic deformation recovers instantaneously, the radial displacement is equivalent to the total viscous displacement, which is negative in the ice-covered region and positive in the ice-free region. As in hydrostatic equilibrium, the values are governed by the surface density and are therefore largest for earth model S. Fig. 5b–f shows the relaxation process following load removal. In the ice-covered region, the depressed surface starts to rise while, in the region far away from the ice sheet, the upwarped surface starts to subside. The behaviour in the intermediate region is characterized by the interference of both types of radial motion. However, the uplift dominates and is particularly pronounced just outside the ice margin, where a peripheral bulge develops (Fig. 5b and c). The points between uplift and subsidence are near $\theta = 76°$ for earth model H, near $\theta = 80°$ for earth model C and near $\theta = 85°$ for earth model S. As relaxation proceeds, the peripheral bulge reaches its maximum and begins to collapse. The critical times are about $t = 1700$ a for earth model H, $t = 1800$ a for earth model C and $t = 2000$ a for earth model S (Fig. 5b–d). The peak amplitudes of the bulge located near $\theta = 16°$ are 74.445 m for earth model H, 89.538 m for earth model C and 161.78 m for earth model S. 10000 a after load removal, the residual radial displacements on the load axis are -9.4533 m for earth model H, -14.488 m for earth model C and -50.343 m for earth model S, which corresponds to about 1.29 per cent, 1.73 per cent and 4.01 per cent, respectively, of the maximum subsidence (Fig. 5f). This shows that, in agreement with Fig. 3, the relaxation proceeds more slowly for earth models C and S than for earth model H.

Next, we compare the geoid heights for earth models H, C and S (Fig. 6). At $t = 0^+$ (Fig. 6a) and in the ice-covered region, the strongly negative geoid height reflects the mass deficit associated with the surface de-

Figure 4. (a) Radial surface displacement, (b) geoid height and (c) free-air gravity anomaly as functions of angular distance from the load axis. The free-air gravity anomaly is shown above and below the surface load. Results apply to the hydrostatic limit and to earth models H, C and S.

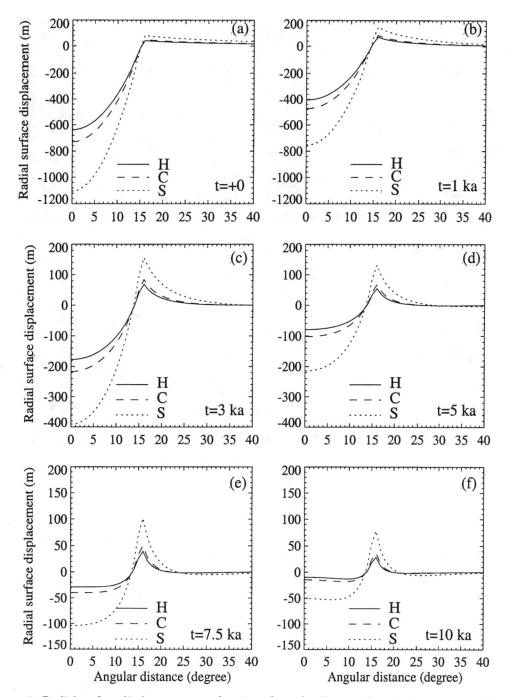

Figure 5. Radial surface displacement as a function of angular distance from the load axis for different times after load removal. Results apply to earth models H, C and S.

pression. In the ice-free region, the slightly positive geoid height is produced by the excess mass related to the upwarped surface. A small contribution is also due to the warping of the core–mantle boundary. The intercept points are near $\theta = 30°$ for earth models H and C and near $\theta = 29°$ for earth model S. As relaxation proceeds, the geoid height shows the recovery of the surface depression in the ice-covered region and the development of the peripheral bulge outside the ice margin (Fig. 7c–e). Whereas the central geoid low weakens gradually, the peripheral geoid high is a transient feature, which ultimately collapses (Fig. 7e and f). An

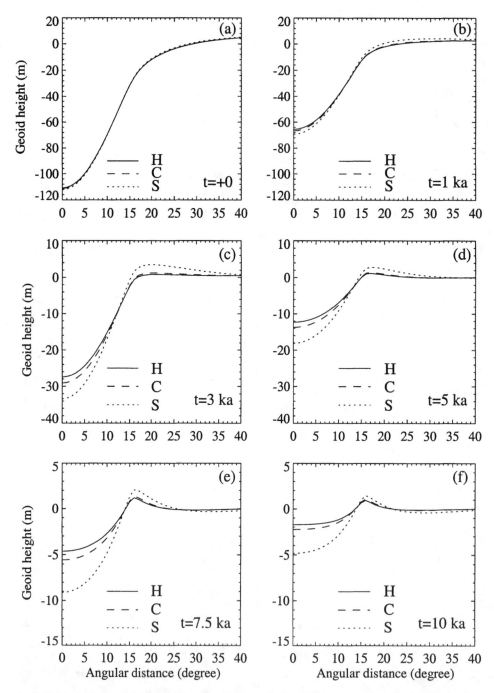

Figure 6. Geoid height as a function of angular distance from the load axis for different times after load removal. Results apply to earth models H, C and S.

exception is earth model S, for which the high decreases continuously. We note that the geoid height is smaller in magnitude by about a factor of ten than the radial surface displacement. Initially, the values are nearly identical for the three earth models, because the surface masses produced by the warped surface are nearly

the same. As time proceeds, the geoid height relaxes fastest for earth model H followed by earth models C and H in accordance with the differences in relaxation times for the three models.

Fig. 7 shows the differences between the free-air gravity anomalies for the three earth models. The sources

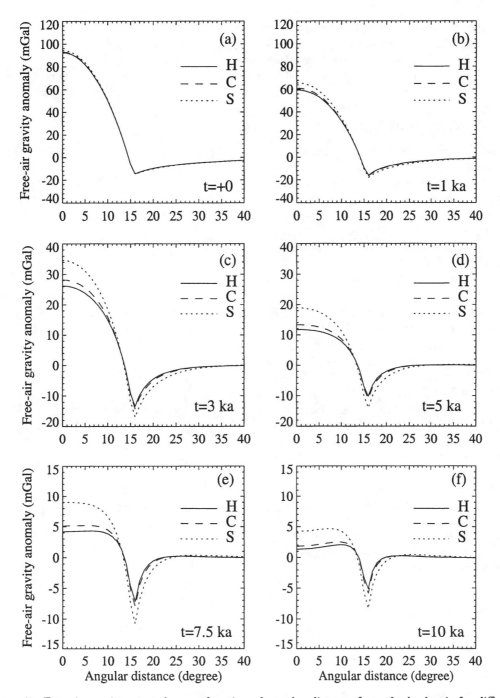

Figure 7. Free-air gravity anomaly as a function of angular distance from the load axis for different times after load removal. Results apply to earth models H, C and S.

of the anomalies agree with those responsible for the perturbations of the geoid. In addition, an effect arises from referring the gravity anomaly to the local geoid. This gives a positive contribution in the ice-covered region and a negative contribution in the ice-free region. However, the free-air gravity anomaly is negative in the ice-covered region and positive in the ice-free region. As for the geoid height, the differences between the three earth models are initially small (Fig. 7a and b), but the relative differences continuously increase as time proceeds due to the differences in relaxation times mentioned before (Fig. 7c–f). Of interest is the narrow positive ridge in the region of the peripheral bulge, where the following maximum values occur: 16.207 mGal near $t = 900$ a for earth model H, 16.646 mGal near $t = 1000$ a for earth model C and 18.271 mGal near $t = 1400$ a for earth model S.

6. CONCLUDING REMARKS

We have studied the problem of load-induced, gravitational–viscoelastic perturbations of a compressible earth model initially in hydrostatic equilibrium. In particular, we have derived an explicit solution to the perturbation equations for a spherical earth model consisting of a compositionally homogeneous mantle surrounding a fluid core. Novel features considered in our solution are the following.

1. The initial density stratification applies to compositionally homogeneous shells and is given by Darwin's law, which can be shown to satisfy the field equations governing the initial state and, in particular, be consistent with the assumption of compressibility. This feature distinguishes our study from those by *Plag and Jüttner* [1995], *Vermeersen et al.* [1996b], *Hanyk et al.* [1999], *Wieczerkowski* [1999] and *Vermeersen and Mitrovica* [2000], who solved the problem for an earth model consisting of shells with homogeneous density.

2. We have assumed that the gravitational–viscoelastic perturbations are compressible. The perturbed state is thus consistent with the initial state and no instabilities arise in the solution to the perturbation equations. In contrast to this, solutions for compressible perturbations of earth models consisting of homogeneous-density shells may show such instabilities, which was discussed in detail in the investigations referenced above.

3. The sole approximation admitted in our study is that the compressible perturbations are constrained by the assumption of local incompressibility. This approximation is almost perfect near the fluid limit of the material but less satisfactory near the elastic limit.

4. Our study represents a generalization of the solution derived by *Wolf and Kaufmann* [2000] for a plane earth model without IGF. The generalized problem for a spherical earth model with IGF and consisting of an arbitrary number of compositionally homogeneous, compressible shells is solved in the paper by *Martinec et al.* [2001] by means of propagator matrices.

APPENDIX A: MATRICES IN THE GENERAL SOLUTION

The explicit forms of matrices D and $F(r)$ are as follows:

$$D = \begin{bmatrix} 1 & 1 & 1 \\ D_{21} & D_{22} & D_{23} \\ \frac{l-\beta+3}{l(l+1)} & \frac{2-l-\beta}{l(l+1)} & \frac{\nu+\tau+2-\beta}{l(l+1)} \\ 1+\frac{l-\beta+3}{l+1} & 2+\frac{\beta-2}{l} & D_{43} \\ 0 & 0 & 0 \\ 0 & 0 & 0 \end{bmatrix}$$

$$\begin{matrix} 1 & \frac{\beta A}{\alpha_1 \alpha_2} & \frac{\beta A}{\gamma_1 \gamma_2} \\ D_{24} & D_{25} & D_{26} \\ \frac{\nu-\tau+2-\beta}{l(l+1)} & \frac{l+3-2\beta}{l(l+1)}\frac{\beta A}{\alpha_1 \alpha_2} & \frac{2-2\beta-l}{l(l+1)}\frac{\beta A}{\gamma_1 \gamma_2} \\ D_{44} & D_{45} & D_{46} \\ 0 & 1 & 1 \\ 0 & l & -(l+1) \end{matrix} \Bigg], \quad (A1)$$

$$\begin{aligned} F(r) &= \text{diag}\,[F_1(r), F_2(r), F_3(r), \\ & \quad F_4(r), F_5(r), F_6(r)] \\ &= \text{diag}\,[r^{l+\beta}, r^{-l+\beta-1}, r^{\nu+\tau+\beta-1}, \\ & \quad r^{\nu-\tau+\beta-1}, r^l, r^{-(l+1)}], \end{aligned} \quad (A2)$$

where

$$D_{21} = e_1 D_{11} + e_3 D_{31} - (l+3) D_{41}, \quad (A3)$$

$$D_{22} = e_1 D_{12} + e_3 D_{32} + (l-2) D_{42}, \quad (A4)$$

$$D_{23} = e_1 D_{13} + e_3 D_{33} - (\nu+\tau+2) D_{43}, \quad (A5)$$

$$D_{24} = e_1 D_{14} + e_3 D_{34} - (\nu-\tau+2) D_{44}, \quad (A6)$$

$$\begin{aligned} D_{25} &= e_1 D_{15} + e_3 D_{35} + (\beta-3-l) D_{45} \\ & \quad -A D_{55}, \end{aligned} \quad (A7)$$

$$\begin{aligned} D_{26} &= e_1 D_{16} + e_3 D_{36} + (l+\beta-2) D_{46} \\ & \quad -A D_{56}, \end{aligned} \quad (A8)$$

$$D_{43} = 1 + \frac{(\nu+\tau+2-\beta)(\nu+\tau-1)}{l(l+1)}, \quad (A9)$$

$$D_{44} = 1 + \frac{(\nu-\tau+2-\beta)(\nu-\tau-1)}{l(l+1)}, \quad (A10)$$

$$D_{45} = \left[1 + \frac{(l+3-2\beta)(l-\beta)}{l(l+1)}\right]\frac{\beta A}{\alpha_1 \alpha_2}, \quad (A11)$$

$$\begin{aligned} D_{46} &= \left[1 + \frac{(l+2\beta-2)(l+\beta+1)}{l(l+1)}\right] \\ & \quad \times \frac{\beta A}{\gamma_1 \gamma_2}, \end{aligned} \quad (A12)$$

$$\nu = \frac{1}{2}(\beta-3), \tau = \frac{1}{2}[(\beta-1)^2 + 4l(l+1)]^{\frac{1}{2}}, \quad (A13)$$

$$\alpha_1 = (l-\beta)(l-\beta+1) - l(l+1), \quad (A14)$$

$$\alpha_2 = \frac{(l-\beta+1)(l-2\beta+4)+2-\beta}{l(l+1)} - 1, \quad (A15)$$

$$\gamma_1 = (l + \beta)(l + \beta + 1) - l(l + 1), \quad \text{(A16)}$$

$$\gamma_2 = \frac{(l + \beta)(l + 2\beta - 3) + 2 - \beta}{l(l + 1)} - 1, \quad \text{(A17)}$$

$$e_1 = 2(\beta - 3), \quad e_3 = 4l(l + 1) - 2. \quad \text{(A18)}$$

APPENDIX B: MATRIX B IN THE INTERFACE CONDITIONS

The explicit form of matrix $B(\mu(s), \chi_1, \chi_2, \chi_3, \chi_4)$ is as follows:

$$B = \begin{bmatrix} B_{11} & B_{12} & B_{13} \\ B_{21} & B_{22} & B_{23} \\ D_{41}F_1(a) & D_{42}F_2(a) & D_{43}F_3(a) \\ D_{41}F_1(b) & D_{42}F_2(b) & D_{43}F_3(b) \\ k_3 D_{21}F_1(a) & k_3 D_{22}F_2(a) & k_3 D_{23}F_3(a) \\ k_4 D_{21}F_1(b) & k_4 D_{22}F_2(b) & k_4 D_{23}F_3(b) \end{bmatrix}$$

$$\begin{bmatrix} B_{14} & B_{15} & B_{16} \\ B_{24} & B_{25} & B_{26} \\ D_{44}F_4(a) & D_{45}F_5(a) & D_{46}F_6(a) \\ D_{44}F_4(b) & D_{45}F_5(b) & D_{46}F_6(b) \\ k_3 D_{24}F_4(a) & B_{55} & B_{56} \\ k_4 D_{24}F_4(b) & B_{65} & B_{66} \end{bmatrix}, \quad \text{(B1)}$$

where

$$B_{1j} = [\chi_1 \mu(s) D_{2j} + \chi_2 k_1 D_{1j}] F_j(a), \quad \text{(B2)}$$

$$B_{2j} = \{\chi_3 \mu(s)[D_{2j} + (\delta_{5j} + \delta_{6j})k_5 D_{5j}] \\ -\chi_4 k_2 D_{1j}\} F_j(b), \quad \text{(B3)}$$

$$B_{55} = [k_3 D_{25} + D_{65} + (l + 1)D_{55}] F_5(a), \quad \text{(B4)}$$

$$B_{56} = [k_3 D_{26} + D_{66} + (l + 1)D_{56}] F_6(a), \quad \text{(B5)}$$

$$B_{65} = [k_4 D_{25} + D_{65} + (k_6 - l)D_{55}] F_5(b), \quad \text{(B6)}$$

$$B_{66} = [k_4 D_{26} + D_{66} + (k_6 - l)D_{56}] F_6(b), \quad \text{(B7)}$$

$$k_1 = \rho^{(0)}(a^-)g^{(0)}(a)a, \quad \text{(B8)}$$

$$k_2 = \delta\rho^{(0)}(b)g^{(0)}(b)b, \quad \text{(B9)}$$

$$k_3 = 4\pi G \frac{a^{1-\beta}}{g^{(0)}(a)}, \quad \text{(B10)}$$

$$k_4 = 4\pi G \frac{b^{1-\beta}}{g^{(0)}(b)}, \quad \text{(B11)}$$

$$k_5 = \rho^{(0)}(b^-)b^\beta, \quad \text{(B12)}$$

$$k_6 = 4\pi G \frac{\rho^{(0)}(b^-)b}{g^{(0)}(b)}. \quad \text{(B13)}$$

Acknowledgments. We would like to thank Bert Vermeersen and an anonymous reviewer for constructive comments.

REFERENCES

Amelung, F., and D. Wolf, Viscoelastic perturbations of the earth: significance of the incremental gravitationalforce in models of glacial isostasy, *Geophys. J. Int., 117*, 864–879, 1994.

Bullen, K.E., *The Earth's Density*, 420 pp., Chapman and Hall, London, England, 1975.

Darwin, G.H., On the figure of equilibrium of a planet of heterogeneous density, *Proc. R. Soc. London, 36*, 158–166, 1884.

Dziewonski, A.M., and D.L. Anderson, Preliminary reference earth model, *Phys. Earth Planet. Inter., 25*, 297–356, 1981.

Greff-Lefftz, M., and H. Legros, Some remarks about the degree-one deformation of the earth, *Geophys. J. Int., 131*, 699–723, 1997.

Hanyk, L., C. Matyska, and D.A. Yuen, Secular gravitational instability of a compressible viscoelastic sphere, *Geophys. Res. Lett., 26*, 557–560, 1999.

Li, G., and D.A. Yuen, Viscous relaxation of a compressible shell, *Geophys. Res. Lett., 14*, 1227–1230, 1987.

Martinec, Z., Spectral–finite element approach to three-dimensional viscoelastic relaxation in a spherical earth, *Geophys. J. Int., 142*, 117–141, 2000.

Martinec, Z., and D. Wolf, Explicit form of the propagator matrix for a multi-layered, incompressible viscoelastic sphere, *Sci. Techn. Rep., GFZ Potsdam, STR98/08*, 13 pp., 1998.

Martinec, Z., M. Thoma, and D. Wolf, Material versus local incompressibility and its influence on glacial–isostatic adjustment, *Geophys. J. Int., 144*, 136–156, 2001.

Munk, W.H., and G.J.F. MacDonald, *The Rotation of the Earth: a Geophysical Discussion*, 323 pp., Cambridge University Press, Cambridge, England, 1960.

Plag, H.-P., and H.-U. Jüttner, Rayleigh–Taylor instabilities of a self-gravitating earth, *J. Geodyn., 20*, 267–288, 1995.

Sabadini, R., D.A. Yuen, and E. Boschi, Polar wandering and forced response of a rotating, multilayered, viscoelastic planet, *J. Geophys. Res., 87*, 2885–2903, 1982.

Spada, G., R. Sabadini, D.A. Yuen, and Y. Ricard, Effects on post-glacial rebound from the hard rheology in the transition zone, *Geophys. J. Int., 109*, 683–700, 1992.

Vermeersen, L.L.A., and J.X. Mitrovica, Gravitational stability of spherical self-gravitating relaxation models, *Geophys. J. Int., 142*, 351–360, 2000.

Vermeersen, L.L.A., and R. Sabadini, A new class of stratified viscoelastic models by analytical techniques, *Geophys. J. Int., 129*, 531–570, 1997.

Vermeersen, L.L.A., R. Sabadini, and G. Spada, Analytical visco-elastic relaxation models, *Geophys. Res. Lett., 23*, 697–700, 1996a.

Vermeersen, L.L.A., R. Sabadini, and G. Spada, Compressible rotational deformation, *Geophys. J. Int., 126*, 735–761, 1996b.

Wieczerkowski, K., Gravito-Viskoelastodynamik für verallgemeinerte Rheologien mit Anwendungen auf den Jupitermond Io und die Erde, *Veröff. Deut. Geod. Komm., Ser. C, 515,* 130 pp., 1999.

Williamson, E.D., and L.H. Adams, Density distribution in the earth, *J. Washington Acad. Sci., 13,* 413–428, 1923.

Wolf, D., The relaxation of spherical and flat Maxwell earth models and effects due to the presence of the lithosphere, *J. Geophys., 56,* 24–33, 1984.

Wolf, D., Lamé's problem of gravitational viscoelasticity: the isochemical, incompressible planet, *Geophys. J. Int., 116,* 321–348, 1994.

Wolf, D., Gravitational viscoelastodynamics for a hydrostatic planet, *Veröff. Deut. Geod. Komm., Ser. C, 452,* 96 pp., 1997.

Wolf, D., Gravitational–viscoelastic field theory, in *Dynamics of the Ice Age Earth: a Modern Perspective,* edited by P. Wu, pp. 55–86, Trans Tech Publications, Uetikon, Switzerland, 1998.

Wolf, D., and G. Kaufmann, Effects due to compressional and compositional density stratification on load-induced Maxwell viscoelastic perturbations, *Geophys. J. Int., 140,* 51–62, 2000.

Wu, J., and D.A. Yuen, Post-glacial relaxation of a viscously stratified compressible mantle, *Geophys. J. Int., 104,* 331–349, 1991.

Wu, P., and W.R. Peltier, Viscous gravitational relaxation, *Geophys. J. R. Astron. Soc., 70,* 435–485, 1992.

Wu, P., and Z. Ni, Some analytical solutions for the viscoelastic gravitational relaxation of a two-layer non-self-gravitating incompressible spherical earth, *Geophys. J. Int., 126,* 413–436, 1996.

D. Wolf and G. Li, GeoForschungsZentrum Potsdam, Kinematics and Dynamics of the Earth, Telegrafenberg, D-14473 Potsdam, Germany. (email: dasca@gfz-potsdam.de)

Glacial Isostatic Adjustment on a Three-dimensional Laterally Heterogeneous Earth: Examples From Fennoscandia and the Barents Sea

Georg Kaufmann

Institut für Geophysik, Universität Göttingen, 37075 Göttingen, Germany

Patrick Wu

*Department of Geology and Geophysics, University of Calgary,
Calgary, Alberta T2N 1N4, Canada*

Fennoscandia and the Barents Sea have both been covered by large ice sheets during the last glacial phases. Evidence of the dynamical response of these regions to changes in the ice sheet distribution has been recorded in numerous observables, e.g. successions of submerged or uplifted strandlines. The time-dependent relaxation of the crust and mantle is still observable today, as seen in the distribution of present-day velocities, free-air gravity anomalies, and the earthquake potential. While the dependence on the radial viscosity profile of glacially-induced observations is well understood, effects of lateral viscosity variations are often neglected in modeling the glacial isostatic adjustment process.

We develop fully three-dimensional earth models with lateral viscosity variations in the lithosphere and asthenosphere, derived from seismological observations and tectonical setting. Based on these earth models, we calculate present-day velocities, free-air gravity anomalies, and the earthquake potential for a three-dimensional ice sheet distribution over Fennoscandia and the Barents Sea. Comparisons with radially symmetric earth models enable us to study the influence of lateral viscosity variations on glacial signatures, which can be significant (e.g. 1-2 mm/yr in present-day velocities, 2-4 mGal for free-air gravity anomalies). Also, the earthquake fault potential in Fennoscandia is significantly influenced by lateral viscosity variations.

1. INTRODUCTION

The Late Pleistocene glacial cycles are responsible for large-scale redistributions of ice and water on the Earth's surface. Both the additional weight of the Late Pleistocene ice sheets such as the Laurentide and Fennoscandian Ice

Sheets and the corresponding mass deficits over the oceans due to falling sea levels have deformed the Earth's surface, a process still observable at present due to the time-dependent viscous relaxation of the Earth's interior.

The adjustment of the Earth to the changing loads is successfully modeled within the framework of global glacial isostatic adjustment theory [*Peltier*, 1974; *Farrell and Clark*, 1976; *Mitrovica et al.*, 1994a; *Milne and Mitrovica*, 1998]. However, while the complex three-dimensional variation of the combined ice and water load is fully taken into account [*Mitrovica et al.*, 1994b; *Peltier*, 1996; *Lambeck et*

Ice Sheets, Sea Level and the Dynamic Earth
Geodynamics Series 29

10.1029/029GD18

al., 1998], the properties of the Earth's crust and mantle are assumed to vary in the radial direction only. This latter assumption, which simplifies the modeling procedure significantly, ignores the complex three-dimensional structure of the Earth's interior as seen by seismic tomographical imaging [*Su and Dziewonski,* 1991; *Li and Romanowicz,* 1996; *Trompert,* 1998].

Several studies have been carried out in the past to infer the effects of lateral variations in the rheology of the Earth's interior on observations related to the glacial isostatic adjustment process. While most of these studies, based on simple axisymmetrical ice load histories, were intended to assess the effects of some three-dimensional variations in the Earth's mantle in general [*Sabadini and Gasperini,* 1989; *Gasperini and Sabadini,* 1989; *Kaufmann et al.,* 1997; *Giunchi et al.,* 1997; *Kaufmann and Wu,* 1998a; *Ni and Wu,* 1998], studies on more realistic, fully three-dimensional ice and earth models are rare [*Kaufmann and Wu,* 1998b; *Wu et al.,* 1998; *Kaufmann et al.,* 2000]. Recently, the theoretical framework for spherical models of glacial isostatic adjustment on a fully three-dimensional earth has been laid out [*Martinec,* 1999, 2000; *Tromp and Mitrovica,* 1999a, 1999b, 2000].

In this paper, we present model calculations for glacially-induced present-day velocities and gravity anomalies for two regions once covered by prominent ice sheets: Fennoscandia and the Barents Sea. In addition the earthquake potential and the fault instability is also calculated for Fennoscandia. As these two regions are adjacent to the continental shelf, the structure of the lithosphere and mantle is expected to vary both in the radial and the lateral direction. We have developed three-dimensional earth models characterizing the uppermost mantle structure of these two regions and compare model predictions for the three-dimensional models with reference models, whose properties vary in the radial direction only. Realistic three-dimensional ice models for both Fennoscandia and the Barents Sea are chosen to represent the deglaciation during the last glacial cycle. Hence, our calculations provide valuable insight into the variability of glacially-induced present-day velocities, gravity anomalies, and fault instability, if a fully three-dimensional structure of the Earth is taken into account.

2. EARTH MODELS

A layered, isotropic, compressible, Maxwell-viscoelastic halfspace with a constant gravitational attraction of $g = 9.81$ m s^{-2} is used to model the glacially-induced perturbations of the solid Earth in the Fennoscandian region. As it has been shown earlier [e.g., *Wolf,* 1984; *Amelung and Wolf,* 1994], the flat-earth approximation is adequate to describe

the glacial isostatic adjustment for an ice model of the size of the former Late Pleistocene ice cover over Fennoscandia. We solve the Boussinesq problem for a layered, viscoelastic halfspace using the commercial finite-element package Abaqus, which has been modified to include pre-stress in order to allow the deformed free surface to return to its initial equilibrium via viscous flow [*Wu,* 1992a,1992b]. Thus, the equation that describes the conservation of momentum is given by:

$$\nabla \cdot \sigma - \rho g \nabla w = 0, \qquad (1)$$

where σ is the incremental stress tensor, ρ the density, g the gravitational acceleration, and w is the vertical displacement. The first term in equation (1), the divergence of stress, describes the surface force deforming the Earth. The second term arises because the undisturbed Earth is assumed to be in hydrostatic equilibrium, with the forces of self–gravitation balanced by the hydrostatic pre–stress. This pre–stress is being advected along with the material when the body deforms either elastically or viscoelastically. Thus, the second term in equation (1) represents the gradient of the advected pre-stress, $\rho g w$. The presence of this term is required in order to provide the buoyancy force that is needed to satisfy the boundary conditions in the fluid limit, and without this term, there would be no viscous gravitational relaxation. The validity of the finite-element model to predict glacial isostatic adjustment has been shown previously [*Wu and Johnston,* 1998].

Earth models consist of an elastic lithosphere, a viscoelastic asthenosphere and an underlying viscoelastic mantle. Density, shear and bulk modulus are volume-averaged values derived from PREM [*Dziewonski and Anderson,* 1981]. Mantle viscosity within each sublithospheric layer is taken as free parameter, as well as lithospheric thickness. We compare three sets of earth models. Set R represents laterally homogeneous reference models with constant viscosity within each layer. For Fennoscandia, set L simulates the laterally varying earth model properties as described in section 3.1. Model B1 is a representation of a simple laterally variable earth model for the Barents Sea region as described in section 4.1.

3. FENNOSCANDIA

3.1. Setting

Fennoscandia is dominated by the Precambrian basement rocks of the Baltic Shield, with younger sequences around the western edges resulting from the Caledonian Orogeny, and the Permo-Carboniferous rocks of the Oslo Graben to the south (Fig. 1a). The Baltic Shield encompasses mainland Norway, Sweden, Finland, the Kola Peninsula, and

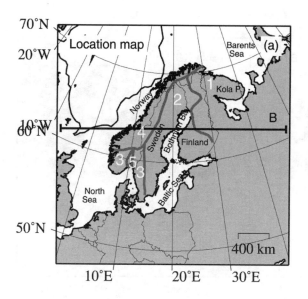 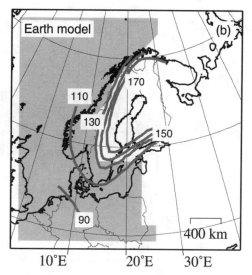

Figure 1. (a) Location map of Scandinavia, including the tectonic provinces **1** Suecokarelides, **2** Suecofennides, **3** Sueconorwegian, and encircled by more recent provinces: **4** Caledonides, **5** Oslo Graben. Profile B is used for model predictions. The thick solid line indicates the 1000 m isobath of the continental margin. (b) Lateral lithosphere and viscosity variation employed for models. The solid lines indicate the lithospheric thickness (in km, from Calcagnile, 1982), the grey-shaded area the asthenospheric low-viscosity zone (from Panza et al., 1980 and Calcagnile, 1982).

Soviet Karelia. Four orogenic episodes between 3100 and 1500 Ma, the Saamian (3100–2900 Ma), Lopian (2900–2650 Ma), Suecofennian (2000-1750 Ma), and Gothian Orogeny (1750–1500 Ma), have formed the stable Baltic Shield, which is characterized by a thick crust and low heatflow [e.g., *Čermák and Ryback*, 1979; *Balling*, 1995]. The shield is zoned with its oldest rocks exposed in Northwest Finland and the Kola Peninsula (Suecokarelides), the younger Suecofennides of western Finland and southeastern Sweden, and the most recent Sueconorwegian of southern Norway. The latter areas have been accreted and reworked during the Gothian and Grenvillian (1250–900 Ma) Orogenies. The closing of the Iapetus between 650–400 Ma resulted in the Caledonian Orogeny, which reworked the westernmost parts of the Baltic Shield and added oceanic fragments and eclogitic rocks along the Norwegian coast (Caledonides). During the last tectonically active phase between 305–245 Ma, the Oslo Graben opened as an intracontinental rift.

As a result of the different orogenies, crustal and subcrustal structure beneath Fennoscandia is expected to vary considerably from the offshore regions of the Norwegian coast towards the central parts of the old Baltic Shield complex. For example, the inference of the lithosphere asthenosphere boundary by *Calcagnile* [1982] shown in Fig. 1b reveils a strong lateral variation from values around 90 km along the southern margins of Fennoscandia and increasing

towards 170 km in the central parts of the Baltic Shield. These inferences are based on the inversion of eleven seismic profiles of Rayleigh-wave dispersion data, with crustal thickness estimates fixed to the *Bungum et al.* [1980] inference, and a variable subcrustal thickness. The increase of lithospheric thickness reveals a sharp gradient from the Caledonian Province towards the Baltic Shield. Additionally, *Calcagnile* [1982] has found indications for a low-viscosity asthenosphere derived from a low-velocity zone present beneath the Caledonides and the Baltic Sea, which is absent underneath the older parts of the Baltic Shield (Suecokarelides and part of Suecofennides). The low-velocity zone is weaker underneath the southern boundary of the shield (Sueconorwegian), an indication for a stronger asthenosphere in that part. *Kukkonen and Peltonen* [1999] revised a geotherm for the Fennoscandian Shield, based on new petrological data on mantle kimberlites and a numerical model of the thermal conditions of the lithosphere. The authors deduced a thermal thickness of the lithosphere around 130 – 185 km for southwestern Finland, and additionally found no evidence for an asthenosphere underneath the central parts of the Baltic Shield. Their results are thus in agreement with earlier inferences from seismic studies. Both the lithospheric thickness variation and the low-viscosity zone underneath central Europe, but absent underneath the Baltic Shield, has been confirmed by a recent study of P-wave tomography using the TOR array [*Arlitt et al., 2000*].

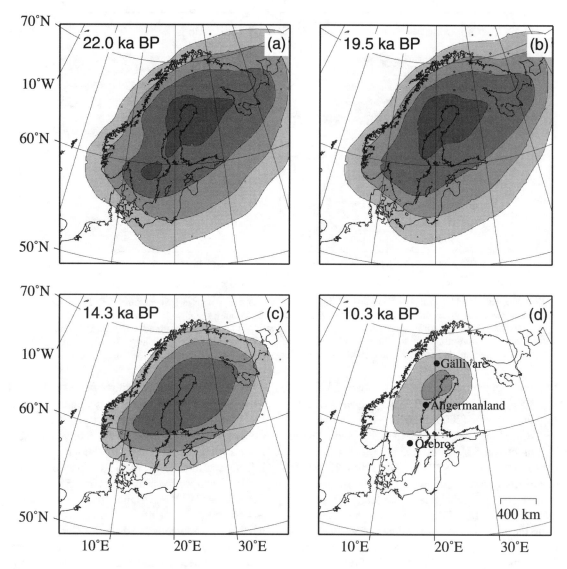

Figure 2. Ice model FBK8 over Scandinavia at four different epochs. Contours are drawn every 500m, starting from the zero contour. In (d), the three locations Gällivare, Ångermanland, and Örebro are shown.

3.2. Ice Model

The ice load model FBK8 (Fig. 2) used in this approach has been derived by *Lambeck et al.* [1998] from forward modeling predictions of relative sea-level change in Northern Europe. The ice model encompasses two glacial cycles and reached its maximum extent at the last glacial maximum (LGM) at around 22 ka BP. The deglaciation history throughout the Late Pleistocene is based on ice retreat isochrons from *Andersen* [1981] and *Pedersen* [1995]. Deglaciation ended around 9 ka BP. Ice thickness is inferred by a least-squares fit of the relative sea-level observations, and is characterized by a relatively modest maximum ice height of around 1800 m at the last glacial maximum. The

ice model has been modified to match the coarser finite-element grid, which has 225 4-node bilinear elements on the surface with dimensions of 200 km over the area of interest. Rigid boundary conditions are applied to the sides and bottom, which surround the area of interest with a margin of 63,500 km on each side. In the finite-element implementation, the ice volume is 3.3×10^6 km^3, corresponding to 9.6 m of eustatic sea-level rise.

3.3. Results

We start establishing differences in model predictions, which arise from radial variations in mantle viscosity alone. Introducing the first set of earth models R1-R3 (table 1),

Table 1. Parameters of the earth models, with h the thickness of a layer and η the viscosity. The subscripts denote lithosphere (lith), asthenosphere (asth), upper (um) and lower (lm) mantle, respectively.

Model	h_{lith} [km]	η_{lith} [Pa s]	h_{asth} [km]	η_{asth} [Pa s]	h_{um} [km]	η_{um} [Pa s]	h_{lm} [km]	η_{lm} [Pa s]
R1	72	∞	-	-	598	10^{21}	2216	10^{21}
R2	72	∞	-	-	598	7.0×10^{20}	2216	7.0×10^{21}
R3	72	∞	-	-	598	3.6×10^{20}	2216	1.7×10^{22}
L1	variable[1]	∞	-	-	598	3.6×10^{20}	2216	1.7×10^{22}
L1'	variable[2]	∞	-	-	598	3.6×10^{20}	2216	1.7×10^{22}
L2	72	∞	100	variable[3]	498	3.6×10^{20}	2216	1.7×10^{22}
L3	variable[1]	∞	100	variable[3]	498	3.6×10^{20}	2216	1.7×10^{22}
B1	72	∞	100	variable[4]	498	3.6×10^{20}	2216	1.7×10^{22}

[1] Lithospheric thickness variation as in Fig. 1b, but reduced by a factor of two (45-85 km).
[2] Lithospheric thickness variation as in Fig. 1b (90-170 km).
[3] Asthenospheric viscosity variation as in Fig. 1b (from 10^{18} Pa s in the grey area to 3.6×10^{20} Pa s elsewhere).
[4] Asthenospheric viscosity variation as in Fig. 8b.

we aim to infer the variability of predictions of present-day velocities and gravity anomalies subject to a uniform mantle (R1), and two layered mantle models with viscosity increasing across the 670 km seismic discontinuity by a factor of 10 (R2) and 47 (R3), respectively. The average mantle viscosity therefore differs significantly between the three models with 10^{21} (R1), 3.4×10^{21} (R2), and 8.3×10^{21} Pa s (R3). We note that the latter two viscosity profiles are in accordance with recent modeling predictions of glacial isostatic adjustment in northern Europe [e.g., *Lambeck et al.,* 1998]. All models R1 to R3 have in common a laterally uniform elastic lithosphere.

We continue to discuss the effects of lateral variations of physical properties in the earth model. We therefore introduce the second set of earth models L1-L3 (table 1), whose upper and lower mantle properties are similar to model R3. Lateral variations are allowed either in lithospheric thickness (L1,L1'), in asthenospheric viscosity (L2), or in both parameters (L3). Variations in lithospheric thickness are based on observational evidence discussed earlier and therefore resemble a reasonable model for the region. However, as the seismological and rheological lithosphere are conceptually different, we have scaled the lithospheric thickness estimates of *Calcagnile* [1982] by 0.5 for model L1 and 1.0 for model L1'. The average lithospheric thickness for these two models is 52 and 104 km, respectively. Since the presence of an asthenospheric layer is less reliable, we adopt a simple model with a 100 km thick low-viscosity asthenosphere underneath the Caledonides and the Sueconorwegian, but which is absent underneath the central Baltic Shield. The

chosen asthenospheric viscosity represents an upper limit (see also section 4.3). Both models of lateral variations follow the pattern shown in Fig. 1b.

3.3.1. Present-day Velocities. In Fig. 3 (left column), both radial and horizontal velocity predictions are shown for the three radially varying models R1-R3. Focusing on the radial velocity patterns first, we observe a pattern of positive uplift rates up to 6 mm yr^{-1} over the Fennoscandian Peninsula for model R1, thus coinciding with the former LGM ice margin. Beyond that area, the surface experiences slight subsidence, not exceeding 2 mm yr^{-1}. In contrast, models R2 and R3 predict a much broader area of positive uplift rates, extending towards the Atlantic Ocean, the Barents Sea, and encompassing Denmark, which is a result of the stiffer mantle rheology, maintaining a higher degree of disequilibrium. In addition, the maximum values are higher by about 1-3 mm yr^{-1}. Uplift rates above 2 mm yr^{-1} are now predicted over the entire Fennoscandian region.

Next, we discuss predictions of present-day velocities for the laterally heterogeneous models L1-L3 (Fig. 3, right column). Comparing radial uplift rates for models L1 and R3, we observe that the two predicted patterns are essentially similar. Closer inspection reveals that contours over Sweden and Finland are almost identical, while a slight shift towards the east is detectable over Norway for model L1. We can confirm these differences by discussing the uplift rates along an W-E profile through the center of the former ice sheet (see line B in Fig. 1a). In Fig. 4a, predictions for models L1 and R3 are indistinguishable over the eastern part of the profile, while small differences are present over the west-

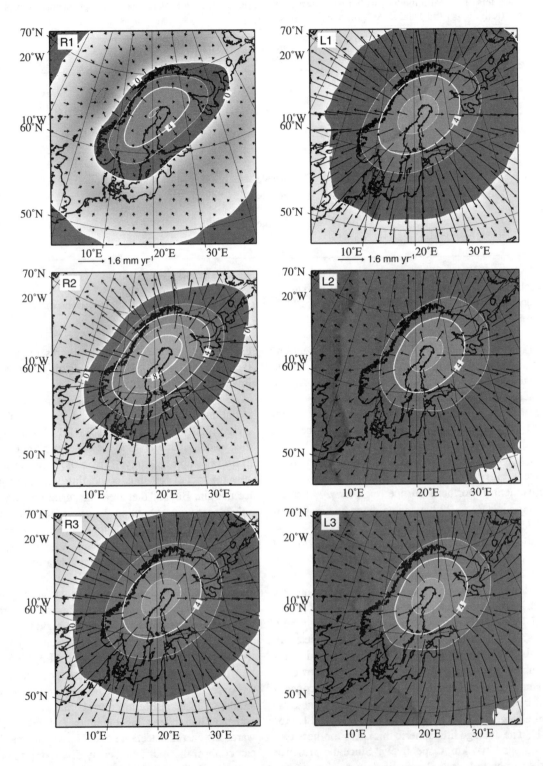

Figure 3. Predicted uplift rates (contour lines, in mm yr^{-1}) and horizontal present-day velocities (arrows) for earth models R1-R3 (left panels) and earth models L1-L3 (right panels).

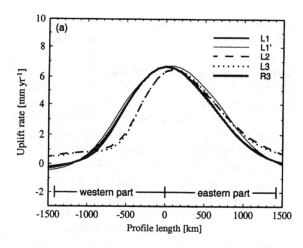
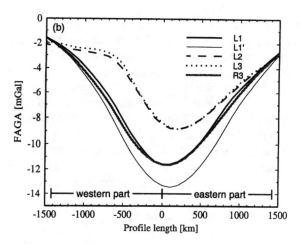

Figure 4. (a) Predicted uplift rates along an W-E profile for different earth models as indicated in the legend. (b) Predicted free-air gravity anomalies along an W-E profile for different earth models as indicated in the legend.

ern section. These differences are the result of the thinner lithosphere underneath that section, which results in smaller uplift rate predictions for model L1. When we assume a generally thicker laterally variable lithosphere as in model L1', differences are more pronounced. In Fig. 4, predictions for models L1' and R3 are identical over the western part of the profile, while differences of up to 1 mm yr^{-1} arise over the eastern section. These differences are the result of the much thicker lithosphere underneath that section for model L1', which results in larger uplift rate predictions.

The influence of a laterally heterogeneous asthenosphere is more readily visible in the comparison of uplift rates predictions between models L2 and R3. While the maximum of both predictions is similar, the distance between uplift contours over Norway and southern Sweden has converged for model L2, but the zero contour line extents around 100 km further to the east over Finland, when compared to model R3. We can quantify the effect by discussing the two corresponding profiles L2 and R3 shown in Fig. 4a. Here, the introduction of an asthenosphere significantly reduces the uplift rate over the western part of the profile by about 2 mm yr^{-1}. Differences over the eastern part are only minor, with slightly larger uplift rates for L2 close to the easternmost end. Finally, allowing both lateral variations in lithospheric thickness and asthenospheric viscosity as in model L3, the resulting uplift pattern reflects both the effects discussed above: The presence of an asthenosphere is seen in the reduced uplift rates over the western part of the profile, which are about 2 mm yr^{-1} smaller, while the thin lithosphere underneath Norway is responsible for the further reduction in uplift rates along the western section of the profile.

Differences in predictions of horizontal velocities are even more pronounced, which is in agreement with published results [*Mitrovica et al.,* 1994b]. While no significant horizontal movement is predicted for model R1, velocities are below 0.4 mm yr^{-1} with convergence dominating the pattern outside the formerly glaciated area, models R2 and R3 predict horizontal movements in excess of 1 mm yr^{-1}, oriented radially away from Bothnian Bay. The extensional deformation is significant far beyond the former ice margins. The predicted patterns for models R2 and R3 are a result of the stiffer mantle rheology, which has maintained a larger degree of disequilibrium than for model R1. In general, all predictions are symmetrical around the former ice margin, as no other ice sheets such as the Laurentide ice sheet are taken into account in this regional study.

Significant differences between the laterally homogeneous model R3 and the laterally heterogeneous models L2 and L3 can be seen in predictions of horizontal velocities (Fig. 3). Here, velocities are reduced by a factor of three or more over the North Sea and the North Atlantic, which can be attributed to the more rapid relaxation of the asthenosphere underneath that region. However, horizontal velocities over Finland and eastern Europe remain almost unchanged. We note that changes in lithospheric thickness as in model L1 have no significant effect on predictions of horizontal velocities.

3.3.2. Present-day Free-air Gravity Anomalies. Predictions of present-day gravity anomalies can be used as indicator of the remaining departure from isostatic equilibrium. As mantle material is still displaced and is moving backwards towards its initial position before the onset of the glacial loading, both the remaining displacement and the gravity anomaly are correlated. Following *Gasperini et al.* [1991], we derive predictions of gravity anomalies by converting the remaining present-day displacement, using a Bouguer correction. We use a gravity-displacement ratio of 0.19 mGal m^{-1}, which is in agreement with a newer inference of 0.2 mGal m^{-1} derived from Fennoscandian uplift and gravity measurements [*Ekman and Mäkinen*, 1996; *Ekman*, 1998]. In Fig. 5 (left column), predictions of gravity anomalies for the three models R1-R3 are shown. The predictions are characterized by a negative pattern with minima around -11 to -17 mGal centered over Bothnian Bay. Again, predictions for models R2 and R3 experience a much broader pattern, encompassing the entire region shown, while the pattern for model R1 is closely following the LGM ice margin. The predicted minima can also be interpreted as remaining land uplift. Using the gravity-displacement ratio introduced above, this results in predictions of remaining land uplift in the range of 55-90 m, which is somewhat lower than the recent estimate of about 90 m by *Ekman and Mäkinen* [1996].

Finally, we discuss effects from lateral variations in lithospheric thickness and asthenospheric viscosity on predictions of the free-air gravity anomaly. The predicted patterns for the three laterally heterogeneous models L1-L3 are shown in Fig. 5 (right column), and predictions along the previously introduced W-E profile are redrawn in Fig. 4b. As a general remark, we observe that the effects are more pronounced than for the present-day velocities. When we compare predictions for models L1 and R3, we observe a similar shape of the free-air gravity anomaly for both models. Again, contour lines have slightly shifted to the east over the Norwegian coast. This can be seen more readily on the profile shown in Fig. 4b, with a slight reduction in free-air gravity anomaly for model L1 over the western section of the profile, when compared to model R3. In contrast, predictions for model L1' significantly differ from the reference model R3 along the entire profile. Here, the much thicker lithosphere underneath the central Baltic Shield results in an 20 percent increase in amplitude, with minimum values around -12 mGal for model R3, but around -13.5 mGal for model L1'. A slight asymmetry in the profile predictions can also be seen, resulting from the asymmetric change in lithospheric thickness. In contrast, the effect of a lateral variation in asthenosphere viscosity modifies both the pattern and the amplitude of the predicted free-air gravity anomaly. Contour lines for model L2 have converged over Norway, but diverge over Finland, when compared to model R3. This can also be seen in the profile prediction, with predictions along the western part of the profile reduced by about 5 mGal, and only reaching a minimum of -9 mGal. Predictions over the eastern section are also reduced, but to a lesser degree.

As expected, the combination of lateral variations in lithospheric thickness and asthenospheric viscosity is a superposition of effects. Differences are present over the western part of the profile. Here, the faster relaxation due to the presence of the asthenosphere obviously dominates the response. Along the eastern section, predictions of models L3 and R3 are almost similar, because the competing effects of lithospheric thickness variation and asthenospheric viscosity variation compensate each other.

3.3.3. Fault Instability. In this section we study the effects of lateral variations in lithospheric thickness and asthenospheric viscosity on glacially-induced fault instability and earthquakes. Recent studies [*Wu*, 1997; *Wu*, 1998; *Johnston et al.*, 1998; *Wu et al.*, 1999] have shown that postglacial faults in Fennoscandia and Eastern Canada are likely to be caused by glacial unloading, which ended around 9 ka BP. Even the intraplate earthquakes in Eastern Canada today are probably triggered by the stresses due to postglacial rebound, although tectonic stress is needed to keep the pre-existing faults near the condition of failure [*Wu and Hasegawa*, 1996]. The rebound stress in Fennoscandia is also large enough to trigger seismicity today, however, more information on local fault and stress properties is required to understand the modern seismicity pattern and mode of failure there [*Wu et al.*, 1999].

In this and the above studies, earthquake or fault potential is related to the state of stress by the Mohr-Coulomb Failure criteria. The quantity calculated is the change in Fault Stability Margin (dFSM) [see *Wu and Hasegawa* [1996] for details of its computation]. This quantity varies in space and time as rebound stress changes. When dFSM is positive, the Mohr circle at that location moves away from failure, thus pre-existing faults which are originally close to failure become stabilized. However, when dFSM is negative, then the Mohr circle moves towards failure and pre-existing faults which were originally close to failure can become reactivated. During glaciation, the normal stress due to the load generally gives positive values of dFSM under the load, i.e. fault stability [*Johnston*, 1987]. However, after deglaciation the values of dFSM become negative and faults become unstable. The time when dFSM at a location first becomes negative is the onset time of faulting. When dFSM is negative, its magnitude is an indicator of the amount of rebound stress available to trigger earthquakes.

The effects of lithospheric thickness, upper and lower mantle viscosities in a laterally homogeneous earth on the onset time and the magnitude of maximum instability in Fennoscandia have been studied by *Wu et al.* [1999]. They found that lower mantle viscosity has little effect on fault stability. However, the onset time is most sensitive to lithospheric thickness and upper mantle viscosity in the north

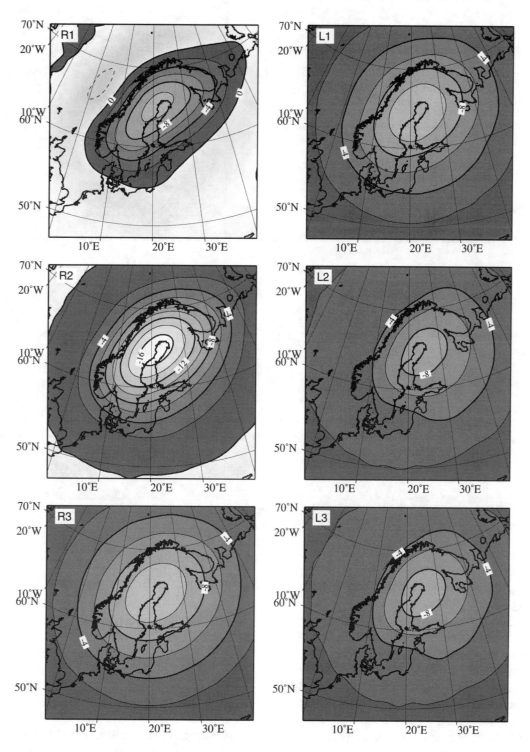

Figure 5. Predicted free-air gravity anomalies (in mGal) for earth models R1-R3 (left panels) and earth models L1-L3 (right panels).

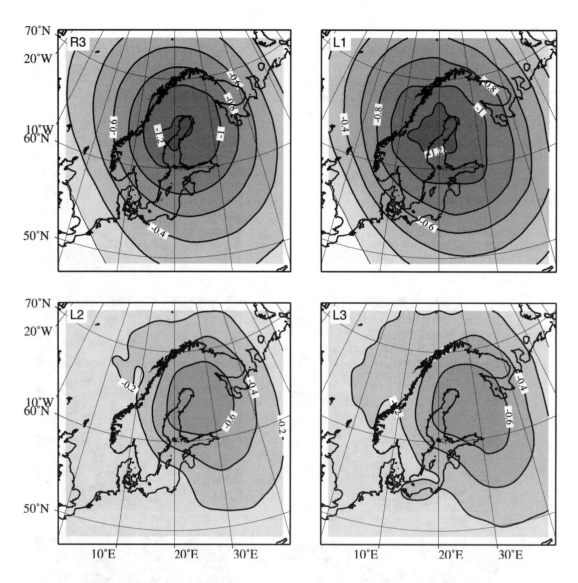

Figure 6. Fault stability margin at the present for four different earth models, R3, L1-L3.

of Fennoscandia near Gällivare, but the minimum dFSM is most sensitive to lithospheric thickness and upper mantle viscosity in Ångermanland near the center of rebound (see Fig. 2d for locations). The effects of lateral variations in lithospheric thickness and/or asthenospheric viscosity on the spatial distribution of dFSM at the present are shown in Fig. 6. For laterally homogenous models such as model R3, the contours are generally symmetrical around the former load margin. Introducing a laterally heterogeneous lithosphere (model L1), the shape of the contours remains symmetrical but the magnitude increases slightly. For a laterally heterogeneous asthenosphere (model L2), the contours become asymmetrical, and contour lines converge over Nor-

way, but diverge over Finland and Russia. When both a laterally heterogeneous asthenosphere and lithosphere are present (model L3), the result is similar to model L2 but the magnitudes are slightly larger, and the zero contour moves west. The contour maps for all these models are very similar at the last glacial maximum (not shown), but start to show different patterns from 10 ka BP on, since as noted before, the load pressure determines the stess field during the glacial phase.

The temporal variations of the dFSM at three representative sites near the center of rebound are shown in Fig. 7. It is evident that the presence of lateral viscosity variations has an insignificant effect on the onset times of faulting in

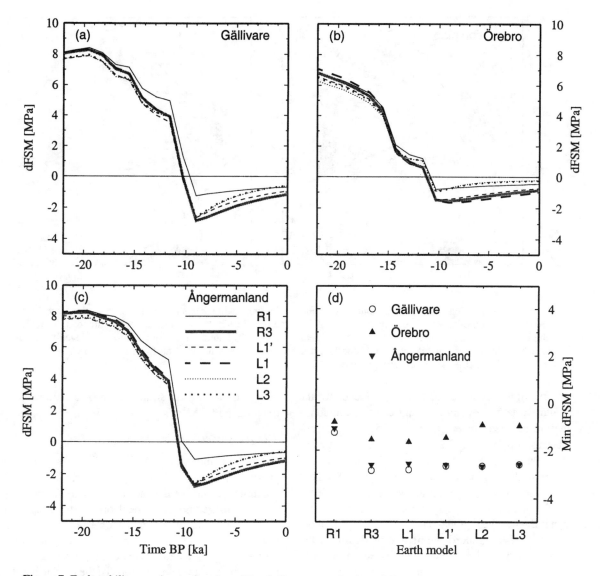

Figure 7. Fault stability margin as a function of time before present for three different locations (a)-(c), and minimum fault stability margin (d).

Ångermanland, Gällivare, and Örebro (i.e. time shifts of less than 1 ka). Changes in the radial viscosity profile are the main contribution to differences in fault instability (models R1 and R3) at Ångermanland and Gällivare, while at Örebro the effect from the low-viscosity asthenosphere on the onset time becomes evident, but is still small (model R3 and models L2, L3). As shown in Fig. 7d, the dependence of the minimum fault instability on earth rheology is minor, with the exception of Örebro, where models L2 and L3 tend to increase the dFSM value. Again, the radial viscosity profile is the main contributor to differences (models R1 and R3). In summary, the location of the zero contour and the magnitude of maximum instability is most sensitive to the presence of a low-viscosity asthenosphere.

4. BARENTS SEA

4.1. Setting

The Barents Sea is a shallow ocean basin in northwestern Europe including the arctic islands of the Svalbard Archipelago, Franz Joseph Land and Novaya Zemlya (Fig. 8). The area was covered by an essentially marine ice sheet during the Late Pleistocene, which reached glacial maximum some 22 ka ago. Although the extent of the ice sheet at the

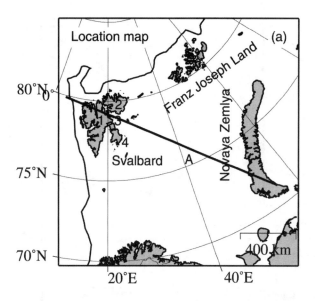
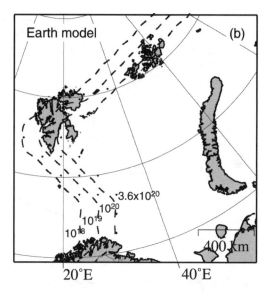

Figure 8. (a) Location map of the Barents Sea, including the four main islands of the Svalbard Archipelago Spitsbergen (1), Nordaustlandet (2), Barentsøya (3), and Edgeøya (4). Profile A is used for model predictions. The thick solid line indicates the 1000 m isobath of the continental margin. (b) Lateral viscosity variation employed for models. The dashed lines separate regions of constant asthenospheric viscosity, starting from 10^{18} Pa s in the west to 3.6×10^{20} Pa s in the east, as assumed for model B1.

last glacial maximum is still controversial, evidence for the northern and western ice margins being close to the continental margin has been given [*Rutter*, 1995]. The ocean-continent transition from the Greenland deep-ocean basin to the shallow Barents Sea continental shelf [*Eldholm et al.*, 1987] occurs in that region, therefore we can expect lateral variations in rheological properties across the Barents Sea.

It is likely that lateral heterogeneities in the uppermost mantle can influence the regional pattern of land uplift in the northern Barents Sea. Indeed, numerical modeling of the deglaciation history of the Barents Sea, based on laterally homogeneous earth models, performed by *Breuer and Wolf* [1995] and *Kaufmann and Wolf* [1996] indicates that land uplift in the northern Barents Sea is best explained by assuming an increase in asthenospheric viscosity towards the Eurasian continent. Later on, *Kaufmann et al.* [1997] have shown that the quantitative inference of lateral viscosity variations required a proper treatment of glacial isostatic adjustment modeling using laterally heterogeneous earth models. This suggestion has been investigated with incompressible earth models in *Kaufmann and Wu* [1998a,1998b], where we have shown that lateral viscosity variations are significant for the predictions of land uplift as well as present-day velocities and gravity anomalies. In this section, we extend the results of *Kaufmann and Wu* [1998a,1998b], and include a more realistic earth model with compressible Maxwell rhe-

ology and a realistic stratification of elastic properties derived from PREM.

4.2. Ice Model

The load model BARENTS-2 (Fig. 9) used in this approach has been developed by *Kaufmann et al.* [1996]. The maximum extent at the last glacial maximum is in accordance with geomorphological observations by *Grosswald* [1980,1993] and numerical reconstructions from *Isaksson* [1992]. The space-time history of the deglaciation largely follows the deglaciation isochrons proposed by *Andersen* [1981], with a rapid deglaciation of the marine ice sheet over the central Barents Sea between 22 ka BP and 17 ka BP, and a slower disintegration of the remaining ice caps over the Svalbard Archipelago, Franz Joseph Land and Novaya Zemlya. These remaining ice caps disappeared at about 6.4 ka BP. In *Kaufmann and Wu* [1998a,1998b], we assumed that the ice model was in isostatic equilibrium prior to the last glacial maximum. In this paper, we extend the ice model to encompass the last two glacial cycles, simulating the slow ice build up over 90,000 years, and the more rapid deglaciation phases of 10,000 years.

The ice model has been modified to match the coarser finite-element grid, with 196 4-node bilinear elements each 121 km wide over the area of interest, and a 20 times larger grid surrounding the area. Both the vertical and bottom

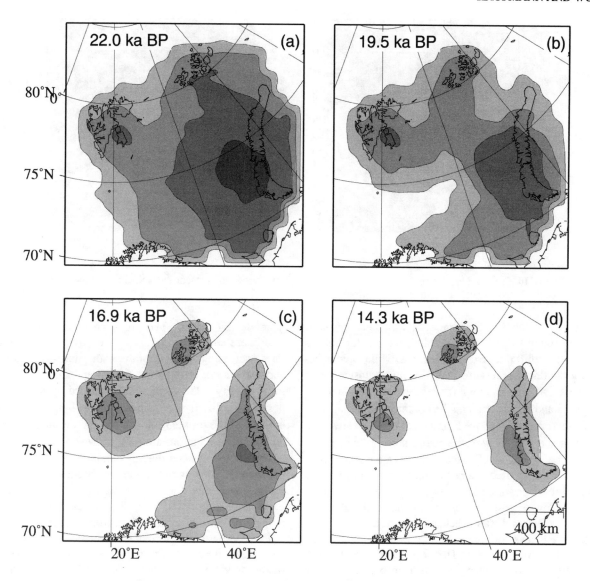

Figure 9. Ice model BARENTS-2 over the Barents Sea at four different epochs. Contours are drawn every 1000m, starting from the zero contour.

boundaries are no-slip boundaries. In the finite-element implementation, the ice volume is 3×10^6 km^3, corresponding to 9 m of eustatic sea-level rise.

4.3. Results

We discuss our results for the reference earth model R3 (table 1) with a variation of viscosity in the radial direction only, and a three-dimensional model B1 including a laterally variable asthenospheric viscosity (10^{18} Pa s to 3.6×10^{20} Pa s) as indicated in Fig. 8b. The lateral asthenospheric viscosity variation is chosen to match the sugges-

tion of *Kaufmann and Wolf* [1996]. However, no attempt has been made so far to correlate the variation to Tertiary surface tectonic features as the Spitsbergen fracture zone [*Maher et al., 1997*]. The lateral variation chosen should also be regarded as an upper limit, smaller variations as discussed in *Wu et al.* [1998] and *Kaufmann and Wu* [1998] result in less prominent differences.

4.3.1. Present-day Velocities. Present-day velocities predicted for the Barents Sea region for earth models R3 and B1 are shown in Fig. 10. The velocities reflect the ongoing rebound of the Earth due to the remaining disequilibrium. We observe a broad area of ongoing uplift indicated by the pos-

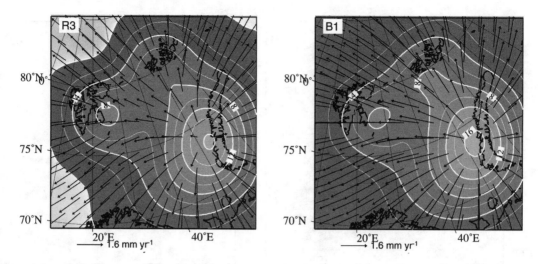

Figure 10. Predicted uplift rates (contour lines, in mm yr^{-1}) and horizontal present-day velocities (arrows) for earth model R3 (left panel) and earth model B1 (right panel).

itive uplift rate for both earth models, which encompasses the entire Barents Sea region. At the center of the former ice sheet the uplift rates exceed 16 mm/a, while regions beyond the continental margin experience minor subsidence up to 2 mm/a. Differences in uplift rates between the two earth models R3 and B1 occur along the continental margin, with significant reductions (above 1 mm/yr) of uplift rate over the western area of Spitsbergen, the main island of the Svalbard Archipelago. For model R3, the entire area covering Spitsbergen is being uplifted at a rate of more than 4 m/yr, while for model B1 rates are well below 4 mm/yr. Hence, the presence of the low-viscosity region around the continental margin results a faster relaxation beneath that area.

Next, we discuss uplift rates along a profile in the Barents Sea (line A in Fig. 8a). The profile is chosen to sample the area, in which changes of asthenospheric viscosity

in model B1 occur. The predicted uplift rates along the profile are shown in Fig. 11a. We observe positive uplift rates increasing from 1 mm/a at the northwestern end to about 16 mm/a at the southeastern end for model R3. A low-viscosity asthenosphere as proposed by model B1 reduces the uplift rates by about 1 mm/a along the northwestern part of the profile (200-600 km). Over the southeastern end, model predictions are essentially similar, as both models resemble the same rheology underneath the central Barents Sea. We conclude that the measurement of present-day uplift rates can be a useful tool to discriminate between the earth models proposed here, if the measurements are taken along profiles transsecting the continental shelf area.

Proceeding to the horizontal velocities, both models are characterized by flow directed radially away from the center of the previously glaciated area, with a westerly to north-

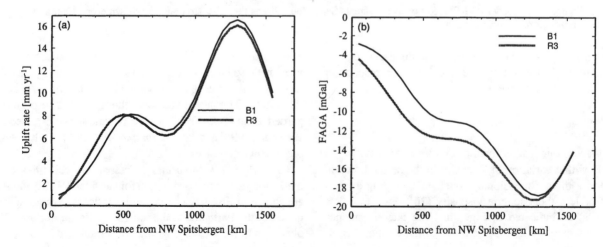

Figure 11. (a) Predicted uplift rates along a profile for different earth models as indicated in the legend. (b) Predicted free-air gravity anomalies along a profile for different earth models as indicated in the legend.

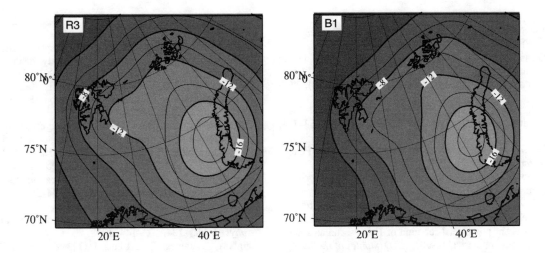

Figure 12. Predicted free-air gravity anomalies (in mGal) for earth model R3 (left panel) and earth model B1 (right panel).

westerly flow across the Svalbard Archipelago of about 1.5 to 2 mm/a. For model B1 the flow field over the Svalbard Archipelago and Franz Joseph Land is somewhat intensified.

4.3.2. Present-day Free-air Gravity Anomalies. In Fig. 12, we discuss present-day free air gravity anomalies for earth models R3 and B1. For model R3, we observe negative values with a minimum of about −18 mGal located over the center of the former ice sheet. Except for the westernmost part of Spitsbergen Island, the entire Svalbard Archipelago experiences negative free-air gravity anomalies below −8 mGal. In contrast, upon the introduction of a low-viscosity asthenosphere underneath that area (model B1) the entire Svalbard Archipelago is characterized by free-air gravity anomalies above −8 mGal.

This can also be seen on profile A. In Fig. 11b, free air gravity anomalies for profile A are shown, with values ranging from −4 mGal at the northwestern end to −19 mGal at the southeastern end for the laterally homogeneous model R3. For the heterogeneous model B1, the signal is reduced by 2 − 3 mGal between 0 and 1000 km, as the region has relaxed faster than in model R3. Over the southeastern end, predictions for both models are similar.

5. CONCLUSIONS

For the two regions Fennoscandia and the Barents Sea, we have developed three-dimensional viscoelastic earth models, which are based on seismological and tectonical field evidence. We have shown that these earth models with their three-dimensional structure can significantly influence glacial isostatic adjustment, when compared to the generally used radially symmetric earth models.

Glacial signatures such as present-day velocities, free-air gravity anomalies and the earthquake potential are strongly

affected by the presence of a low-viscosity asthenosphere underneath the Atlantic and central Europe, which peters out towards the central parts of the European continent. Changes in lithospheric thickness result in less pronounced differences. Hence, modern space-geodetical techniques, such as GPS and satellite-gravity measurements, have the potential to discriminate between simplified radially symmetric and more realistic three-dimensional earth models.

In addition, repeated measurements with absolute gravimeters can reduce the uncertainty in solid-surface gravity anomalies to less than 0.2 μGal/yr [e.g., *Ekman and Mäkinen,* 1996; *Lambert et al.,* 1996]. As our models result in secular rates for the solid-surface gravity up to 1 − 3 μGal/yr in amplitude, we would be able to discriminate between different earth models on the basis of repeated gravity measurements.

A low-viscosity asthenosphere underneath the Caledonides and central Europe also has the potential to reduce the fault instability in Denmark and southern Norway and Sweden, after the Weichselian Ice Sheet has melted completely.

Acknowledgments. We would like to thank Kurt Lambeck for providing the Fennoscandian ice model FBK8. The figures in this paper are drawn using the GMT graphics package [*Wessel and Smith,* 1991].

REFERENCES

Amelung, F. and Wolf, D., Viscoelastic perturbations of the Earth: significance of the incremental gravitational force in models of glacial isostasy, *Geophys. J. Int.,* **117 (3)**, 864–879, 1994.

Andersen, B. G., Barents Sea and Arctic islands, in: *The Last Great Ice Sheets,* edited by G. H. Denton and T. J. Hughes, pp. 41–45, John Wiley and Sons, New York, 1981.

Arlitt, R., Kissling, E., Ansorge, J., and Group, T. W., P-wave velocity structure of the lithosphere-asthenosphere system across the TESZ in Denmark, *Geophys. Res. Abtracts,* **2,** 5, 2000.

Balling, N., Heat flow and thermal structure of the lithosphee across the Baltic Shield and the northern Tornquist Zone, *Tectonophys.*, **244**, 13–50, 1995.

Breuer, D. and Wolf, D., Deglacial land emergence and lateral upper–mantle heterogeneity in the Svalbard Archipelago – I. first results for simple load models, *Geophys. J. Int.*, **121 (3)**, 775–788, 1995.

Bungum, H., Pirhonen, S. E., and Husebye, E. S., Crustal thickness in Fennoscandia, *Geophys. J. R. Astron. Soc.*, **63**, 759–774, 1980.

Calcagnile, G., The lithosphere-asthenosphere system in Fennoscandia, *Tectonophys.*, **90**, 19–35, 1982.

Čermák, V. and Ryback, L. *Terrestrial Heat Flow in Europe*, Springer, Berlin, 1979.

Dziewonski, A. M. and Anderson, D. L., Preliminary reference Earth model. *Phys. Earth Planet. Int.*, **25**, 279–356, 1981.

Ekman, M., Recent postglacial rebound of Fennoscandia: a short review and some numerical results, in: *Dynamics of the Ice Age Earth: A Modern Perspective*, edited by P. Wu, editor, pp. 383–392, Trans Tech Pub., Zürich, Switzerland, 1998.

Ekman, M. and Mäkinen, J., Recent postglacial rebound, gravity change and mantle flow in Fennoscandia, *Geophys. J. Int.*, **126**, 229–234, 1996.

Eldholm, O., Faleide, J. I., and Myhre, A. M., Continent–ocean transition at the western Barents–Sea/Svalbard continental margin, *Geology*, **15**, 1118–1122, 1987.

Farrell, W. E. and Clark, J. A., On postglacial sea level, *Geophys. J. R. astr. Soc.*, **46**, 647–667, 1976.

Gasperini, P. and Sabadini, R., Lateral heterogeneities in mantle viscosity and post-glacial rebound, *Geophys. J.*, **98**, 413–428, 1989.

Gasperini, P., Sabadini, R., and Yuen, D. A., Deep continental roots: the effects of lateral variations of viscosity on post–glacial rebound, in: *Glacial Isostasy, Sea–Level and Mantle Rheology*, edited by R. Sabadini, K. Lambeck, and E. Boschi, pp. 21–32. Kluwer Acad. Pub., Dordrecht, 1991.

Giunchi, C., Spada, G., and Sabadini, R., Lateral viscosity variations and post–glacial rebound: effects on present–day VLBI baseline deformations, *Geophys. Res. Lett.*, **24 (1)**, 13–16, 1997.

Grosswald, M. G., Late Weichselian ice sheet of northern Eurasia, *Quat. Res.*, **13**, 1–32, 1980.

Grosswald, M. G., Extent and melting history of the late Weichselian ice sheet, the Barents–Kara continental margin, in: *Ice in the Climate System*, edited by W. R. Peltier, pp. 1–20, Springer–Verlag, Berlin, Heidelberg, 1993.

Isaksson, E., The western Barents Sea and the Svalbard Archipelago 18000 years ago – a finite–difference computer model reconstruction, *J. Glaciology*, **38 (129)**, 295–301, 1992.

Johnston, A. C., Suppression of earthquakes by large continental ice sheets, *Nature*, **330**, 467–469, 1987.

Johnston, P., Wu, P., and Lambeck, K., Dependence of horizontal stress magnitude on load dimension in glacial rebound models, *Geophys. J. Int.*, **132**, 41–60, 1998.

Kaufmann, G. and Wolf, D., Deglacial land emergence and lateral upper–mantle heterogeneity in the Svalbard Archipelago – II. Extended results for high-resolution load models, *Geophys. J. Int.*, **127 (1)**, 125–140, 1996.

Kaufmann, G. and Wu, P., Upper mantle lateral viscosity variations and postglacial rebound: application to the Barents Sea, in: *Dynamics of the Ice Age Earth: A Modern Perspective*, edited

by P. Wu, pp. 583–602, Trans Tech Pub., Zürich, Switzerland, 1998a.

Kaufmann, G. and Wu, P., Lateral asthenospheric viscosity variations and postglacial rebound: a case study for the Barents Sea, *Geophys. Res. Lett.*, **25 (11)**, 1963–1966, 1998b.

Kaufmann, G., Wu, P. and Li, G., Glacial isostatic adjustment in Fennoscandia for a laterally heterogeneous Earth, *Geophys. J. Int.*, **143 (1)**, 262–273, 2000.

Kaufmann, G., Wu, P., and Wolf, D., Some effects of lateral heterogeneities in the upper mantle on postglacial land uplift close to continental margins, *Geophys. J. Int.*, **128**, 175–187, 1997.

Kukkonen, I. T. and Peltonen, P., Xenolith-controlled geotherm for the central Fennoscandian Shield: implications for lithosphere-asthenosphere relations, *Tectonophys.*, **304**, 301–315, 1999.

Lambeck, K., Smither, C., and Johnston, P., Sea-level change, glacial rebound and mantle viscosity for northern Europe, *Geophys. J. Int.*, **134**, 102–144, 1998.

Lambert, A., James, T. S., Liard, J. O., and Courtier, N., The role and capability of absolute gravity measurements in determining the temporal variations in the Earth's gravity field, in: *Global gravity field and its temporal variations*, edited by R. H. Rapp, A. A. Cazenave, and R. S. Nerem, pp. 20–29, Springer, New York, 1996.

Li, X. D. and Romanowicz, B., Global mantle shear velocity model developed using nonlinear asymptotic coupling theory, *J. Geophys. Res.*, **101 (B10)**, 22245–22272, 1996.

Maher Jr., H. D., Bergh, S., Braathen, A., and Ohta, Y., Svartfjella, Eidembukta, and Daudmannsodden lineament: Tertiary orogen-parallel motion in the crystalline hinterland of Spitsbergen's fold-thrust belt, *Tectonics*, **16 (1)**, 88–106, 1997.

Martinec, Z., Spectral, initial value approach for viscoelastic relaxation of a spherical earth with a three-dimensional viscosity – I. Theory, *Geophys. J. Int.*, **137**, 469–488, 1999.

Martinec, Z., Spectral-finite element approach to three-dimensional viscoelastic relaxation in a spherical earth, *Geophys. J. Int.*, **142**, 117–141, 2000.

Milne, G. A. and Mitrovica, J. X., Postglacial sea–level change on a rotating Earth, *Geophys. J. Int.*, **133**, 1–19, 1998.

Mitrovica, J. X., Davis, J. L., and Shapiro, I. I., A spectral formalism for computing three–dimensional deformations due to surface loads 1. Theory, *J. Geophys. Res.*, **99 (B4)**, 7057–7073, 1994a.

Mitrovica, J. X., Davis, J. L., and Shapiro, I. I., A spectral formalism for computing three-dimensional deformations due to surface loads 2. Present–day glacial isostatic adjustment, *J. Geophys. Res.*, **99 (B4)**, 7075–7101, 1994b.

Ni, Z. and Wu, P., Effects of removing concentric positioning on postglacial vertical displacement in the presence of lateral variation in lithospheric thickness, *Geophys. Res. Lett.*, **25 (16)**, 3209–3212, 1998.

Pedersen, S. S., Israndslinier i norden. *Tech. Rep.*, Danm. Geol. Unders, 1995.

Peltier, W. R., The impulse response of a Maxwell Earth, *Rev. Geophys. Space Sci.*, **12 (4)**, 649–669, 1974.

Peltier, W. R., Mantle viscosity and ice-age ice sheet topography. *Science*, **273**, 1359–1364, 1996.

Rutter, N., Problematic ice sheets, *Quat. Int.*, **28**, 19–37, 1995.

Sabadini, R. and Gasperini, P., Glacial isostasy and the interplay between upper and lower mantle lateral viscosity heterogeneities, *Geophys. Res. Lett.*, **16 (5)**, 429–432, 1989.

Su, W. and Dziewonski, A. M., Predominance of long–wavelength heterogeneity in the mantle, *Nature*, **352**, 121–126, 1991.

Tromp, J. and Mitrovica, J. X., Surface loading of a viscoelastic earth - I. General theory, *Geophys. J. Int.*, **137** (3), 847–855, 1999a.

Tromp, J. and Mitrovica, J. X., Surface loading of a viscoelastic earth - II. Spherical models, *Geophys. J. Int.*, **137** (3), 856–872, 1999b.

Tromp, J. and Mitrovica, J. X., Surface loading of a viscoelastic earth - III. Aspherical models, *Geophys. J. Int.*, **140**, 425–441, 2000.

Trompert, J., Global seismic tomography: the inverse problem and beyond, *Inverse Problems*, **14** (3), 371–385, 1998.

Wessel, P. and Smith, W. H. F., Free software helps map and display data, *EOS*, **72**, 441–446, 1991.

Wolf, D., The relaxation of spherical and flat Maxwell Earth models and effects due to the presence of the lithosphere, *J. Geophys.*, **56**, 24–33, 1984.

Wu, P., Deformation of an incompressible viscoelastic flat Earth with power–law creep: a finite element approach, *Geophys. J. Int.*, **108**, 35–51, 1992a.

Wu, P., Viscoelastic versus viscous deformation and the advection of pre–stress, *Geophys. J. Int.*, **108**, 136–142, 1992b.

Wu, P., Effect of viscosity structure on fault potential and stress orientations in Eastern Canada, *Geophys. J. Int.*, **130**, 365–382, 1997.

Wu, P., Intraplate earthquakes and postglacial rebound in Eastern Canada and Northern Europe. in: *Dynamics of the Ice Age Earth: A Modern Perspective*, edited by P. Wu, editor, pp. 603–628, Trans Tech Pub., Zürich, Switzerland, 1998.

Wu, P. and Hasegawa, H. S., Induced stresses and fault potential in Eastern Canada due to a disc load: a preliminary analysis, *Geophys. J. Int.*, **125**, 415–430, 1996.

Wu, P. and Johnston, P., Validity of using flat-earth finite element models in the study of postglacial rebound, in: *Dynamics of the Ice Age Earth: A Modern Perspective*, edited by P. Wu, editor, pp. 191–202, Trans Tech Pub., Zürich, Switzerland, 1998.

Wu, P., Ni, Z., and Kaufmann, G., Postglacial rebound with lateral heterogeneities: from 2D to 3D modeling, in: *Dynamics of the Ice Age Earth: A Modern Perspective*, edited by P. Wu, editor, pp. 557–582. Trans Tech Pub., Zürich, Switzerland, 1998.

Wu, P., Johnston, P., and Lambeck, K., Postglacial rebound and fault instability in Fennoscandia, *Geophys. J. Int.*, **139**, 657–670, 1999.

Georg Kaufmann, Institut für Geophysik, Universität Göttingen, Herzberger Landstrasse 180, 37075 Göttingen, Germany (email: gkaufman@uni-geophys.gwdg.de).

Patrick Wu, Department of Geology and Geophysics, University of Calgary, Calgary, Alberta T2N 1N4, Canada (email: ppwu@acs.ucalgary.ca).